Ergebnisse der Mathematik und ihrer Grenzgebiete

Band 55

Herausgegeben von
P. R. Halmos · P. J. Hilton · R. Remmert · B. Szőkefalvi-Nagy

Unter Mitwirkung von
L. V. Ahlfors · R. Baer · F. L. Bauer · R. Courant
A. Dold · J. L. Doob · S. Eilenberg · M. Kneser · G. H. Müller
M. M. Postnikov · B. Segre · E. Sperner

Geschäftsführender Herausgeber: P. J. Hilton

Wolfgang Walter

Differential and Integral Inequalities

Translated by
Mrs. Lisa Rosenblatt
and
Prof. Lawrence Shampine
(Technical Advisor)

With 18 Figures

Springer-Verlag Berlin Heidelberg New York 1970

Prof. Dr. rer. nat. Wolfgang Walter
Mathematisches Institut der Universität Karlsruhe

Translation of
Differential- und Integral-Ungleichungen
1964
(Springer Tracts in Natural Philosophy, Vol. 2)

AMS Subject Classifications (1970):

Primary 34-02, 34 A 10, 34 A 35, 34 A 40, 34 A 45, 34 G 05, 35-02, 35 A 35, 35 B 05, 35 B 30, 35 B 35, 35 B 45, 35 K 05, 35 K 10, 35 K 15, 35 K 20, 35 K 55, 35 L 10, 35 L 15, 35 L 60, 45-02, 45 D 05, 45 G 99, 45 L 05, 65 M 15, 65 M 20, 65 N 15

Secondary 34 A 15, 34 A 20, 34 A 50, 34 D 20, 34 J 10, 35 J 15, 35 J 25, 35 J 60, 35 J 70, 35 K 45, 35 K 50, 35 L 05, 45 E 10, 45 L 10, 45 N 05, 65 L 05

ISBN 978-3-642-86407-0 ISBN 978-3-642-86405-6 (eBook)
DOI 10.1007/978-3-642-86405-6

To Irmgard

Preface

In 1964 the author's monograph "Differential- und Integral-Ungleichungen," with the subtitle "und ihre Anwendung bei Abschätzungs- und Eindeutigkeitsproblemen" was published. The present volume grew out of the response to the demand for an English translation of this book. In the meantime the literature on differential and integral inequalities increased greatly. We have tried to incorporate new results as far as possible. As a matter of fact, the Bibliography has been almost doubled in size.

The most substantial additions are in the field of existence theory. In Chapter I we have included the basic theorems on Volterra integral equations in Banach space (covering the case of ordinary differential equations in Banach space). Corresponding theorems on differential inequalities have been added in Chapter II. This was done with a view to the new sections, dealing with the line method, in the chapter on parabolic differential equations. Section 35 contains an exposition of this method in connection with estimation and convergence. An existence theory for the general nonlinear parabolic equation in one space variable based on the line method is given in Section 36. This theory is considered by the author as one of the most significant recent applications of inequality methods. We should mention that an exposition of Krzyżański's method for solving the Cauchy problem has also been added.

The numerous requests that the new edition include a chapter on elliptic differential equations have been satisfied to some extent. A survey of the most important theorems on elliptic differential inequalities, with brief proofs, is given in an appendix. We note that all new material has been incorporated in such a way as to leave unchanged the numbers of sections, subsections, formulas, . . . appearing in the German edition.

I am most grateful to the translator, Lisa Rosenblatt, for her reliable collaboration. My thanks also go to Professor L. F. Shampine for his valuable assistance in resolving technical difficulties. I am greatly indebted to Professor B. Noble, Professor G. H. Knightly and Mr. Wickwire, who read parts of the manuscript and made valuable suggestions, and to Dr. H. Becker, Dr. K. Deimling, Dr. E. Mues, Dr. G.

Schleinkofer and Mr. A. Voigt for their assistance in proof-reading. Finally, the publisher, who expedited the completion of this work in every way with patience in the face of many problems, deserves my special thanks.

Karlsruhe, August 1970 Wolfgang Walter

Preface to the German Edition

The theory of differential and integral inequalities has been greatly enriched in the past fifteen years by new understanding and knowledge. This book is the first comprehensive presentation of these developments. It covers Volterra integral equations (in one and several variables) as well as ordinary, hyperbolic, and parabolic differential equations.

The present volume of "Springer Tracts" is simultaneously textbook, guide to the literature, and research monograph. My intention of writing a self-contained textbook that can also be read by advanced students was consistent with the subject. The theory of differential and integral inequalities is basically of an elementary nature and no special preparation is needed for its understanding. Relatively much effort is spent on one-dimensional problems in the first two chapters to show the methods to be used. Thus the essential ideas of the proofs are first clearly worked out in very elementary cases. The central core of the book is basically concerned with partial differential equations, and particularly with parabolic equations, to which the last chapter, by far the most extensive, is devoted.

My special thanks are due to Professor K. Nickel for many fruitful discussions and suggestions. His advice was indispensable in writing the section on boundary layer theory, in which I see the most important and most elegant application of the theory to date.

My gratitude also goes to Dr. H. Brakhage, Dr. P. Werner, and Mr. H. Weigel for their valuable help in proofreading and for their criticisms.

The Editor, Professor L. Collatz, furthered the progress of this work by his continued interest during the entire time of its preparation. I wish to thank him, as well as Springer-Verlag, for their willing cooperation and the excellent production of this book.

Karlsruhe, March 1964 Wolfgang Walter

Contents

Chapter III

Volterra Integral Equations in Several Variables
Hyperbolic Differential Equations

Chapter IV

Parabolic Differential Equations

Appendix

Notation

Formulas are numbered with Arabic numerals; theorems, definitions, remarks, etc., are given Roman numbers in sequence. Thus 8 X (α) means hypothesis (α) of Theorem X of Section 8, (27.2) means formula (2) of Section 27. References within a section omit the section number. Literature references include the author and the publication date in parentheses; if this is ambiguous, an identifying letter is added, e.g., Ciliberto (1956 b).

The following notational principles will be used (see also the list of notations at the end of the bool). Independent variables are denoted by $t, \tau \in E^1$ and $x=(x_1, \ldots, x_m)$, $\xi=(\xi_1, \ldots, \xi_m) \in E^m$ (not bold-face), while $z \in E^1$ and $\mathbf{z}=(z_1, \ldots, z_n) \in E^n$ (bold-face) denote a function and a system of functions, respectively; n is the number of equations in a system of differential equations or integral equations. The function classes $\mathscr{D}, \mathscr{E}, \mathscr{H}, \mathscr{P}$ refer to the "right hand sides" of differential equations or kernels of integral equations, which are generally denoted by f, k, or ω. The domain of definition of a function f is denoted by $D(f)$; $D(f)$ can have a very different meaning depending, for example, on whether $f=f(t,z)$ or $f=f(t,x,z,p,r)$. The classes $Z, Z_0, Z_c, \ldots$ ("admissible" functions) refer to solutions or approximate solutions of the problems under consideration. These symbols also have different meanings in different chapters, yet are consistent. Thus, for instance, for ordinary differential equations Z is the class of functions $\varphi(t)$ which are continuous for $0 \leqq t \leqq T$ and differentiable for $0 < t \leqq T$, while for parabolic differential equations Z stands for the class of functions $\varphi=\varphi(t,x)$ described in detail in 23 III. Yet for a function $\varphi=\varphi(t)$ which is independent of x, Definitions 5 I and 23 III are equivalent. For the most part, no assumptions will be made concerning the domain of definition $D(f)$ of the right hand side f of a differential equation. The function classes $Z(f), Z_0(f), \ldots$ take this into account. If, for example, an ordinary differential equation $u'=f(t,u)$ is given for $0<t \leqq T$, then $u \in Z(f)$ means, first, that u is in the class Z defined above and, second, that u "can be substituted" in f, i.e., $(t, u(t)) \in D(f)$ for $0<t \leqq T$.

The concept of monotonicity (and of quasimonotonicity) is defined in 6 II in the weak sense, i.e., with equality permitted. If equality is excluded, we speak of strict monotonicity. Thus a real-valued function

$\varphi(t)$ of a real variable t is said to be monotone increasing if $\varphi(t_1) \leqq \varphi(t_2)$ for $t_1 \leqq t_2$, and strictly monotone increasing if $\varphi(t_1) < \varphi(t_2)$ for $t_1 < t_2$.

The symbols used are listed at the end of the book.

Introduction

I. General and Historical Remarks. In this book, differential and integral *inequalities* are treated with a view to the theory of differential and integral *equations* rather than for their own sake. We are primarily concerned with two types of problem: integral equations and initial or boundary value problems for differential equations. These present the mathematician with a series of problems, most of which fall into one of the following categories (the dividing lines are not always sharp): (α) the existence problem; (β) the uniqueness problem and the related problem of continuous dependence on the various given data; (γ) qualitative and quantitative properties of the solution (e.g., statements about monotonicity or convexity of the solution, validity of the maximum principle, ..., numerical determination of the solution).

A great variety of methods is available today for handling such problems. It is characteristic of the procedures of this book that *inequalities* play an essential role, and in particular those inequalities to which the determining *equations* in the problem under consideration are related in a simple fashion. One could speak of a "method of inequalities" in this as-yet vague sense (it will be made more precise by the contents of the book!).

To this end let us consider, as an elementary example, the equation $u^2 = a$ $(a > 0)$ for a positive real number u. Problems (α), (β), and (γ) can be solved in very different ways (one need only compare the possibilities presented in analysis texts). A procedure in the sense of the "method of inequalities" would consist of considering positive numbers v with $v^2 < a$ ("lower solution") and positive numbers w with $w^2 > a$ ("upper solution"). The proof of existence is then carried out by letting $u = \inf w$ or $\bar{u} = \sup v$, while the fact that the inequality $v < u < w$ holds for every solution u can be used both for uniqueness and the numerical determination of bounds. Methods of this kind are as old as mathematics itself. As the first "modern" application relevant to the problems of interest here we note the solution of the integration problem due to Perron (1914). If $f(t)$ is a real function defined for $0 \leqq t \leqq T$, then the integral $u(t) = \int_0^t f(\tau)\,d\tau$ is to be defined appropriately. The method given by Perron, today associated with the name Perron integral, proceeds as follows (omitting technical details like replacement of the derivative by a Dini-derivate). First the problem is rewritten as a differential equation $u' = f(t)$ with $u(0) = 0$. Then we consider the functions $v(t)$ with $v(0) = 0$, $v' \leqq f(t)$ (subfunctions) and the functions $w(t)$ with $w(0) = 0$, $w' \geqq f(t)$ (super-functions). Again we have $v \leqq w$, and the solution is obtained from $\underline{u} = \inf w$ or $\overline{u} = \sup v$ (integrability is defined by $\underline{u} = \overline{u}$ and is thus identical with uniqueness).

This example of the definite integral demonstrated for the first time the simplicity and elegance of the method as well as its extraordinary scope (it is well known that the Perron integral is more general than the Lebesgue integral). Soon after, Perron (1915) published a new proof of Peano's existence theorem, again based on the same idea. This approach to existence proofs, known now as the Perron method, turned out to be quite efficient for the first boundary value problem for the potential equation and the heat equation [Perron (1923) and Sternberg (1929), respectively], and subsequently for a great number of problems. The report of Rosenbloom (1958) gives a concise summary and a survey of the more recent literature.

The cited works deal primarily with the problem of existence. The systematic use of inequalities in problems (β) and (γ) is more recent. An early significant result in the area of partial differential equations was obtained by Hopf (1927): the proof of the strong maximum (minimum) principle for nonlinear elliptic differential equations. The result, already known in part, was of even greater interest because of its surprising new and completely elementary method of proof [part of which can be traced back to Paraf (1892)]. The historical development of the method of inequalities, which leads from ordinary differential equations to partial differential equations of first order and then to parabolic and elliptic differential equations, will be discussed in the appropriate places. Here we name only two mathematicians, Nagumo and Ważewski, who together with their collaborators and students, did authoritative work in the development of this theory.

II. Monotonicity. Many of the theorems in this book, although derived from very different sources, have related structures. This structural similarity is expressed aptly in the terminology "*problems of monotonic type*" coined by Collatz (1952). Such a problem is present (one can think of an integral equation or a differential equation with boundary conditions) if it can be written in the form $Tu = 0$ with the help of an operator T, and if

(α) $Tv < Tw$ *implies* $v \leqq w$.

Such a statement naturally assumes that order relations have been assigned to v, w as well as to Tv, Tw.

Originally, problems of monotonic type were characterized by Collatz (1952) by the sharper requirement

(β) $Tv \leqq Tw$ *implies* $v \leqq w$.

We call a theorem with the structure given in (α) or (β) a *monotonicity theorem.* It should be noted, however, that some authors prefer the name "comparison theorem" (this terminology is in wide use in connection with boundary value problems for ordinary differential equations of second order).

Here is an example. The initial value problem $u' = f(t, u)$ for $0 < t \leqq T$, $u(0) = \eta$ can be written in the form $Tu = 0$ with the help of $Tv = (v' - f(t, v), v(0) - \eta)$, where $0 =$ (the function $\varphi(t) \equiv 0$, the number 0). We write $\varphi < \psi$ if $\varphi(t) < \psi(t)$ for $0 \leqq t \leqq T$, and $(\varphi, a) < (\psi, b)$ if $\varphi < \psi$ and $a < b$. Then with no further assumptions on f we have a monotonicity theorem of type (α): by Theorem 8 V, $Tv < Tw$ indeed implies $v < w$.

A monotonicity theorem of the type (β) implies, in particular, uniqueness for the problem $Tu = 0$. For suppose u and $\bar{u}$ are two solutions. Then $Tu \leqq T\bar{u}$ and $Tu \geqq T\bar{u}$; therefore $u \leqq \bar{u}$ and $u \geqq \bar{u}$, i.e., $u = \bar{u}$. Thus, for example, in the initial value problem $u' = \sqrt{|u|}, u(0) = 0$, monotonicity holds in the sense of (α), but not in the sense of (β) (it has infinitely many solutions).

Closely related to monotonicity theorems is the concept of

III. Quasi-Monotonicity. As has already been said, the initial value problem for an ordinary differential equation of first order is of monotonic type in the sense of II (α). This is not true in general for a system of differential equations $\boldsymbol{u}' = \boldsymbol{f}(t, \boldsymbol{u})$, but only when the right hand side $\boldsymbol{f} = (f_1, \ldots, f_n)$ has a certain monotonicity property which was discovered by Müller (1926); it will be more carefully defined in 6 II. Since we are dealing with a very central concept, whose significance for systems of ordinary differential equations was clearly recognized by Kamke (1932) and which was further developed by Szarski (1947, 1950) and Ważewski (1950) among others, we have given it the new name "quasi-monotonicity". More recently Mlak (1957) discovered that this same monotonicity property also characterizes problems of monotonic type for systems of parabolic differential equations.

IV. Superfunctions and Subfunctions. These concepts were first used by Perron (1914) in developing the integration theory named after him; see I.

As its name indicates, a superfunction is a majorant for the solution (in case of non-uniqueness: for all solutions) of a problem. Here it is essential that the superfunctions are defined by *inequalities* which arise naturally from the *equations* which define the problem (say, a differential equation plus boundary conditions). Here there is a connection with problems of monotonic type. If u is a solution of a problem $Tu=0$ and if $Tv<0$, $Tw>0$, then v is a sub- and w a superfunction (i.e., $v \leqq u \leqq w$) precisely when the problem is of monotonic type.

To avoid a lengthy enumeration of definitions tailored to the problem at hand, we agree to the following, in contrast to the normal usage in the literature: *A function v (resp. w) is called a subfunction (resp. superfunction) with respect to a given problem if it is defined in the appropriate domain and if $v \leqq u$ (resp. $u \leqq w$) for every solution u of the problem (correspondingly $\mathbf{v} \leqq \mathbf{u}$, resp. $\mathbf{u} \leqq \mathbf{w}$ for systems of equations).*

Thus we call *every* upper bound a superfunction and for individual problems give criteria in the form of inequalities for a given function to be a superfunction. On the contrary, most authors designate as superfunctions (sometimes also as upper solutions) only those upper bounds which satisfy such special criteria.

V. Monotone Increasing Operators. Volterra integral equations in one or several variables, particularly the initial value problems of Chapters I and III for ordinary and hyperbolic differential equations, can be put into the form $u=Su$, where S is an integral operator. We say S is a monotone increasing operator if, roughly speaking, $\varphi \leqq \psi$ implies $S\varphi \leqq S\psi$ (note the precise Definition 1 IV, or 17 II). We have the following relation to problems of monotonic type: If S is a monotone increasing operator and we let $T\varphi = \varphi - S\varphi$, then a monotonicity theorem holds for T, i.e., super- and subfunctions can be characterized by the inequalities $v<Sv$, $w>Sw$ (cf. 1 VI and 17 III).

The integral operator corresponding to a Volterra integral equation is in general *not* monotone increasing (this is the case only if its kernel has certain monotonicity properties). Thus super- and subfunctions cannot in general be determined in the proposed simple fashion. For this reason it is important that

VI. Simultaneous Determination of Super- and Subfunctions for Non-Monotonic Problems is possible in many cases. The corresponding theorems for a problem $u=Su$ are of the form: Suppose v and $w>v$ are two functions such that for all functions φ with $v \leqq \varphi \leqq w$ the inequality $v<S\varphi<w$ holds; then v is a subfunction and w is a superfunction (see 4 VII and 17 X).

Of special significance is the case when S is a *monotone decreasing operator* (i.e., $-S$ is a monotone increasing operator). The assumption is then simply $v<Sw$, $w>Sv$.

The first such simultaneous bounding from below and above for a non-monotone problem was carried out by Müller (1927) for systems of ordinary differential equations. (Although the *differential* equation stands in the foreground in this theorem — cf. 12 IV — the same fundamental idea is involved.) The application of this principle to hyperbolic differential equations can be found in the book by Hukuhara and Satō (1957, Ch. 6) in several places; see 21 XIII.

VII. Successive Approximation. For a problem $u = Su$, the method of successive approximation consists of determining a sequence of functions u_n by $u_{n+1} = Su_n$, with u_0 given. The solution is then obtained as the limit of this sequence. If S is a monotone increasing operator and $u_0 \geqq Su_0$, then we have immediately

$$u_0 \geqq u_1 \geqq u_2 \geqq u_3 \geqq \cdots;$$

if S is a monotone decreasing operator and $u_0 \geqq Su_1 \geqq u_1 = Su_0$, then it follows that

$$u_1 \leqq u_3 \leqq u_5 \leqq \cdots \leqq u_4 \leqq u_2 \leqq u_0 .$$

Thus we have monotone convergence or alternating convergence. Hence, not only is the problem of convergence cleared up (usually the most difficult part of an existence proof), but the successive approximations simultaneously yield bounds for the solution which can be used in the numerical determination of the solution.

Iteration procedure of this and similar kinds were investigated from the point of view of error bounds by Weissinger (1952) and especially thoroughly by Collatz and Schröder. Their work, in particular Collatz (1952, 1960), Collatz and Schröder (1959) and Schröder (1958/59, 1959, 1959/60, 1960, 1961, 1961 a), shows that some of the theorems presented in this book can also be proved by a consideration of suitable iteration procedures and by reduction to fixed point theorems in suitably chosen general spaces.

There is an extensive literature on successive approximation, especially on the convergence problem and its connection with the uniqueness problem. Ordinary differential equations are treated in papers by Nagumo (1930), Dieudonné (1945), Wintner (1946), Coddington and Levinson (1952), Viswanatham (1952), Baiada and Lorefice (1957), Brauer and Sternberg (1958, 1959), Luxemburg (1958), Brauer (1959, 1959a) among others, hyperbolic differential equations by Walter (1959a, 1965) and Kisyński (1960); see also the literature cited in 7 XII.

Approximation procedures are not a primary concern of this book. However, we have included a convergence theorem for successive approximation since a monotonicity argument plays the fundamental role in its proof (7 XII). The so-called method of Chaplygin (it can be described as a combination of Newton's method and the regula falsi in an abstract version) is not considered here. We note only that monotonicity theorems are an important tool in the development of Chaplygin's method.

VIII. Error Bounds and Defect. An essential part of this book is concerned with theorems on error estimation. They are all formulated according to a uniform scheme. We prove an inequality $|v-u| \leqq \varrho$, where u is a solution of a given problem, say $Tu=0$, and v is an approximate solution for whose "defect" Tv we know a "defect bound" $|Tv| \leqq \delta$. The bound ϱ is determined from a problem with a similar structure, $\Omega\varrho=0$. This new problem, called the "comparison problem," is essentially determined by two quantities, first, the defect of v, and second, a condition for the original problem, say, for the right hand side of a differential equation or the kernel of an integral equation, which reduces to a Lipschitz condition in the simplest case.

We illustrate this in more detail by the example of an ordinary differential equation $u'=f(t,u), u(0)=\eta$. For an explicitly given approximate solution $v(t)$, both of the quantities $v'-f(t,v)$ and $v(0)-\eta$ which determine the defect can be obtained without knowing the solution u. In order to be able to draw a conclusion about the behavior of $|v-u|$ from this, we also need information about the growth of $f(t,z)$ in z, which is most often given as a Lipschitz condition:

$$|f(t,z)-f(t,\bar{z})| \leqq L|z-\bar{z}|. \tag{1}$$

If the defect bound

$$|v'-f(t,v)| \leqq \delta, \quad |v(0)-\eta| \leqq \varepsilon, \tag{2}$$

holds for v, then ϱ is a bound for $|v-u|$, i.e., $|v-u| \leqq \varrho$, provided ϱ is the solution of the linear differential equation (the "comparison problem")

$$\varrho' = L\varrho + \delta \quad \text{with} \quad \varrho(0)=\varepsilon. \tag{3}$$

This state of affairs is not connected with the linearity of the right hand side of (1), i.e., of the initial value problem (3). If f satisfies a nonlinear condition

$$|f(t,z)-f(t,\bar{z})| \leqq \omega(t,|z-\bar{z}|), \tag{1'}$$

then (3) must be replaced by

$$\varrho' > \omega(t,\varrho)+\delta, \quad \varrho(0)=\varepsilon \tag{3'}$$

and once again a bound is obtained for $|v-u|$[1]. From (3) we see that the Lipschitz constant L is crucial to the growth of ϱ and thus to the sharpness of the bound. This point will be examined in somewhat greater detail.

First we note the following. The inequality $|v-u| \leqq \varrho$ can be proved in another way, by using an appropriate criterion to show that $v-\varrho$ is a subfunction and $v+\varrho$ a superfunction (for the original problem). Now the comparison problem fades into the background. We do *not* prove

[1] We can also determine ϱ from the corresponding differential equation, but then we have to take the maximal solution.

estimation theorems in this fashion here, even though it would be possible in many problems, for the following reason. We can proceed in this manner only if the *original* problem is of monotonic type, while in our treatment the *comparison* problem must be of monotonic type. This means, for example, that the proof using the comparison problem can be extended to systems of differential equations, while the method of proof of the original problem, using super- and subfunctions, cannot. (More precisely, it can be extended only to systems which present problems of monotonic type, i.e., for which the right hand side is quasi-monotone increasing.)

IX. Remarks on the Lipschitz Condition, in Particular on One-Sided Conditions. According to (1) the Lipschitz constant L is a bound for the absolute value of $f_z(t, z)$. Now it is well known that the behavior of a solution u with respect to neighboring solutions is completely different depending on whether $f(t, z)$ is monotone increasing or decreasing in z. In the first case the solutions diverge (Fig. 1), in the second case they converge (Fig. 2). But the estimate (1) suppresses these differences. Let

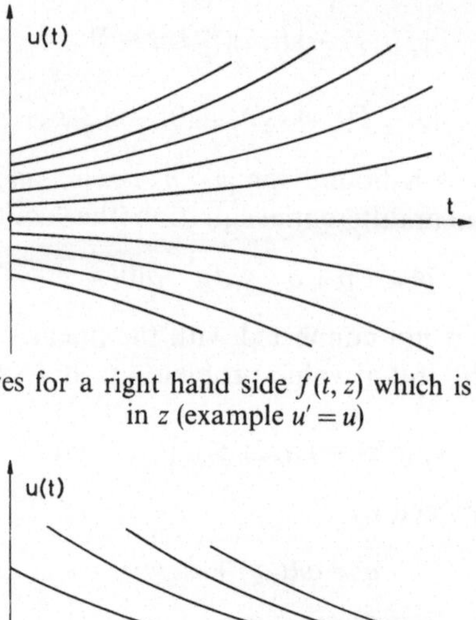

Fig. 1. Solution curves for a right hand side $f(t, z)$ which is monotone increasing in z (example $u' = u$)

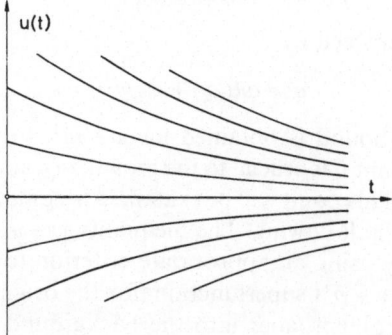

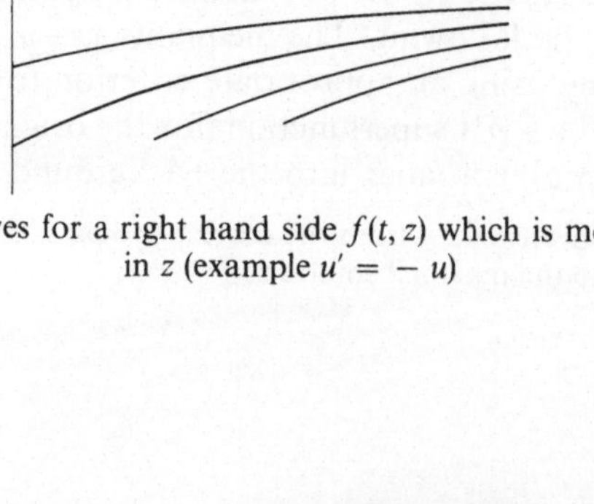

Fig. 2. Solution curves for a right hand side $f(t, z)$ which is monotone decreasing in z (example $u' = -u$)

us consider a simple example. Suppose

$$u' = u\,, \quad u(0) = \eta\,;$$

then $f(t, z) = z$ and $L = 1$. If v is also a solution of the differential equation and $|v(0) - \eta| \leqq \varepsilon$, then from (3) we obtain $|v - u| \leqq \varepsilon e^t$, an obviously optimal bound. But if $f(t, z) = -z$, then $L = 1$, and

$$u' = -u, \qquad u(0) = \eta, \qquad v' = -v, \qquad |v(0) - \eta| \leqq \varepsilon,$$

so that according to (3) we have the same estimate $|v - u| \leqq \varepsilon e^t$ even though the difference actually behaves like e^{-t} rather than e^t (see 9 V).

The condition (1) or (1'), which connects the original problem for u with the comparison problem for ϱ, is therefore not appropriate here. The difference of the two solutions (and similarly for the difference between solution and approximate solution) does not depend on the absolute value of f_z, but on f_z itself; indeed, it increases as f_z increases. This behavior can be deduced not from (1), but from the one-sided condition

$$f(t, z) - f(t, \bar{z}) \leqq L(z - \bar{z}) \quad \text{for} \quad z \geqq \bar{z} \tag{4}$$

or the more general condition

$$f(t, z) - f(t, \bar{z}) \leqq \omega(t, z - \bar{z}) \quad \text{for} \quad z \geqq \bar{z} \tag{4'}$$

where we allow L and ω to be negative. This *one-sided* condition on f is actually sufficient for the *two-sided* bound $|v - u| \leqq \varrho$; in the estimate given above in VIII, the condition (1) (resp. (1')) can be replaced by (4) (resp. (4')). Thus this estimate now becomes quite useful numerically.

We will pay particular attention to the extent to which one-sided conditions on f can be used for bounds in the problems treated here (ordinary, hyperbolic and parabolic differential equations and systems of such differential equations). It has long been known that only a one-sided condition as in (4'), rather than one like (1'), is required in the familiar uniqueness theorems for a first order differential equation; this can already be found in Tonelli (1925) and Iyanaga (1928). Müller (1927) showed that a condition which is in a certain sense one-sided will also work for systems of ordinary differential equations. This principle of one-sided conditions was extended to estimation theorems in the more recent work of Eltermann (1955) and Uhlmann (1957, 1957a). One-sided conditions for hyperbolic differential equations first appeared in Zwirner (1952), and for parabolic differential equations in Giuliano (1952); the appropriate generalization for parabolic systems is given here (in the German edition) for the first time.

X. Further Remarks on the Lipschitz Condition. For some very simple nonlinear ordinary differential equations, the right hand side does not satisfy a Lipschitz condition. If, say, $u' = u^2$, then

$$f(t,z) - f(t,\bar{z}) = z^2 - \bar{z}^2 = (z+\bar{z})(z-\bar{z}), \tag{5}$$

which is not $\leqq L(z-\bar{z})$ for $z \geqq \bar{z}$. A well-known device helps in this case. First we seek a superfunction $w(t)$ for the solution of the initial value problem (if this is impossible, we can set $w = \eta + M, M > 0$, if $u(0) = \eta$ and thus at least have a bound for small t). Once such an upper bound $w(t)$ has been found, we consider only the points (t,z) with $z \leqq w(t)$, i.e., we consider $f(t,z)$ on the set $D(f)$ defined by the inequalities $0 \leqq t \leqq T$, $z \leqq w(t)$. Then, for example, the following Lipschitz condition

$$f(t,z) - f(t,\bar{z}) \leqq 2w(t)(z-\bar{z}) \quad \text{for} \quad z \leqq \bar{z} \quad \text{and} \quad (t,z),(t,\bar{z}) \in D(f) \tag{6}$$

holds for the example in (5) (it is needed only in this form for the estimation theorem). This procedure of first determining crude upper and/or lower bounds and then suitably restricting the domain of definition of f, is usually necessary in nonlinear problems (see Example 9 VIII). For this reason, our theorems will be formulated in such a way as to give greatest possible latitude to the choice of $D(f)$.

We can also handle these difficulties by putting the functions u, v in the inequality (4′) instead of free variables $z, \bar{z}$, i.e.,

$$f(t,v) - f(t,u) \leqq \omega(t, v-u). \tag{4''}$$

In fact such a condition in connection with (2) and (3′) is sufficient for the validity of the inequality $v - u < \varrho$. In this form the condition is still not usable, since the unknown solution u appears in it. Rather, it must be rewritten so that only known quantities, the approximation v and the bound ϱ, occur. Then (as in 9 I)

$$f(t,v) - f(t,v-\varrho) \leqq \omega(t,\varrho). \tag{4'''}$$

Whenever possible, we will give this central inequality, which relates the original problem to the comparison problem, in a realistic form, i.e., so that on the one hand we assume as little as possible and on the other hand the unknown solution is not explicitly involved.

This point of view is even more important for partial differential equations. Suppose we are given the nonlinear heat equation $u_t = \varkappa(u)u_{xx}$ (thus $f(t,x,z,p,r) = \varkappa(z)r$). Then the Lipschitz condition required in the estimation theorem is

$$[\varkappa(z) - \varkappa(\bar{z})]r \leqq L(z-\bar{z}) \quad \text{for} \quad z \geqq \bar{z},$$

which cannot be satisfied in this form since r is arbitrary. In order to use the theorem, it is therefore necessary to know that the derivative v_{xx} of the approximate solution v can be used in place of the free variable r, i.e., that simply the condition

$$[\varkappa(z)-\varkappa(\bar{z})]v_{xx} \leqq L(z-\bar{z}) \quad \text{for} \quad z \geqq \bar{z}$$

is needed (see 25 VI).

Rather than a Lipschitz condition as in (4), a more general condition as in (4′) is the foundation for our estimation theorems. There are essentially two reasons for this. First, the simple Lipschitz condition is generally too crude for numerical purposes. However, in almost all cases one can work with a generalized Lipschitz condition (for ordinary di-ferential equations with $\omega(t,z)=l(t)\,z$). In this case the new problem for the bound ϱ is still linear, and hence explicitly solvable. Second, we would like to keep the estimation theorems general enough so that they yield the known uniqueness theorems as special cases. *Uniqueness theorems will always be obtained here by specialization of estimation theorems. This has the great advantage that we also obtain information about the continuous dependence of the solution on the initial values or the boundary values and on the right hand side of the differential equation.*

XI. Stability Problems. The problems that go by this name for ordinary and recently also for parabolic differential equations will be mentioned only in passing (see 1 X, 11 VII, 28 XII). This restriction was imposed by the size of the book. For nonlinear parabolic differential equations many authors use methods close to those in this book. For ordinary differential equations the closest connections are with the so-called second method of Lyapunov; a survey of the very rich literature in this field can be found in Cesari (1959) and Hahn (1963).

XII. Existence Problems. Existence theorems for Volterra integral equations are presented in Sections 2, 3, 7, and 18. These theorems are relatively easy to prove with the help of the Banach or Schauder fixed point theorem. (At the time the German edition was published they had not yet been treated in textbooks and some were new). These theorems apply in particular to ordinary and hyperbolic differential equations.

Boundary value problems for parabolic differential equations from the point of view of existence are handled at the end of Chapter IV. The line method, which is used there, turns out to be a powerful new method for linear and nonlinear problems.

It should finally be noted that existence problems are deliberately separated from estimation and uniqueness problems. The latter will be handled by elementary methods, in particular without using existence results. That part of the theory where existence and estimation theorems are combined will be presented as late as possible.

For example, we first prove, in a quite elementary way, the uniqueness theorem 10 I for ordinary differential equations. The uniqueness theorem of Kamke, whose proof depends on the existence theorem of Peano, is obtained later as a special case, in 10 VII. As long as we have only ordinary differential equations in mind, this remark is not of importance since Peano's existence theorem is easily accessible. On the other hand, one learns in such a situation how to attack the uniqueness problem if a general existence theorem is not known, as say for hyperbolic differential equations.

For the above reasons we first obtain the bound ϱ in estimation theorems from a (differential or integral) *inequality*. The analogous theorem, where ϱ is the maximal solution of a corresponding *equality*, appears as a special case (for which additional hypotheses are needed).

In a recent report (1962; reprint of a lecture at the DMV-GAMM Meeting 1962 in Bonn), the author presented the essential ideas underlying this book. Consideration of elliptic differential equations can be found in a survey by Redheffer (1962a). The problems treated here are also discussed briefly in the monograph of Beckenbach and Bellman (1961) and extensively in Szarski's book on differential inequalities (1965) which appeared after the German edition of the present book. In the elegant book *Maximum Principles in Differential Equations* by Protter and Weinberger (1967) elliptic, parabolic and hyperbolic differential equations are treated from a viewpoint which has much in common with ours.

CHAPTER I

Volterra Integral Equations

1. Monotone Kernels

In this chapter we investigate operator equations and inequalities for functions of one real variable. Our particular objective here is nonlinear Volterra integral equations and ordinary differential equations. Unless explicitly stated otherwise, the Lebesgue concept of integral is always presupposed. As for the results for ordinary differential equations, they are obtained via the corresponding integral equations and therefore also hold for differential equations in the generalized sense of Carathéodory. In the second chapter we will handle ordinary differential equations with a different method, which puts *differential* equations in the foreground. Then the concept of a classical solution is basic. In this case the differential equations and differential inequalities under consideration must be satisfied at *every* point (and not just almost everywhere).

We start with some frequently needed

I. Notation ($J, J_0, C, L, E^n, Z_c(k), D(k)$, *monotonicity*). J is the interval $0 \leqq t \leqq T$ with $0 < T < \infty$, J_0 is the interval $0 < t \leqq T$. We use E^n to denote n-dimensional Euclidean space, i.e., the set of all n-tuples of real numbers. If G is a measurable point set in E^n, then $C(G)$ stands for the class of (real-valued) functions continuous on G, $L(G)$ for the class of (real-valued) functions integrable (in the Lebesgue sense) over G. In these and other classes to be introduced later which depend on a basic domain G, G will not be explicitly mentioned when there is no danger of misunderstanding.

First we consider Volterra integral equations of the form

$$u(t) = g(t) + \int_0^t k(t, \tau, u(\tau))\, d\tau \qquad (t \in J). \tag{1}$$

The kernel $k(t, \tau, z)$ is defined as a real-valued function of the three real variables t, τ, z on a certain subset $D(k)$ of E^3. The class $Z_c(k)$ of functions "admissible" for the kernel k consists of the functions $\varphi(t) \in C(J)$ with the property that $k(t, \tau, \varphi(\tau))$ is defined and L-integrable in the interval $0 \leqq \tau \leqq t$ for every $t \in J$.[1] Thus the admissible functions are those continuous functions for which the integral in (1) is defined. We shall speak of a "kernel which is monotone increasing in the variable z" or simply of a "monotone increasing kernel" if[2]

$$k(t, \tau, z) \leqq k(t, \tau, \bar{z}) \quad \text{for} \quad z \leqq \bar{z} \quad \text{and} \quad (t, \tau, z), (t, \tau, \bar{z}) \in D(k). \tag{2}$$

Monotone increasing kernels play a special role in the determination of bounds and uniqueness. For these kernels it is easy to obtain bounds for a solution u of the integral equation (1) according to the following theorem.

II. Theorem. *Suppose $k(t, \tau, z)$ is a monotone increasing kernel, $v(t)$ and $w(t)$ are functions of the class $Z_c(k)$, $g(t)$ is a function defined in J, and in J*

$$v(t) \leqq g(t) + \int_0^t k(t, \tau, v(\tau))\, d\tau, \qquad w(t) \geqq g(t) + \int_0^t k(t, \tau, w(\tau))\, d\tau,$$

where for each t equality holds in at most one place. Then

$$v < w \quad in \quad J.$$

The basic idea of the following simple proof will be seen often below. For $t=0$ it follows from the hypothesis that $v(0) \leqq g(0)$, $w(0) \geqq g(0)$, where there cannot be equality in both places; hence $v(0) < w(0)$. If the assertion were false, there would be a first point $t_0 \in J_0$ such that $v(t_0) = w(t_0)$ and $v < w$ for $0 < t < t_0$. On the other hand, because of the monotonicity of k,

$$v(t_0) \leqq g(t_0) + \int_0^{t_0} k(t_0, \tau, v)\, d\tau \leqq g(t_0) + \int_0^{t_0} k(t_0, \tau, w)\, d\tau \leqq w(t_0),$$

where there is strict inequality in at least one position. The contradiction thus obtained proves the validity of the theorem.

As a first application of II we prove the well known

III. Gronwall's Inequality. *Let $g(t), v(t) \in C(J)$, $0 \leqq h(t) \in L(J)$ and*

$$v(t) \leqq g(t) + \int_0^t h(\tau) v(\tau)\, d\tau \quad in \quad J. \tag{3}$$

---

[1] Here by definition $\int_0^0 k(t, \tau, \varphi(\tau))\, d\tau = 0$ (independent of a possible singularity of k at $t=0$). The subscript c of $Z_c(k)$ is a reminder that – in contrast to other classes of functions needed later – we are dealing with *continuous* functions.

[2] For example, the kernel $k(t, \tau, z) = -1/z$ is a monotone increasing kernel on the set $D(k)$ defined by $0 \leqq \tau \leqq t \leqq T$, $z > 0$, but not on the set $D(k)$ defined by $0 \leqq \tau \leqq t \leqq T$, $z \neq 0$.

Then in J

$$v(t) \leqq g(t) + \int_0^t g(\tau)h(\tau)e^{H(t)-H(\tau)}d\tau = e^{H(t)}\left[g(0) + \int_0^t g'(\tau)e^{-H(\tau)}d\tau\right] \quad (4)$$

with

$$H(t) = \int_0^t h(\tau)d\tau$$

(the second form of the bound holds only if $g(t)$ is absolutely continuous in J).

The proof makes use of the fact that the function

$$w(t) = \bar{g}(t) + \int_0^t \bar{g}(\tau)h(\tau)e^{H(t)-H(\tau)}d\tau$$

is a solution of the integral equation

$$w(t) = \bar{g}(t) + \int_0^t h(\tau)w(\tau)d\tau$$

(the equality of both integrals is most easily obtained by differentiation). If we choose $\bar{g} > g$, then besides (3) we have the inequality

$$w(t) > g(t) + \int_0^t h(\tau)w(\tau)d\tau\,,$$

and hence by II [with $k(t, \tau, z) = h(\tau)z$], $v < w$. The first expression in (4) is obtained for $\bar{g} \to g$, while the second expression represents only a transformation by integration by parts.

A special form of this inequality goes back to Gronwall (1918/19)[1]. Bellman (1943), Candirov (1958), and others have generalized it; see also the bibliography in Sansone-Conti (1956, p. 15), the literature given in X and Džabbarov (1964), Chu and Metcalf (1967).

We mention briefly some other forms of Gronwall's inequality and generalizations of a different kind.

(α) Neglecting the term $H(\tau)$ in (4), one gets the simpler bound

$$v(t) \leqq g(t) + e^{H(t)}\int_0^t g(\tau)h(\tau)d\tau\,. \quad (4')$$

[1] There the lemma reads as follows: "When, for $x_0 \leqq x \leqq x_0 + h$, the continuous funtion $z = z(x)$ satisfies the inequality

$$0 \leqq z \leqq \int_{x_0}^x (Mz + A)dx\,,$$

where the constants M and A are positive or zero, then

$$0 \leqq z \leqq Ahe^{Mh}, \quad (x_0 \leqq x \leqq x_0 + h)."$$

(β) If, in addition to the above regularity assumptions, $0 \leqq k(t) \in L(J)$ and

$$v(t) \leqq g(t) + k(t) \int_0^t h(\tau) v(\tau) d\tau ,$$

then

$$v(t) \leqq g(t) + k(t) \int_0^t g(\tau) h(\tau) d\tau \left(\exp \int_0^t h(\tau) k(\tau) d\tau \right).$$

This reduces to (α) for $k=1$. The general case is easily obtained from this particular one (thus it is not a generalization of the original statement (3) (4)). If we put $\bar{k}(t) = k(t) + \varepsilon$ ($\varepsilon > 0$), $\bar{v} = v/\bar{k}$, $\bar{g} = g/\bar{k}$, $\bar{h} = h\bar{k}$, then (3) holds with $\bar{v}, \bar{g}, \bar{h}$. Using the bound given in (α) and letting $\varepsilon \to 0$, the result follows.

This form of Gronwall's inequality was given by Willett (1965; Theorem 0). See also the literature cited there.

(γ) Willett (1965) also gives explicit bounds for v under more general assumptions, e.g.,

$$v(t) \leqq g(t) + \sum_{i=1}^{n} k_i(t) \int_0^t h_i(\tau) v(\tau) d\tau .$$

We will not reproduce the resulting inequality which is somewhat lengthy.

(δ) Gronwall's inequality has been generalized to functions in L_p by Willett (1964). Let v, g, h be nonnegative functions of class $L_p(J)$ ($1 \leqq p < \infty$) and

$$v(t) \leqq g(t) + h(t) \left(\int_0^t v^p d\tau \right)^{1/p} .$$

Then

$$\left(\int_0^t v^p d\tau \right)^{1/p} \leqq \frac{\left(\int_0^t g^p \varepsilon(\tau) d\tau \right)^{1/p}}{1 - \{1 - \varepsilon(t)\}^{1/p}} ,$$

where

$$\varepsilon(t) = \exp\left(- \int_0^t h^p(\tau) d\tau \right).$$

See also Willett and Wong (1965).

As first noted by K. Nickel (1961a), Theorem II can be generalized to certain operators:

IV. Definition (*Volterra operator = operator, monotone increasing operator, $Z_c(K)$*). Here we consider only "Volterra operators." A Volterra operator K with domain of definition $Z_c(K) \subset C(J)$ associates with every function $\varphi(t) \in Z_c(K)$ which is continuous in J a real function $K\varphi$ defined in J. The essential thing is that the value of the function $K\varphi$ at a point $t_0 \in J$, denoted by $(K\varphi)(t_0)$, depends only on the values of φ in the interval $0 \leqq t \leqq t_0$, i.e., $(K\varphi)(t_0) = (K\psi)(t_0)$ if $\varphi, \psi \in Z_c(K)$ and $\varphi(t) = \psi(t)$ for $0 \leqq t \leqq t_0$. Since all the operators we consider are Volterra operators in this sense, we simply call them "operators" (see VIII (δ) below).

An operator K is called a "monotone increasing operator" if it has the following property:

If $\varphi, \bar{\varphi} \in Z_c(K)$ and if for some $t_0 \in J_0$ the inequality $\varphi(t) < \bar{\varphi}(t)$ holds in the interval $0 < t < t_0$, then

$$(K\varphi)(t_0) \leqq (K\bar{\varphi})(t_0).$$

An important example is the integral operator

$$K\varphi = \int_0^t k(t, \tau, \varphi(\tau))\, d\tau. \tag{5}$$

For this operator the class $Z_c(K)$ agrees with the class $Z_c(k)$ defined above, and furthermore K is a monotone increasing operator if the kernel k is monotone increasing in z.

Before generalizing Theorem II to operators, we introduce new notation which will be needed in some applications. In a sequence of theorems where, as in II, an inequality $v < w$ is proved, we need to assume $v(0) < w(0)$. A glance at the proofs shows that one can often substitute the weaker hypothesis "$v < w$ for $0 < t < \delta, \delta > 0$." This fact motivates the following

V. Definition. ($\varphi(a+) < \bar{\varphi}(a+)$). If both functions $\varphi, \bar{\varphi}$ are defined in an interval $a < t < a + \varepsilon$ ($\varepsilon > 0$) and if there exists a $\delta > 0$ such that $\varphi < \bar{\varphi}$ for $a < t < a + \delta$, then we write simply

$$\varphi(a+) < \bar{\varphi}(a+).$$

VI. Theorem. *For a monotone increasing operator K and two functions $v, w \in Z_c(K)$, suppose we have*

(α) $v(0+) < w(0+)$;

(β) $v - Kv < w - Kw$ *in* J_0.

Then

$$v < w \quad \text{in} \quad J_0.$$

The assumption (α) can be discarded if K has the property that $(K\varphi)(0) = 0$ for all $\varphi \in Z_c(K)$ and if (β) also holds for $t = 0$ (indeed it then follows from (β)).

The proof proceeds in essence exactly as in II. If the assertion is false, then there exists a $t_0 \in J_0$ with the properties noted in the proof of II. Then because of the monotonicity of K

$$(Kv)(t_0) \leqq (Kw)(t_0),$$

whence, with the aid of (β),

$$v = (v - Kv) + Kv < (w - Kw) + Kv \leqq w$$

at the point $t = t_0$, so that we have arrived at a contradiction to the assumption $v(t_0) = w(t_0)$.

Now we consider the operator equation

$$u = g + Ku; \tag{6}$$

the function $g(t)$ is defined on J. By a *solution* we will always mean a function continuous in J which satisfies Eq. (6) in J. If the operator K is monotone increasing, then from VI we can easily obtain conditions for a function to be a superfunction or a subfunction relative to Eq. (6). The hypothesis (β) is actually a consequence of (β'):

(β') $v \leqq g + Kv$, $w \geqq g + Kw$ in J_0, where equality holds in at most one place [1].

VII. Superfunctions, Subfunctions. The function $v \in Z_c(K)$ is a subfunction, the function $w \in Z_c(K)$ is a superfunction relative to Eq. (6) with a monotone increasing operator K if

(α) $v(0+) < u(0+) < w(0+)$ for every solution u;

(β) $v < g + Kv$, $w > g + Kw$ in J_0.

For two functions v, w satisfying these assumptions and an arbitrary solution u we then have

$$v < u < w \quad \text{in} \quad J_0 \tag{7}$$

(the two assertions about v, or w, are independent of each other).

The operator Eq. (6) thus represents a problem of monotonic type in the sense of the introduction (II) if K is a monotone increasing operator. In order to be consistent with the notation of the introduction, we need only introduce the operator T given by $T\varphi = \varphi - g - K\varphi$. Likewise, Theorem VI is a monotonicity theorem for T.

If in particular we have an integral equation (1), i.e., if K is the operator defined by (5) and if $k(t, \tau, z)$ is a monotone increasing kernel, then condition (β) becomes

$$(\beta') \quad v(t) < g(t) + \int_0^t k(t, \tau, v(\tau))d\tau, \qquad w(t) > g(t) + \int_0^t k(t, \tau, w(\tau))d\tau.$$

---

[1] Theorem VI is only apparently more general with hypothesis (β) than with (β'). Indeed, if (β) is satisfied, then so is (β') with $g = v - Kv$.

If we also require both of these inequalities to hold in J (instead of J_0), (α) is automatically satisfied, since $v(0) < g(0) = u(0) < w(0)$ for every solution u[1]. The corresponding remark holds for every operator K with the property that $(K\varphi)(0) = 0$ for $\varphi \in Z_c(K)$.

A monotone increasing operator which does *not* possess this last property is the mean-value operator

$$K\varphi = \frac{1}{t}\int_0^t \varphi(\tau)\,d\tau \quad \text{in} \quad J_0, \qquad (K\varphi)(0) = \varphi(0).$$

VI would be false for it without the assumption (α), as can be seen immediately[2].

A further example is provided by the following operator defined by a Stieltjes integral:

$$K\varphi = \int_0^t k(t, \tau, \varphi(\tau))\,d\alpha(\tau).$$

If $\alpha(t)$ is monotone increasing in t and $k(t, \tau, z)$ is monotone increasing in z, then K is obviously a monotone increasing operator. Here $(K\varphi)(0) = 0$ holds only if $\alpha(t)$ is continuous at zero.

VIII. Generalizations and Remarks. (α) Frequently $k(t, \tau, z)$ will be defined for all points of the set $\{0 \leqq \tau \leqq t \leqq T, |z| < \infty\}$ and will be assigned properties such that the integral (5) exists for all $\varphi \in C(J)$. However it will sometimes be useful to restrict the domain of definition of K artificially. We give an example. We have two functions v, w for which VII (β') holds in J. Suppose the kernel $k(t, \tau, z)$ is monotone increasing not for all z, but only for those z with $v(\tau) \leqq z \leqq w(\tau)$. Then the bound (7) holds for all solutions of Eq. (1). If we take the set of all (t, τ, z) with $0 \leqq \tau \leqq t \leqq T$, $v(\tau) \leqq z \leqq w(\tau)$ for the domain of definition of k, then VII applies. All solutions u lie in this domain, which we see as follows: Since $v(0) < u(0) < w(0)$, there is for u a greatest positive number $T_1 \leqq T$ such that $v \leqq u \leqq w$ in the interval $0 \leqq t \leqq T_1$. But then VII also implies that $v < u < w$ in this interval and hence $T_1 = T$.

(β) The functions u, v, w are allowed to have singularities at the origin. Thus we consider functions from $C(J_0)$ and operators K which map a subset of the space $C(J_0)$ onto the set of functions defined in J_0. In this case VI and VII also hold.

---

[1] The example $g = u = 0, v = t, k = 2z/\tau$ shows that (7) is false in general if the inequality (β') holds only in J_0.

[2] Here too $k(t, \tau, z) = z/t$ is a monotone increasing kernel. The new situation arises only because the convention $(K\varphi)(0) = 0$ usually met for integral equations is here set aside; cf. Footnote 1, page 14.

(γ) It is not necessary for the operator equation to be linear in u and Ku. We can treat the equation

$$h(t, u, Ku) = 0 \tag{8}$$

in exactly the same way; here $h(t, z, p)$ is defined on $J \times E^2$ and is monotone decreasing in p. In Theorem VI we then have to replace (β) by

$$h(t, v, Kv) < h(t, w, Kw); \tag{9}$$

the modifications in the proof are trivial. The corresponding substitution $h(t, v, Kv) < 0 < h(t, w, Kw)$ is needed in VII (β).

At first glance it looks as if this also covers integral equations *of the first kind*. However, this is not so. For if we set $h(t, z, p) = h_1(t) - p$, then $Kv > Kw$ follows from (9). On the other hand, from VII (α), $v < w$ and thus $Kv \leqq Kw$ for small positive T. Thus in this case there do not exist any super- and subfunctions in the sense of VII. This was first noted by Nickel (1961a, p. 166).

(δ) For an operator K, the value of $K\varphi$ at the point t_0 depends only on the values assumed by $\varphi(t)$ in the interval $0 \leqq t \leqq t_0$. We speak of operators "of Volterra type." Such operators were first investigated by Volterra (see the book by Volterra (1927), which contains details on the older literature) and Tonelli (1928). A more recent bibliography can be found in Zitarosa (1954).

IX. Allowing Equality Signs in Inequalities. In using VI and VII, inequalities must be satisfied in which no equality signs are permitted. This is often inconvenient in practice. A satisfactory general answer to the question of when equality signs are permissible cannot be given; this is a problem which has to be solved for each case. The considerations needed in this regard are mostly based on one of the following three principles.

(α) The first concerns the operator property of being "*strictly monotone increasing*." This means: if $\varphi < \bar{\varphi}$ for $0 < t < t_0$, then

$$(K\varphi)(t_0) < (K\bar{\varphi})(t_0).$$

There can be equality signs in VI (β), VII (β) – but not in (α) – for strictly monotone increasing operators K. The proof of VI remains the same, except that now the symbol $<$ has a different location in that chain of inequalities.

(β) The second principle is most closely related to the concepts of *maximal and minimal solution*, which will be investigated in the next section.

If K is a monotone increasing operator and $(K\varphi)(0)=0$ for all $\varphi \in Z_c(K)$, then we have the following theorem:

VI: *$v \leqq g+Kv, w>g+Kw$ in J implies $v<w$ in J.*

This can be sharpened into a theorem

VIa: *$v \leqq g+Kv, u^*=g+Ku^*$ in J implies $v \leqq u^*$ in J,*

if for u^* there exists a sequence of functions w_n with $w_n>g+Kw_n$ and $w_n \to u^*$ $(n\to\infty)$. Then by VI we have $v<w_n$, which, for $n\to\infty$, implies the assertion $v \leqq u^*$. Such a sequence w_n exists under conditions which will be investigated in more detail in 2 IV when u^* is the maximal solution of Eq. (6). A similar result holds for the minimal solution u_* of Eq. (6) under corresponding conditions:

$$w \geqq g+Kw \text{ in } J \text{ implies } u_* \leqq w \text{ in } J.$$

This was proved for integral equations by Satō and Iwasaki (1955).

(γ) We now come to the third and most important principle. We look for hypotheses under which the following theorem holds:

VIb: *$v \leqq g+Kv, w \geqq g+Kw$ in J implies $v \leqq w$ in J.*

Obviously K needs to be a monotone increasing operator here, $v, w \in Z_c(K)$ and $(K\varphi)(0)=0$ for $\varphi \in Z_c(K)$. For the validity of VIb, it suffices to have a uniqueness condition for K of the type considered in Section 4; more precisely, a bound $|K\varphi - K\bar{\varphi}| \leqq \Omega(|\varphi-\bar{\varphi}|)$ for φ, $\bar{\varphi} \in Z_c(K)$ by a monotone increasing operator $\Omega \in \mathscr{E}$ (see 4 II, III). For if $\varrho(t)$ is a function which is positive and continuous in J and $\varrho > \Omega\varrho$ with $w \geqq g+Kw$, then

$$w+\varrho \geqq g+Kw+\varrho > g+K(w+\varrho),$$

since $K(w+\varrho)-Kw \leqq \Omega\varrho < \varrho$. By VI (applied to v and $w+\varrho$) we also have $v < w+\varrho$ and hence $v \leqq w$, since by the definiton of $\mathscr{E}$ there exist arbitrarily small functions ϱ with these properties. Of course in the hypotheses we also have to include $w+\varrho \in Z_c(K)$.

On the other hand, uniqueness for the equation $u=Ku+g$ immediately follows from Theorem VI b. For if $u, \bar{u}$ are two solutions we then get $u \leqq \bar{u}$ as well as $\bar{u} \leqq u$ from VI b. So, we have the following situation: *If* VI b *holds, then uniqueness prevails; if on the contrary K satisfies the uniqueness condition above, then* VI b *holds.* This shows that the given hypotheses are appropriate here and are not merely required for technical reasons in the proof. Another advantage is that we have here an explicitly verifiable condition which is independent of existence theory.

Of course this also sharpens Theorem VII: under the hypotheses of VI b we have $v \leqq u \leqq w$ for the uniquely determined solution u of (6).

X. Application to Stability Theory. Gronwall's inequality has significant applications in the theory of stability of ordinary differential equations; see Bellmann (1943, 1953, especially p. 35), Levinson (1946), Weyl (1946), Szmydt (1955), Bihari (1956, 1957), Golomb (1958), Cesari (1959, especially § 3, § 6). The point of departure for this is the following well known fact.

Suppose we are given a system of n linear differential equations (see 6 I for the vector notation)

$$\boldsymbol{u}'(t) = A(t)\boldsymbol{u}(t) + \boldsymbol{b}(t) \tag{10}$$

($A(t)$ is an $n \times n$ matrix, $\boldsymbol{u}$ and $\boldsymbol{b}$ are column vectors, A and $\boldsymbol{b}$ are continuous in J) and $Y(t)$ is the fundamental system of n solutions of the homogeneous equation $\boldsymbol{u}' = A(t)\boldsymbol{u}$ with $Y(0) = E$ (identity matrix). Then the solution of Eq. (10) with initial value $\boldsymbol{u}(0) = \boldsymbol{\eta}$ is

$$\boldsymbol{u}(t) = Y(t)\boldsymbol{\eta} + \int_0^t Y(t)Y^{-1}(\tau)\boldsymbol{b}(\tau)d\tau . \tag{11}$$

In the special case where A is a constant matrix, $Y(t)\,Y^{-1}(\tau) = Y(t-\tau) = e^{A(t-\tau)}$. If we use the compatible norms $\|\boldsymbol{z}\|$ for vectors $\boldsymbol{z}$ and $\|M\|$ for $n \times n$ matrices M (i.e., such that $\|M\boldsymbol{z}\| \leqq \|M\|\,\|\boldsymbol{z}\|$), then for a solution $\boldsymbol{u}$ of the equation

$$\boldsymbol{u}' = A(t)\,\boldsymbol{u} + \boldsymbol{g}(t, \boldsymbol{u}) \tag{12}$$

we obtain the integral equation (11) with $\boldsymbol{g}(\tau, \boldsymbol{u}(\tau))$ in place of $\boldsymbol{b}(\tau)$ and hence

$$\|\boldsymbol{u}(t)\| \leqq \|Y(t)\|\,\|\boldsymbol{\eta}\| + \int_0^t \|Y(t)Y^{-1}(\tau)\|\,\|\boldsymbol{g}(\tau, \boldsymbol{u}(\tau))\|\,d\tau , \tag{13}$$

which, together with a Lipschitz condition on $\boldsymbol{g}$

$$\|\boldsymbol{g}(t, \boldsymbol{z})\| \leqq h(t)\,\|\boldsymbol{z}\|$$

leads directly to an inequality of the form (3). These considerations also hold for complex differential equations.

Example. If A is a constant matrix whose eigenvalues all have negative real part, then $\|Y(t)\| \leqq ce^{-\alpha t}$ for positive constants c and α. Thus from (13),

$$\|\boldsymbol{u}(t)\| \leqq c\,\|\boldsymbol{\eta}\|\,e^{-\alpha t} + \int_0^t e^{-\alpha(t-\tau)}h(\tau)\,\|\boldsymbol{u}(\tau)\|\,d\tau .$$

For $v(t) = e^{\alpha t}\,\|\boldsymbol{u}(t)\|$ we then have

$$v(t) \leqq c\,\|\boldsymbol{\eta}\| + \int_0^t h(\tau)v(\tau)d\tau$$

and hence by Gronwall's inequality

$$v(t) = e^{\alpha t}\|\boldsymbol{u}(t)\| \leqq c\|\boldsymbol{\eta}\| e^{\int_0^t h(\tau)d\tau}$$

From this it is easy to derive the stability theorems which go back to Lyapunov and which are basic for nonlinear systems of the form (12) with constant A. Compare the above proof with Bellman (1953, "first proof" of the "fundamental stability theorem"), Cesari (1959, p. 92), Coddington and Levinson (1955, p. 314ff).

2. Remarks on the Existence Problem. Maximal and Minimal Solutions

An existence theorem for the integral equation

$$u = g + Ku \tag{1}$$

with

$$Ku = \int_0^t k(t, \tau, u(\tau))d\tau \tag{2}$$

is easily obtained with help of the Schauder fixed point theorem. This path was taken by Satō (1953) among others. Since a proof with sufficiently general hypotheses is not to be found in the textbooks and in order to enable the reader to proceed independently of original sources, we go briefly into the proof.

I. Definition ($\mathscr{K}$, $\mathscr{K}_0$). Suppose that for fixed $(t, z) \in J \times E^1$ the kernel $k(t, \tau, z)$ is measurable in the interval $0 \leqq \tau \leqq t$, for fixed $\tau \in J$ it is continuous for $(t, z) \in \{\tau \leqq t \leqq T\} \times E^1$ (the latter need hold only for almost all $\tau \in J$), and for every number $M > 0$ there exists a function $h(\tau) \in L(J)$ with the property that

$$|k(t, \tau, z)| \leqq h(\tau) \tag{3}$$

for $0 \leqq \tau \leqq t \leqq T$, $|z| \leqq M$. These hypotheses are summarized as $k \in \mathscr{K}_0$, while $k \in \mathscr{K}$ means that $k \in \mathscr{K}_0$ and there exists an $h \in L(J)$ such that (3) holds for all z.

It will be shown that if g is continuous and K is generated by a kernel $k \in \mathscr{K}$, the mapping $\varphi \to g + K\varphi$ maps a suitable convex and compact subset Φ of $C(J)$ into itself and this mapping is continuous (i.e., that the functions from Φ are uniformly bounded and equicontinuous and that $K\varphi_n - K\varphi$ converges to zero in J uniformly with $\varphi_n - \varphi$). Then according to Schauder's fixed point theorem there exists a fixed point u, and this function u represents a solution of the integral equation (1) (2).

In the proof we shall need the two constants

$$M=\int_0^T h(\tau)d\tau, \qquad M_1=M+\max_J |g(t)|.$$

Let

$$\delta(\tau,\varepsilon)=\sup\{|k(\bar{t},\tau,\bar{z})-k(t,\tau,z)|;\ \bar{t}-t+|z-\bar{z}|\leqq\varepsilon,\ |z|\leqq M_1\}, \quad (4)$$

where, in addition, the variables are restricted by $0\leqq\tau\leqq t\leqq\bar{t}\leqq T$. This function is measurable in τ, continuous and monotone decreasing in ε for $\varepsilon\geqq 0$, and by (3) we have $0\leqq\delta(\tau,\varepsilon)\leqq 2h(\tau)$ and $\delta(\tau,0)=0$ in J. By the theorem on dominated convergence [Kamke (1956, p. 146)] we then have

$$\int_0^T \delta(\tau,\varepsilon)d\tau\to 0 \quad \text{as} \quad \varepsilon\to 0. \quad (5)$$

As a subset Φ of $C(J)$ we choose the set of all continuous functions $\varphi(t)$ for which $|\varphi(t)-g(t)|\leqq M$ and

$$|\varphi(\bar{t})-\varphi(t)|\leqq\int_0^T\delta(\tau,\bar{t}-t)d\tau+\int_t^{\bar{t}}h(\tau)d\tau+|g(\bar{t})-g(t)| \quad \text{for} \quad 0\leqq t<\bar{t}\leqq T.$$

The subset Φ is convex, closed with respect to uniform convergence and equicontinuous, thus compact.

The continuity of the operator K follows immediately from the inequalities

$$|(K\varphi)(t)-(K\bar{\varphi})(t)|\leqq\int_0^t|k(t,\tau,\varphi)-k(t,\tau,\bar{\varphi})|d\tau\leqq\int_0^T\delta(\tau,\max_J|\varphi-\bar{\varphi}|)d\tau$$

and (5). That Φ is mapped into itself by the mapping $\varphi\to g+K\varphi$ is a consequence of the following bound which holds for $0\leqq t\leqq\bar{t}\leqq T$:

$$\begin{aligned}|(K\varphi)(\bar{t})-(K\varphi)(t)|&\leqq\int_0^t|k(\bar{t},\tau,\varphi)-k(t,\tau,\varphi)|d\tau+\int_t^{\bar{t}}|k(\bar{t},\tau,\varphi)|d\tau\\&\leqq\int_0^T\delta(\tau,\bar{t}-t)d\tau+\int_t^{\bar{t}}h(\tau)d\tau.\end{aligned}$$

We combine this in the following

II. Existence Theorem. *If* $g(t)\in C(J)$, $k(t,\tau,z)\in\mathscr{K}$, *then the integral equation* (1) (2) *has at least one solution* $u(t)\in C(J)$.

Theorem II was proved (even for bounded, measurable g) by Kanazawa and Murakami (1955); see also Tonelli (1927/28), Cameron and Shapiro (1955). Using II it is not difficult to obtain the usual extensions of existence theory. These concern the case where $k(t,\tau,z)$ is defined only on a point set $J\times D$ (D *is a domain of the* (τ,z) *space*) and a bound (3) is valid only locally. They culminate in the theorem that a solution can be continued to the right up to the boundary of the domain of definition. This was carried out by Satō (1953) for continuous kernels. If for example $k \in \mathscr{K}_0$, then existence can be obtained with the help of well known devices. First we restrict the variable z to $|z| \leqq M$, i.e., instead of $k(t, \tau, z)$ we consider the kernel $k(t, \tau, [z]_M)$ where $[z]_M = z$ for $|z| \leqq M$ and $= M \operatorname{sgn} z$ for $|z| > M$. For every M we have $k(t, \tau, [z]_M) \in \mathscr{K}$, and the solution u of the integral equation with the modified kernel is also a solution of the original equation as long as $|u| \leqq M$. In this way, if M is taken sufficiently large, we obtain existence in an interval $0 \leqq t \leqq T_1$ $(\leqq T)$. Now we consider the integral equation

$$u(t) = \bar{g}(t) + \int_{T_1}^{t} k(t, \tau, u(\tau)) d\tau \quad \text{with} \quad \bar{g}(t) = g(t) + \int_{0}^{T_1} k(t, \tau, u(\tau)) d\tau$$

in the interval $T_1 \leqq t \leqq T$ and proceed analogously, etc. This type of reasoning is well known for ordinary differential equations [Kamke (1945; p. 75f), Coddington-Levinson (1955; Chap. 1, Sec. 4)].

III. Maximal Solution, Minimal Solution. The following definition suggests itself: "the solution $u^* \in Z_c(K)$ of Eq. (1) is a maximal solution if for every other solution $u \in Z_c(K)$ the inequality $u \leqq u^*$ in J holds." However, for the integral equation (1) (2) with $J = [0, 1]$, $g = 0$, $k(t, \tau, z) = 2z/\tau(1-\tau)$, it would have the paradoxical consequence that $u = 0$ is a maximal solution since all other solutions $u = Ct^2/(1-t)^2$ (C arbitrary) fail to exist in all of J. With regard to such examples, we therefore also consider solutions which exist only in an interval $0 \leqq t < T_1 (\leqq T)$ (this is possible only if K is an operator of Volterra type, i.e., if $(K\varphi)(t_0)$ depends only on the values of φ in the interval $0 \leqq t \leqq t_0$). If $u^* \in Z_c(k)$ is a solution of (1) and if, when u is an arbitrary continuous solution of (1) in an interval $0 \leqq t < T_1 \leqq T$, we have $u \leqq u^*$ for $0 \leqq t < T_1$, then we say u^* is a maximal solution (in J). If the solution u^* exists only in $J^*: 0 \leqq t < T^*$ and if for an arbitrary solution u existing, say, in the interval $0 \leqq t \leqq T_1$ we have $u \leqq u^*$ for $0 \leqq t < \min(T_1, T^*)$, then we speak of a maximal solution in J^*. Corresponding definitions hold for the minimal solution.

Now we shall consider two methods for constructing maximal and minimal solutions with various ranges of application (a further method is given in the following section).

The first method is suggested by 1 VI; it can be used if K is a monotone increasing operator and an existence theorem holds. We consider a sequence of positive numbers $\varepsilon_n \searrow 0$ and an associated sequence of solutions $u = u_{\varepsilon_n}$ of the equation

$$u = g + \varepsilon_n + Ku\,. \tag{6}$$

If $\varepsilon_n > \varepsilon_{n+1} > 0$ and u is a solution of (1), then by 1 VI we have $u < u_{\varepsilon_{n+1}} < u_{\varepsilon_n}$ (we assume that $(K\varphi)(0) = 0$). Therefore the u_{ε_n} converge to a function

$u^* \geqq u$ which is a solution of (1) and hence the desired maximal solution, if the passage to the limit $n \to \infty$ in (6) can be carried inside the operator K. This train of reasoning leads us to

IV. Theorem. *Suppose the monotone increasing operator K has the properties*

(α) *there is a $\delta > 0$ such that the equation ($g \in C(J)$)*

$$u = g + \varepsilon + Ku$$

has at least one solution $u_\varepsilon \in Z_c(K)$ for all ε with $|\varepsilon| < \delta$;

(β) $(Ku_\varepsilon)(0) = 0$ *for all solutions u_ε.*

Suppose further that one of the following hypotheses (γ_1), (γ_2) holds.

(γ_1) *The solutions u form an equicontinuous set of functions; if a sequence u_{ε_n} of solutions converges uniformly in J to a function $\bar{u}$, then $\bar{u} \in Z_c(K)$ and*[1] *$Ku_{\varepsilon_n} \to K\bar{u}$ as $n \to \infty$.*

(γ_2) *If a sequence of solutions u_{ε_n} increases or decreases and converges to a bounded function $\bar{u}$, then $K\bar{u}$ exists, and*[1] *$Ku_{\varepsilon_n} \to K\bar{u}$ as $n \to \infty$, $K\bar{u} \in C(J)$.*

Then Eq. (1) *has a maximal solution u^* and a minimal solution u_* in J, and*

(δ) *u_* (resp. u^*) is the limit of a monotone increasing sequence of subfunctions v_n (resp. a monotone decreasing sequence of superfunctions w_n), i.e.,*

$$v_n < g + Kv_n, \qquad w_n > g + Kw_n, \qquad v_n \nearrow u_*, \qquad w_n \searrow u^* .$$

In case (γ_1) we note that a monotone, bounded sequence converges uniformly due to the equicontinuity, while in case (γ_2) the continuity of u^* is obtained only *after* passing to the limit in the equation $u^* = g + Ku^*$ and using the continuity of both functions on the right hand side. The procedure is the same for the minimal solution, except for the use of a sequence of negative numbers $\varepsilon_n \nearrow 0$.

It is precisely property (δ) which is essential for our purposes. As discussed in 1 IX (β), this property is needed in the proof of the theorem that the inequality $v \leqq u^*$ follows from $v \leqq g + Kv$. The integral equation (1) (2) represents the most important application of IV. *If $k \in \mathscr{K}$ is a monotone increasing kernel, then a maximal and a minimal solution exist and they have the property* IV (δ).

This is obtained, as has been said, from IV; the hypotheses (α), (β), (γ_2) are easily verified. The important property IV (δ) is lost in the second method of proof, which we now outline. Here it is essential that along with u and $\bar{u}$ the functions $\max(u, \bar{u})$ and $\min(u, \bar{u})$ are also solutions of (1) (ordinary differential equations, for example, possess this property).

[1] The following limit should hold for every $t \in J$ (not necessarily uniformly).

V. Theorem. *Suppose the operator K has the properties*

(α) *The solutions of* (1) *form a nonempty set of equicontinuous functions;*

(β) *along with u and $\bar{u}$ the functions* $\max(u, \bar{u})$ *and* $\min(u, \bar{u})$ *are also solutions;*

(γ) $(Ku)(0) = 0$ *for all solutions;*

(δ) *the limit of a uniformly convergent sequence of solutions is also a solution.*

Then Eq. (1) *has a maximal and a minimal solution.*

If, say, $u^*(t) = \sup\{u(t): u \text{ a solution}\}$ and if $\{t_k\}$ is a countable dense point set in J, then for every k there exists a sequence of solutions u_{kn} $(k, n = 1, 2, \ldots)$ with the property

$$u_{kn}(t_k) \to u^*(t_k) \quad \text{as} \quad n \to \infty .$$

By (β), the functions $u_m = \max(u_{kn}: 1 \leqq k, n \leqq m)$ are solutions, and we have

$$\lim_{m \to \infty} u_m(t) = u^*(t)$$

for all $t = t_k$. Because of (α) this limit exists and is moreover uniform for all $t \in J$; because of (δ), u^* is a solution.

The method described in V applies to ordinary differential equations (5 II). See also Viswanatham (1965).

VI. Difference Kernels ($\mathscr{K}_1$). The class of functions $\mathscr{K}$ is not suitable for an important class of kernels, the so-called difference kernels. For if $k(t, \tau, z) = k_1(t - \tau, z)$ is a difference kernel and the function $k_1(t, z)$ is unbounded in a neighborhood of a point (t, z), then k certainly does not lie in $\mathscr{K}$. In order to be able to deal with such kernels as well, it is useful to define a class $\mathscr{K}_1$ as follows. Let $k(t, \tau, z)$ be measurable in $(t, \tau) \in \{0 < \tau < t \leqq T\}$ for fixed $z \in E^1$, continuous in $z \in E^1$ for fixed (t, τ). If

$$(\alpha) \quad \int_0^t \sup_z |k(t, \tau, z)| \, d\tau \leqq M < \infty \quad \text{for} \quad t \in J$$

and if to a given N there corresponds a modulus of continuity $\delta = \delta_N$ such that for $0 < \tau < t < \bar{t} \leqq T$ and $|z|, |\bar{z}| \leqq N$

$$(\beta) \quad \int_t^{\bar{t}} \sup_z |k(t, \tau, z)| \, d\tau \leqq \delta(\bar{t} - t)$$

$$(\gamma) \quad \int_0^t \sup_z |k(\bar{t}, \tau, z) - k(t, \tau, z)| \, d\tau \leqq \delta(\bar{t} - t)$$

$$(\delta) \quad \int_0^t \sup_{|z - \bar{z}| \leqq \varepsilon} |k(t, \tau, \bar{z}) - k(t, \tau, z)| \, d\tau \leqq \delta(\varepsilon),$$

then we say that $k \in \mathscr{K}_1$.

The existence theorem II *also holds if the kernel $k(t, \tau, z)$ belongs to the class $\mathscr{K}_1$.*

For the proof we take M from (α) and choose the set Φ of all functions φ in $C(J)$ satisfying

$$|\varphi(t)-g(t)| \leqq M \quad \text{and} \quad |\varphi(\bar{t})-\varphi(t)| \leqq |g(\bar{t})-g(t)|+2\delta(\bar{t}-t)$$

where $\delta=\delta_N$, $N=M+\max|g|$. The operator $\varphi \rightarrow g+K\varphi$ maps Φ into itself since for $\varphi \in \Phi$, $\psi=K\varphi$ we have $|\psi| \leqq M$ from (α) and $|\psi(\bar{t})-\psi(t)| \leqq 2\delta(\bar{t}-t)$ from (β) (γ). Furthermore, K is continuous by virtue of (δ). According to Schauder's fixed point theorem we again obtain the existence of a solution of the integral equation (1) (2).

For example, the kernel

$$k(t,\tau,z)=p(t,\tau,z)\,q(t-\tau) \tag{7}$$

belongs to the class $\mathscr{K}_1$ if $q \in L(J)$ and if p is a bounded function which is measurable in τ and continuous in (t,z) uniformly on bounded subsets of $\{0 \leqq \tau \leqq t \leqq T\} \times E^1$. In this example the proof of (α) (β) (δ) is simple, while the proof of (γ) uses the fact that

$$\int_0^t |q(s+\varepsilon)-q(s)|\,ds \rightarrow 0 \quad \text{for} \quad \varepsilon \rightarrow 0 \quad \text{uniformly in } t\,.$$

The case where p is independent of t and continuous, $q(t)=t^\alpha$, $\alpha>-1$, was investigated in detail by Dinghas (1958).

As above, we get a local existence theorem if we dispense with (α). Then the integral u exists possibly only in a smaller interval $0 \leqq t < T_1$ ($\leqq T$). Observe that a local version of (α) is contained in (β) for $t=0$.

The assumptions (α)—(δ) are similar to those made by Li Mun-Su (1965; p. 294) (he treats a more general case; see 6 XII).

Picone (1960a) investigated difference kernels and, in a further work (1960), kernels of another special form. An interesting nonlinear heat conduction problem was solved by Padmavally (1958) also by reduction to a nonlinear Volterra integral equation with difference kernel.

There is an extensive literature on integral equations of "Hammerstein type" whose kernels are of the special form

$$k(t,\tau,z)=h(t,\tau)\,f(\tau,z)\,. \tag{8}$$

The result presented above for the class $\mathscr{K}_1$ immediately yields an existence theorem for equations of Hammerstein type under very general assumptions. Equations with kernels of the form (7) or (8) are treated, among others, by Corduneanu (1963, 1963a, 1965), Friedman (1963), Reichert (1962, 1965), Vinokurov (1967).

3. Generalization of the Monotonicity Concept

This section can be omitted in a first reading. All the investigations up to now have been based on a stringent monotonicity hypothesis. The question arises as to whether we can avoid this condition by using another method of proof, or at least weaken it. This is possible in certain circumstances, as was shown by Satō (1953) for integral equations with continuous kernel. The results of Satō will now be presented from a somewhat more general point of view. As an application we shall show a third way of obtaining maximal and minimal solutions.

I. Definition ($g_\alpha(t\,|\,\bar{t})$, *α-monotonicity*). If $g(t)$ is a function defined in J, α is a real number with $0 \leqq \alpha \leqq 1$ and $0 \leqq t < \bar{t} \leqq T$, then we define

$$g_\alpha(t\,|\,\bar{t}) = g(\bar{t}) - \alpha g(t)\,.$$

Thus, for example, $k_\alpha(t\,|\,\bar{t}, \tau, z) = k(\bar{t}, \tau, z) - \alpha k(t, \tau, z)$ and $(K\varphi)_\alpha(t\,|\,\bar{t}) = (K\varphi)(\bar{t}) - \alpha(K\varphi)(t)$. The kernel k is said to be α-monotone if

$$k_\alpha(t\,|\,\bar{t}, \tau, z) \leqq k_\alpha(t\,|\,\bar{t}, \tau, \bar{z}) \quad \text{for} \quad z \leqq \bar{z}$$

and $0 \leqq \tau \leqq t < \bar{t} \leqq T$ (as long as the arguments lie in the domain of definition of k). (Moreover, it suffices that this inequality hold for those t of a left-sided neighborhood of $\bar{t}$). A 0-monotone kernel is thus monotone in the ordinary sense of 1 I.

II. Theorem. *Suppose K is an operator, $v(t)$ and $w(t)$ are functions in $Z_c(K)$, $g(t)$ and $d(t)$ are functions defined in J, α is a number between* 0 *and* 1 *and*

(α) $v(0+) < w(0+)$;

(β) $v_\alpha(t\,|\,\bar{t}) \leqq g_\alpha(t\,|\,\bar{t}) + (Kv)_\alpha(t\,|\,\bar{t})$, $\quad w_\alpha(t\,|\,\bar{t}) \geqq g_\alpha(t\,|\,\bar{t}) + d_\alpha(t\,|\,\bar{t}) + (Kw)_\alpha(t\,|\,\bar{t})$ *for all $t < \bar{t} \in J_0$ sufficiently close to $\bar{t}$;*

(γ) $(K\varphi)_\alpha(t\,|\,\bar{t}) \leqq (K\bar{\varphi})_\alpha(t\,|\,\bar{t}) + d_\alpha(t\,|\,\bar{t})$ *(for $\alpha = 0$ we need the $<$ symbol here) for all $t < \bar{t}$ sufficiently close to $\bar{t}$, if $\varphi, \bar{\varphi} \in Z_c(K)$ and $\varphi < \bar{\varphi}$ for $0 < t < \bar{t} \leqq T$, $\varphi(\bar{t}) = \bar{\varphi}(\bar{t})$.*

Then

$$v < w \quad \textit{in} \quad J_0\,.$$

For $\alpha = d = 0$ this theorem (with a slight change) becomes 1 VI, and the proof also follows the previous pattern. From the assumption that the assertion is false we find in a familiar way that $v(\bar{t}) = w(\bar{t})$ and $v < w$ for $0 < t < \bar{t}$ for suitable $\bar{t} \in J_0$. The contradiction which is obtained here with the help of (β) (γ) is contained in the following chain of inequalities ($t<\bar{t}$)

$$0<\alpha(w(t)-v(t))=v_\alpha(t\,|\,\bar{t})-w_\alpha(t\,|\,\bar{t})$$
$$\leqq (Kv)_\alpha(t\,|\,\bar{t})-(Kw)_\alpha(t\,|\,\bar{t})-d_\alpha(t\,|\,\bar{t})\leqq 0$$

(when $\alpha=0$ there is an equality sign at the beginning and then a $<$ in the middle).

A specialization to integral equations gives the

III. Consequence. Suppose the kernel $k(t,\tau,z)$ is α-monotone (where $\alpha\in[0,1]$) and the following condition is satisfied:

(α) there is a function $\delta(\tau)\in L(J)$ such that if $(t_0,z_0)\in J_0\times E^1$,

$$k(t_0,\tau,z)-k(t_0,\tau,\bar{z})\leqq\delta(\tau)\quad\text{for}\quad t_0-\varepsilon\leqq\tau\leqq t_0,\ z_0-\varepsilon\leqq z\leqq\bar{z}\leqq z_0+\varepsilon$$

for suitable $\varepsilon=\varepsilon(t_0,z_0)>0$ (as long as the arguments are in the domain of definition of k).

Then if $u,v,w\in Z_c(k)$ and u, v and w are respectively solutions of the integral equation

$$\varphi(t)=g(t)+d(t)+\int_0^t k(t,\tau,\varphi(\tau))\,d\tau \tag{1}$$

with

$$\left.\begin{array}{c} d(t)=0,\quad d(t)=-\beta-\int_0^t\delta(\tau)\,d\tau,\quad d(t)=\beta+\int_0^t\delta(\tau)\,d\tau \\ (\beta>0)\end{array}\right\} \tag{2}$$

then we have the inequalities

$$v<u<w\quad\text{in}\quad J.$$

We reduce the proof of the inequality $u<w$ to II (with u instead of v). II (α) is obtained from $u(0)=g(0)<g(0)+\beta=w(0)$; II ($\beta$) is satisfied with the equality sign for all $t<\bar{t}$. The proof of II (γ) follows from the identity

$$\left.\begin{array}{r}(K\varphi)_\alpha(t\,|\,\bar{t})-(K\bar{\varphi})_\alpha(t\,|\,\bar{t})=\int_0^t[k_\alpha(t\,|\,\bar{t},\tau,\varphi)-k_\alpha(t\,|\,\bar{t},\tau,\bar{\varphi})]\,d\tau \\ +\int_t^{\bar{t}}[k(\bar{t},\tau,\varphi)-k(\bar{t},\tau,\bar{\varphi})]\,d\tau\,.\end{array}\right\} \tag{3}$$

Here the first integrand is $\leqq 0$ since $\varphi<\bar{\varphi}$ and k is α-monotone. The second integrand is $\leqq\delta(\tau)$, at least for t near $\bar{t}$, since by (α) there is an $\varepsilon=\varepsilon(\bar{t},\varphi(\bar{t}))>0$ such that this inequality holds as long as $\bar{t}-\varepsilon<\tau<\bar{t}$ and $\varphi(\bar{t})-\varepsilon\leqq\varphi(t)\leqq\bar{\varphi}(\tau)\leqq\varphi(\bar{t})+\varepsilon$. The left hand side of (3) is thus (for t near $\bar{t}$)

$$\leqq\int_t^{\bar{t}}\delta(\tau)\,d\tau=d(\bar{t})-d(t)\leqq d(\bar{t})-\alpha d(t)=d_\alpha(t\,|\,\bar{t})\,,$$

which was the only thing left to prove (for $\alpha=0$ the $<$ sign actually holds since $\delta(\tau)\geqq 0$ and hence $d(t)\geqq\beta>0$).

If we apply the result just proved to the functions $\bar{u}, \bar{w}, \bar{g}$, where $\bar{u}=v, \bar{w}=u, \bar{g}=g-d$, then it follows that $v<u$ and we have the rest of the assertion.

We can use III to show the existence of maximal and minimal solutions for integral equations with α-monotone kernels. Let us assume that the kernel $k(t, \tau, z)$ is α-monotone and that for every integer n there is a function $\delta(\tau)$, denoted by $\delta_n(\tau)$, such that III (α) holds, and the integral $\int_0^T \delta_n(\tau)d\tau < 1/n$.

Suppose further that the integral equation (1) for the functions

$$d(t)=d_n(t)=\frac{1}{n}+\int_0^t \delta_n(\tau)d\tau \tag{4}$$

has solutions w_n which are equicontinuous. Then there exists a uniformly convergent subsequence with limit $u^*(t) \in C(J)$. If the hypotheses on k are such that the passage to the limit in (1) under the integral sign is allowed, then u^* is the desired maximal solution of (1) with $d=0$. The corresponding arguments with $-d_n(t)$ instead of $d_n(t)$ lead to the minimal solution.

If $k(t, \tau, z)$ is continuous in all three variables, then III (α) holds for every function $\delta(\tau)=\text{const.}>0$, and the above reasoning goes through without difficulty. (If we set $\delta_n=1/n$ here, we even obtain that the w_n decrease monotonically.) If more generally $k \in \mathscr{K}$, then it follows from (2.4) (2.5) that functions $\delta_n(\tau)$ with the above properties exist. We summarize the main result:

IV. Maximal and Minimal Solutions *for integral equations of the form* (2.1) (2.2) *exist if* $g \in C(J)$ *and the kernel* $k(t, \tau, z) \in \mathscr{K}$ *is* α*-monotone* (*for* α *between* 0 *and* 1). *If* $k \in \mathscr{K}_0$ *and* k *is* α*-monotone, then there is a maximal solution* $u^*(t)$ *either in* J *or in a largest interval* $0 \leqq t < T_1 \leqq T$, *where* $\limsup_{t\to T_1} |u^*(t)| = \infty$ *(and similarly for the minimal solution).*

Thus if the kernel is monotone in the ordinary sense ($\alpha=0$), the maximal solution is very simple to approximate. According to 2 IV (δ) this can be done by functions w_n which satisfy the inequality $w_n > g + K w_n$. But if k is α-monotone ($\alpha>0$), we need solutions of the integral equation $w_n = g + d_n + K w_n$, where d_n depends on the kernel in a complicated fashion; see III (α) and (4). (The situation is of course simpler for continuous kernels since we can set $\delta_n = \text{const.}$) The special significance, expressed in 1 IX (β), of the maximal solution relative to inequalities depended essentially on 2 IV (δ). An analog of this is furnished by the following

V. Theorem. Suppose the operator K is generated by an α-monotone kernel $k \in \mathscr{K}, u_*$ and u^* are respectively the minimal and maximal solutions of the integral equation $u=g+Ku$, and v and w are continuous functions in J for which

$$v_\alpha(t|\bar{t}) \leqq g_\alpha(t,\bar{t}) + (Kv)_\alpha(t|\bar{t}), \quad w_\alpha(t|\bar{t}) \geqq g_\alpha(t|\bar{t}) + (Kw)_\alpha(t|\bar{t}).$$

Then

$$v \leqq u^* \quad \text{and} \quad w \geqq u_* \quad \text{in} \quad J.$$

For the proof we approximate u^* in the fashion described above by w_n and apply II to v and w_n (remembering that II (γ) follows from III (α)). We proceed analogously with the second inequality.

The α-monotonicity can be generalized in the sense that we allow a function $\alpha(t)$ instead of a constant α. Then the kernel is said to be α-monotone if the function $k_\alpha(t|\bar{t}, \tau, z) = k(\bar{t}, \tau, z) - \alpha(t)k(t, \tau, z)$ is monotone increasing in z; here $\alpha = \alpha(t)$ is a function defined in J and $0 \leqq \alpha(t) \leqq 1$. It is left to the reader to show that the method of proof still works with this concept of monotonicity.

4. Estimates and Uniqueness Theorems

With the methods developed so far, we can handle only operator equations such as

$$u = g + Ku \tag{1}$$

for which K has certain monotonicity properties. Now we shall give bounds which hold for quite general operators K. The fundamental idea is to associate the *arbitrary* operator K with a *monotone increasing* operator Ω by means of an inequality such as

$$|K\varphi - K\bar{\varphi}| \leqq \Omega(|\varphi - \bar{\varphi}|). \tag{2}$$

This idea of passing from the original problem to a "*comparison problem*" of similar type for which simple bounds are available proves most fruitful here and in all later chapters. In the case under consideration, Eq. (1) is the original problem, while the comparison problem is an operator equation

$$\sigma = d + \Omega\sigma \tag{3}$$

for a function $\sigma \in Z_c(\Omega)$. This approach was first developed by Bompiani (1925) [see also Tonelli (1925), Perron (1926)] for the uniqueness problem for ordinary differential equations. While earlier uniqueness criteria, i.e., that of Lipschitz (1880), Osgood (1898), Rosenblatt (1909), Montel (1926), always work with a particular explicit bound, Bompiani characterizes bounds which are sufficient for uniqueness by properties of the corresponding "comparison problem."

I. Theorem. *Assume that when $\varphi, \bar{\varphi} \in Z_c(K)$ and $|\varphi - \bar{\varphi}| \in Z_c(\Omega)$, the bound (2) holds for the operator K and the monotone increasing operator Ω. Suppose that the functions u, v, ϱ, d, g are defined in J and that $u, v \in Z_c(K)$, $|v-u|, \varrho \in Z_c(\Omega)$. Then*

(α) $|v-u|(0+) < \varrho(0+)$;

(β) $u = g + Ku, |v - g - Kv| \leqq d$ *in* J_0;

(γ) $\varrho > d + \Omega\varrho$ *in* J_0

imply the inequality

$$|v-u| < \varrho \quad in \quad J_0 .$$

If the operators K and Ω have the property that $K\varphi$ and $\Omega\varphi$ vanish at zero and (β) (γ) also hold for $t=0$, then (α) is superfluous since then at zero we have $|v-u| = |v-g| \leqq d < \varrho$.

If, moreover, the operator Ω satisfies the hypotheses of 2 IV, $d \in C(J)$ and $\sigma(t) \in Z_c(\Omega)$ is the maximal solution of Eq. (3) (it exists, for example, if Ω is generated by a monotone increasing kernel $\omega(t, \tau, z) \in \mathscr{K}$), then ($\beta$) alone implies

$$|u-v| \leqq \sigma \quad \text{in} \quad J .$$

The theorem follows from 1 VI if we replace the quantities v, w, K by $|v-u|$, ϱ, Ω. Then both hypotheses (α) coincide and so do both assertions. The hypothesis 1 VI (β) becomes $|v-u| - \Omega(|v-u|) \leqq d < \varrho - \Omega\varrho$; it can be obtained from (β) (γ) and (2):

$$|v-u| = |v - g - Kv + Kv - Ku| \leqq d + |Kv - Ku| \leqq d + \Omega(|v-u|) .$$

Finally we obtain the estimate with the maximal solution by approximating σ by functions ϱ for which the theorem applies; this is possible by 2 IV (δ).

Sharp uniqueness statements can be obtained from I by specialization. We introduce the following concepts to expedite their formulation.

II. Definition. ($\mathscr{E}, \mathscr{E}_1$). The monotone increasing operator Ω belongs to the class $\mathscr{E}$ [resp. $\mathscr{E}_1$] if all continuous and nonnegative functions φ in J for which $\varphi(0) = 0$ [resp. $\varphi(t) = o(t)$ for $t \to +0$] belong to $Z_c(\Omega)$ and if for every $\varepsilon > 0$ there exist a function $\varrho \in Z_c(\Omega)$ and a $\delta > 0$ such that

$$\varrho > \Omega\varrho \quad \text{and} \quad \delta \leqq \varrho \leqq \varepsilon \text{ [resp. } \delta t \leqq \varrho \leqq \varepsilon] \quad \text{in} \quad J_0 .$$

If Ω is an integral operator generated by the kernel $\omega(t, \tau, z)$, we thus require that $\omega(t, \tau, z)$ be defined and monotone increasing in z for $0 \leqq \tau \leqq t \leqq T$, $z \geqq 0$ and that the inequalities

$$\varrho(t) > \int_0^t \omega(t, \tau, \varrho) d\tau \quad \text{and} \quad \delta \leqq \varrho \leqq \varepsilon \text{ [resp. } \delta t \leqq \varrho \leqq \varepsilon]$$

can be satisfied in J_0. In this case we also write $\omega \in \mathscr{E}$ [*resp.* $\in \mathscr{E}_1$].

For example, $\Omega \in \mathscr{E}$ if (besides the monotonicity and the hypothesis on the domain of definition) the equation $\sigma = \Omega\sigma$ has a maximal solution in J with the property 2 IV (δ) which vanishes identically. In particular we have $\omega(t, \tau, z) \in \mathscr{E}$ if ω is a monotone increasing kernel from the class $\mathscr{K}$ and the function $\sigma(t) \equiv 0$ is the maximal solution (which surely exists in J) of the integral equation

$$\sigma(t) = \int_0^t \omega(t, \tau, \sigma(\tau))d\tau .$$

The same holds for $\omega \in \mathscr{K}_0$: *If the maximal solution* $\sigma(t)$ *of the above integral equation exists in all of* J *and if* $\sigma(t) \equiv 0$ *in* J, *then* $\omega \in \mathscr{E}$ *(note Definition* 2 III*)*.

The latter is easy to see. The function $\bar{\omega}$ defined by $\bar{\omega}(t, \tau, z) = \omega(t, \tau, \min(z, 1))$ for $z \geqq 0$, $\bar{\omega} = 0$ for $z < 0$ belongs to $\mathscr{K}$. The maximal solution $\sigma(t)$ of the equation with $\bar{\omega}$ instead of ω exists in all of J and vanishes there. This follows easily from the fact that this function σ is also a solution of the original equation in a possibly smaller interval, as long as it is $\leqq 1$. According to 2 IV (δ), this maximal solution $\sigma(t) \equiv 0$ can be approximated from above by functions ϱ which satisfy the requirements imposed above (first with respect to $\bar{\omega}$ and therefore, provided they are $\leqq 1$, with respect to ω).

III. Uniqueness Theorem. *Suppose* $(K\varphi)(0) = 0$ *for all* $\varphi \in Z_c(K)$. *If the bound* (2) *with* $\Omega \in \mathscr{E}$ *holds for all* $\varphi, \bar{\varphi} \in Z_c(K)$, *then the operator equation* (1) *has at most one solution.*

If Eq. (1) *has a maximal solution or if it is true that with* u *and* $\bar{u}$ *the function* $\max(u, \bar{u})$ *is also a solution of* (1), *then instead of* (2) *the following weaker condition suffices for uniqueness: There is an* $\Omega \in \mathscr{E}$ *such that*

(α) *for all* $\varphi, \bar{\varphi} \in Z_c(K)$ *which satisfy the inequality* $\varphi \leqq \bar{\varphi}$,

$$K\bar{\varphi} - K\varphi \leqq \Omega(\bar{\varphi} - \varphi) . \tag{4}$$

The first part of the theorem is just a special case of Theorem I. If u and v are two solutions and ϱ is determined according to II, then by I we have $|v - u| < \varrho \leqq \varepsilon$ where $\varepsilon > 0$ is arbitrary, and hence $u = v$. In the second part of the theorem we again start with two solutions u, v, but may now assume $u \leqq v$ in the proof of $u = v$. As in the proof of I, we refer back to 1 VI and replace v, w, K, g in 1 VI (β') by $v - u, \varrho, \Omega, 0$. Since from (4)

$$v - u = Kv - Ku \leqq \Omega(v - u) ,$$

we obtain $0 \leqq v - u < \varrho \leqq \varepsilon$, i.e., $v = u$.

From III via specialization we obtain

IV. Uniqueness Assertions for Integral Equations, which will be given special attention because of their importance. *If K is an integral operator*

$$K\varphi = \int_0^t k(t, \tau, \varphi(\tau))d\tau\,,$$

then the integral equation (1) *has at most one solution if we can find a function* $\omega \in \mathscr{E}$ *such that for* $(t, \tau, z), (t, \tau, \bar{z}) \in D(k)$,

$$|k(t, \tau, z) - k(t, \tau, \bar{z})| \leqq \omega(t, \tau, |z - \bar{z}|)\,; \tag{2'}$$

if Eq. (1) *has a maximal solution, then the condition*

$$k(t, \tau, z) - k(t, \tau, \bar{z}) \leqq \omega(t, \tau, z - \bar{z}) \quad \textit{for} \quad z \geqq \bar{z} \tag{4'}$$

suffices for uniqueness.

These uniqueness results, which generalize a theorem of Satō and Iwasaki (1955), can be sharpened if the kernel $k(t, \tau, z)$ is continuous at the point $(0, 0, g(0))$. Then for every solution u

$$\frac{1}{t}\int_0^t k(t, \tau, u(\tau))d\tau \to k(0, 0, g(0))\,,$$

the limiting value being independent of u. The difference of two solutions is then $o(t)$ as $t \to +0$. Thus the hypothesis I (α) is satisfied not only for $\varrho(0) > 0$ but also for ϱ such that $\varrho(t) \geqq \delta t\,(\delta > 0)$.

If $k(t, \tau, z)$ *is continuous at the point* $(0, 0, g(0))$, *uniqueness is also ensured as long as a condition* (2') *or* (4') *with a function* $\omega \in \mathscr{E}_1$ *holds.*

V. Examples. (α) Of particular importance for numerical applications is the (generalized) *Lipschitz condition*

$$|k(t, \tau, z) - k(t, \tau, \bar{z})| \leqq l(\tau)|z - \bar{z}|\,, \tag{5}$$

where $l(\tau) \in L(J)$ is assumed. Suppose $u, v \in Z_c(K), d \in C(J)$ and

$$u(t) = g(t) + \int_0^t k(t, \tau, u)d\tau, \quad \left|v(t) - g(t) - \int_0^t k(t, \tau, v)d\tau\right| \leqq d(t)$$

in J. Then we get the bound

$$\left.\begin{aligned} |v - u| &\leqq d(t) + \int_0^t d(\tau)l(\tau)e^{L(t) - L(\tau)}d\tau \\ &= e^{L(t)}\left[d(0) + \int_0^t d'(\tau)e^{-L(\tau)}d\tau\right], \end{aligned}\right\} \tag{6}$$

where

$$L(t) = \int_0^t l(\tau) d\tau$$

[the second line in (6) holds only for absolutely continuous d].

This is a special case of I. The right hand side of (6) is the maximal solution of Eq. (3), where Ω is now an integral operator with the kernel $\omega(t, \tau, z) = l(\tau)z$. This was shown in the proof of 1 III. The inequality (6) contains a uniqueness theorem as well as a theorem on the continuous dependence of the solution on g and k if k satisfies a Lipschitz condition. It was given by Carathéodory (1918) for kernels of the form $k(t, \tau, z) = f(\tau, z)$.

(β) Condition of Osgood (1898), Tonelli (1925)[1] and Montel (1926). It is given by

$$\omega(t, \tau, z) = l(\tau)\psi(z) \in \mathscr{E},$$

where $0 \leqq l(\tau) \in L(J)$ while $\psi(z)$ is continuous and monotone increasing for $z \geqq 0$, $\psi(0) = 0$, $\psi(z) > 0$ for $z > 0$ and the integral

$$\int_0^1 \frac{dz}{\psi(z)}$$

is divergent.

(γ) Condition of Nagumo (1926)[1]:

$$\omega(t, \tau, z) = \frac{z}{\tau} \in \mathscr{E}_1.$$

(δ) For $0 \leqq l(\tau) \in L(J)$ we have

$$\omega(t, \tau, z) = \left[\frac{1}{\tau} + l(\tau)\right] z \in \mathscr{E}_1.$$

For the examples (β) – (δ) it is easy to find functions ϱ as required in II. In the case (β) we can choose, say, functions $\varrho(t)$ defined by the relation

$$\int_\varrho^\varepsilon \frac{dz}{\psi(z)} = \int_t^T l(\tau) d\tau;$$

they satisfy the integral equation

$$\varrho(t) = \varrho(0) + \int_0^t l(\tau)\psi(\varrho(\tau)) d\tau, \quad \varrho(0) > 0.$$

In the case (δ) we can use the functions

$$\varrho(t) = Cte^{L(t)+t}, \quad C > 0,$$

[1] The cited authors considered only the special case of ordinary differential equations.

for which $\varrho'(t)=[l(t)+1+1/t]\,\varrho(t)>[l(t)+1/t]\,\varrho(t)$ holds almost everywhere, while (γ) is a special case of (δ) (note that the functions ϱ must satisfy an integral inequality in J_0 with equality not allowed).

(ε) Suppose the function $\omega(\tau,z)$, which is independent of t, is defined in $J_0\times\{z\geqq 0\}$, is measurable in τ for fixed z, and is continuous and monotone increasing in z for fixed τ. Suppose further that $\omega(\tau,0)=0$ and $\omega(\tau,\delta)\in L(t_0\leqq\tau\leqq T)$ for a fixed $\delta>0$ and every $t_0\in J_0$. Let the function $\sigma(t)\equiv 0$ be the only continuous nonnegative function in J which satisfies the integral equation

$$\sigma(t)=\int_0^t \omega(\tau,\sigma(\tau))\,d\tau \quad \text{in} \quad J. \tag{7}$$

Then $\omega\in\mathscr{E}$.

This theorem can be found (for ordinary differential equations and with additional assumptions) in Coddington-Levinson (1955), pp. 49–51[1]. We obtain appropriate functions ϱ by setting $\omega=0$ for $z<0$, defining $\varrho(T)=\varepsilon$ $(0<\varepsilon<\delta)$, and continuing $\varrho(t)$ as a C-solution[2] of the differential equation $\varrho'(t)=\omega(t,\varrho)$ in J towards the *left* up to zero (see the following section). This is obviously possible; $\varrho(t)$ is monotone increasing in t and $\varrho(0)>0$ (otherwise we would have a nontrivial solution of (7)), and thus

$$\varrho(t)>\int_0^t \omega(\tau,\varrho)\,d\tau \quad \text{in} \quad J.$$

VI. Remark. Tonelli (1925) showed that V(β) gives a uniqueness condition for ordinary differential equations even without the monotonicity assumption on $\psi(z)$. It does not seem possible to prove this fact with the methods used here. A direct proof goes as follows: If u and v are two C-solutions (see Section 5), then so are $\bar u=\min(u,v)$ and $\bar v=\max(u,v)$. If there were a point t_1 where $d=\bar v-\bar u$ is positive and if t_0 were the first zero of $d(t)$ to the left of t_1, then $d'=\bar v'-\bar u'$ $=f(t,\bar v)-f(t,\bar u)\leqq l(t)\,\psi(d)$ and hence, for $t_0<t\leqq t_1$,

$$\int_t^{t_1} d'(\tau)\,[\psi(d(\tau))]^{-1}d\tau=\int_{d(t)}^{d(t_1)}[\psi(s)]^{-1}ds\leqq\int_t^{t_1} l(\tau)\,d\tau.$$

As $t\to t_0$ we now get a contradiction since the middle integral approaches ∞ but the right hand side remains bounded.

VII. Super- and Subfunctions for General Operators. If the operator K is not monotone increasing, then super- and subfunctions can be determined according to the procedure of the Introduction VI. *If v and w are two functions from the class $Z_c(K)$ and*

(α) $v(0+)<u(0+)<w(0+)$ *for every solution u of Eq.* (1);

(β) $v(t_0)<g(t_0)+(K\varphi)(t_0)<w(t_0)$ *for* $t_0\in J_0$ *and all* $\varphi\in Z_c(K)$ *with* $v<\varphi<w$ *in* $0<t<t_0$,

then

$$v<u<w \quad \text{in} \quad J_0 \quad \text{for every solution } u \text{ of Eq. (1).}$$

[1] We mean the theorem stated in Coddington-Levinson, p. 51, at the end of the proof of Theorem 2.2. This theorem refers to systems of differential equations; it requires, when specialized to the case of *one* differential equation, a bound of the absolute value as in (2').

[2] Translator's note: C-solution; see 5 I.

From the assumption that (for a fixed solution u) t_0 is the first point at which one of the two inequalities of the assertion is false, a contradiction is easily obtained. If, say, $v(t_0)=u(t_0)$ and $v<u<w$ for $0<t<t_0$, then $v(t_0)<w(t_0)$ follows immediately from (β) for $u=\varphi$.

The assumption (β) is simplified for

VIII. Monotone Decreasing Operators. These are operators K such that $-K$ is monotone increasing. If K is monotone decreasing and $(K\varphi)(0)=0$ for all $\varphi \in Z_c(K)$, then the two inequalities

(β') $v<g+Kw, \quad w>g+Kv$ in J

are sufficient for v to be a subfunction and w a superfunction.

It should be noted that equality signs are permissible in (β') as well as in VII (β) if the operator K satisfies a uniqueness condition (i.e., the hypothesis of 4 III). Justification for this is almost word for word the same as in 1 IX (γ).

If the operator K is monotone decreasing, then successive approximation is particularly effective. If we start with a function v_0 and define a sequence v_n by $v_{n+1}=g+Kv_n$ and if $v_0 \leqq v_1$ and $v_0 \leqq v_2$, then, by what has just been proved, we obtain $v_0 \leqq u \leqq v_1$, provided the operator K satisfies a uniqueness condition [u is the uniquely determined solution of (1)]. Then by induction we easily obtain

$$v_0 \leqq v_2 \leqq v_4 \leqq \cdots \leqq u \leqq \cdots \leqq v_5 \leqq v_3 \leqq v_1 . \tag{8}$$

The approximations are alternately super- and subfunctions, and in many cases they can be used in existence proofs (see the introduction (VII)).

As an example we consider the initial value problem $y'''+yy''=0$ for $y=y(t), 0 \leqq t<\infty, y(0)=y'(0)=0, y''(0)=1$. It is equivalent to the following integral equation for $u(t)=y''(t)$:

$$u(t)=Ku=e^{-\frac{1}{2}\int_0^t (t-\tau)^2 u(\tau)\,d\tau} \tag{9}$$

The operator K is monotone decreasing and the sequence v_n, if we begin with $v_0=0$, has the properties noted above, $v_0 \leqq v_1, v_0 \leqq v_2$, i.e. (8) holds. The proof that K satisfies a uniqueness condition and that the v_n converge is very simple. Thus we have both an existence proof for the solution in the interval $0 \leqq t<\infty$ as well as a practical numerical procedure. The problem was solved in this fashion by H. Weyl (1942) (it arises in flow over a flat plate). The transformation of the problem to the integral equation (9) and the inequalities (8) were first given by Mohr (1937; p. 490–491).

5. Ordinary Differential Equations (in the Sense of Carathéodory)

I. Definition (*initial value problem, solution, C-solution, Z, Z_{ac}*). By an initial value problem for an ordinary differential equation of first order we mean the problem of finding a function u such that

$$u'(t) = f(t, u(t)) \tag{1}$$

and $u(0) = \eta$, where $f(t, z)$ is a real valued function defined on a certain set $D(f)$ of the (t, z)-plane and η is a given real number. In the following we distinguish between two concepts of solution. The function class Z comprises all functions u which are continuous in J and differentiable in J_0, while the class Z_{ac} consists of all functions u which are absolutely continuous in J. If, moreover, $(t, u(t)) \in D(f)$ for $t \in J_0$ (resp. for almost all $t \in J$), then $u \in Z(f)$ (resp. $u \in Z_{ac}(f)$). A function $u \in Z(f)$ which satisfies the differential equation (1) in J_0 is said to be a *solution* of this equation, while a *solution in the sense of Carathéodory* or C-*solution* belongs to the class $Z_{ac}(f)$ and satisfies the differential equation almost everywhere in J. We will speak of a solution or a C-solution of the initial value problem if the initial condition $u(0) = \eta$ is also satisfied.

In this section we will be concerned with C-solutions, on which Carathéodory's theory was based. The theory in the area of *solutions* requires other methods of proof and also leads to other results, as we shall see in Chapter II. The function u is a C-solution of the initial value problem if and only if $u \in Z_c(f)$[1] and

$$u(t) = \eta + \int_0^t f(\tau, u(\tau))\,d\tau \quad \text{in} \quad J. \tag{2}$$

Thus the initial value problem, in the context of C-solutions, is reduced to a special class of integral equations, namely those with a kernel $k(t, \tau, z) = f(\tau, z)$[2] which does not depend on t, and the earlier theorems can be applied immediately to this case. It should be noted that not every solution is also a C-solution (a differentiable function is not necessarily absolutely continuous!); of course every continuously differentiable solution is a C-solution.

II. Existence and Uniqueness Assertions. *If $f \in \mathscr{K}$, then according to the results of Section* 2, *the initial value problem has a* C-*solution for every η, and by* 2 V *there is a maximal and a minimal solution. Both of*

[1] According to 1 I, this means that $f(t, u(t)) \in L(J)$ and that u is continuous in J. Since the right hand side of (2) is absolutely continuous, we automatically have $u \in Z_{ac}(f)$.

[2] However in the text, except in integrals, we write $f(t, z)$ instead of $f(\tau, z)$ and later $\omega(t, z)$.

these C-solutions can be approximated in the sense of 2 IV (δ) *if* f *is monotone increasing in* z. The same holds in a possibly smaller interval if $f \in \mathscr{K}_0$; or if f is given only in a neighborhood of the point $(0, \eta)$ and is there measurable in t and continuous in z, and can be bounded as in (2.3). Moreover, the kernel f is 1-monotone in the sense of 3 I, so that the existence of the maximal and minimal solutions also follows from the entirely different method of proof of 3 IV.

The uniqueness theorem also holds in the sharper form (4.4′): *A one-sided condition*

$$f(t, z) - f(t, \bar{z}) \leqq \omega(t, z - \bar{z}) \quad \text{for} \quad z \geqq \bar{z}$$

with $\omega \in \mathscr{E}$ [*resp.* $\omega \in \mathscr{E}_1$ *if* f *is continuous at the point* $(0, \eta)$] *is sufficient for uniqueness.* In particular the uniqueness criteria 4 V (α)–(ε) can be used.

Of course Theorem 1 VI can be specialized immediately to differential equations. This gives a theorem in which $f(t, z)$ must be assumed to be monotone increasing in z. However, an essentially sharper result can be formulated.

III. Theorem. *Suppose* $v, w \in Z_{ac}(f)$ *and*

(α) $v(0+) < w(0+)$

(β) $v' \leqq f(t, v)$, $w' \geqq f(t, w)$ *a.e. in* J.

Then

$$v < w \quad \text{in} \quad J_0$$

if f *also satisfies the following condition:*

(γ) *For every* $t_0 \in J_0$ *there is an* $\omega(t, z) \in \mathscr{E}$ *such that in a left-sided neighborhood of* t_0

$$f(t, z) - f(t, \bar{z}) \geqq -\omega(t_0 - t, z - \bar{z}) \quad \text{for} \quad z \geqq \bar{z}$$

(*provided the arguments lie in* $D(f)$).

The condition 5 III (γ) is actually a uniqueness condition "to the left," i.e., a condition to ensure that when we move from left to right, two solutions do not meet. We not that the following assumption suffices instead of (β):

(β′) *if* $v < w$ *for* $0 < t < t_0$, $v(t_0) = w(t_0)$, *then* $v' \leqq f(t, v)$, $w' \geqq f(t, w)$ *for almost all* t *of a left-sided neighborhood of* t_0.

The proof starts with the assumption that the assertion is false. Then there is a first point $t_0 \in J_0$ where $v = w$. For the functions $\bar{v}(t) = v(t_0 - t)$, $\bar{w}(t) = w(t_0 - t)$ we then have

$$\bar{v}(0) = \bar{w}(0), \quad \bar{v}' \geqq \bar{f}(t, \bar{v}), \bar{w}' \leqq \bar{f}(t, \bar{w}) \quad \text{with} \quad \bar{f}(t, z) = -f(t_0 - t, z)$$

for small positive t. Then from the inequality in (γ) we obtain for the difference $d=\bar{w}-\bar{v}$

$$d'=\bar{w}'-\bar{v}'\leqq\bar{f}(t,\bar{w})-\bar{f}(t,\bar{v})\leqq\omega(t,d) \quad\text{and}\quad d(0)=0\,,$$

i.e., in the notation of 4 II $d\leqq\Omega d$, while by 4 II there are arbitrarily small functions ϱ for which $\varrho>\Omega\varrho$ and $\varrho(0)>0$. By 1 VI then $d<\varrho\leqq\varepsilon$, i.e., $d\equiv 0$ in a right-sided neighborhood of the point $t=0$ or $v=w$ in a left-sided neighborhood of t_0, which contradicts our choice of t_0.

IV. Super- and Subfunctions. Let f satisfy the condition III (γ). The function $v\in Z_{ac}(f)$ is a subfunction and the function $w\in Z_{ac}(f)$ a superfunction relative to a given initial value problem if:

(α) $v(0+)<u(0+)<w(0+)$ for every C-solution u of the initial value problem;

(β) $v'\leqq f(t,v)$, $w'\geqq f(t,w)$ a.e. in J.

Then $v<u<w$ in J_0. The condition (α) is satisfied, for example, if $v(0)<\eta<w(0)$. (The bounds are independent of each other.)

V. Theorem. *Suppose $u,v\in Z_{ac}(f)$, ϱ and $\bar{\varrho}$ are two functions $\in Z_{ac}(\omega)$ positive in J_0, where $\omega(t,z)$ is defined for $z\geqq 0$ and satisfies the condition III (γ), $\delta(t)$ and $\bar{\delta}(t)$ are two functions defined on J and*

(α) $-\bar{\varrho}(0+)<(v-u)(0+)<\varrho(0+)$;

(β) $u'=f(t,u)$, $-\bar{\delta}(t)\leqq v'-f(t,v)\leqq\delta(t)$ *a.e. in* J;

(γ) $\varrho'\geqq\delta(t)+\omega(t,\varrho)$, $\bar{\varrho}'\geqq\bar{\delta}(t)+\omega(t,\bar{\varrho})$ *a.e. in* J;

(δ) $f(t,z)-f(t,\bar{z})\leqq\omega(t,z-\bar{z})$ *for* $z\geqq\bar{z}$ *and* $(t,z),(t,\bar{z})\in D(f)$.

Then

$$-\bar{\varrho}<v-u<\varrho \quad\textit{in}\quad J_0\,.$$

First we verify the inequality $v-u<\varrho$ by going back to III. If the difference $d=v-u$ is not smaller than ϱ in all of J, then there exists a first point $t_0\in J_0$ with $d(t_0)=\varrho(t_0)>0$. We choose a point $t_1, 0<t_1<t_0$, such that $d\geqq 0$ in $t_1\leqq t\leqq t_0$ and use III on the interval $[t_1,t_0]$ instead of J and with the functions $d,\varrho,\delta(t)+\omega(t,z)$ instead of $v,w,f(t,z)$. From (β) and (δ) follows

$$d'=v'-u'\leqq\delta(t)+f(t,v)-f(t,u)\leqq\delta(t)+\omega(t,d)\,.$$

This together with the first inequality (γ) and the inequality $d(t_1)<\varrho(t_1)$ shows that the hypotheses (α) and (β) of III hold in the case under consideration and consequently $d<\varrho$ for $t_1\leqq t\leqq t_0$, contradicting the assumption $d(t_0)=\varrho(t_0)$. The inequality $u-v<\bar{\varrho}$ is proved in exactly the same way.

VI. Definition ($\mathscr{E}_2,\mathscr{E}_3$). The function $\omega(t,z)$ defined on $J_0\times\{z\geqq 0\}$ belongs to the class $\mathscr{E}_2$ [resp. $\mathscr{E}_3$] if it has the property III (γ) and if for every $\varepsilon>0$ a function ϱ which is absolutely continuous in J and a number $\delta>0$ exist such that

$$\varrho' \geqq \omega(t,\varrho) \text{ a.e. in } J \text{ and } \delta \leqq \varrho \leqq \varepsilon \;[\text{resp. } \delta t \leqq \varrho \leqq \varepsilon] \text{ in } J.$$

VII. Uniqueness Theorem. *The initial value problem for the equation* (1) *has at most one* C*-solution if f satisfies the condition* V (δ) *with a function $\omega \in \mathscr{E}_2$ [resp. $\omega \in \mathscr{E}_3$ if f is continuous at the point* $(0,\eta)$*].*

The manner in which VII is reduced to V is immediately evident (see the proof of 4 III).

VIII. Example and Remarks. (α) *Lipschitz Condition.* Suppose that in its domain of definition the function f satisfies a one-sided Lipschitz condition

$$f(t,z)-f(t,\bar{z}) \leqq l(t)(z-\bar{z}) \quad \text{for} \quad z \geqq \bar{z},$$

where $l(t) \in L(J)$. Suppose further that V (β) holds for the functions $u, v \in Z_{ac}(f)$ with $\delta(t), \bar{\delta}(t) \in L(J)$ and $-\bar{\varepsilon} \leqq v(0)-u(0) \leqq \varepsilon, \varepsilon \geqq 0, \bar{\varepsilon} \geqq 0$. Then, letting $L(t)=\int_0^t l(\tau)\,d\tau$, we have

$$-e^{L(t)}\left[\bar{\varepsilon}+\int_0^t \bar{\delta}(\tau)e^{-L(\tau)}d\tau\right] \leqq v(t)-u(t) \leqq e^{L(t)}\left[\varepsilon+\int_0^t \delta(\tau)e^{-L(\tau)}d\tau\right],$$

as long as the expression on the left hand side is $\leqq 0$ and that on the right hand side $\geqq 0$.

This bound, easily obtained from V, differs from 4 V primarily in that the Lipschitz condition is only one-sided and that the sign of $l(t)$ is not restricted. Thus in some cases it is possible to arrive at essentially more precise bounds by using *negative* Lipschitz "constants" $l(t)$.

(β) The essential point in Theorem V is the inequality (δ): if we want to obtain sharp bounds, we have to determine a suitable function ω. Here it is useful to note that this inequality certainly need not hold for *all* $z \geqq \bar{z}$. As the proof shows, it is sufficient to have

$$f(t,v)-f(t,u) \leqq \omega(t,v-u) \quad \text{if} \quad 0<v-u<\varrho,$$

$$f(t,u)-f(t,v) \leqq \omega(t,u-v) \quad \text{if} \quad 0<u-v<\bar{\varrho}.$$

Almost the same result can be obtained directly from V (δ) by restricting (this is permissible) the domain of definition of f from the beginning to the set $-\bar{\varrho}(t) \leqq v(t)-z \leqq \varrho(t)$. (Indeed, the proof of V shows that a solution u cannot leave this domain even under the weakened hypothesis (δ).) Finally it should also be noted that we can get by without the positivity of the two bounds $\varrho, \bar{\varrho}$, but then (δ) has to be modified.

(γ) The assertions III–VII are related to the earlier 1 VI–VII, 4 I–III. However, they are not special cases of those theorems. What is essentially new is that in place of a strong monotonicity condition we use the much weaker requirement III (γ). In fact III (γ) is satisfied in most cases which are encountered in practice, e.g., if a partial derivative $f_z(t,z)$ exists which is bounded from below. The following example shows that, moreover, Theorem III is false without the hypothesis (γ), and indeed becomes false as soon as $\mathscr{E}$ is replaced by $\mathscr{E}_1$: $T=2, v\equiv 0$, $w=1-t$, $f(t,z)=z/(t-1)$ for $t\neq 1$, $f(t,z)=0$ for $t=1$. The hypotheses III (α) (β) are satisfied; despite this we do not have $v<w$ for $0\leqq t\leqq 2$. Here III (γ) cannot be satisfied at the point $t_0=1$; this would however be possible with the Nagumo function $\omega(t,z)=z/t\in\mathscr{E}_1$. The theorem cannot even be saved — still without III (γ) — by requiring a genuine $<$ or $>$ in III (β); to see this we need only set $f=2z/(t-1)$ in the given example.

The question arises, can we, by a similar procedure, get around monotonicity in the earlier theorems as well? A glance at the proof of the two decisive theorems 1 II (or 1 VI) and III shows this is rather improbable. The new idea in III consists of taking as starting point for the integral equation not the point 0, but the point t_0, and reasoning from there "to the left." Such a rewriting of the integral equation on another fixed limit of integration turns out to be possible just for those integral equations equivalent to a differential equation. We note that there is a conclusion (although of a different form) "from right to left" in the uniqueness theorem of Kamke (1945, p. 139f).

IX. Integral Inequalities. Gronwall's inequality 1 III is the most important special case of Theorem 1 VI. The upper bound w is here obtained as a solution of a linear differential equation. Further special cases with an explicitly computable upper bound are contained in the following theorem.

If $f(t,z)\in\mathscr{K}$ is monotone increasing in z, $g(t)$ is absolutely continuous in J and $u^*\in Z_{ac}(f)$ is the maximal solution of the initial value problem

$$u'=g'+f(t,u)\quad \text{a.e. in } J,\quad u(0)=g(0),$$

then for every function $v\in Z_c(f)$ satisfying the inequality

$$v(t)\leqq g(t)+\int_0^t f(\tau,v(\tau))d\tau$$

the inequality

$$v\leqq u^*\quad \text{in}\quad J$$

holds.

By 2 IV (δ) there are functions w arbitrarily close to u^* which satisfy

$$w(t)>g(t)+\int_0^t f(\tau,w(\tau))d\tau\,.$$

Then the inequality $v < w$ follows from 1 VI, and hence also the assertion in the usual way.

It is not difficult to extend the theorem to the case where f is defined in a domain $D(f)$ containing the point $(0, g(0))$ and is such that a local existence theorem holds.

As has been said, the case $f(t, z) = l(t)z$ leads to Gronwall's inequality; the case $f(t, z) = l(t)\psi(z)$ was given by Bihari (1956) and the general case (with sharper hypotheses) by Babkin (1953).

X. Another Theorem on Integral Inequalities. Simple counterexamples (say $f(t, z) = -z, g(t) = 1, u = e^{-t}, v = t$) show that the monotonicity of f is indispensable in the above theorem. The following similar theorem, however, does not contain any monotonicity condition.

Suppose $f(t, z) \in \mathscr{K}, v \in Z_c(f)$ and $u^* \in Z_{ac}(f)$ is the maximal solution of the initial value problem $u' = f(t, u), u(0) = \eta$. Then $v(0) \leqq \eta$ and

$$v(\bar{t}) - v(t) \leqq \int_t^{\bar{t}} f(\tau, v(\tau))d\tau \qquad (0 \leqq t < \bar{t} \leqq T)$$

imply the inequality

$$v \leqq u^* \quad \text{in} \quad J.$$

If v is absolutely continuous, then the integral inequality in the hypotheses can be replaced by

$$v' \leqq f(t, v) \quad \text{a.e. in } J.$$

This theorem, which was proved by Baiada (1947) and Cafiero (1948), is contained in 3 V (letting $k = f$ and $\alpha = 1$ there); see also Lakshmikantham (1957), Olech and Opial (1960).

6. Systems of Integral Equations

In this section we extend the previous results to systems of Volterra integral equations

$$u_\nu(t) = g_\nu(t) + \int_0^t k_\nu(t, \tau, u_1(\tau), \ldots, u_n(\tau))d\tau \qquad (\nu = 1, \ldots, n) \tag{1}$$

and more generally to systems of operator equations. Consequently we shall make use of vector notation and give some

I. Definitions ($z, z < \bar{z}, |z|, \varphi \in C(J), \ldots$). Let $z = (z_1, \ldots, z_n), n \geqq 1$, be a point (or vector) of E^n. By $|z|$ we mean the vector

$$|z| = (|z_1|, \ldots, |z_n|).$$

Inequality between vectors $\boldsymbol{y}=(y_1,\ldots,y_n)$ and $\boldsymbol{z}$ is defined as

$$\boldsymbol{y}<\boldsymbol{z} \quad \text{equivalent to} \quad y_\nu<z_\nu \quad \text{for} \quad \nu=1,\ldots,n$$

(and similarly for the symbols $\leqq, >, \geqq$). If we are given a function $\boldsymbol{\varphi}(t)=(\varphi_1(t),\ldots,\varphi_n(t))$ depending on a real variable t, then statements like "$\boldsymbol{\varphi}$ is continuous in J", "$\boldsymbol{\varphi}\in L(J)$", ..., mean that every component $\varphi_\nu(t)$ has the corresponding property. Phrases like „vector function $\boldsymbol{\varphi}(t)$", "scalar function $\varphi(t)$" will occasionally be used for clarity.

II. Definitions ($\boldsymbol{K}$, $Z_c(\boldsymbol{K})$, *operator, monotone increasing operator*, $\mathscr{K}^n$, $\mathscr{K}_0^n$, $\boldsymbol{\varphi}(a+)<\overline{\boldsymbol{\varphi}}(a+)$, *monotonicity in* $\boldsymbol{z}$, *quasi-monotonicity in* $\boldsymbol{z}$). *A* Volterra operator $\boldsymbol{K}$ is a mapping which associates a vector function $\boldsymbol{K\varphi}$ defined in J with every vector function $\boldsymbol{\varphi}(t)$ from a certain set $Z_c(\boldsymbol{K})\subset C(J)$. It has the property that for every fixed t, $(\boldsymbol{K\varphi})(t)\in E^n$ depends only on the values of $\boldsymbol{\varphi}$ in $[0,t]$. The monotone increasing operators are defined as in 1 IV but with bold-face $\boldsymbol{\varphi}, \overline{\boldsymbol{\varphi}}, \boldsymbol{K}$. If in particular we are given an integral operator with a kernel $\boldsymbol{k}(t,\tau,\boldsymbol{z})=(k_1(t,\tau,z_1,\ldots,z_n),\ldots,k_n(t,\tau,z_1,\ldots,z_n))$, then we define the class $Z_c(\boldsymbol{k})$ as in 1 I ($\boldsymbol{k}$, $\boldsymbol{z}$ bold-face), and the classes $\mathscr{K}^n$ and $\mathscr{K}_0^n$ as in 2 I ($\boldsymbol{k}, \boldsymbol{z}, \boldsymbol{h}, \boldsymbol{M}>0$ bold-face)[1]. Similarly $\boldsymbol{\varphi}(a+)<\overline{\boldsymbol{\varphi}}(a+)$ is defined as for scalar functions in 1 V.

A vector function $\boldsymbol{F}(x,\boldsymbol{z})=(F_1(x,\boldsymbol{z}),\ldots,F_n(x,\boldsymbol{z}))$, defined in $D(\boldsymbol{F})$ and depending on $\boldsymbol{z}=(z_1,\ldots,z_n)$ and possibly also other variables combined in the symbol x, is said to be "*monotone increasing in* $\boldsymbol{z}$" if

$$\boldsymbol{F}(x,\boldsymbol{z})\leqq\boldsymbol{F}(x,\overline{\boldsymbol{z}}) \quad \text{for} \quad \boldsymbol{z}\leqq\overline{\boldsymbol{z}},$$

and "*quasi-monotone increasing in* $\boldsymbol{z}$" if for $\nu=1,\ldots,n$

$$F_\nu(x,\boldsymbol{z})\leqq F_\nu(x,\overline{\boldsymbol{z}}) \quad \text{for} \quad \boldsymbol{z}\leqq\overline{\boldsymbol{z}},\ z_\nu=\overline{z}_\nu,$$

(provided $(x,\boldsymbol{z}), (x,\overline{\boldsymbol{z}})\in D(\boldsymbol{F})$).

If $\boldsymbol{F}(x,\boldsymbol{z})$ with x fixed is defined for all $\boldsymbol{z}$ (or more generally for those $\boldsymbol{z}$ of a convex set of E^n), then $\boldsymbol{F}$ is monotone increasing in $\boldsymbol{z}$ if and only if every component $F_\nu(x,z_1,\ldots,z_n)$ is monotone increasing in each of the variables z_μ in the usual sense, and is quasi-monotone increasing in $\boldsymbol{z}$ if and only if every component F_ν is monotone increasing in each of the variables z_μ, $\mu\neq\nu$, in the usual sense. The reader should be aware that this assertion is false in general if the domain of definition is not a convex set. A more detailed discussion of this fact and critical remarks on earlier work can be found in Ważewski (1950). The significance of the monotonicity property which is called *quasi-monotonicity* here was

[1] Equivalently, $|k_\nu(t,\tau,\boldsymbol{z})|\leqq h(\tau)\in L(J)$ for $0\leqq\tau\leqq t\leqq T$, $|z_\mu|<M$, $\mu,\nu=1,\ldots,n$ (case $\mathscr{K}_0^n$).

first recognized by Müller (1926, 1927) and Kamke (1932) in carrying over fundamental theorems from *one* differential equation to *systems* of differential equations. Every function is quasi-monotone for $n=1$.

Below we shall be concerned with the operator equation

$$\boldsymbol{u} = \boldsymbol{g} + \boldsymbol{K}\boldsymbol{u} \tag{2}$$

or the integral equation (1), which can also be written in the form

$$\boldsymbol{u}(t) = \boldsymbol{g}(t) + \int_0^t \boldsymbol{k}(t, \tau, \boldsymbol{u}(\tau))d\tau . \tag{1'}$$

The formal correspondence with the one-dimensional case, emphasized by our choice of notation, covers a large part of the theory up to the present. Let us begin with Section 1. Theorem 1 VI holds without modification other than the bold-face, and will be stated again because of its importance.

III. Theorem. *If* $\boldsymbol{v}, \boldsymbol{w} \in Z_c(\boldsymbol{K})$ *and* $\boldsymbol{K}$ *is a monotone increasing operator, then from*

(α) $\boldsymbol{v}(0+) < \boldsymbol{w}(0+)$;

(β) $\boldsymbol{v} - \boldsymbol{K}\boldsymbol{v} < \boldsymbol{w} - \boldsymbol{K}\boldsymbol{w}$ *in* J_0

it follows that

$$\boldsymbol{v} < \boldsymbol{w} \quad \text{in} \quad J_0 .$$

Likewise, 1 VII *remains valid word for word.*

The proof of 1 VI is slightly modified. The point t_0 now has to be determined so that $\boldsymbol{v}(t) < \boldsymbol{w}(t)$ for $0 < t < t_0$ and $u_\nu(t_0) = w_\nu(t_0)$ for at least one ν. Everything else remains the same.

It is obvious that the generalizations 1 VIII also hold.

From this we easily prove that Theorem 5 IX is true for systems of ordinary differential equations ($\boldsymbol{f}(t, z)$, $\boldsymbol{g}$, $\boldsymbol{u}$, $\boldsymbol{v}$, $\boldsymbol{u}^*$ bold-face). It was given by Opial (1957) and Ważewski (1957); see also Olech (1967).

Section 2 offers no difficulties. We can simply take over the existence Theorem 2 II, the definition of maximal and minimal solution, and Theorem 2 IV (in 2 IV (α), for example, we use $\boldsymbol{\delta} > \boldsymbol{0}$, $|\boldsymbol{\varepsilon}| < \boldsymbol{\delta}$ and $\boldsymbol{u} = \boldsymbol{g} + \boldsymbol{\varepsilon} + \boldsymbol{K}\boldsymbol{u}$). On the other hand, a translation of 2 V is not meaningful. In other words: *The first of the two methods described in Section 2 for constructing maximal and minimal solutions can be carried over to systems of integral equations (where the kernel* $\boldsymbol{k}$ *must be monotone increasing in* z*); the second cannot.* It is not surprising that the second method cannot be carried over, since we know that in general no maximal solution exists for systems of ordinary differential equations.

Thus we have the

IV. Theorem. *If* $\boldsymbol{g} \in C(J)$, $\boldsymbol{k} \in \mathscr{K}^n$, *then a solution of the integral equation* (1) *exists. If* $\boldsymbol{k}$ *is a monotone increasing kernel, i.e., if* $\boldsymbol{k}(t, \tau, \boldsymbol{z}) \leqq \boldsymbol{k}(t, \tau, \bar{\boldsymbol{z}})$ *for* $\boldsymbol{z} \leqq \bar{\boldsymbol{z}}$, *then there exist a maximal and a minimal solution. These assertions hold for* $\boldsymbol{k} \in \mathscr{K}_0$ *either in* J *or in an interval* $0 \leqq t < T_1$, $T_1 \in J_0$. *The maximal solution can be approximated by superfunctions, and the minimal solution by subfunctions in the sense of* 2 IV (δ).

A carry-over of Section 3 leads to lengthy considerations and will therefore not be made. On the other hand the theorems of Section 4 present no problems.

V. Theorem and Definition ($\mathscr{E}^n$, $\mathscr{E}_1^n$). The estimation theorem 4 I still holds for systems ($\boldsymbol{K}, \boldsymbol{\Omega}, \boldsymbol{\varphi}, \bar{\boldsymbol{\varphi}}, \boldsymbol{u}, \boldsymbol{v}, \boldsymbol{\varrho}, \boldsymbol{g}, \boldsymbol{d}, \boldsymbol{\sigma}, \boldsymbol{\omega}, \boldsymbol{z}$ bold-face). The classes $\mathscr{E}^n$, $\mathscr{E}_1^n$ are defined verbatim like the classes $\mathscr{E}$, $\mathscr{E}_1$ in 4 II except for bold-face $\boldsymbol{\Omega}, \boldsymbol{\varphi}, \boldsymbol{\varrho}, \boldsymbol{\varepsilon} > \boldsymbol{0}, \boldsymbol{\delta} > \boldsymbol{0}, \boldsymbol{\omega}(t, \tau, \boldsymbol{z})$. Then the uniqueness theorem 4 III and the uniqueness assertions for integral equations 4 IV also hold for the Eqs. (2) and (1) (with appropriate bold-face).

For example the uniqueness condition (4.2') for systems of integral equations now reads as follows

$$|\boldsymbol{k}(t, \tau, \boldsymbol{z}) - \boldsymbol{k}(t, \tau, \bar{\boldsymbol{z}})| \leqq \boldsymbol{\omega}(t, \tau, |\boldsymbol{z} - \bar{\boldsymbol{z}}|)$$

or, in detail,

$$|k_\nu(t, \tau, z_1, \dots, z_n) - k_\nu(t, \tau, \bar{z}_1, \dots, \bar{z}_n)| \leqq \omega_\nu(t, \tau, |z_1 - \bar{z}_1|, \dots, |z_n - \bar{z}_n|)$$

($\nu = 1, \dots, n$). The one-sided conditions 4 III (α) and (4.4') can likewise be read vectorially (the hypothesis is the existence of a maximal solution).

The question of non-uniqueness is discussed in Šatalov (1967).

VI. Examples. (α) *Lipschitz Condition.* The function $\boldsymbol{\omega}(t, \tau, \boldsymbol{z})$ with components

$$\omega_\nu(t, \tau, \boldsymbol{z}) = l(\tau)(z_1 + \cdots + z_n) \qquad (\nu = 1, \dots, n)$$

belongs to the class $\mathscr{E}^n$ provided $0 \leqq l(t) \in L(J)$, as is easily seen with the help of the functions

$$\varrho_\nu(t) = C e^{nL(t)}, \qquad L(t) = \int_0^t l(\tau)\, d\tau, \qquad C > 0 \qquad (\nu = 1, \dots, n).$$

(β) *Generalized Nagumo Condition.* Suppose the $n \times n$ matrix $A = (a_{\mu\nu})$ has only nonnegative elements, $a_{\mu\nu} \geqq 0$, and is irreducible. Then by a theorem of Frobenius, A has a positive eigenvalue λ and an associated positive eigenvector $\boldsymbol{\alpha} = (\alpha_1, \dots, \alpha_n) > \boldsymbol{0}$. The eigenvalue λ exceeds the real part of every other eigenvalue [see e.g. Gantmacher (1959), Chapter XIII]. If $\lambda \leqq 1$, then

$$\omega_\nu(t, \tau, \boldsymbol{z}) = \frac{1}{\tau}(a_{\nu 1} z_1 + \dots + a_{\nu n} z_n) \qquad (\nu = 1, \dots, n)$$

and also

$$\omega_\nu(t, \tau, z) = \frac{1}{\tau}(a_{\nu 1} z_1 + \cdots + a_{\nu n} z_n) + l(\tau)(z_1 + \cdots + z_n) \qquad (\nu = 1, \ldots, n)$$

belong to the class $\mathscr{E}_1^n$ $[0 \leqq l(t) \in L(J)]$.

For the proof of this assertion we consider the vector function $\boldsymbol{\varrho}$ with the components

$$\varrho_\nu = C\alpha_\nu t\varphi(t) \quad \text{with} \quad C > 0, \quad \varphi(t) = e^{\beta L(t)}, \quad \beta = 2\max_\nu \frac{\alpha_1 + \cdots + \alpha_n}{\alpha_\nu}.$$

We have

$$\begin{aligned}\varrho_\nu' = C\alpha_\nu\varphi(t)[1 + \beta t l(t)] &\geqq C\varphi(t)[a_{\nu 1}\alpha_1 + \cdots + \alpha_{\nu n}\alpha_n + \alpha_\nu \beta t l(t)] \\ &> \frac{1}{t}(a_{\nu 1}\varrho_1 + \cdots + a_{\nu n}\varrho_n) + l(t)(\varrho_1 + \cdots + \varrho_n),\end{aligned}$$

whence follows the assertion. Compare Example 14 III (γ).

Now we come to Section 5. The initial value problem for a system of differential equations

$$\boldsymbol{u}' = \boldsymbol{f}(t, \boldsymbol{u}), \quad \boldsymbol{u}(0) = \boldsymbol{\eta} \tag{3}$$

as well as the classes $Z(\boldsymbol{f})$, $Z_{ac}(\boldsymbol{f})$ and the various notions of solution are defined as in 5 I. A general existence theorem is contained in IV. The assertions found there concerning maximal and minimal solutions will be sharpened later. A bound

$$|\boldsymbol{f}(t, \boldsymbol{z}) - \boldsymbol{f}(t, \bar{\boldsymbol{z}})| \leqq \boldsymbol{\omega}(t, |\boldsymbol{z} - \bar{\boldsymbol{z}}|) \tag{4}$$

with $\boldsymbol{\omega} \in \mathscr{E}^n$ [or $\boldsymbol{\omega} \in \mathscr{E}_1^n$ if $\boldsymbol{f}$ is continuous at the point $(0, \boldsymbol{\eta})$] is sufficient for uniqueness[1].

We encounter somewhat more trouble from 5 III; in particular, from the condition 5 III (γ). Its n-dimensional analog is contained in the following

VII. Theorem. *Suppose $\boldsymbol{f}(t, \boldsymbol{z})$ is quasi-monotone increasing in $\boldsymbol{z}$, that $\boldsymbol{v}, \boldsymbol{w} \in Z_{ac}(\boldsymbol{f})$ and*

(α) $\boldsymbol{v}(0+) < \boldsymbol{w}(0+)$;

(β) $\boldsymbol{v}' \leqq \boldsymbol{f}(t, \boldsymbol{v})$, $\boldsymbol{w}' \geqq \boldsymbol{f}(t, \boldsymbol{w})$ *a.e. in* J.

Then

$$\boldsymbol{v} < \boldsymbol{w} \quad \text{in} \quad J_0$$

if $\boldsymbol{f}$ also satisfies the following condition:

(γ) *The component f_ν has the property* 5 III (γ) *relative to z_ν, i.e., for every $t_0 \in J_0$ and every ν there is an $\omega(t, z) \in \mathscr{E}$ such that in a left-sided neighborhood of t_0*

$$f_\nu(t, \boldsymbol{z}) - f_\nu(t, \bar{\boldsymbol{z}}) \geqq -\omega(t_0 - t, z_\nu - \bar{z}_\nu) \quad \textit{for} \quad z_\nu \geqq \bar{z}_\nu,\ z_\mu = \bar{z}_\mu\ (\mu \neq \nu)$$

(provided the arguments lie in $D(\boldsymbol{f})$).

[1] We call attention to the notation $\boldsymbol{f}(t, \boldsymbol{z})$ and $\boldsymbol{\omega}(t, \boldsymbol{z})$ in place of $\boldsymbol{f}(\tau, \boldsymbol{z})$ and $\boldsymbol{\omega}(\tau, \boldsymbol{z})$; it was already used in Section 5.

The proof of 5 III must be modified. The point t_0 is chosen so that $\boldsymbol{v}<\boldsymbol{w}$ for $0<t<t_0$ and $v_\nu(t_0)=w_\nu(t_0)$ for at least one ν. Letting $\bar{\boldsymbol{v}}(t)=\boldsymbol{v}(t-t_0)$, $\bar{\boldsymbol{w}}(t)=\boldsymbol{w}(t-t_0)$, $\bar{\boldsymbol{f}}(t,\boldsymbol{z})=-\boldsymbol{f}(t-t_0,\boldsymbol{z})$, $d(t)=\bar{w}_\nu(t)-\bar{v}_\nu(t)$, we have

$$\begin{aligned} d' = \bar{w}'_\nu - \bar{v}'_\nu &\leqq \bar{f}_\nu(t,\bar{\boldsymbol{w}}) - \bar{f}_\nu(t,\bar{\boldsymbol{v}}) \\ &\leqq \bar{f}_\nu(t,\bar{v}_1,\ldots,\bar{v}_{\nu-1},\bar{w}_\nu,\bar{v}_{\nu+1},\ldots,\bar{v}_n) - \bar{f}_\nu(t,\boldsymbol{v}) \leqq \omega(t,d), \end{aligned}$$

where the monotonicity of $\boldsymbol{f}$ was used. All the rest is obtained as in 5 III. We remark that hypotheses (β) (γ) were used only in the following weaker form:

(β') If $\boldsymbol{v}<\boldsymbol{w}$ in an interval $0<t<t_0$ and $v_\nu(t_0)=w_\nu(t_0)$, then

$$v'_\nu \leqq f_\nu(t,\boldsymbol{v}),\ w'_\nu \geqq f_\nu(t,\boldsymbol{w}) \quad \text{a.e. in} \quad t_0-\varepsilon<t<t_0 \qquad (\varepsilon>0);$$

(γ') $g_\nu(t,z)=f_\nu(t,v_1,\ldots,v_{\nu-1},z,v_{\nu+1},\ldots,v_n)$ has the property 5 III (γ) ($\nu=1,\ldots,n$).

The theorem under consideration is, like 5 III, fundamental. The essential improvement over III is expressed in the condition (γ), which appears in place of monotonicity. However the difference is not as striking as for *one* differential equation, since for the component f_ν the monotonicity requirement in the variables z_μ with $\mu\neq\nu$ still holds. Now 5 IV–VIII can also be carried over in sharpened form. We dwell on this briefly.

Super- and subfunctions can be defined as in 5 IV if $\boldsymbol{f}$ is quasi-monotone increasing in $\boldsymbol{z}$ and satisfies the condition VII (γ). Now 5 V becomes the following

VIII. Theorem. *Suppose the function $\boldsymbol{\omega}(t,\boldsymbol{z})$ is defined on $J\times\{\boldsymbol{z}\geqq\boldsymbol{0}\}$ and is quasi-monotone increasing in $\boldsymbol{z}$; suppose further that it satisfies the condition* VII (γ). *For the functions $\boldsymbol{u},\boldsymbol{v}\in Z_{ac}(\boldsymbol{f})$, $\boldsymbol{\varrho}\in Z_{ac}(\boldsymbol{\omega})$ and $\boldsymbol{\delta}(t)$ suppose*

(α) $|\boldsymbol{v}-\boldsymbol{u}|(0+)<\boldsymbol{\varrho}(0+)$;

(β) $\boldsymbol{u}'=\boldsymbol{f}(t,\boldsymbol{u})$, $|\boldsymbol{v}'-\boldsymbol{f}(t,\boldsymbol{v})|\leqq\boldsymbol{\delta}(t)$ *a.e. in* J;

(γ) $\boldsymbol{\varrho}'\geqq\boldsymbol{\delta}(t)+\boldsymbol{\omega}(t,\boldsymbol{\varrho})$ *a.e. in* J;

(δ) $\left.\begin{array}{l} f_\nu(t,\boldsymbol{v})-f_\nu(t,\boldsymbol{v}-\boldsymbol{z}) \\ f_\nu(t,\boldsymbol{v}+\boldsymbol{z})-f_\nu(t,\boldsymbol{v}) \end{array}\right\} \leqq \omega_\nu(t,|\boldsymbol{z}|)$ *for* $|\boldsymbol{z}|\leqq\boldsymbol{\varrho}$, $0\leqq z_\nu\leqq\varrho_\nu$.

Then

$$|\boldsymbol{u}-\boldsymbol{v}|<\boldsymbol{\varrho} \quad \text{in} \quad J_0.$$

We give only an indication of the proof. First, from VII we obtain [with $\boldsymbol{v}\equiv\boldsymbol{0}$, $\boldsymbol{w}=\boldsymbol{\varrho}$ and $\boldsymbol{f}=\boldsymbol{\omega}$] that $\boldsymbol{\varrho}>\boldsymbol{0}$ [according to (δ) we have $\boldsymbol{\omega}(t,\boldsymbol{0})\geqq\boldsymbol{0}$]. The proof proper is carried out by reduction to VII with

$|\boldsymbol{v}-\boldsymbol{u}|, \boldsymbol{\varrho}, \boldsymbol{\delta}(t)+\boldsymbol{\omega}(t, z)$ instead of $\boldsymbol{v}, \boldsymbol{w}, \boldsymbol{f}(t, z)$, where that theorem is used with the hypothesis (β')[1].

VIII contains a

IX. Uniqueness Theorem. *The initial value problem* (3) *has at most one C-solution if* $\boldsymbol{f}(t, z)$ *satisfies an inequality*

$$f_\nu(t, z) - f_\nu(t, \bar{z}) \leqq \omega_\nu(t, |z-\bar{z}|) \quad \textit{for} \quad z_\nu \geqq \bar{z}_\nu \quad \textit{and} \quad (t, z), (t, \bar{z}) \in D(\boldsymbol{f})$$

$(\nu = 1, \ldots, n)$ *with* $\boldsymbol{\omega} \in \mathscr{E}_2^n$ [*resp.* $\boldsymbol{\omega} \in \mathscr{E}_3^n$ *provided* $\boldsymbol{f}$ *is continuous at the point* $(0, \boldsymbol{\eta})$].

Here both the classes $\mathscr{E}_2^n, \mathscr{E}_3^n$ are defined like the corresponding classes in 5 VI ($\boldsymbol{\omega}(t, z), \boldsymbol{\varrho}, \boldsymbol{\varepsilon} > \mathbf{0}, \boldsymbol{\delta} > \mathbf{0}$), but with VII (γ) instead of 5 III (γ) as well as the requirement that $\boldsymbol{\omega}$ be quasi-monotone increasing in z.

As a last application of VII we give a theorem on maximal and minimal solutions. If the function $\boldsymbol{f}(t, z)$, quasi-monotone increasing in z, satisfies the condition VII (γ) and if for a sequence $\boldsymbol{\varepsilon}_n$ of vectors with the properties $\boldsymbol{\varepsilon}_n > \boldsymbol{\varepsilon}_{n+1} > \mathbf{0}$, $\boldsymbol{\varepsilon}_n \to \mathbf{0}$ the initial value problems

$$\boldsymbol{u}' = \boldsymbol{f}(t, \boldsymbol{u}), \ \boldsymbol{u}(0) = \boldsymbol{\eta} + \boldsymbol{\varepsilon}_n$$

are solvable (let $\boldsymbol{u}_n$ be a solution), then VII implies that $\boldsymbol{u}_n > \boldsymbol{u}_{n+1}$ and $\boldsymbol{u}_n > \boldsymbol{u}$ for every solution $\boldsymbol{u}$ of the initial value problem (3). If we write these initial value problems as integral equations, then we can carry out the passage to the limit $n \to \infty$ under the integral sign under very general hypotheses on $\boldsymbol{f}$. In this way it is shown that the sequence $\boldsymbol{u}_n$ converges to a solution $\boldsymbol{u}^*$, namely, the desired maximal solution of the initial value problem (3). One proceeds similarly for the minimal solution.

X. Theorem. *If* $\boldsymbol{f} \in \mathscr{K}^n$ *is quasi-monotone increasing in* z *and has the property* VII (γ), *then the initial value problem* (3) *has a maximal solution* $\boldsymbol{u}^*$ *and a minimal solution* $\boldsymbol{u}_*$ *in* J. *If* $\boldsymbol{v}\,[\boldsymbol{w}] \in Z_{ac}(\boldsymbol{f})$ *and*

(α) $\boldsymbol{v}(0) \leqq \boldsymbol{\eta} \quad [\boldsymbol{w}(0) \geqq \boldsymbol{\eta}]$,

(β) $\boldsymbol{v}' \leqq \boldsymbol{f}(t, \boldsymbol{v}) \quad [\boldsymbol{w}' \geqq \boldsymbol{f}(t, \boldsymbol{w})]$ *a.e. in* J

then

$$\boldsymbol{v} \leqq \boldsymbol{u}^* \quad [\boldsymbol{w} \geqq \boldsymbol{u}_*] \quad \textit{in } J.$$

If $\boldsymbol{f}$ is only in $\mathscr{K}_0$, then the maximal solution exists either in J or in a largest interval $0 \leqq t < T^* \leqq T$; in the latter case we have

$$\limsup_{t \to T^*} |u_\nu(t)| = \infty$$

for at least one ν (and similarly for the minimal solution).

[1] The above theorem is distinguished from the corresponding one-dimensional Theorem 5 V by the fact that the very same bound is used for bounding above and below. A more general theorem for systems which does not have this deficiency will be proved in 13 V (it is easily carried over to C-solutions). The remarks relating to 13 V nevertheless show that such a theorem is of only slight practical significance.

The method of proof used above is precisely the "first method" described in 2 III–IV. We did not prove the assertion concerning $\boldsymbol{v}$, but it follows immediately since $\boldsymbol{u}_n > \boldsymbol{v}$ by VII. However in this proof we need the additional assumption that $\boldsymbol{f}$ satisfy VII (γ).

Let us sketch another proof which dispenses with VII (γ). It is in the spirit of the "second method" of Section 2. We define $\boldsymbol{u}^* = \sup\{\boldsymbol{v}(t): \boldsymbol{v} \in V\}$, where V is the class of all functions absolutely continuous in J which satisfy (α) (β). The following two properties are easily proved using the quasimonotonicity of $\boldsymbol{f}$. If $\boldsymbol{v} = \max(\boldsymbol{v}_1, \boldsymbol{v}_2)$ (componentwise), where $\boldsymbol{v}_1, \boldsymbol{v}_2 \in V$, then $\boldsymbol{v} \in V$. The operator T_i defined by $\boldsymbol{w} = T_i \boldsymbol{v} = (v_1, \ldots, v_{i-1}, \varphi, v_{i+1}, \ldots, v_n)$, where φ is the maximal solution of

$$\varphi' = f_i(t, v_1, \ldots, v_{i-1}, \varphi, v_{i+1}, \ldots, v_n), \qquad \varphi(0) = \eta_i,$$

maps V into V, and we have $\boldsymbol{v} \leqq T_i \boldsymbol{v}$ (see 5 X for the last inequality).

Let $\boldsymbol{h}(t) \in L(J)$ be such that $|\boldsymbol{f}(t, z)| \leqq \boldsymbol{h}(t)$. Because of the second property it is sufficient to consider the set V' of those elements of V which satisfy $\boldsymbol{v}(0) = \boldsymbol{\eta}$ and $|\boldsymbol{v}'(t)| \leqq \boldsymbol{h}(t)$. Since the set V' is equicontinuous and equibounded, we find by the method of 2 V a sequence of functions belonging to V' and converging uniformly to $\boldsymbol{u}^*$. Thus $\boldsymbol{u}^* \in V'$. We have $\boldsymbol{u}^* \leqq T_i \boldsymbol{u}^*$ by the second property proved above and $\boldsymbol{u}^* \geqq T_i \boldsymbol{u}^*$ by definition, hence $\boldsymbol{u}^* = T_i \boldsymbol{u}^*$ for each i. This shows that $\boldsymbol{u}^*$ is a solution of (3) and indeed the desired maximal solution.

Szarski proved Theorem X in his book (1965; Theorem 16.2) (even for infinite systems) using a method due to Mlak and Olech (1963). It is somewhat different from our procedure.

XI. Super- and Subfunctions can also be characterized for arbitrary right hand sides $\boldsymbol{f}$ by differential inequalities. However we encounter the difficulty that we have to specify a super- and a subfunction simultaneously. The function $\boldsymbol{v} \in Z_{ac}(\boldsymbol{f})$ is a subfunction and the function $\boldsymbol{w} \in Z_{ac}(\boldsymbol{f})$ a superfunction relative to the initial value problem (3) if

(α) $\boldsymbol{v}(0+) < \boldsymbol{u}(0+) < \boldsymbol{w}(0+)$ for every solution $\boldsymbol{u}$;
(β) $\boldsymbol{v}'(t) \leqq \boldsymbol{f}(t, z) \leqq \boldsymbol{w}'(t)$ *for* $\boldsymbol{v}(t) \leqq z \leqq \boldsymbol{w}(t)$ a.e. in J.

If $\boldsymbol{f}$ satisfies the condition VII (γ), then (β) can be replaced by

(β') $v_\nu'(t) \leqq f_\nu(t, z)$ for $\boldsymbol{v}(t) \leqq z \leqq \boldsymbol{w}(t), v_\nu(t) = z_\nu$ a.e. in J
$w_\nu'(t) \geqq f_\nu(t, z)$ for $\boldsymbol{v}(t) \leqq z \leqq \boldsymbol{w}(t), w_\nu(t) = z_\nu$ a.e. in J.

In both cases

$$\boldsymbol{v} < \boldsymbol{u} < \boldsymbol{w} \quad \text{in } J_0 \text{ for every solution } \boldsymbol{u} \text{ of the initial value problem (3).}$$

Let us assume for the proof that one of these inequalities is false and that $\boldsymbol{v} < \boldsymbol{u} < \boldsymbol{w}$ for $0 < t < t_0$, $u_\nu(t_0) = w_\nu(t_0)$. In the first case we then have $\boldsymbol{u}' \leqq \boldsymbol{w}'$ in the entire interval $0 \leqq t \leqq t_0$, whence a contradiction follows immediately. In the second case for $0 \leqq t \leqq t_0$

$$u_\nu' = g(t, u_\nu),\ w_\nu' = g(t, w_\nu) \quad \text{with} \quad g(t, z) = f_\nu(t, u_1, \ldots, u_{\nu-1}, z, u_{\nu+1}, \ldots, u_n).$$

Hence we arrive at a contradiction to the equality $u_\nu(t_0)=w_\nu(t_0)$ with the aid of 5 III, applied to u_ν, w_ν, g instead of v, w, f [note that g has the property 5 III (γ)].

The results analogous to XI for differential equations in the classical sense were found by M. Müller (1927).

XII. Integral Equations with Discontinuous Kernels. The concepts of a Volterra integral equation and of an ordinary differential equation have been generalized in various ways. We mention briefly one such generalization. The main thing is that the continuity of the kernel $\boldsymbol{k}(t,\tau,z)$ with respect to z is dropped. Instead one assumes that $\boldsymbol{k}$ is measurable in z. With $\boldsymbol{k}$ we associate two kernels

$$\boldsymbol{k}^*(t,\tau,z)=\limsup_{\zeta\to z}\boldsymbol{k}(t,\tau,\zeta),\qquad \boldsymbol{k}_*(t,\tau,z)=\liminf_{\zeta\to z}\boldsymbol{k}(t,\tau,\zeta),$$

where the lim sup is taken componentwise and defined as

$$\limsup_{\zeta\to z}=\lim_{\varrho\to 0}\Big\{\operatorname*{ess\,sup}_{\|z-\zeta\|\leqq\varrho}\Big\}.$$

A function $\boldsymbol{u}$ is called a solution of the integral equation (1) if $\boldsymbol{u}$ is continuous in J and

$$\boldsymbol{g}(t)+\int_0^t\boldsymbol{k}_*(t,\tau,\boldsymbol{u}(\tau))\,d\tau\leqq\boldsymbol{u}(t)\leqq\boldsymbol{g}(t)+\int_0^t\boldsymbol{k}^*(t,\tau,\boldsymbol{u}(\tau))\,d\tau\ \text{ in } J. \tag{5}$$

Under conditions similar to those given in 2 VI for the class $\mathscr{K}_1$ an existence theorem can be proved. Use is made here of the fact that the kernel

$$\boldsymbol{k}_\varrho(t,\tau,z)=\frac{1}{V_\varrho}\int_{K_\varrho}\boldsymbol{k}(t,\tau,z+\zeta)\,d\zeta\qquad(\varrho>0)$$

is continuous in z (K_ϱ is the ball $\|\zeta\|\leqq\varrho$, V_ϱ its volume). The solution $\boldsymbol{u}$ of (5) is constructed from the solutions $\boldsymbol{u}_\varrho$ of the equation with the kernels $\boldsymbol{k}_\varrho$, using a familiar compactness argument. For the details we refer to the literature given below.

The theorem on integral inequalities III is easily extended to the present case:

Suppose $\boldsymbol{v},\boldsymbol{w},\boldsymbol{g}\in C(J)$. Let $\boldsymbol{k}(t,\tau,z)$ be increasing and measurable in z and such that $\boldsymbol{k}^*(t,\tau,\boldsymbol{v}(\tau))$ and $\boldsymbol{k}_*(t,\tau,\boldsymbol{w}(\tau))$ are integrable on $(0,t)$ for every $t\in J$. Then the inequality

$$\boldsymbol{v}(t)-\boldsymbol{g}(t)-\int_0^t\boldsymbol{k}^*(t,\tau,\boldsymbol{v}(\tau))\,d\tau<\boldsymbol{w}(t)-\boldsymbol{g}(t)-\int_0^t\boldsymbol{k}_*(t,\tau,\boldsymbol{w}(\tau))\,d\tau\quad\text{in}\quad J$$

implies

$$\boldsymbol{v}<\boldsymbol{w}\quad\text{in}\quad J.$$

The proof follows the same lines as in III. The difficulty that now two kernels are involved is easily overcome by using the following property: if $z<\bar{z}$, then $\boldsymbol{k}^*(t,\tau,z)\leqq\boldsymbol{k}_*(t,\tau,\bar{z})$.

The theory sketched above was developed by Li Mun Su (1965) (1965a); see also Li Mun Su and Ragimhanov (1967). For other works on differential and integral equations with discontinuities see Matrosov (1967, 1967a).

7. Bounds for Systems Using *K*-Norms

A second method for extending Theorem 4 I and similar theorems to systems consists of going from the vector function $\boldsymbol{v}(t)-\boldsymbol{u}(t)$ [let $\boldsymbol{u}(t)$ be a solution and $\boldsymbol{v}(t)$ an approximate solution of Eq. (6.2)] by means of a norm to the scalar function $\|\boldsymbol{v}(t)-\boldsymbol{u}(t)\|$ and estimating this scalar function with the help of the appropriate "scalar" theorems. Thus we obtain scalar bounds, in contrast to Section 6, where the bound was an n-dimensional vector. In principle intermediate stages are also possible between these two extremes. The bound is then a vector of lower dimension, and the components of the function to be bounded are divided into groups where each group is bounded by a scalar function. We let this matter rest and proceed to the second method.

I. Definition (*norm, K-norm,* $\|\boldsymbol{z}\|_e$). A real-valued scalar function $\|\boldsymbol{z}\|$ defined for $\boldsymbol{z}\in E^n$ is called a K-norm [or according to Hukuhara (1941a) a "Kamke function"] if it has both of the properties

(α) $\|\boldsymbol{y}+\boldsymbol{z}\|\leqq\|\boldsymbol{y}\|+\|\boldsymbol{z}\|$ (triangle inequality)
(β) $\|\alpha\boldsymbol{z}\|=\alpha\|\boldsymbol{z}\|$ for real $\alpha\geqq 0$,

while a norm in the usual sense is defined by the three properties (α),

(β') $\|\alpha\boldsymbol{z}\|=|\alpha|\,\|\boldsymbol{z}\|$ for real α
(γ) $\|\boldsymbol{z}\|>0$ for $\boldsymbol{z}\neq 0$.

Every norm is a K-norm, but not vice-versa.

In textbooks one finds most often in the treatement of systems of ordinary differential equations the norm

(δ) $\|\boldsymbol{z}\|=|z_1|+\cdots+|z_n|$,

or occasionally the Euclidean distance

(ε) $\|\boldsymbol{z}\|_e=\sqrt{z_1^2+\cdots+z_n^2}$.

Of the two examples

(ζ) $\|\boldsymbol{z}\|=\max_\nu|z_\nu|$
(η) $\|\boldsymbol{z}\|=\max_\nu z_\nu$

the first is a norm, the second a K-norm.

The fact that the theorems under consideration here are quite independent of a special norm, and moreover can be stated for K-norms, was brought to light by the work of Kamke (1930a) [Müller (1927) had previously proved uniqueness theorems with the K-norm (η)].

II. Some Properties of K-norms. (α) The inequality

$$\|\boldsymbol{y}\| - \|\boldsymbol{z}\| \leqq \|\boldsymbol{y} - \boldsymbol{z}\|$$

follows from I (α) if we replace $\boldsymbol{y}$ by $\boldsymbol{y} - \boldsymbol{z}$.

(β) For every K-norm there is a constant $N > 0$ such that

$$|\|\boldsymbol{z}\|| \leqq N \sum_\nu |z_\nu| .$$

For if we use $\boldsymbol{e}_\nu \in E^n$ to denote the vector which has a 1 in the νth position and zeros everywhere else, then $\boldsymbol{z} = \sum_\nu z_\nu \boldsymbol{e}_\nu$, and thus by I ($\alpha$) ($\beta$)

$$\|\boldsymbol{z}\| \leqq \sum_\nu \|z_\nu \boldsymbol{e}_\nu\| = \sum_\nu |z_\nu| \, \|\pm \boldsymbol{e}_\nu\| \leqq N \sum_\nu |z_\nu|$$

(where the positive or negative sign is used depending on whether z_ν is positive or negative). It further follows from I (α), with $\boldsymbol{y} = -\boldsymbol{z}$, that

$$0 \leqq \|-\boldsymbol{z}\| + \|\boldsymbol{z}\|, \quad \text{thus} \quad -\|\boldsymbol{z}\| \leqq \|-\boldsymbol{z}\| \leqq N \sum_\nu |z_\nu| .$$

(γ) From (α) and (β) we obtain the fact (which is well known for norms) that every K-norm is a continuous function. Furthermore, $\|\boldsymbol{\varphi}(t)\|$ is a continuous (resp. absolutely continuous) scalar function if $\boldsymbol{\varphi}(t)$ is a continuous (resp. absolutely continuous) vector function.

(δ) If the vector function $\boldsymbol{\varphi}(t)$ is differentiable from the left [right] at the point t, then

$$D^- \|\boldsymbol{\varphi}(t)\| \leqq \|\boldsymbol{\varphi}'_-(t)\| \quad \text{resp.} \quad D^+ \|\boldsymbol{\varphi}(t)\| \leqq \|\boldsymbol{\varphi}'_+(t)\| .$$

According to (α) and I (β) for $t < \bar{t}$ we have

$$\frac{\|\boldsymbol{\varphi}(\bar{t})\| - \|\boldsymbol{\varphi}(t)\|}{\bar{t} - t} \leqq \left\| \frac{\boldsymbol{\varphi}(\bar{t}) - \boldsymbol{\varphi}(t)}{\bar{t} - t} \right\| ,$$

and for $t \to \bar{t} - 0$ [resp. $\bar{t} \to t + 0$] the assertion follows because of the continuity of the K-norm.

(ε) By (γ) and (δ), $\|\boldsymbol{\varphi}(t)\|$ is therefore absolutely continuous along with $\boldsymbol{\varphi}(t)$, and

$$\|\boldsymbol{\varphi}(t)\|' \leqq \|\boldsymbol{\varphi}'(t)\| \quad \text{a.e.}.$$

(ζ) If $\boldsymbol{\varphi}(t)$ is L-integrable over an interval $a \leqq t \leqq b$, then the same holds for $\|\boldsymbol{\varphi}(t)\|$, and

$$\left\| \int_a^b \boldsymbol{\varphi}(t) dt \right\| \leqq \int_a^b \|\boldsymbol{\varphi}(t)\| dt$$

(likewise for multidimensional integrals). For the proof we note that by I (α) (β)

$$\left\| \sum_i \varepsilon_i \boldsymbol{\varphi}(t_i) \right\| \leqq \sum_i \varepsilon_i \|\boldsymbol{\varphi}(t_i)\| \quad \text{for positive } \varepsilon_i .$$

But these sums are Riemann approximation sums for the above integrals for suitably chosen ε_i and t_i. In this way we first obtain the above inequality in the case of Riemann integrability and then the general case in the usual way.

A more precise investigation of K-norms was made by M. Hukuhara (1941a). Among other things, in sharpening (δ) he proved that the right (left) handed derivative of $\|\boldsymbol{\varphi}(t)\|$ exists if this holds for $\boldsymbol{\varphi}(t)$. The property (γ) seems to have eluded him, since he includes it in the axioms. A K-norm is nothing but a "subadditive, positively homogeneous functional" occuring in many places in functional analysis, e.g. in the Hahn-Banach theorem; see Banach (1932, p. 28f).

After these preliminaries, Theorem 4 I will be carried over to a system of integral equations or more generally to an operator equation

$$\boldsymbol{u} = \boldsymbol{g} + \boldsymbol{K}\boldsymbol{u} . \tag{1}$$

We assume that as in (4.2), an inequality

$$\|\boldsymbol{K}\boldsymbol{\varphi} - \boldsymbol{K}\overline{\boldsymbol{\varphi}}\| \leqq \Omega(\|\boldsymbol{\varphi} - \overline{\boldsymbol{\varphi}}\|) \tag{2}$$

holds. Here $\boldsymbol{K}$ is now a vector operator in the sense of 6 II, Ω a scalar operator in the sense of 1 IV.

III. Theorem. *Let $\boldsymbol{K}$ be a (vector) operator, Ω a monotone increasing (scalar) operator and $\|\cdot\|$ a K-norm. Suppose that the bound* (2) *holds if $\boldsymbol{\varphi}, \overline{\boldsymbol{\varphi}} \in Z_c(\boldsymbol{K})$ and $\|\boldsymbol{\varphi} - \overline{\boldsymbol{\varphi}}\| \in Z(\Omega)$. Let the functions $\boldsymbol{u}, \boldsymbol{v}, \varrho, d, \boldsymbol{g}$ be defined in J and $\boldsymbol{u}, \boldsymbol{v} \in Z_c(\boldsymbol{K})$, $\|\boldsymbol{v} - \boldsymbol{u}\|, \varrho \in Z_c(\Omega)$. Then*

(α) $\|\boldsymbol{v} - \boldsymbol{u}\|(0+) < \varrho(0+)$;
(β) $\boldsymbol{u} = \boldsymbol{g} + \boldsymbol{K}\boldsymbol{u}$, $\|\boldsymbol{v} - \boldsymbol{g} - \boldsymbol{K}\boldsymbol{v}\| \leqq d$ *in* J_0;
(γ) $\varrho > d + \Omega\varrho$ *in* J_0

imply the inequality

$$\|\boldsymbol{v} - \boldsymbol{u}\| < \varrho \quad \text{in} \quad J_0 .$$

If the operator Ω satisfies the hypotheses of 2 IV and if $\sigma(t)$ is the maximal solution of the equation

$$\sigma = d + \Omega\sigma , \tag{3}$$

then it follows from $\|\boldsymbol{v}(0) - \boldsymbol{u}(0)\| \leqq \sigma(0)$ and (β) that

$$\|\boldsymbol{v} - \boldsymbol{u}\| \leqq \sigma \quad \text{in} \quad J .$$

The proof of 4 I can be transferred; only the absolute value bars are to be replaced by double bars and u, v, g, K by the corresponding bold-face letters.

IV. Uniqueness Theorem. *Let* $(\boldsymbol{K}\boldsymbol{\varphi})(0)=0$ *for all* $\boldsymbol{\varphi}\in Z_c(\boldsymbol{K})$. *Suppose further that there is an operator* $\Omega\in\mathscr{E}$ *such that for all* $\boldsymbol{\varphi}, \bar{\boldsymbol{\varphi}}\in Z_c(\boldsymbol{K})$ *the inequality* (2) *holds, and indeed with a K-norm which has the property that* $\|\boldsymbol{z}\|=\|-\boldsymbol{z}\|=0$ *only for* $\boldsymbol{z}=0$. *Then the operator equation* (1) *has at most one (continuous) solution.*

It follows from the previous theorem just as in the proof of 4 III that if $\boldsymbol{u}$ and $\boldsymbol{v}$ are two solutions then $\|\boldsymbol{v}-\boldsymbol{u}\|\leqq 0$ and, since $\boldsymbol{u}$ and $\boldsymbol{v}$ can be interchanged, $\|\boldsymbol{u}-\boldsymbol{v}\|\leqq 0$. However, since $\|\boldsymbol{z}\|+\|-\boldsymbol{z}\|\geqq 0$ as we saw in II (β), we have $\|\boldsymbol{v}-\boldsymbol{u}\|=\|\boldsymbol{u}-\boldsymbol{v}\|=0$ and hence $\boldsymbol{u}=\boldsymbol{v}$.

V. Systems of Integral Equations. If $\boldsymbol{K}$ and Ω are integral operators with the kernels $\boldsymbol{k}(t,\tau,\boldsymbol{z})$ and $\omega(t,\tau,z)$, the statement under 4 IV holds mutatis mutandis. Because of II (ζ), the condition (2) is a consequence of

$$\|\boldsymbol{k}(t,\tau,\boldsymbol{z})-\boldsymbol{k}(t,\tau,\bar{\boldsymbol{z}})\|\leqq\omega(t,\tau,\|\boldsymbol{z}-\bar{\boldsymbol{z}}\|)\,. \tag{2'}$$

If such an inequality (in the domain of definition of $\boldsymbol{k}$) holds with a function $\omega\in\mathscr{E}$ and a K-norm which has the property designated in IV, then we have uniqueness. If $\boldsymbol{k}$ is continuous at the point $(0,0,\boldsymbol{g}(0))$, then even $\omega\in\mathscr{E}_1$ suffices.

In applications, $\boldsymbol{k}$ often satisfies a Lipschitz condition

$$\|\boldsymbol{k}(t,\tau,\boldsymbol{z})-\boldsymbol{k}(t,\tau,\bar{\boldsymbol{z}})\|\leqq l(\tau)\,\|\boldsymbol{z}-\bar{\boldsymbol{z}}\|\quad\text{with}\quad l(\tau)\geqq 0\,.$$

The bound ϱ in III can then be given explicitly. We need not state the formulation of this important special case, since it coincides word for word (except for bold-face and double bars) with 4 V (α).

As in Section 5, it is useful to consider the initial value problem for a system of equations

$$\boldsymbol{u}'=\boldsymbol{f}(t,\boldsymbol{u}),\qquad \boldsymbol{u}(0)=\boldsymbol{\eta} \tag{4}$$

not simply as a special case of Eq. (1). By going back to Section 5 we obtain sharper theorems which are free of monotonicity requirements.

VI. Estimation Theorem for Systems of Differential Equations. *Suppose* $\varrho(t)$ *is absolutely continuous in* J *and positive in* J_0, $\omega(t,z)$ *is defined for* $0\leqq z\leqq\varrho(t)$ *and the property* 5 III (γ) *holds for* ω. *Suppose further that* $\boldsymbol{u},\boldsymbol{v}\in Z_{ac}(\boldsymbol{f})$, $\delta(t)$ *is a function defined in* J, *and that*

(α) $\|\boldsymbol{v}-\boldsymbol{u}\|(0+)<\varrho(0+)$;
(β) $\boldsymbol{u}'=\boldsymbol{f}(t,\boldsymbol{u})$, $\|\boldsymbol{v}'-\boldsymbol{f}(t,\boldsymbol{v})\|\leqq\delta(t)$ *a.e. in* J;
(γ) $\varrho'\geqq\delta(t)+\omega(t,\varrho)$ *a.e. in* J;
(δ) $\|\boldsymbol{f}(t,\boldsymbol{v})-\boldsymbol{f}(t,\boldsymbol{v}+\boldsymbol{z})\|\leqq\omega(t,\|\boldsymbol{z}\|)$ *for* $0\leqq\|\boldsymbol{z}\|\leqq\varrho(t)$, $(t,\boldsymbol{v}+\boldsymbol{z})\in D(\boldsymbol{f})$.

Then

$$\|\boldsymbol{v}-\boldsymbol{u}\|<\varrho \quad in \quad J_0 .$$

This theorem can be reduced to 5 III if we replace v, w, f there by $\|\boldsymbol{v}-\boldsymbol{u}\|, \varrho, \delta+\omega$. Then ($\alpha$) becomes 5 III ($\alpha$). Further, by II ($\varepsilon$), $\|\boldsymbol{v}-\boldsymbol{u}\|$ is absolutely continuous, and it follows from $\|\boldsymbol{v}-\boldsymbol{u}\|<\varrho$ for $0<t<t_0$, $\|\boldsymbol{v}-\boldsymbol{u}\|=\varrho$ for $t=t_0$ that $0\leqq\|\boldsymbol{v}-\boldsymbol{u}\|\leqq\varrho$ in a left-sided neighborhood of t_0; i.e., by (β) (δ)

$$\|\boldsymbol{v}-\boldsymbol{u}\|'\leqq\|\boldsymbol{v}'-\boldsymbol{u}'\|=\|\boldsymbol{v}'-\boldsymbol{f}(t,\boldsymbol{v})+\boldsymbol{f}(t,\boldsymbol{v})-\boldsymbol{f}(t,\boldsymbol{u})\|\leqq\delta(t)+\omega(t,\|\boldsymbol{v}-\boldsymbol{u}\|).$$

Thus 5 III (β') also holds with the new quantities.

From VI it is very easy to obtain the following

VII. Uniqueness Theorem for Systems of Differential Equations. *Let* $\|\cdot\|$ *be a K-norm for which* $\|\boldsymbol{z}\|=\|-\boldsymbol{z}\|$ *holds only for* $\boldsymbol{z}=0$. *The initial value problem* (4) *has at most one* C-*solution if the right hand side of the differential equation permits a bound*

$$\|\boldsymbol{f}(t,\boldsymbol{z})-\boldsymbol{f}(t,\bar{\boldsymbol{z}})\|\leqq\omega(t,\|\boldsymbol{z}-\bar{\boldsymbol{z}}\|) \quad for \quad \|\boldsymbol{z}-\bar{\boldsymbol{z}}\|\geqq 0 \quad and \quad (t,\boldsymbol{z}),(t,\bar{\boldsymbol{z}})\in D(\boldsymbol{f})$$

with a function $\omega\in\mathscr{E}_2$. *If* $\boldsymbol{f}$ *is continuous at the point* $(0,\boldsymbol{\eta})$, then $\omega\in\mathscr{E}_3$ *suffices.*

VIII. Remark. Except for trivialities, the theorems proved here differ particularly in three points from corresponding well known theorems as far as they relate to systems of differential equations in the sense of Carathéodory. They hold not only for norms but even for K-norms; the bound of $\boldsymbol{f}$ by ω need not hold for all arguments but only for $\|\boldsymbol{z}-\bar{\boldsymbol{z}}\|\geqq 0$ in, say, the uniqueness theorem; finally, a monotonicity requirement for ω is superfluous. Of these generalizations, both of the first two can be arrived at without complications in the proof worth mentioning, while the third goes back to 5 III and its different type of proof. Of course no generalization is produced by adding "for $\|\boldsymbol{z}-\bar{\boldsymbol{z}}\|\geqq 0$" for norms; on the other hand it can be of considerable significance for K-norms. Thus the uniqueness theorem 5 VII has now become a special case of VII, and indeed in its full scope including the one-sided condition 5 V (δ). To see this we need only set $n=1$ and $\|z\|=z$ in VII.

IX. Example. We consider the system of linear integral equations

$$\boldsymbol{u}(t)=\boldsymbol{b}(t)+\int_0^t A(t,\tau)\boldsymbol{u}(\tau)d\tau . \tag{5}$$

Here let $A(t,\tau)$ be an $n\times n$ matrix in the class $L(0\leqq\tau\leqq t)$ for every $t\in J$; the functions $\boldsymbol{u}(t)$, $\boldsymbol{b}(t)\in C(J)$ are to be taken as column vectors. If $\|\boldsymbol{z}\|$ and $\|A\|$ are norms for column vectors $\boldsymbol{z}$ and $\boldsymbol{n}\times\boldsymbol{n}$ matrices A such that $\|A\boldsymbol{z}\|\leqq\|A\|\,\|\boldsymbol{z}\|$ (e.g., $\|\boldsymbol{z}\|=|z_1|+\cdots+|z_n|$, $\|A\|=\max|a_{\mu\nu}|$), then (5) implies the inequality

$$\|\boldsymbol{u}(t)\| \leqq g(t) + \int_0^t h(t,\tau)\|\boldsymbol{u}(\tau)\|\,d\tau$$

with

$$g(t) \geqq \|\boldsymbol{b}(t)\|, \qquad h(t,\tau) \geqq \|A(t,\tau)\|\,.$$

If $g(t)\in C(J)$, $h(t,\tau)\in L(0\leqq\tau\leqq t)$ for every $t\in J$ and if $h(t,\tau)$ is monotone increasing in t, then we have the inequality

$$\|\boldsymbol{u}(t)\| \leqq g(t) + \int_0^t g(\tau)h(t,\tau)e^{H(t,t)-H(t,\tau)}\,d\tau$$

with

$$H(t,\tau) = \int_0^\tau h(t,s)\,ds\,,$$

and in particular for $g(t)=\text{const.}$,

$$\|\boldsymbol{u}(t)\| \leqq g(0)e^{H(t,t)}.$$

For the special case $t=T$ this is obtained immediately from 1 III if we replace the function $h(\tau)$ there by $h(T,\tau)$. Since this conclusion is also correct in every subinterval $0\leqq t\leqq T^* < T$, the assertion follows for arbitrary $t=T^*\in J$.

This estimate generalizes a result of Satō (1952); see also Butlewski (1948, 1959) and Gheorghiu (1965). Similar estimates for non-linear equations are given by Bantas (1965).

X. Integral and Differential Equations in Banach Spaces. The theorems of this section can be carried over without difficulties to functions whose values belong to a Banach space. We outline one of the several possibilities for the translation which are motivated by the various topologies and concepts of integral in a Banach space.

We consider a Banach space B with the norm $\|\cdot\|$ and functions $\varphi(t)$ whose values are in B for every $t\in J$. We take the norm topology (strong topology) as the basis for all limit processes. Thus the function $\varphi(t)$ is continuous at the point t_0 (resp. differentiable with derivative $\varphi'(t_0)$) if

$$\|\varphi(t_0+h)-\varphi(t_0)\|\to 0 \quad \text{resp.} \quad \left\|\frac{1}{h}[\varphi(t_0+h)-\varphi(t_0)]-\varphi'(t_0)\right\|\to 0$$

as $h\to 0$. By $C(J)$ we mean the set of functions continuous in J in this sense, and by a Volterra operator $\boldsymbol{K}$ a mapping of a set $Z_c(\boldsymbol{K})\subset C(J)$ into $C(J)$ with the property that $(\boldsymbol{K}\varphi)(t_0)=(\boldsymbol{K}\psi)(t_0)$ if $\varphi(t)=\psi(t)$ for $0\leqq t\leqq t_0$.

The estimation theorem III *and the uniqueness theorem* IV *hold without change.* The proofs also carry over.

The application to integral equations is based on the Bochner integral (see Hille-Phillips (1957, Ch. 3.1)). The kernel $\boldsymbol{k}(t,\tau,\boldsymbol{z})(0\leqq\tau\leqq t \leqq T, \boldsymbol{z}\in B, \boldsymbol{k}\in B)$ generates the integral operator

$$(\boldsymbol{K}\varphi)(t)=\int_0^t \boldsymbol{k}(t,\tau,\varphi(\tau))d\tau\,;$$

here $\varphi\in Z_c(\boldsymbol{k})$ means that $\varphi\in C(J)$ and that this Bochner integral exists for every $t\in J$ [i.e., $\|\boldsymbol{k}(t,\tau,\varphi(\tau))\|$ is integrable in the usual sense and $\boldsymbol{k}(t,\tau,\varphi)$ is measurable; see Hille-Phillips, Theorem 3.7.4].

In particular, these theorems can be applied to ordinary differential equations in a Banach space

$$\boldsymbol{u}'=\boldsymbol{f}(t,\boldsymbol{u})\,. \tag{6}$$

In the definition of the class Z_{ac} we have to note that the absolute continuity of the function $\|\boldsymbol{u}(t)\|$ generally does *not* imply the existence of the derivative $\boldsymbol{u}'(t)$ (a.e. in J). Thus we define Z_{ac} as the class of functions $\boldsymbol{u}(t)$ which are continuous and a.e. differentiable in J and for which $\|\boldsymbol{u}(t)\|$ is absolutely continuous. The fundamental theorem of differential and integral calculus $\boldsymbol{u}(t)=\boldsymbol{u}(0)+\int_0^t \boldsymbol{u}'(\tau)d\tau$ holds for $\boldsymbol{u}\in Z_{ac}$ (Hille-Phillips, Theorem 3.8.6). In this class of solutions — it is natural to speak of solutions in the sense of Carathéodory — the initial value problem for Eq. (6) can be transformed into an integral equation. *Theorems* V *and* VI, *and in particular the special uniqueness criteria* 4 V, *then hold.*

In recent years an extensive literature on differential equations in Banach spaces has appeared. The following, among others, have connections in subject matter and methods with the problems raised here: Krasnosel'skii and Krein (1955), Kato (1956), Corduneanu (1957), Kisyński (1959), Mlak (1959), Pulvirenti (1960, 1961), Lakshmikantham (1962a, 1964), Jedryka (1965).

XI. Existence and Successive Approximation. Summing up, we may state that our theory of uniqueness and error estimation carries over to equations in Banach space without difficulty; quite in contrast the existence theory offers new aspects. Among others we have the surprising fact that the initial value problem for Eq. (6) has in general no solution if $\boldsymbol{f}$ is merely bounded and continuous. Here we shall give an existence proof for the integral equation

$$\boldsymbol{u}(t)=\boldsymbol{g}(t)+\int_0^t \boldsymbol{k}(t,\tau,\boldsymbol{u}(\tau))d\tau \quad \text{in} \quad J\,, \tag{7}$$

written $\boldsymbol{u}=\boldsymbol{g}+\boldsymbol{K}\boldsymbol{u}$ for short. The proof is based on successive approximation and uses monotonicity arguments in an essential way. The main assumption is an inequality

$$\|k(t, \tau, z) - k(t, \tau, \bar{z})\| \leqq \omega(t, \tau, \|z - \bar{z}\|) \tag{2'}$$

familiar in connection with uniqueness questions. As is well known, successive approximation means that we construct a sequence (u_n) by $u_{n+1} = g + Ku_n$, where u_0 is a given zero-approximation.

The integral equation (7) will be considered as an equation in a real Banach space under "Carathéodory hypotheses", that is, the integral is taken in the Bochner sense.

XII. Existence and Convergence Theorem. *Suppose* $k \in \mathscr{K}$, i.e. $k(t, \tau, z)$ *is defined in* $\{0 \leqq \tau \leqq t \leqq T\} \times B$, *where B is a real Banach space, measurable in* τ *(in the Bochner sense) for fixed* (t, z), *continuous in* (t, z) *for fixed* τ, *and* $\|k(t, \tau, z)\| \leqq h(\tau) \in L(J)$. *Suppose further that* k *satisfies the condition* (2'), *where* ω *has the properties*

(α) $\omega(t, \tau, z)$ *is real-valued, monotone increasing in* z *and belongs to* $\mathscr{K}_0$;

(β) *the function* $\sigma(t) \equiv 0$ *is the maximal solution in* J *of the equation*

$$\sigma(t) = \int_0^t \omega(t, \tau, \sigma(\tau)) d\tau;$$

(γ) *for each* $C > 0$ *there is a function* $\varrho \in C(J)$ *satisfying*

$$\varrho(t) \geqq C \quad \text{and} \quad \varrho(t) \geqq \int_0^t \omega(t, \tau, \varrho(\tau)) d\tau.$$

Then Eq. (7) *has one and only one continuous solution* u *in* J *and the sequence* (u_n), *obtained by successive approximation from an arbitrary zero-approximation* $u_0 \in C(J)$, *converges to* u *uniformly in* J.

In the proof we use the abbreviation $\Omega\varrho$ for the integral in (γ). Let (u_n) be an approximating sequence. Since k is bounded in norm by $h(\tau)$, $\|u_n(t)\| \leqq C$ for appropriate C. We define a sequence (ϱ_n) by $\varrho_{n+1} = \Omega\varrho_n$, where ϱ_0 satisfies the inequalities

$$\varrho_0 \geqq 2C \quad \text{and} \quad \varrho_0(t) \geqq \Omega\varrho_0 = \varrho_1.$$

This is possible by (γ). Since $\omega(t, \tau, z)$ is increasing in z and $\omega(t, \tau, 0) = 0$, it is easily seen (by induction) that the sequence (ϱ_n) is decreasing and that the ϱ_n are nonnegative and equicontinuous functions. Let $\sigma = \lim \varrho_n \in C(J)$. From the uniform convergence (or, alternatively, from the fact that $\omega(t, \tau, \varrho_n(\tau))$ is decreasing in n) it follows that σ is a solution of $\sigma = \Omega\sigma$, hence $\sigma \equiv 0$ by (β). Thus the sequence (ϱ_n) converges uniformly to zero.

Let us now prove the inequality

$$\|u_{n+m}(t)-u_n(t)\| \leqq \varrho_n(t) \quad \text{in } J \text{ for } \quad n, m=0,1,2,\ldots$$

It holds for $n=0$ and arbitrary m. A proof by induction (in n) is simple:

$$\|u_{n+m+1}-u_{n+1}\| = \|Ku_{n+m}-Ku_n\| \leqq \Omega(\|u_{n+m}-u_n\|) \leqq \Omega\varrho_n = \varrho_{n+1}.$$

Here we used (2) and the monotonicity of ω in z. Thus the sequence (u_n) is a Cauchy sequence with respect to uniform convergence. Its limit u is a solution of (7), as is seen by passing to the limit in the equation $u_{n+1}=g+Ku_n$.

The uniqueness assertion of XII is contained in X, since we showed in 4 II that $\omega \in \mathscr{E}$. Uniqueness can also be proved directly in the following manner. If u and $\bar{u}$ are two solutions and $\varrho_0 \geqq \|u-\bar{u}\|$, $\varrho_0 \geqq \Omega\varrho_0$, then the inequality $\|u-\bar{u}\| \leqq \varrho_n (n=1,2,\ldots)$ follows by induction. This together with $\varrho_n \to 0$ yields $u=\bar{u}$.

We note the most important special case of Theorem XII: *Existence, uniqueness and convergence of successive approximation hold if* **k** *satisfies a generalized Lipschitz condition*

$$\|k(t,\tau,z)-k(t,\tau,\bar{z})\| \leqq l(\tau)\,\|z-\bar{z}\| \quad \textit{with} \quad l(\tau) \in L(J). \tag{8}$$

In conclusion we remark that there is a nice device which gives a very short proof of Theorem XII under the condition (8). The space $C(J)$ is a Banach space with the norm

$$|u|^* = \max_J \|u(t)\|\, e^{-L(t)}, \qquad L(t) = \int_0^t l(\tau)\,d\tau.$$

With respect to the norm $|\cdot|^*$, the map $u \to g+Ku$ is a contraction, and Theorem XII follows from the Banach fixed point theorem (contractive mapping principle). Indeed

$$\begin{aligned}\|(Ku-Kv)(t)\| &\leqq \int_0^t \|k(t,\tau,u)-k(t,\tau,v)\|\,d\tau \\ &\leqq \int_0^t l(\tau)\,\|u(\tau)-v(\tau)\|e^{L(\tau)}e^{-L(\tau)}d\tau \\ &\leqq \ |u-v|^* \int_0^t l(\tau)\,e^{L(\tau)}d\tau = |u-v|^*(e^{L(t)}-1).\end{aligned}$$

Multiplying by $e^{-L(t)}$ we get

$$|Ku-Kv|^* \leqq |u-v|^* \max_J (1-e^{-L(t)}) = q|u-v|^* \quad \text{with} \quad q = 1-e^{-L(T)} < 1,$$

i.e., $\boldsymbol{K}$ is contractive. This idea of working with a weighted maximum norm proves useful in other instances; see, e.g. Walter (1965). To the author's knowledge it goes back to Morgenstern (1952).

Naturally Theorem XII applies to the case $B = E^n$, i.e., to systems of Volterra integral equations of the form (6.1) and especially to systems of ordinary differential equations in the sense of Carathéodory (6.3) and in the classical sense (Chapter II). It contains cum grano salis all known convergence assertions on this type of equation. Since there are theorems in the literature which seem more general at first glance, some remarks are required.

Suppose $\boldsymbol{k}$ satisfies the hypotheses of Theorem XII. Let $M \geqq \|g(t)\| + \int_0^t h(\tau)d\tau$ in J and

$$\bar{\omega}(t, \tau, s) = \sup\{\|\boldsymbol{k}(t, \tau, z) - \boldsymbol{k}(t, \tau, \bar{z})\| \mid \|z - \bar{z}\| \leqq s;\ \|z\|, \|\bar{z}\| \leqq M\}.$$

If the condition (2') holds with a function ω, then it also holds with $\bar{\omega}$, and $\bar{\omega} \leqq \omega$. Furthermore, if ω satisfies (α) (β), then according to 1 IX (β) the same is true for $\bar{\omega}$. But $\bar{\omega}$ also satisfies (γ) since $\bar{\omega}(t, \tau, z) \leqq 2h(\tau)$. Therefore, when comparing theorems which are based on a condition (2') one should always take into account that ω can be replaced by $\bar{\omega}$. This remark applies especially to the older literature which treats systems of ordinary differential equations with a continuous right hand side. If for such an equation the inequality (2') holds with a function $\omega = \omega(\tau, z)$ of the Kamke type (see 10 VII), then the same is true with a function $\bar{\omega}$ of the Perron type (see 10 IV (δ)). This result was proved by Olech (1960); see also the remarks in Walter (1965; no. 5) and in 14 IV.

There is an extensive literature on successive approximation; see, for instance, the bibliography given in Walter (1965), the works cited in the introduction (VII), Kwapisz (1962), Ziebur (1962) (1965), Rama Mohana Rao (1963), Wazewski (1965), for more general Volterra functional equations Mamedov (1964), Kwapisz (1967).

CHAPTER II

Ordinary Differential Equations

8. Basic Theorems on Differential Inequalities

Many theorems of this chapter are very similar to the corresponding theorems of the preceding chapter. If we nevertheless devote a new chapter to an old subject, this will be justified by a new method. This method deals with *differential* equations and inequalities, whereas earlier the corresponding *integral* equations stood in the foreground. This is not done merely to give the old theorems a second proof. Rather, both of these first two chapters furnish, besides their concrete assertions, a methodology for later problems connected with partial differential equations. Chapter III, devoted to hyperbolic differential equations, contains in essence a translation of Chapter I into several dimensions, while the theory of the parabolic equations in Chapter IV is closely connected to the present Chapter II.

Most of this chapter can be read independently of Chapter I; an exception is provided by the existence statements 2 II. On the other hand, we do go back to the earlier notation and definitions, in particular, 1 I, V, 2 I, III, 5 I, 6 I, II, 7 I, II.

I. Definition (*φ', φ'_+, φ'_-, $D^+\varphi$, $D_+\varphi$, $D^-\varphi$, $D_-\varphi$, rules for $\pm\infty$*). The derivative of a function $\varphi(t)$ is denoted by $\varphi'(t)$, the right-sided derivative by $\varphi'_+(t)$, the right-sided (Dini) derivates by $D^+\varphi(t)$, $D_+\varphi(t)$, and a $-$ instead of a $+$ sign for the corresponding left-sided concepts. Thus for example

$$\varphi'_-(t) = \lim_{\bar{t}\to t-0} \frac{\varphi(t)-\varphi(\bar{t})}{t-\bar{t}}, \qquad D^+\varphi(t) = \limsup_{\bar{t}\to t+0} \frac{\varphi(\bar{t})-\varphi(t)}{\bar{t}-t}.$$

If we are speaking of the derivative or one-sided derivative of a function, it will be implicit that we are dealing with a *finite* value of the derivative; however, for Dini derivatives the values $\pm\infty$ are permitted. Here the usual computational rules hold for ∞, in particular,

$$a<\infty \; [a>-\infty] \quad \text{means:} \quad a = \text{real number or} \quad a=-\infty \; [+\infty]\,.$$

II. Lemma. *Let $\varphi(t), \psi(t)$ be two continuous functions in J_0 with the property: If $\varphi=\psi$ at a point $t_0 \in J_0$, then*

$$D_-\varphi < D_-\psi \quad \text{or} \quad D^-\varphi < D^-\psi$$

at this point t_0 (thus $\varphi'(t_0) < \psi'(t_0)$ if the derivatives exist).

Then exactly one of the following two cases holds:

(α) $\varphi < \psi$ in J_0;

(β) there is a $\bar{t} > 0$ with $\varphi \geqq \psi$ for $0 < t \leqq \bar{t}$.

The proof is simple. If the case (α) does not hold, then there is a $\bar{t} \in J_0$ with $\varphi(\bar{t}) \geqq \psi(\bar{t})$. We want to show that then there does not exist a $t_1 < \bar{t}$ with $\varphi(t_1) < \psi(t_1)$. For if this inequality were true for a $t_1 \in J_0$, there would be a first point to the right of t_1 at which $\varphi = \psi$, i.e., a $t_0 > t_1$ with the properties

$$\varphi(t_0) = \psi(t_0), \qquad \varphi < \psi \quad \text{for} \quad t_1 \leqq t < t_0 .$$

This would imply

$$\frac{\varphi(t)-\varphi(t_0)}{t-t_0} > \frac{\psi(t)-\psi(t_0)}{t-t_0} \quad \text{for} \quad t_1 \leqq t < t_0$$

(the denominators are negative!) and, finally, by passing to lim sup or lim inf for $t \to t_0 - 0$,

$$D_-\varphi(t_0) \geqq D_-\psi(t_0) \quad \text{and} \quad D^-\varphi(t_0) \geqq D^-\psi(t_0)$$

contrary to the hypothesis.

This lemma is fundamental for all further considerations. We can say, although this oversimplifies the situation somewhat, that all error estimates and uniqueness theorems of this chapter represent nothing but the application of this lemma to appropriate functions. The proofs for the most part consist of verifying that the hypotheses of II hold and that the case (β) cannot occur. In connection with this we note that (β) is incompatible with each of the following three properties:

(γ_1) $\varphi(0+) < \psi(0+)$;

(γ_2) $\varphi(0) < \psi(0)$;

(γ_3) $\varphi(0) = \psi(0)$, $D_+\varphi(0) < D_+\psi(0)$ or $D^+\varphi(0) < D^+\psi(0)$.

In (γ_2) and (γ_3) we assume $\varphi, \psi \in C(J)$.

III. The Initial Value Problem for an ordinary differential equation of first order consists of finding a function u which for a real valued function $f(t, z)$ defined on a point set $D(f) \subset E^2$ and a given initial value η satisfies the differential equation

$$u' = f(t, u) \quad \text{in} \quad J_0 \tag{1}$$

and the initial condition

$$u(0) = \eta . \tag{2}$$

A *solution* is a function $u \in Z(f)$ with both of these properties[1]. This concept of solution, which was already introduced in 5 I, deviates from the usual to the extent that the existence of $u'_+(0)$ and the validity of the differential equation for $t=0$ are not required. This generalization is not very important since in most cases f is continuous at the point $(0, \eta)$ and thus (1) is automatically satisfied for $t=0$. On the other hand it does not cause complications in the following, and now and then problems arise in which f has a singularity at the point $t=0$. Also, in view of parabolic differential equations, our concept of solution is the more natural one; there initial values are prescribed for $t=0$ but the differential equation is in general satisfied only in the interior of the domain, and thus not for $t=0$.

Most of the following theorems hold without any regularity assumption on f; one will be explicitly formulated whenever it is needed.

IV. Definition *(Defect P)*. By the defect Pv of a function $v(t)$ (with respect to the differential equation (1)) we mean the function

$$Pv \equiv Pv(t) \equiv v'(t) - f(t, v(t)) .$$

The defect is defined only where v is differentiable. As we shall see, the known theorems on error bounds involving the defect can, however, be extended to continuous functions v if Pv is replaced by, say, $D^- v(t) - f(t, v)$.

V. Theorem. *Suppose the following relationships hold for two functions $v, w \in Z(f)$:*

(α) $v(0+) < w(0+)$;

(β) $Pv < Pw$ in J_0.

Then

$$v(t) < w(t) \quad \textit{in} \quad J_0 .$$

For the proof we return to Lemma II, where we now have v and w in place of φ and ψ. It follows from $v(t_0) = w(t_0)$ that $f(t_0, v) = f(t_0, w)$ and then, by (β), that $v'(t_0) < w'(t_0)$; thus the hypothesis of Lemma II is satisfied. Further, since the case II (β) is eliminated because of (α), II (α) holds, as asserted.

The proof shows that the hypothesis (β) can be weakened as follows and the theorem is then true even for $v, w \in C(J_0)$:

(β') $D^- v(t_0) - f(t_0, v) < D^- w(t_0) - f(t_0, w)$ (or the corresponding hypothesis with D_-) holds for all $t_0 \in J_0$ with $v(t_0) = w(t_0)$.

[1] Phrases like "solution for $0 \leqq t < T_0 (\leqq T)$" are quite self-explanatory; here continuity in $[0, T_0)$, differentiability in $(0, T_0)$ are assumed.

VI. Super- and Subfunctions. Both of these concepts were originally introduced by Perron for the initial value problem (1) (2). A function $w(t)$ which is continuous and left and right differentiable in J is called, according to Perron (1915), a superfunction for the initial value problem (1) (2) if $w(0)=\eta$ and the inequalities $w'_{\pm}(t)>f(t,w)$ hold in J.

From V we can immediately read off a more general criterion.[1] The function w is a superfunction [the function v is a subfunction] if it is continuous in J and has both of the properties

(α) $w(0+)>u(0+)$ $[v(0+)<u(0+)]$ *for every solution* u *of the initial value problem;*

(β) $D^{-}w(t)>f(t,w)$ $[D_{-}v(t)<f(t,v)]$ *in* J_0.

Then according to V,

$$v<u<w \quad \text{in } J_0 \text{ for every solution } u\,.$$

Often (α) will be in the form

(α_1) $w(0)>\eta$ $[v(0)<\eta]$.

If the differential equation (1) holds also for $t=0$, then (α) is a consequence of

(α_2) $w(0)=\eta$ *and* $D^{+}w(0)>f(0,\eta)$ $[v(0)=\eta$ *and* $D_{+}v(0)<f(0,\eta)]$.

Because $u'(0)=f(0,\eta)$, we then have $u(0)=w(0)$, $D^{+}u(0)<D^{+}w(0)$; compare with II (γ_3).

There is a series of relevant facts which can be clarified only in conjunction with estimation and existence theorems. Therefore it is necessary to dwell briefly on existence theory.

VII. Remarks on the Existence Problem $(\mathscr{F},\mathscr{F}_0)$. The right hand side $f(t,z)$ of the differential equation belongs to the class $\mathscr{F}$ if $D(f)$ is the intersection of an open set of E^2 with the strip $J\times E^1$ and if f is continuous on $D(f)$. According to the classical existence theorem of Peano (1890), the initial value problem has at least one solution if $f\in\mathscr{F}$ and $(0,\eta)\in D(f)$. This solution $u(t)$ generally does not exist in the entire interval, but it can be "continued to the right up to the boundary of $D(f)$." More precisely: the solution u exists in J or in an interval $0\leqq t<T_1\leqq T$; in the latter case, either $\limsup_{t\to T_1}|u(t)|=\infty$ or, if $\delta(t)$ is the (Euclidean) distance of the point $(t,u(t))$ from the complementary set $E^2-D(f)$, $\lim_{t\to T_1}\delta(t)=0$.

The function $f(t,z)$ belongs to the class $\mathscr{F}_0\supset\mathscr{F}$ if $D(f)$ has the properties above, if f is continuous at those points of $D(f)$ with $t>0$ and if for every η with $(0,\eta)\in D(f)$ there exists a $\delta>0$ and a function $h(t)\in L(J)$ such that

$$|f(t,z)|\leqq h(t) \quad \text{for} \quad 0<t<\delta, \quad |z-\eta|<\delta$$

[1] Note that the concepts "superfunction" and "subfunction" have a meaning here different from that of Perron; see Introduction IV.

(for sufficiently small δ these points lie in $D(f)$)[1]. The above existence results also hold for the class $\mathscr{F}_0$. This follows in familiar fashion from the discussions of Section 2 on the existence of C-solutions, since for continuous f a C-solution is itself continuously differentiable and is thus a solution.

VIII. Theorem. *If $f \in \mathscr{F}_0$ and $u \in Z(f)$ is a uniquely determined*[2] *solution of the initial value problem* (1) (2) *existing in J, then for sufficiently small values of $|\varepsilon|$ and $|\delta|$ the solutions of the initial value problem*

$$\varphi' = f(t, \varphi) + \varepsilon, \qquad \varphi(0) = \eta + \delta \tag{3}$$

exist in all of J, and they converge uniformly in J to $u(t)$ as $\delta \to 0, \varepsilon \to 0$.

For the proof let $f \in \mathscr{F}$ and let G be the set of all points in $J \times E^1$ which are at a distance $\leqq \alpha\,(\alpha > 0)$ from a point of the curve $(t, u(t))$. For sufficiently small α, $G \subset D(f)$. If f is extended from its values in G as a continuous and bounded function on the entire strip $J \times E^1$, then the initial value problems (3) with the new f are surely solvable in J. If $|\varepsilon| + |\delta| \leqq 1$, these solutions form a set of equi-continuous functions. If (δ_k), (ε_k) are two null sequences and if the sequence (φ_k) of the associated solutions is convergent (thus uniformly convergent!), then the limit φ of this sequence represents a solution of (1) with the new f. This is shown in the usual way by using the integral representation. As long as this solution remains in G, it is identical with u since u is uniquely determined. From this it is easy to see that it cannot leave the domain G, and thus coincides with u in J. From the equi-continuity of the solutions of (3) and the fact just proved that the limit is uniquely determined, both assertions of the theorem follow for the new f. But since the solutions remain in G for small $|\varepsilon|$ and $|\delta|$ because of the uniform convergence, they also hold for the old f.

Only trivial alterations are needed in the proof for $f \in \mathscr{F}_0$.

IX. Maximal and Minimal Solution. In the following, j, j^* are subintervals of J which contain the origin. A solution u^* of the initial value problem (1) (2) which exists in j^* is called a maximal solution if for an arbitrary solution u, existing in j,

$$u(t) \leqq u^*(t) \quad \text{in} \quad j \cap j^*.$$

On the basis of VII the maximal solution is easily constructed. If $f \in \mathscr{F}_0$ and if δ_k, ε_k are two strictly decreasing null sequences and $\varphi = w_k$ are solutions of Eq. (3) with $\delta = \delta_k$, $\varepsilon = \varepsilon_k$, then by V we have

$$u < w_{k+1} < w_k$$

[1] The values of f for $t = 0$ are irrelevant.

[2] This means: If $\bar{u}$ is a solution of (1) in $0 \leqq t < T_1 \leqq T$ which satisfies the initial condition (2), then $\bar{u} = u$ in this interval.

for every solution u (where these solutions exist). The w_k converge to the desired maximal solution u^*. The w_k and thus also u^* will in general not exist in the entire interval J. However, like every other solution, the maximal solution can be continued to the right up to the boundary of $D(f)$.

As in VIII, it can be shown that: *If $f \in \mathscr{F}_0$ and the maximal solution of the initial value problem* (1) (2) *exists in J, then the solutions φ of* (3) *also exist in all of J for sufficiently small $\varepsilon > 0, \delta > 0$. These functions φ are superfunctions which converge uniformly in J to the maximal solution as $\varepsilon \to +0, \delta \to +0$.*

Obviously the following form of this theorem is thus also proved: *If $f \in \mathscr{F}_0$ and the maximal solution exists in the interval $0 \leqq t < T_1$, then it can be uniformly approximated in every interval $0 \leqq t \leqq T_1 - \delta$ $(\delta > 0)$ by superfunctions φ which are solutions of* (3) *with positive ε and δ.*

The corresponding results hold for the minimal solution. It is interesting to contrast this behavior with that corresponding to C-solutions. For continuous f the monotonicity theorem V holds without further assumptions on f. Using this theorem we approximate the maximal solution "from above" by superfunctions w which satisfy the corresponding differential inequality

$$w' > f(t, w).$$

This method of proof is exactly the "first method" described in 2 IV. Under Carathéodory hypotheses, say $f \in \mathscr{K}$, the monotonicity theorem, as given in 5 III, holds only with the additional assumption 5 III (γ). In this case the "first method" is applicable. The general case in the context of C-solutions, however, needs a proof based on a quite different foundation. The maximal solution is constructed "from below" as the supremum of all solutions of the initial value problem. This is the "second method" treated in 2 V and applied to the initial value problem for C-solutions in 5 II and 6 X (for systems). We recall that a third method is available by extending the theory of α-monotonicity as originated by Satō (1953) to Carathéodory hypotheses; see 3 III, IV.

It should be noted, however, that the fundamental property of the maximal solution, as expressed in Theorem X below, is also valid in those cases where the first method does not apply; see 5 X for one equation and 6 X for systems of equations in the Carathéodory sense.

The two following theorems result from the application of the two principles given in 1 IX (β) (γ) to the initial value problem under consideration.

X. Theorem. *If $f(t, z) \in \mathscr{F}_0$ and u^* [resp. u_*] $\in Z(f)$ is the maximal [resp. minimal] solution of the initial value problem* (1) (2) *— suppose it exists in J — and if for a function v [resp. w] $\in Z(f)$*

(α) $v(0) \leqq \eta$ $[w(0) \geqq \eta]$;
(β) $v' \leqq f(t, v)$ $[w' \geqq f(t, w)]$ *in* J_0,
then

$$v \leqq u^* \quad [w \geqq u_*] \quad in \quad J.$$

The following weaker hypothesis suffices in place of (β) (the theorem then also holds for continuous functions v or w): For a $\delta > 0$ we have

(β') $D_-v(t) \leqq f(t, v)$, if $t \in J_0$ and $u^*(t) < v(t) < u^*(t) + \delta$
$[D^-w(t) \geqq f(t, w)$, if $t \in J_0$ and $u_*(t) - \delta < w(t) < u_*(t)]$.

For the proof we let G denote the point set of the (t, z) plane defined by the inequalities $0 < t \leqq T, u^*(t) < z < u^*(t) + \delta$. According to the procedure of IX, for every positive $\varepsilon < \delta$ there is a superfunction $\bar{w} < u^* + \varepsilon$ existing in J which therefore remains entirely in G. Then V applied to v and $\bar{w}$ implies $v < \bar{w} \leqq u^* + \varepsilon$ and hence the assertion $v \leqq u^*$. The other part of the theorem is proved in the same way.

Corduneanu (1964) gives a similar theorem, assuming (β') with D_-, D^- replaced by D_+, D^+. Using Lemma IIa formulated in XII (ε) at the end of this section, we can prove it in roughly the same way as the above theorem.

XI. Theorem. *If f satisfies a uniqueness condition* (10.3) *with* $\omega \in \mathscr{E}_5$ *and if for two functions* $v, w \in Z(f)$ *we have*
(α) $v(0) \leqq w(0)$;
(β) $Pv \leqq Pw$ *in* J_0,
then

$$v(t) \leqq w(t) \quad in \quad J.$$

This theorem is derived from V by the method given in 1 IX (γ). If $\varrho' > \omega(t, \varrho)$ and $0 < \delta \leqq \varrho \leqq \varepsilon$, then V can be applied to the functions v and $\bar{w} = w + \varrho$. The hypothesis V (β) follows from (β) and the inequalities

$$P\bar{w} - Pw = \varrho' - f(t, w + \varrho) + f(t, w) \geqq \varrho' - \omega(t, \varrho) > 0.$$

Hence we have $v < \bar{w} = w + \varrho \leqq w + \varepsilon$, and thus $v \leqq w$.

Thus for the case where f satisfies a uniqueness condition, Theorem VI can be replaced by a theorem

VIa: *v is a subfunction if* $v(0) \leqq \eta$, $v' \leqq f(t, v)$;
w is a superfunction if $w(0) \geqq \eta$, $w' \geqq f(t, w)$.

The continuity of both functions also suffices here if we replace the derivative by D_-v or D^-w respectively.

XII. Remarks. (α) The essential theorems of this section are well known, mostly under sharper hypotheses; they can be readily found in very general form — except for more recent work — in Hukuhara (1940) (the same is true for the next section). Ważewski (1952) first called attention to the fact that the differential inequality X (β) does not have to be satisfied for all $t \in J$ but only when the point

$(t, v(t))$ lies in the set G defined in the proof of X. He calls this set "épiderme supérieur" (loc. cit.); the "épiderme supérieur" is constructed somewhat more generally with (in our notation) a function $\delta(t)$ which is continuous and positive for $0<t<T$ instead of with the constant δ. It should be noted that the theorem formulated by Ważewski is true only under the additional hypothesis that $\delta(t) \geqq \delta_1 > 0$ in a right sided neighborhood of the origin [however $\delta(t) \to 0$ as $t \to T$ is permissible]. Counterexample[1]: $f(t, z) \equiv 0$, $u^*(t) \equiv 0$, $\delta(t) = t$, $v(t) = 2t$.

(β) We should again point out the difference between Theorems V, VI and the corresponding theorems for absolutely continuous functions, 5 III and 5 IV. The theorems of this section hold for completely arbitrary functions f. This generalization is obtained at the expense of a sharpened requirement (β), which must now hold at *every* point and with the exclusion of the equality sign. These theorems become false if equality holds in (β) for even *one* t value[2]. If for example $f(t, z) = 4\sqrt{|z|}$, $u(t) \equiv 0$, $v(t) = (t-1)^3$, $T = 2$, then $v(0) < 0$, $v' = 3(t-1)^2 < 4\sqrt{|v|}$ for $t \neq 1$, but v is not a subfunction since $v > u$ for $1 < t \leqq 2$.

(γ) In the formulation of the initial value problem on which our presentation is based, it is assumed that the point t_0 at which initial values are prescribed always coincides with the origin. Translating the theorem to intervals $t_0 \leqq t \leqq T$, $t_0 \leqq t < T$, $t_0 \leqq t < \infty$ is trivial; it is likewise easy to do this for $T \leqq t \leqq t_0$ $(-\infty < T < t_0)$ via the substitution $\bar{t} = t_0 - t$. Our choice of a t interval closed on the right is to some extent arbitrary. We can come up with a physical reason for this which will become still more significant in the treatment of parabolic differential equations: The variable t often represents time in applications, and we would like to determine the behavior from a known initial situation $(t=0)$ to the time $t = T$.

(δ) More than once it was shown by corollaries that the theorems first formulated for differentiable functions also hold for continuous functions if the derivative is replaced by a left-sided Dini derivate. In particular, we can replace derivatives by left-sided derivatives in all assumptions. Accordingly, the class $Z(f)$ can be enlarged (and with this the concept of solution generalized) by requiring only left-sided differentiability in J_0.

(ε) A further peculiarity of our presentation rests on the fact that we work with left-sided derivates or derivatives in II and correspondingly in all later theorems. As long as we consider only ordinary differential equations there is no compelling reason to do this. Indeed, Lemma II has the following counterpart.

IIa. Lemma. Suppose $\varphi(t)$, $\psi(t)$ are two continuous functions in J_0 with the property: If $\varphi = \psi$ at a point $t_0 \in J_0$, then

$$D^+\varphi < D^+\psi \quad \text{or} \quad D_+\varphi < D_+\psi$$

at this point t_0.

Then exactly one of the two cases occurs:

(α) $\varphi \leqq \psi$ in J_0;
(β) there is a $\bar{t} > 0$ with $\varphi > \psi$ for $0 < t < \bar{t}$.

Assuming, for the proof, that (α) is not true, we start out from a point $\bar{t} \in J_0$ with $\varphi(\bar{t}) > \psi(\bar{t})$ and look for the first point t_0 to the left of $\bar{t}$ with $\varphi(t_0) = \psi(t_0)$. If such a $t_0 > 0$ existed, we would have $[\varphi(t) - \varphi(t_0)]/(t - t_0) > [\psi(t) - \psi(t_0)]/(t - t_0)$ for $t_0 < t < \bar{t}$, whence, however, we obtain a contradiction to the hypothesis as in II. Other versions of Lemma IIa are given by Bebernes and Meisters (1967).

[1] See also the review by Stewart in *Mathematical Reviews* **15**, 704 (1954).
[2] In Kamke (1959) p. 11, 2.7 (d), the symbols $\leqq$, $\geqq$ must be replaced by $<$, $>$.

With this it is possible to express the theorems of this chapter for right-sided derivates also. Since (α) now reads $\varphi \leqq \psi$ and not $\varphi < \psi$ as before, the equality sign must then be added in the assertions of several of the theorems. This asymmetry between left and right sided derivates becomes stronger for parabolic differential equations in Chapter IV. There we are *forced* to consider left sided derivatives in the t direction, and it is impossible to express the corresponding theorems for right sided derivatives. It should be noted that frequently differential inequalities are required for the left *and* right sided derivatives in the literature on bounds for ordinary differential equations. This is, however, unnecessary.

9. Estimates for the Initial Value Problem for an Ordinary Differential Equation of First Order

Theorem V of the preceding section is suited for the approximate determination of a solution $u(t)$ of the initial value problem for the differential equation

$$u' = f(t, u) \tag{1}$$

only when super- and subfunctions can be explicitly given. For the most part we will arrive at an "approximate solution" (either by trying with a functional form which appears suitable or by using one of the familiar procedures for numerical integration of the differential equation) for which the defect Pv is indeed small, but not of one sign. The problem then consists of making assertions about the goodness of the approximation from a knowledge of v [and thus of Pv and $v(0) - \eta$]. If we have found an approximate solution v for which the expressions Pv and $v(0) - \eta$ are only a "little" different from zero, then we will be inclined to assume that the same also holds for the difference $v(t) - u(t)$. Whether and how much this assumption is justified depends crucially on the properties of the function f and in every case necessitates a painstaking analysis. Our theorems are based on inequalities such as

$$|f(t, z) - f(t, \bar{z})| \leqq \omega(t, |z - \bar{z}|) \tag{2}$$

which can be considered as generalizations of the classical Lipschitz condition. But since the quality of a bound for $v - u$ stands or falls with the quality of an estimation of f by ω, great care must be taken with this point. It turns out that instead of (2) a one-sided inequality

$$f(t, z) - f(t, \bar{z}) \leqq \omega(t, z - \bar{z}) \quad \text{for} \quad z \geqq \bar{z} \tag{3}$$

often suffices, and that the set of values $z, \bar{z}$ which are to be considered can sometimes be reduced further. Another refinement is contained in the first two theorems. Two different functions $\omega, \bar{\omega}$ are used for bounds from above and below.

If in a theorem we list $\varrho \in Z(\omega)$, say, as the only assumption about ϱ and ω, then according to 5 I this signifies: ϱ is continuous in J, differentiable in J_0, $\omega(t, z)$ is defined as a real valued function on a point set $D(\omega) \subset E^2$, and $(t, \varrho(t)) \in D(\omega)$ for $t \in J_0$.

I. Theorem. *For the functions* $u, v \in Z(f)$, $\varrho \in Z(\omega)$, $\bar{\varrho} \in Z(\bar{\omega})$, *assume the relations*

(α) $-\bar{\varrho}(0+) < (v-u)(0+) < \varrho(0+)$;

(β) $u' = f(t, u)$ *and* $-\bar{\delta}(t) \leqq Pv \equiv v' - f(t, v) \leqq \delta(t)$ *in* J_0;

(γ) $\varrho' > \omega(t, \varrho) + \delta(t)$ *and* $\bar{\varrho}' > \bar{\omega}(t, \bar{\varrho}) + \bar{\delta}(t)$ *in* J_0

hold with suitable functions $\delta(t), \bar{\delta}(t)$ *defined in* J_0. *Further suppose* f *satisfies the conditions*

(δ) $f(t, v) - f(t, v-\varrho) \leqq \omega(t, \varrho)$ *if* $t \in J_0$, $(t, v-\varrho) \in D(f)$,
$f(t, v+\bar{\varrho}) - f(t, v) \leqq \bar{\omega}(t, \bar{\varrho})$ *if* $t \in J_0$, $(t, v+\bar{\varrho}) \in D(f)$.

Then

$$-\bar{\varrho}(t) < v(t) - u(t) < \varrho(t) \quad in \quad J_0 .$$

Actually we have here two theorems which are independent of each other. If only the inequalities involving quantities with bars [resp. without bars] hold, then the left [resp. right] hand side of the assertion holds. Further, it suffices for $v, \varrho, \bar{\varrho}$ to be continuous in J and for the inequalities (β) (γ) to hold with $D^- v, D^- \varrho, D_- \bar{\varrho}$ or $D_- v, D_- \varrho, D^- \bar{\varrho}$ instead of $v', \varrho', \bar{\varrho}'$ (this remark is useful if we have a piecewise linear curve as approximating function).

We use 8 II with $\varphi = v - u$, $\psi = \varrho$ for the proof of the inequality $v - u < \varrho$. It then suffices to show that $v' - u' < \varrho'$ for all t with $v - u = \varrho$. In fact for these t by (β) – (δ)

$$v' - u' \leqq \delta(t) + f(t, v) - f(t, u) = \delta(t) + f(t, v) - f(t, v - \varrho)$$

$$\leqq \delta(t) + \omega(t, \varrho) < \varrho' ,$$

whereby this part of the assertion has been proved. We proceed in exactly the same way in the second part.

The inequalities in (γ) are often inconvenient in applications. It is sometimes easier to determine functions $\sigma, \bar{\sigma}$ which satisfy the corresponding equations. The following theorem gives conditions under which this is permitted.

II. Theorem. Suppose the functions $\delta, \bar{\delta}$ are continuous in J_0 and integrable over J, the functions $\omega, \bar{\omega}$ are from $\mathscr{F}_0$, and for two functions $u, v \in Z(f)$ the following hold:

(α) $-\bar{\varepsilon} \leqq v(0) - u(0) \leqq \varepsilon$;

(β) $u' = f(t, u)$ and $\bar{\delta}(t) \leqq Pv \leqq \delta(t)$ in J_0.

Suppose further that $\sigma(t) \in Z(\omega)$, $\bar{\sigma}(t) \in Z(\bar{\omega})$ are the maximal solutions[1] of

(γ) $\sigma' = \omega(t, \sigma) + \delta(t)$, $\sigma(0) = \varepsilon$ and $\bar{\sigma}' = \bar{\omega}(t, \bar{\sigma}) + \bar{\delta}(t)$, $\bar{\sigma}(0) = \bar{\varepsilon}$ resp.

Suppose there is a positive number α with the property that

(δ) $f(t, v) - f(t, v - z) \leqq \omega(t, z)$ if $\sigma(t) < z < \sigma(t) + \alpha$, $(t, v - z) \in D(f)$,
$f(t, v + z) - f(t, v) \leqq \bar{\omega}(t, z)$ if $\bar{\sigma}(t) < z < \bar{\sigma}(t) + \alpha$, $(t, v + z) \in D(f)$.

Then

$$-\bar{\sigma}(t) \leqq v(t) - u(t) \leqq \sigma(t) \quad \text{in} \quad J.$$

First we note that the arguments (t, z) occurring in (δ) are in $D(\omega)$ and $D(\bar{\omega})$ respectively if α is sufficiently small since these sets are open and contain the points (t, σ) and $(t, \bar{\sigma})$ respectively. The assertion is obtained in a well-known way from I. For an arbitrary given positive $\beta < \alpha$ there are, according to 8 IX, functions $\varrho, \bar{\varrho}$ which satisfy the inequalities I (γ) and

$$\sigma < \varrho < \sigma + \beta, \qquad \bar{\sigma} < \bar{\varrho} < \bar{\sigma} + \beta$$

in J. Since I (δ) is a consequence of (δ) for these functions $\varrho, \bar{\varrho}$, the assertion of I holds, whence, since β is arbitrary, the present assertion follows.

In the special case $\delta = \bar{\delta}$, $\omega = \bar{\omega}$, Theorems I, II take on the following simpler form.

III. Consequence. *Suppose that for $u, v \in Z(f)$, $\varrho \in Z(\omega)$*

(α) $|v - u|\,(0+) < \varrho(0+)$;

(β) $u' = f(t, u)$ *and* $|Pv| \equiv |v' - f(t, v)| \leqq \delta(t)$ *in* J_0;

(γ) $\varrho' > \omega(t, \varrho) + \delta(t)$ *in* J_0;

(δ) $\left.\begin{array}{l} f(t, v + \varrho) - f(t, v) \\ f(t, v) - f(t, v - \varrho) \end{array}\right\} \leqq \omega(t, \varrho)$ *if* $(t, v \pm \varrho) \in D(f)$.

Then

$$|v - u| < \varrho \quad \text{in} \quad J_0 .$$

If $\delta \in C(J_0)$ and $\in L(J)$, $\omega \in \mathscr{F}_0$ and further if $\sigma \in Z(\omega)$ is the maximal solution of an initial value problem

(γ') $\sigma' = \omega(t, \sigma) + \delta(t)$ *with* $\sigma(0) \geqq |v(0) - u(0)|$,

existing in J, and (β) holds as well as (for an $\alpha > 0$)

(δ') $\left.\begin{array}{l} f(t, v+z) - f(t, v) \\ f(t, v) - f(t, v-z) \end{array}\right\} \leqq \omega(t, z)$ *if* $\sigma(t) < z < \sigma(t) + \alpha$ *and* $(t, v \pm z) \in D(f)$,

then

$$|u - v| \leqq \sigma \quad \text{in} \quad J.$$

This is obtained directly from I, II since the inequalities I (δ) and II (δ) respectively follow from the inequality (δ) if $\omega = \bar{\omega}$.

From the two hypotheses (δ), (δ') we see that much less is necessary for bounds than the inequality (2) (for arbitrary $z, \bar{z}$). (δ) *as well as* (δ') *is a consequence of the one-sided condition* (3).

[1] It is thus assumed that these maximal solutions exist in J.

IV. Generalizations and Remarks. (α) One might ask why the bound ϱ in III is positive; no assumption was made about this. By III (δ) we have $\omega(t, 0) \geqq 0$ provided $\varrho(t) = 0$. Consequently, because of III (γ) and because $\delta(t) \geqq 0$, we have $\varrho' > 0$ if $\varrho = 0$. Thus $\varrho > 0$ by Lemma 8 II.

(β) The generalizations given in connection with I can also be carried out for II and III with slight alterations. In particular one of the expressions $D_- v - f(t, v)$, $D^- v - f(t, v)$ can be put in place of Pv in I–III (with otherwise unchanged hypotheses) if v is continuous.

(γ) For *two-sided* bounds on the difference $v - u$ it is sufficient, according to III, for f to allow a very special one-sided condition III (δ) or III (δ'). Each of these two hypotheses is a consequence of the one-sided condition (3). If, for example, it is known that in $D(f)$ an inequality $A(t) \leqq f_z(t, z) \leqq B(t)$ holds [see also the introduction (IX)], then the function $B(t)z$ can be chosen for ω in III (for negative B too). On the other hand if in III we had to satisfy a condition on the absolute value of f as in (2), as is found in many places in the literature, then we would have to set $\omega(t, z) = C(t)z$ with $C = \max(|A|, |B|)$. Hence the error bound would be made significantly worse under certain circumstances. This will be illustrated by the transparent example V.

(δ) In Theorems I and II negative bounds $\varrho, \bar{\varrho}, \sigma, \bar{\sigma}$ are also allowed. Here it is important to note that in the case of negative bounds the hypotheses (δ) of these theorems do *not* follow from (3) (this is the case for positive bounds). In practical problems it can certainly happen that starting with an approximation function v, we obtain by Theorem I two bounds $\varrho, \bar{\varrho}$ of different sign and with this a "strip" $v(t) - \varrho(t) < z < v(t) + \bar{\varrho}(t)$ in the (t, z) plane which does *not* contain the function v and in which the solution u must remain. Example VIII is of this type.

For bounds of arbitrary sign we can also find a tractable, easily applicable formulation. For simplicity suppose $\omega_1(t, z)$, $\omega_2(t, z)$ are two functions defined on $J \times \{z \geqq 0\}$ which vanish for $z = 0$. If in its domain of definition the function f satisfies the condition

$$\omega_1(t, z - \bar{z}) \leqq f(t, z) - f(t, \bar{z}) \leqq \omega_2(t, z - \bar{z}) \quad \text{for} \quad z \geqq \bar{z}, \tag{4}$$

then, when we set

$$\omega(t, z) = \bar{\omega}(t, z) = \begin{cases} \omega_2(t, z) & \text{for} \quad z \geqq 0 \\ -\omega_1(t, -z) & \text{for} \quad z < 0, \end{cases}$$

conditions I (δ) and II (δ) are surely satisfied. Thus in the case of a negative bound ϱ or $\bar{\varrho}$ we need a bound on the f-difference "from below."

V. Example. Suppose the initial value problem is $u' = u$, $u(0) = \eta$, and the approximate solution $v(t)$ satisfies $|Pv| = |v' - v| \leqq \delta$, $|v(0) - \eta| \leqq \varepsilon$. Then it follows from III with $\omega(t, z) = z$ that

$$|v - u| \leqq (\varepsilon + \delta)e^t - \delta.$$

If on the other hand the initial value problem is $u' = -u, u(0) = \eta$ and again $|Pv| = |v' + v| \leqq \delta$, $|v(0) - \eta| \leqq \varepsilon$, then in a theorem on error bounds which uses (2) we would obtain the same error bound, whereas the significantly better bound

$$|v - u| \leqq (\varepsilon - \delta)e^{-t} + \delta$$

follows from III with $\omega(t, z) = -z$.

In the following example, the Lipschitz condition just used in V will be discussed in detail.

VI. The Lipschitz Condition. (α) *One-sided Lipschitz conditions.* Suppose that if $(t, z), (t, \bar{z}) \in D(f)$ the function $f(t, z)$ satisfies a one-sided Lipschitz condition

$$f(t, z) - f(t, \bar{z}) \leqq l(t)(z - \bar{z}) \quad \text{for} \quad z \geqq \bar{z}.$$

Then the bounds of Theorem II read

$$\sigma(t) = e^{L(t)}\left[\varepsilon + \int_0^t \delta(\tau)e^{-L(\tau)}d\tau\right],$$

$$\bar{\sigma}(t) = e^{L(t)}\left[\bar{\varepsilon} + \int_0^t \bar{\delta}(\tau)e^{-L(\tau)}d\tau\right],$$

where $L(t) = \int_0^t l(\tau)d\tau$.

Here we must have $\sigma(t) \geqq 0, \bar{\sigma}(t) \geqq 0$. Apart from the fact that now $\delta(t), \bar{\delta}(t), l(t)$ are functions which are continuous in J_0 and integrable over J, the result is identical with 5 VIII.

(β) *Lipschitz conditions from above and below.* We now drop the condition $\sigma, \bar{\sigma} \geqq 0$ and handle the Lipschitz condition from the more general point of view IV (δ). Suppose we have the condition

$$\bar{l}(t)(z - \bar{z}) \leqq f(t, z) - f(t, \bar{z}) \leqq l(t)(z - \bar{z}) \quad \text{for} \quad z \geqq \bar{z},$$

provided $(t, z), (t, \bar{z}) \in D(f)$, and suppose that

$$\omega(t, z) = \bar{\omega}(t, z) = \begin{cases} l(t)z & \text{for} \quad z \geqq 0 \\ \bar{l}(t)z & \text{for} \quad z < 0 \end{cases}$$

Then relations II (α) (β) imply the inequalities

$$-\bar{\sigma}(t) \leqq v(t) - u(t) \leqq \sigma(t) \quad \text{in} \quad J,$$

if $\sigma(t), \bar{\sigma}(t)$ are the solutions of the initial value problems II (γ). What is new compared to (α) is that here no assumptions on the sign of $\varepsilon, \bar{\varepsilon}, \sigma(t)$, $\bar{\sigma}(t)$ are needed. The integrals $\sigma, \bar{\sigma}$ can easily be given in closed form with the help of the familiar formula for integration of linear differential equations. It should be noted that this entire family of problems can also be handled with the methods of Section 5, so that even the hypothesis $\delta, \bar{\delta}, l, \bar{l} \in L(J)$ (without continuity in J_0) is sufficient.

It should be mentioned that the regularity hypotheses imposed on $\delta, \bar{\delta}$ and l in (α), which are stronger here than in Section 5, can also be made considerably milder in the framework of the present theory. If $\delta(t)$ has the following property:

(γ) $\delta(t)$ is finite in J_0 and L-integrable over the interval $\alpha \leqq t \leqq T$ for every positive α, where the limit $\int_0^T \delta(t)\,dt = \lim_{\alpha \to +0} \int_\alpha^T \delta(t)\,dt$ exists, and

$$\delta^*(t) = \limsup_{h \to +0} \frac{1}{h} \int_{t-h}^{t} \delta(\tau)\,d\tau \geqq \delta(t) \quad \text{in} \quad J_0$$

(the value $+\infty$ is allowed for δ^*); and if the other functions $\bar{\delta}(t)$, $l(t)$ also have this property, then the above considerations hold. Under these hypotheses, for the function $\sigma(t)$ of (α) we have

$$D^- \sigma(t) = \delta^*(t) + l^*(t)\sigma(t) \geqq \delta(t) + l(t)\sigma(t).$$

If we replace the function $\delta(t)$ by $\delta(t) + \beta$ ($\beta > 0$) in the formula for σ, then we have a function, denoted by ϱ, for which I (γ) holds with $D^- \varrho$ instead of ϱ'. We then obtain the bound of (α) for $\beta \to 0$ from Theorem I.

VII. On the Construction of Bounds. (α) In most nontrivial (and thus nonlinear) cases we can give only a very bad Lipschitz constant or none at all. Thus we must first suitably bound the domain of definition $D(f)$. If we have already found a superfunction w and a subfunction v of the initial value problem under consideration, then we consider only the point set delimited by $v(t) \leqq z \leqq w(t)$ [see the introduction (X)]. If this is not the case, then we can proceed as follows. Suppose we have an approximate solution $v(t)$ which is to be bounded according to VI (α), and let $v(0) = u(0)$. We take two arbitrary constants $\alpha, \bar{\alpha} > 0$ and determine the Lipschitz "constant" $l(t) = l(t; \alpha, \bar{\alpha})$ relative to the domain $v(t) - \alpha \leqq z \leqq v(t) + \bar{\alpha}$ as well as the bounds $\sigma(t; \alpha, \bar{\alpha})$ and $\bar{\sigma}(t; \alpha, \bar{\alpha})$ according to the formulas of VI (α). The bound $v - \sigma \leqq u \leqq v + \bar{\sigma}$ thus found surely holds provided the functions $v - \sigma$ and $v + \bar{\sigma}$ stay in the delimited strip, i.e., they hold to the right as long as $\sigma(t; \alpha, \bar{\alpha}) \leqq \alpha$ and $\bar{\sigma}(t; \alpha, \bar{\alpha}) \leqq \bar{\alpha}$. We can then proceed by beginning with small numbers and enlarging them when $\sigma > \alpha$ or $\bar{\sigma} > \bar{\alpha}$ (sometimes the bounds are of such a simple form that one can easily give the best values of $\alpha, \bar{\alpha}$ for every point t).

(β) There is yet another way of avoiding these difficulties. For this we refer to Theorem I and assume that the partial derivative $f_z(t, z)$ exists and is monotone increasing [decreasing] in z; moreover let $\varrho \geqq 0, \bar{\varrho} \geqq 0$. Because of

$$f(t, v) - f(t, v - \varrho) \leqq \varrho(t) f_z(t, v) \; [\text{resp.} \leqq \varrho(t) f_z(t, v - \varrho)]$$
$$f(t, v + \bar{\varrho}) - f(t, v) \leqq \bar{\varrho}(t) f_z(t, v + \bar{\varrho}) \; [\text{resp.} \leqq \bar{\varrho}(t) f_z(t, v)]$$

the conditions I (δ) hold if, when f_z is monotone increasing, we set

$$\omega(t, z) = l(t) z \quad \text{with} \quad l(t) = f_z(t, v(t))$$
$$\bar{\omega}(t, z) = f_z(t, v(t) + z) z,$$

and when f_z is monotone decreasing, we set

$$\omega(t, z) = f_z(t, v(t) - z)z$$
$$\bar{\omega}(t, z) = l(t)z \quad \text{with} \quad l(t) = f_z(t, v(t)).$$

With this the earlier considerations on the domain of definition have become completely superfluous. We should note however that for one of the bounds we obtain a linear differential equation while for the other, on the contrary, we obtain a differential inequality which is in general *nonlinear*. Similar considerations can be made if one of the bounds is negative.

A simple

VIII. Example will elucidate what has been said. Suppose we are given the initial value problem

$$u' = u^2 - t, \qquad u(0) = 1 \tag{5}$$

[this is a special Riccati equation which is not integrable in elementary terms; see Kamke (1945), p. 37f]. A simple subfunction is $v(t) = 1 + t$, a superfunction is obtained, for example, by discarding the term $-t$ in the differential equation and solving this problem; we obtain $w(t) = 1/(1-t)$. For the solution $u(t)$ we thus have

$$v = 1 + t < u(t) < 1/(1-t) = w.$$

With this we can bound $f(t, z) = z^2 - t$ in the domain of definition given by $1 + t \leqq z \leqq 1/(1-t)$ above and below according to

$$2(1+t)(z - \bar{z}) \leqq f(t, z) - f(t, \bar{z}) \leqq 2(z - \bar{z})/(1-t) \quad \text{for} \quad z \geqq \bar{z}.$$

Now the superfunction w will be considered as an *approximate solution* and a bound

$$-\bar{\sigma} \leqq w - u \leqq \sigma, \quad \text{i.e.,} \quad w - \sigma \leqq u \leqq w + \bar{\sigma}$$

derived with the help of Theorem I or II. The bound σ is determined according to Theorem II from the differential equation

$$\sigma' = t + \frac{2}{1-t}\sigma \quad \text{with} \quad \sigma(0) = 0;$$

hence

$$\sigma(t) = \frac{t^2}{12(1-t)^2}(6 - 8t + 3t^2).$$

Now we shall try to improve the upper bound w, i.e., to find a negative $\bar{\sigma}$. For this we need — see IV (δ) and VI (β) — a Lipschitz bound "from below"; a corresponding Lipschitz "constant" is, as we have seen, $2(1+t)$. By II we have

$$\bar{\sigma}' = -t + 2(1+t)\bar{\sigma}, \qquad \bar{\sigma}(0) = 0,$$

whence follows

$$\bar{\sigma}(t) = -\int_0^t \tau e^{2t - 2\tau + t^2 - \tau^2} d\tau \leqq -\int_0^t \tau e^{(2+t)(t-\tau)} d\tau$$
$$= \frac{-1}{(2+t)^2}\left[e^{t(2+t)} - (1+t)^2\right] = -\left(\frac{t^2}{2} + \frac{t^3}{3} + \cdots\right).$$

Thus two new bounds

$$v_1 = w - \sigma \leqq u \leqq w_1 = w + \bar{\sigma}$$

are obtained.

Now we want to forget, so to speak, the preceding and demonstrate how one can proceed without a knowledge of bounds. We find $u'(0)=1$ from the differential equation; thus the simplest „Ansatz" is $v(t)=1+t$ with $Pv=-t(1+t)$. Now, starting from this approximate solution (which is simultaneously a subfunction), an upper bound $v+\bar{\sigma}$ will be found according to Theorem II. According to the procedure of VII (α) we choose $\alpha=0, \bar{\alpha}>0$ and obtain $l(t;0,\bar{\alpha})=2(1+\bar{\alpha}+t)$ and thus

$$\bar{\sigma}'=t(1+t)+2(1+\bar{\alpha}+t)\bar{\sigma} \quad \text{with} \quad \bar{\sigma}(0)=0\,,$$

i.e.,

$$\bar{\sigma}(t;0,\bar{\alpha})=\int_0^t \tau(1+\tau)e^{2(1+\bar{\alpha})(t-\tau)+t^2-\tau^2}d\tau<\int_0^t \tau(1+\tau)e^{2A(t-\tau)}d\tau$$

$$=\frac{t^2}{2}+\frac{1+A}{4A^3}(e^{2At}-1-2At-2A^2t^2)=\frac{t^2}{2}+2(1+A)t^3\sum_{\nu=0}^{\infty}\frac{(2At)^\nu}{(\nu+3)!}\,,$$

where we have set $A=1+\bar{\alpha}+t$.

This bound $u\leqq v+\bar{\sigma}=w_2$ holds to the right as long as $\bar{\sigma}(t;0,\bar{\alpha})\leqq\bar{\alpha}$. In the numerical evaluation we begin with a small $\bar{\alpha}$, say $\bar{\alpha}=1/2$, and then increase $\bar{\alpha}$ until $\bar{\sigma}>1/2$, etc.

Let us now make use of the procedure in VII (β). We take, say, the approximate solution $w(t)=1/(1-t)$ and determine corresponding functions ϱ, $\bar{\varrho}$. For ϱ we arrive at precisely the above differential equation; for the negative bound $\bar{\varrho}$, on the other hand, from

$$f(t,w+\bar{\varrho})-f(t,w)\leqq\bar{\varrho}f_z(t,w+\bar{\varrho})=2\bar{\varrho}(w+\bar{\varrho})$$

we can, according to Theorem I, obtain the conditions

$$\bar{\varrho}'>2\bar{\varrho}\left(\bar{\varrho}+\frac{1}{1-t}\right)-t, \qquad \bar{\varrho}(0)=0$$

(we do not have precisely the case discussed in VII. since we want a $\bar{\varrho}<0$) and hence, for example, we obtain the function

$$\bar{\varrho}(t)=\frac{-0.4t^2}{1-t}\,,$$

which is better than the earlier bound. We should keep in mind, however, that finding $\bar{\varrho}$ is now a more difficult matter which depends on experience (and luck), while earlier a linear differential equation, and with it a systematic method, was given. We also notice that, from another point of view, the present procedure is subordinate to the method of super- and subfunctions; in fact $w_3=w+\bar{\varrho}=(1-2t^2/5)/(1-t)$ represents nothing but a new superfunction which could have been found as easily (or with as much trouble) from the original differential equation (thus according to 8 VI)[1].

[1] Naturally the other bounds are also super- or subfunctions, as indeed the function $v-\varrho$ in Theorem I represents a subfunction and the function $v+\bar{\varrho}$ a superfunction in the sense of 8 VI. Seen from this viewpoint, Theorem I is nothing but a recipe for finding sub- and superfunctions. This interpretation, where Theorem I is seen as a special case of 8 VI (we could also adapt the proof accordingly), will *not* be deliberately emphasized here, because its translation to systems of differential equations is not possible.

We can obtain subfunctions very simply in our example from the power series expansion of u since only positive coefficients occur. We find for example

$$v_2 = 1 + t + \frac{t^2}{2} + \frac{2t^3}{3} + \frac{7t^4}{12} + \frac{33t^5}{60}.$$

It is often possible to convert the differential equation by a small alteration into a form integrable by elementary means. In the present case we might consider the initial value problems

$$u' = u^2 + a, \qquad u(0) = 1.$$

They have elementary solutions if a is a constant, and thus if $a = a(t)$ is a step function. If we choose a step function $a(t) \geqq -t$ or $a(t) \leqq -t$, we obtain a super- or a subfunction respectively. Thus we can in principle approximate the solution in closed form with any desired accuracy.

As for the subfunction, this procedure allows an important improvement. Namely, it is more useful to start out from the corresponding Volterra integral equation

$$u(t) = 1 + \int_0^t [u^2(\tau) - \tau]\,d\tau$$

and then to apply the methods of Chapter I. If we want to determine a subfunction in the interval $0 \leqq t \leqq T$ as a solution of $v' = v^2 - \alpha^2$ with constant α, then according to 1 VII the condition is

$$v(t) = 1 + \int_0^t [v^2(\tau) - \alpha^2]\,d\tau < 1 + \int_0^t [v^2(\tau) - \tau]\,d\tau \quad \text{for} \quad 0 \leqq t \leqq T,$$

and thus $\alpha^2 > t/2$. In other words: According to 1 VII we obtain a subfunction in the interval $0 \leqq t \leqq T$ from

$$v' = v^2 - \alpha^2 \quad \text{with} \quad \alpha^2 = T/2, \tag{6}$$

while we must set $\alpha^2 = T$ in connection with 8 VI.

In problems which can be written as an integral equation with a monotone increasing kernel, the transition to the integral equation always brings advantages of this kind.

The solution of (6) with the initial value 1 is

$$v(t) = \alpha \frac{\coth(\alpha t) - \alpha}{\alpha \coth(\alpha t) - 1}.$$

It represents a subfunction for $0 \leqq t \leqq T$ if $\alpha^2 = T/2$, in particular for $t = T$. Since this holds for every positive T, we thus obtain the subfunction $v_3(t)$:

$$v_3(2\alpha^2) = \alpha \frac{\coth(2\alpha^3) - \alpha}{\alpha \coth(2\alpha^3) - 1}.$$

For the value of the asymptote T_1 ($u(t) \to \infty$ as $t \to T_1$) we obtain from w and v_3 the bounds

$$1 < T_1 < 1.59 .$$

Figure 3 shows several bounds together with an approximate solution $u^*(t)$ computed by the Runge-Kutta method[1].

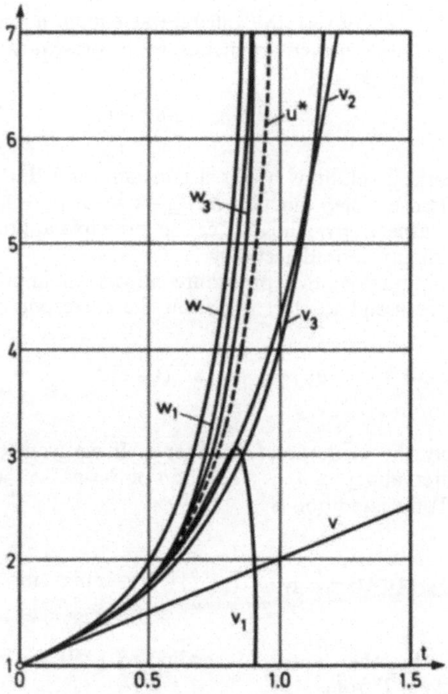

Fig. 3. Several bounds for the problem (5). The solution computed by the Runge-Kutta method is dotted

10. Uniqueness Theorems

In this section we shall show how Theorem 9 III can be used in deriving uniqueness criteria. If u and v are two solutions of one and the same initial value problem

$$u' = f(t, u) \quad \text{in} \quad J_0, \qquad u(0) = \eta , \tag{1}$$

then according to Definition 8 III the difference $d(t) = v(t) - u(t)$ has the property $d(0) = 0$. If we consider, as will often be the case, solutions

[1] The solution was computed on the electronic computer Z 23 of the Universität Karlsruhe using the library program available there (step size 0.05).

which for $t=0$ are also differentiable and satisfy the differential equation[1], then we have

$$d(0)=d'_+(0)=0, \quad \text{i.e.,} \quad d(t)=o(t) \quad \text{for} \quad t\to+0. \tag{2}$$

For solutions in this latter sense the uniqueness criteria can be sharpened.

We now give several classes of functions $\omega(t,z)$ for which uniqueness follows from a one-sided condition

$$f(t,z)-f(t,\bar z)\leqq\omega(t,z-\bar z) \quad \text{for} \quad z\geqq\bar z \quad \text{and} \quad (t,z),(t,\bar z)\in D(f). \tag{3}$$

I. Definition ($\mathscr{E}_4, \mathscr{E}_5, \mathscr{E}_6$). The class $\mathscr{E}_i$ $(i=4,5,6)$ consists of all functions $\omega(t,z)$ defined in $J_0\times\{z\geqq 0\}$ with the property: For every $\varepsilon>0$ there is a number $\delta>0$ and a function $\varrho(t)\in Z(\omega)$ such that $0<\varrho(t)\leqq\varepsilon$ in J_0 and

$$\mathscr{E}_4\colon \varrho'>\omega(t,\varrho)+\delta \quad \text{and} \quad \varrho(t)\geqq\delta \quad \text{in} \quad J_0;$$

$$\mathscr{E}_5\colon \varrho'>\omega(t,\varrho) \quad \text{and} \quad \varrho(t)\geqq\delta \quad \text{in} \quad J_0;$$

$$\mathscr{E}_6\colon \varrho'>\omega(t,\varrho) \quad \text{in } J_0, \quad \varrho(t_k)\geqq\delta t_k \quad \text{for a sequence} \quad t_k\to+0.$$

The second condition in $\mathscr{E}_6$ can also be written as "$\varrho(t)\neq o(t)$ for $t\to+0$". Furthermore, it would suffice in all three cases to require that ϱ be continuous and the inequality hold with $D^-\varrho$ instead of ϱ'.

II. Uniqueness Theorem. *The initial value problem* (1) *has at most one solution, and it depends continuously on the initial value [resp. on the initial value and the right hand side of the differential equation] if f satisfies a condition* (3) *with $\omega\in\mathscr{E}_5$ [resp. $\omega\in\mathscr{E}_4$]. If we consider solutions which also satisfy the differential equation for $t=0$*[1]*, then $\omega\in\mathscr{E}_6$ is sufficient for uniqueness.*

Here the continuous dependence on the initial value [resp. on the initial value and the right hand side] is defined as follows: For every $\varepsilon>0$ there is a $\delta>0$ such that $|u-v|<\varepsilon$ in J if $u\in Z(f)$ denotes the solution of the initial value problem (1) and $v\in Z(f)$ denotes a solution of the differential equation $Pv=0$ in J_0 with $|v(0)-\eta|<\delta$ [resp. v denotes a function from $Z(f)$ with $|Pv|<\delta$ in J_0 and $|v(0)-\eta|<\delta$].

For the proof we first consider the case $\omega\in\mathscr{E}_6$. Let u and v be two solutions of (1) and ϱ be a function with the properties given in I. Then 9 III, where of course $\delta(t)=0$ now, cannot be used immediately since 9 III (α) in general is not satisfied. This difficulty necessitates a small additional consideration. Since $|v(t)-u(t)|=o(t)$ and $\varrho(t)\neq o(t)$ for $t\to+0$, there are arbitrarily small positive numbers t_k with $|v(t_k)-u(t_k)|<\varrho(t_k)$. Then 9 III can be applied to the interval $t_k\leqq t\leqq T$. We obtain

[1] This is automatically the case if f is continuous at the point $(0,\eta)$.

$|v-u|<\varrho$ in the interval $t_k \leqq t \leqq T$ and thereby in J_0, whence the theorem is proved since there is a suitable function ϱ below every positive bound ε. The proof becomes simpler in the cases $\mathscr{E}_4$ and $\mathscr{E}_5$ since we can apply 9 III to J immediately.

III. Remark ($\mathscr{E}[o(g(t))], \ldots$). The uniqueness assertion depends on what we can assert about the behavior of the difference of two solutions $d(t)$ for small positive t. In this sense the two classes $\mathscr{E}_5$ and $\mathscr{E}_6$ represent two more or less arbitrarily chosen special cases of the following general principle. If we know that two solutions have the property $|v(t)-u(t)| = o(g(t))$ or $O(g(t))$ for $t \to +0$, then we define a corresponding class $\mathscr{E}[o(g(t))]$ or $\mathscr{E}[O(g(t))]$ respectively as in I but with

$$\mathscr{E}[o(g(t))]\colon \varrho' > \omega(t,\varrho) \quad \text{in } J_0, \quad \varrho(t) \neq o(g(t)) \quad \text{for} \quad t \to +0\,;$$

$$\mathscr{E}[O(g(t))]\colon \varrho' > \omega(t,\varrho) \quad \text{in } J_0, \quad \varrho(t) \neq O(g(t)) \quad \text{for} \quad t \to +0\,,$$

and we can then prove the corresponding uniqueness theorem[1]:

If f satisfies a condition (3) *with $\omega \in \mathscr{E}[o(g(t))]$ or $\omega \in \mathscr{E}[O(g(t))]$ respectively and if u and v are two solutions of* (1) *for which $|v(t)-u(t)| = o(g(t))$ or $= O(g(t))$ respectively, then these two solutions are identical.*

Example: Suppose the differential equation is not satisfied for $t=0$ but f is bounded in a neighborhood of the point $(0,\eta)$, and thus $|v-u| = O(t)$. For example, the criterion given by Rosenblatt (1909), $\omega(t,z) = \beta z/t, 0<\beta<1$, is allowable for this case; $\beta z/t \in \mathscr{E}[O(t)]$. On the other hand, the Nagumo function $\omega(t,z)=z/t$ does not guarantee uniqueness. To see this we consider the following initial value problem: $\eta=0$, $f(t,z)=0$ for $z \leqq 0$, $f(t,z)=\min(1,z/t)$ for $z>0, t>0$. All functions $u(t)=Ct$, $0 \leqq C \leqq 1$, are solutions, although f is bounded and satisfies a Nagumo condition.

Similar extensions of the uniqueness theorem have been given by Brauer (1959a), Walter (1960a) and Lakshmikantham (1962). In addition the work of Olech (1960) and Walter (1964) should be noted, where connections between the different uniqueness criteria for continuous f are investigated.

IV. Examples and Remarks. Suppose

$$\omega(t,z) = l(t)\psi(z)\,, \tag{4}$$

where $l(t)$ is continuous in J_0 (or instead of this, satisfies 9 VI (γ)) and nonnegative, $\psi(z)$ is continuous for $z \geqq 0$, positive for $z>0$, $\psi(0)=0$ and

$$\int_0^1 \frac{ds}{\psi(s)} = \infty\,.$$

[1] In this notation $\mathscr{E}_6 = \mathscr{E}[o(t)]$ and $\mathscr{E}_5 \subset \mathscr{E}[o(1)]$.

We want to determine functions $\varrho(t)$ with the properties required in I as solutions of

$$\varrho' = (l(t)+\gamma)\psi(\varrho) \quad \text{in } J_0 \text{ with } \varrho(T)=\varepsilon \quad (\gamma>0). \tag{5}$$

This ensures that $\varrho(t) \leqq \varepsilon$ and $\varrho' > \omega(t, \varrho)$. The solution $\varrho(t)$ of (5) is obtained from

$$\int_{\varrho(t)}^{\varepsilon} \frac{ds}{\psi(s)} = \int_t^T l(s)ds + \gamma(T-t). \tag{6}$$

It is uniquely determined, can be continued to the left up to $t=0$ and is positive in J_0 [from $\varrho(t_0)=0, t_0>0$ we obtain a contradiction to (6) since for $t \to t_0+0$ the right hand side of (6) remains bounded but the left hand side, on the contrary, grows beyond all bounds].

(α) If — besides the above hypotheses — $l(t) \in L(J)$, then $\omega \in \mathscr{E}_4$. In this case the derivation just outlined is also true for $t_0=0$, i.e., $\varrho(0)>0$ and thus $\varrho(t) \geqq \alpha > 0$ in J. Furthermore, from this we have

$$\varrho' = (l(t)+\gamma)\psi(\varrho) > l(t)\psi(\varrho) + \delta,$$

if, say, $\delta = \beta\gamma/2$ and β is the (positive) lower bound of $\psi(\varrho(t))$ in J.

Thus we have now shown that the condition of Osgood-Montel-Tonelli (see 4 V (β)), in particular the generalized Lipschitz condition, implies uniqueness and the continuous dependence on the initial value and the right hand side in the initial value problem (1).

(β) From $\delta t \leqq \varrho(t)$ follows

$$\int_{\delta t}^{\varepsilon} \frac{ds}{\psi(s)} \geqq \int_{\varrho(t)}^{\varepsilon} \frac{ds}{\psi(s)}$$

and conversely. Thus we have $\omega \in \mathscr{E}_6$ if for every $\varepsilon>0$ there exist two positive numbers γ, δ such that (compare with (6))

$$\int_{\delta t}^{\varepsilon} \frac{ds}{\psi(s)} \geqq \int_t^T l(s)ds + \gamma(T-t) \tag{7}$$

at least for the values $t=t_k$ of a sequence with $t_k \to +0$. Each of the two relations

$$\limsup_{t\to+0} \int_t^{\alpha} \left[\frac{1}{\psi(s)} - l(s)\right] ds = \infty \quad (\text{for an } \alpha>0) \tag{7'}$$

$$\limsup_{t\to+0} \int_t^{\alpha} \left[\frac{1}{\psi(s)} - l(s)\right] ds > -\infty \quad \text{and} \quad \psi(s) \leqq s \tag{7''}$$

is sufficient for the validity of (7).

Both of these sufficient conditions for $l(t)\psi(z) \in \mathscr{E}_6$ were given by Filippov (1948); see also La Salle (1949). That (7) follows from (7′) is immediately obvious;

in the case of (7″) we must note that the integral

$$\int_{\delta t}^{t} \frac{ds}{\psi(s)} \geqq \ln \frac{t}{\delta t} = \ln \frac{1}{\delta},$$

and thus can be made arbitrarily large by a suitable choice of $\delta > 0$.

(γ) In the linear case $\omega(t, z) = l(t)z$, we obtain from (7″) the condition for $\omega \in \mathscr{E}_6$

$$\liminf_{t \to +0} \left[\log t + \int_t^T l(s) ds \right] < \infty$$

given also by Wintner (1956a). Lakshmikantham (1964a) showed that in this condition $\log t$ can be replaced by $-\log(1 + l(t))$.

(δ) The following nonlinear example is also of historical interest. If $\omega(t, z)$ is continuous in $J \times \{z \geqq 0\}$, $\omega(t, 0) = 0$ and if $\sigma' = \omega(t, \sigma)$ in J, $\sigma(0) = 0$ implies $\sigma = 0$ in J, then $\omega \in \mathscr{E}_4$.

This is the first general uniqueness criterion and was given by Bompiani (1925) and Perron (1926) (Bompiani made the additional assumption that ω is monotone increasing in z). The proof is similar to that given in Example VII below. One constructs the minimal solution σ of $\sigma' = \omega(t, \sigma)$, $\sigma(T) = \varepsilon > 0$ to the left, which is positive in J and can be approximated from below by subfunctions ϱ which are positive in J and have the desired property $\varrho' = \omega(t, \varrho) + \delta (\delta > 0)$ (note that in the application of 8 IX to the present situation we have a positive δ in the approximation from below since we are dealing with an interval to the left of the initial point $t = T$).

The uniqueness theorem can still be generalized. For this we consider the following class of functions.

V. Definition ($\mathscr{E}_7$). We say $\omega \in \mathscr{E}_7$ if the following holds: For every $\varepsilon > 0$ there is a $\delta > 0$, a sequence of numbers $\tau_i \to +0$ $(i \to \infty)$, $\tau_i > 0$, and a sequence of functions $\varrho_i(t)$ which are differentiable in $\tau_i \leqq t \leqq T$, satisfy the inequalities $0 \leqq \varrho_i \leqq \varepsilon$, and for which

$$\varrho_i' > \omega(t, \varrho_i) \quad \text{in} \quad \tau_i \leqq t \leqq T, \qquad \varrho_i(\tau_i) \geqq \delta \tau_i .$$

VI. Uniqueness Theorem. *The initial value problem has at most one solution which also satisfies the differential equation for $t = 0$ if f satisfies a condition* (3) *with $\omega \in \mathscr{E}_7$.*

Since the difference of the two solutions $|v - u| = o(t)$, we have $|v(\tau_i) - u(\tau_i)| < \varrho_i(\tau_i)$ for all large i. Therefore we can derive $|v - u| < \varrho_i$ from 9 III for $\tau_i \leqq t \leqq T$, whence follows the uniqueness.

VII. Example. Suppose $\omega(t, z)$ is continuous in $J_0 \times \{z \geqq 0\}$, $\omega(t, 0) = 0$, $\omega(t, z) \geqq 0$ for $z \geqq 0$, and for every $t_0 \in J_0$ suppose the function $\varphi(t) \equiv 0$ is the only differentiable function in $0 < t \leqq t_0$ for which $\varphi' = \omega(t, \varphi)$ in $0 < t \leqq t_0$ and $\varphi(t) = o(t)$ for $t \to 0$. Then $\omega \in \mathscr{E}_7$.

We indicate briefly how we establish the existence of the functions ϱ_i of Definition V in this case. By familiar theorems there is a minimal solution $\sigma(t)$ of the differential equation $\sigma' = \omega(t, \sigma)$ through the point (T, ε) which can be continued to the left up to $t = 0$ and has the properties $0 \leqq \sigma \leqq \varepsilon$ in J_0, $\sigma(t) \neq o(t)$. Thus there is a null sequence τ_i and a $\delta > 0$ such that $\sigma(\tau_i) > \delta \tau_i$. The minimal solution σ can be approximated from below in every interval $\tau_i \leqq t \leqq T$ by a function $\varrho_i(t)$ which has the required properties [one should note that now $\varrho_i' > \omega(t, \varrho_i)$ in the approximation from *below*, since we are considering an interval to the *left* of the initial value $t=T$].

The uniqueness theorem corresponding to Example VII was given [with the restriction that there is a $<$ symbol in (3)] by Iyanaga (1928) and [with a condition (9.2) and at the same time for systems] by Kamke (1930). We have introduced the complicated-appearing class $\mathscr{E}_7$ only in order to be able to include the known uniqueness theorem of Kamke in our theory. It might be possible that the class $\mathscr{E}_6$ suffices for this (we have not succeeded in finding either a proof or a counterexample)[1]. Szarski (1962) investigates the condition "$\varphi(t)=o(t)$" in the above example and shows, among other things, that it cannot be replaced by "$\varphi(t)\to 0$, $\varphi'(t)\to 0$ for $t\to +0$". A uniqueness class similar to Kamke's and to $\mathscr{E}_6$ was defined by Shen Xin-yao (1963).

Finally we refer to Remark X of the introduction, which is useful in the application of uniqueness theorems. Furthermore, it is well known that for ordinary differential equations the uniqueness "in the large" follows from the uniqueness "in the small". It is therefore sufficient for the uniqueness that for every $t_0\in J_0$ there be a $\delta>0$ and an ω from one of the given function classes such that

$$f(t,z)-f(t,\bar z)\leqq \omega(t-t_0, z-\bar z) \quad \text{for} \quad t_0<t<t_0+\delta, \quad z\geqq \bar z. \tag{8}$$

11. Systems of Ordinary Differential Equations. Estimation by *K*-Norms

In the following four sections of this chapter we consider the initial value problem for a system of ordinary differential equations

$$u_\nu'=f_\nu(t,u_1,\ldots,u_n),\quad u_\nu(0)=\eta_\nu \quad (\nu=1,\ldots,n;\, n\geqq 1); \tag{1}$$

in vector notation this is written as

$$\boldsymbol{u}'=\boldsymbol{f}(t,\boldsymbol{u}),\quad \boldsymbol{u}(0)=\boldsymbol{\eta}. \tag{2}$$

Here $\boldsymbol{u}=(u_1,\ldots,u_n)$, $\boldsymbol{f}=(f_1,\ldots,f_n)$, $\boldsymbol{\eta}=(\eta_1,\ldots,\eta_n)$ are vectors in E^n. In Sections 6 and 7 we introduced a series of notation and definitions which will be used constantly here. $\boldsymbol{P}\boldsymbol{v}=\boldsymbol{v}'-\boldsymbol{f}(t,\boldsymbol{v})$ is the defect (see 8 IV).

Theorem 9 III on error bounds can be translated verbatim to systems. In place of the absolute value we use an arbitrary K-norm.

I. Theorem. *Suppose that for the vector functions* $\boldsymbol{u},\boldsymbol{v}\in Z(\boldsymbol{f})$, *the scalar functions* $\delta(t)$, $\omega(t,z)$, $\varrho(t)\in Z(\omega)$ *and a K-norm* $\|\cdot\|$ *we have*

(α) $\|\boldsymbol{v}-\boldsymbol{u}\|(0+)<\varrho(0+)$;

(β) $\boldsymbol{u}'=\boldsymbol{f}(t,\boldsymbol{u})$, $\|\boldsymbol{P}\boldsymbol{v}\|=\|\boldsymbol{v}'-\boldsymbol{f}(t,\boldsymbol{v})\|\leqq\delta(t)$ *in* J_0;

(γ) $\varrho'>\omega(t,\varrho)+\delta(t)$ *in* J_0;

(δ) $\|\boldsymbol{f}(t,\boldsymbol{v})-\boldsymbol{f}(t,\boldsymbol{u})\|\leqq\omega(t,\|\boldsymbol{v}-\boldsymbol{u}\|)$ *if* $\|\boldsymbol{v}-\boldsymbol{u}\|=\varrho(t)$.

---

[1] The difficulty is that there is a $<$ sign in the inequalities of 9 III (γ) or I respectively, while we are dealing with solutions of the corresponding differential equation in VII. The relation $\mathscr{E}_7=\mathscr{E}_6$ would be established if we could show: For every function ω which has the properties of VII there is a function $\bar\omega$ which also has these properties and for which $\omega(t,z)<\bar\omega(t,z)$ for $t\in J_0$, $z>0$.

Then

$$\|\boldsymbol{v}(t)-\boldsymbol{u}(t)\|<\varrho(t) \quad in \quad J_0 .$$

If $\omega \in \mathscr{F}_0$, the function $\delta(t)$ is continuous in J_0 and integrable over J and if the maximal solution $\sigma(t) \in Z(\omega)$ of an initial value problem

$$(\gamma') \ \sigma' = \omega(t, \sigma) + \delta(t) \quad with \quad \sigma(0) \geqq \|\boldsymbol{v}(0) - \boldsymbol{u}(0)\|$$

exists in J, then

$$\|\boldsymbol{v}(t)-\boldsymbol{u}(t)\| \leqq \sigma(t) \quad in \quad J,$$

provided that along with (β) the hypothesis (with an $\alpha > 0$)

$$(\delta') \ \|\boldsymbol{f}(t, \boldsymbol{v}) - \boldsymbol{f}(t, \boldsymbol{v}-\boldsymbol{z})\| \leqq \omega(t, \|\boldsymbol{z}\|) \quad for \quad \sigma(t) < \|\boldsymbol{z}\| < \sigma(t) + \alpha$$

holds.

That ω is defined for the arguments involved in (δ') is obtained from the fact that $D(\omega)$ is open and contains the points (t, σ).

The second part of the theorem follows in familiar fashion from the first part by approximating the maximal solution σ from above by functions ϱ. For the proof of the first part we refer back to Lemma 8 II, where now $\varphi = \|\boldsymbol{v}-\boldsymbol{u}\|$, $\psi = \varrho$. We then need to show only that the inequality

$$D^- \|\boldsymbol{v}-\boldsymbol{u}\| < \varrho'$$

holds at every point t where $\|\boldsymbol{v}-\boldsymbol{u}\| = \varrho$. For these t,

$$\begin{aligned} D^- \|\boldsymbol{v}-\boldsymbol{u}\| &\leqq \|\boldsymbol{v}'-\boldsymbol{u}'\| = \|\boldsymbol{v}' - \boldsymbol{f}(t, \boldsymbol{v}) + \boldsymbol{f}(t, \boldsymbol{v}) - \boldsymbol{f}(t, \boldsymbol{u})\| \\ &\leqq \delta(t) + \omega(t, \|\boldsymbol{v}-\boldsymbol{u}\|) = \delta(t) + \omega(t, \varrho) < \varrho' \end{aligned}$$

follows from (β) (γ) (δ) with 7 II (δ) and the triangle inequality 7 I (α). Q.E.D.

While we were always able to formulate the hypothesis (δ) in the case of *one* differential equation in such a way that the solution u is not involved, this cannot be done here in a simple fashion. The additional condition "if $\|\boldsymbol{v}-\boldsymbol{u}\| = \varrho$" in (δ) is of course hard to work with since the solution is unknown. Nevertheless we infer from it that (δ) can be replaced by

$$(\delta'') \ \|\boldsymbol{f}(t, \boldsymbol{v}) - \boldsymbol{f}(t, \boldsymbol{v}-\boldsymbol{z})\| \leqq \omega(t, \|\boldsymbol{z}\|) \quad \text{for} \quad \|\boldsymbol{z}\| = \varrho(t).$$

If the bound ϱ or σ is positive, then we can restrict consideration to those $\boldsymbol{z}$ with $\|\boldsymbol{z}\| > 0$; in particular the condition (3) involved in the following uniqueness theorem is then sufficient[1].

Finally we note that I includes Theorem 9 III. If $n = 1$, we have only to take $\|z\| = z$ and $\|z\| = -z$.

[1] This remark is unnecessary for norms but not for K-norms. Compare with Remark V.

From I in connection with a condition

$$\|f(t,z)-f(t,\bar{z})\| \leqq \omega(t, \|z-\bar{z}\|) \text{ for } \|z-\bar{z}\|>0 \text{ and } (t,z),(t,\bar{z}) \in D(f) \quad (3)$$

as in Section 10, we obtain a

II. Uniqueness Theorem. *The initial value problem* (2) *has at most one solution of class* $Z(f)$ *which also satisfies the differential equation for* $t=0$ *if* f *allows a bound* (3) *with* $\omega \in \mathscr{E}_6$ *or* $\omega \in \mathscr{E}_7$ *and if here* $\|\cdot\|$ *is a K-norm for which* $\|z\| = \|-z\| = 0$ *only for* $z=\mathbf{0}$. *If the bound* (3) *also holds with a norm and with* $\omega \in \mathscr{E}_5$ *or* $\mathscr{E}_4$, *then the solution depends continuously on the initial value or on the initial value and the right hand side of the differential equation respectively (here the differential equation need hold only in* J_0*).*

The continuous dependence is defined as in 10 II: $\|v(0)-\eta\| < \delta$ or $\|v(0)-\eta\| < \delta$ and $\|Pv\| < \delta$ respectively implies $\|u-v\| < \varepsilon$ [1].

As in the proof of 10 II, if $\omega \in \mathscr{E}_6$, we obtain $\|v-u\| \leqq 0$ and, since u and v can be interchanged, also $\|u-v\| \leqq 0$. But since the sum $\|z\| + \|-z\|$ is never negative by the triangle inequality, we have $\|v-u\| = \|u-v\| = 0$, i.e., by the assumptions made on the K-norm, $v-u=\mathbf{0}$. The further assertions can also be derived easily from I.

This theorem includes a series of known uniqueness theorems as special cases. The uniqueness theorem of Kamke (1945, p. 139) is obtained with the norm 7 I (δ) and Example 10 VII; see also Coddington-Levinson (1955, Theorem 2.3). Cafiero (1949) and Zwirner (1950) give further uniqueness criteria. Our presentation is closest to the results of Hukuhara (1941). The theorems of M. Müller (1927) are more special in that only the K-norms 7 I (ζ), (η) are used. They are more general in that the values $z, \bar{z}$ are further restricted in (3), which can be significant in applications and which we shall therefore investigate briefly.

III. Corollary. If we use one of the two-K-norms $\|z\| = \max z_\nu$ or $\|z\| = \max |z_\nu|$ in I or II, then we can replace the hypothesis I (δ) by

$$(\delta^*) \quad f_\nu(t,z) - f_\nu(t,\bar{z}) \leqq \omega(t, \|z-\bar{z}\|) \text{ if } \|z-\bar{z}\| = z_\nu - \bar{z}_\nu = \varrho(t) \text{ and}$$

$$(t,z),(t,\bar{z}) \in D(f) \qquad (\nu = 1, \ldots, n);$$

and the hypothesis (3) by

$$f_\nu(t,z) - f_\nu(t,\bar{z}) \leqq \omega(t, \|z-\bar{z}\|) \text{ for } z_\nu \geqq \bar{z}_\nu \text{ and } (t,z),(t,\bar{z}) \in D(f)\,. \quad (3^*)$$

The proof of I undergoes a small alteration. If the assertion is false, then there is a first point t_0 at which $\|v-u\| = \varrho$ and thus for at least

[1] This definition is independent of the special norm, since for every norm $\|\cdot\|$ there exist two numbers $\alpha, \beta > 0$ with $\alpha\|z\|_e \leqq \|z\| \leqq \beta\|z\|_e$ ($\|z\|_e$ is the Euclidean norm).

one index ν, $v_\nu - u_\nu = \varrho$ or $u_\nu - v_\nu = \varrho$. We obtain a contradiction to this with the help of 8 II by applying this lemma to $\varphi = v_\nu - u_\nu$ (resp. $u_\nu - v_\nu$), $\psi = \varrho$. Establishing that $\varphi' < \psi'$ follows from $\varphi = \psi$ is again done with a chain of inequalities similar to that in the proof of I.

What is new in this procedure is that from the vector $\boldsymbol{v} - \boldsymbol{u}$ we first pass to a component $v_\nu - u_\nu$ and then work with this component. This is possible here only for the two given norms; in the following sections we will encounter it frequently in a somewhat different framework.

IV. Example (Lipschitz condition). Let $\|\cdot\|$ be a K-norm and let $\boldsymbol{f}$ satisfy a Lipschitz condition

$$\|\boldsymbol{f}(t, z) - \boldsymbol{f}(t, \bar{z})\| \leqq l(t)\|z - \bar{z}\| \tag{4}$$

in $D(\boldsymbol{f})$. If $\boldsymbol{u}$ is a solution of the initial value problem (2) and $\boldsymbol{v}$ an approximate solution for which

$$\|\boldsymbol{v}(0) - \boldsymbol{\eta}\| \leqq \varepsilon \quad \text{and} \quad \|\boldsymbol{P}\boldsymbol{v}\| = \|\boldsymbol{v}' - \boldsymbol{f}(t, \boldsymbol{v})\| \leqq \delta(t),$$

then we have the bound

$$\|\boldsymbol{v} - \boldsymbol{u}\| \leqq e^{L(t)}\left[\varepsilon + \int_0^t \delta(\tau) e^{-L(\tau)} d\tau\right] \quad \text{with} \quad L(t) = \int_0^t l(\tau) d\tau. \tag{5}$$

This formula was already derived in 7 V for C-solutions with the hypothesis $\delta(t), l(t) \in L(J)$. [Moreover $l(t) \geqq 0$ was required; this is necessary for integral equations but superfluous for differential equations, as 7 VI shows.] In the present derivation based on I we also require $l, \delta \in C(J_0)$; however the property 9 VI (γ) suffices instead of this. If the bound in (5) is positive, then it is sufficient if (4) holds for $\|z - \bar{z}\| > 0$. If it is positive and one of the K-norms given in III is used, then (4) can be replaced by

$$f_\nu(t, z) - f_\nu(t, \bar{z}) \leqq l(t)\|z - \bar{z}\| \quad \text{for} \quad z_\nu \geqq \bar{z}_\nu \quad (\nu = 1, \ldots, n). \tag{4*}$$

V. Remark. The addition of "for $\|z - \bar{z}\| > 0$" in the condition (3) of the uniqueness theorem is of course unnecessary if we consider only norms. It is however significant for more general K-norms. Thus Theorem II for $n = 1$, $\|z\| = z$ becomes 10 II; in other words: The uniqueness theorem 10 II [with the *one-sided* condition (10.3)!] is a special case of II, which was not true in the restriction to norms.

Similarly, the sharpening obtained via (3*) can be quite important. Let us clarify this in an example ($n = 2$)

$$u_\nu' = g_{\nu 1}(t, u_1) + g_{\nu 2}(t, u_2) \quad (\nu = 1, 2).$$

We choose $\|z\| = \max\{|z_1|, |z_2|\}$ and seek conditions under which (3*) holds with $\omega(t, z) = Lz$. This is the case if the functions g_{12}, g_{21} satisfy a Lipschitz condition and the functions g_{11}, g_{22} satisfy a one-sided

Lipschitz condition

$$g_{\nu\nu}(t,z)-g_{\nu\nu}(t,\bar{z})\leqq L(z-\bar{z}) \quad \text{for} \quad z\geqq\bar{z} \qquad (\nu=1,2).$$

On the other hand in an appeal to (3) we need the full Lipschitz condition for the $g_{\nu\nu}$ also.

VI. General Distances. For some problems it is useful to introduce, in place of a K-norm $\|\boldsymbol{z}\|$, a more general distance $V(t,\boldsymbol{z})$ which depends explicitly on t. A possibility for extending I to this case is:

Suppose the function $V(t,\boldsymbol{z})$ is continuously differentiable in $J\times E^n$. For two functions $\boldsymbol{u},\boldsymbol{v}\in Z(\boldsymbol{f})$ and the scalar functions $\delta(t)$, $\omega(t,z)$, $\varrho(t)\in Z(\omega)$ suppose

(α) $V(0,\boldsymbol{v}(0)-\boldsymbol{u}(0))<\varrho(0)$;

(β) $\boldsymbol{P}\boldsymbol{u}=0$ *and* $V_{\boldsymbol{z}}(t,\boldsymbol{v}-\boldsymbol{u})\cdot\boldsymbol{P}\boldsymbol{v}\leqq\delta(t)$ *in* J_0;

(γ) $\varrho'>\omega(t,\varrho)+\delta(t)$ *in* J_0;

(δ) $V_t(t,\boldsymbol{v}-\boldsymbol{u})+V_{\boldsymbol{z}}(t,\boldsymbol{v}-\boldsymbol{u})\cdot[\boldsymbol{f}(t,\boldsymbol{v})-\boldsymbol{f}(t,\boldsymbol{u})]\leqq\omega(t,V(t,\boldsymbol{v}-\boldsymbol{u}))$

(at least for all t with $V(t,\boldsymbol{v}-\boldsymbol{u})=\varrho(t)$).

Then

$$V(t,\boldsymbol{v}-\boldsymbol{u})<\varrho(t) \quad \textit{in} \quad J_0 .$$

Here $V_{\boldsymbol{z}}$ is the vector grad V and $V_{\boldsymbol{z}}\cdot\boldsymbol{z}$ the scalar product. If for the proof we set $\varphi(t)=V(t,\boldsymbol{v}-\boldsymbol{u})$, $\psi(t)=\varrho(t)$, then from $\varphi=\psi$ follows

$$\begin{aligned}\varphi'&=V_t+V_{\boldsymbol{z}}\cdot(\boldsymbol{v}'-\boldsymbol{u}')\\ &=V_t+V_{\boldsymbol{z}}\cdot[\boldsymbol{f}(t,\boldsymbol{v})-\boldsymbol{f}(t,\boldsymbol{u})]+V_{\boldsymbol{z}}\cdot\boldsymbol{P}\boldsymbol{v}\leqq\delta(t)+\omega(t,\varrho)<\varrho' .\end{aligned}$$

The assertion now follows from 8 II.

Instead of the continuous differentiability of $V(t,\boldsymbol{z})$, it is sufficient if $V(t,\boldsymbol{z})$ is continuous in $J\times E^n$ and if there exist two functions $V_t(t,\boldsymbol{z})$, $V_{\boldsymbol{z}}(t,\boldsymbol{z})$ defined in $J\times E^n$ so that for a function $\boldsymbol{\psi}(t)$ which is differentiable in J the inequality

$$D_-V(t,\boldsymbol{\psi}(t))\leqq V_t(t,\boldsymbol{\psi})+V_{\boldsymbol{z}}(t,\boldsymbol{\psi})\cdot\boldsymbol{\psi}$$

holds[1].

VII. Remark on Stability Theory. The above theorem is closely related to the second method of Lyapunov. In this method results are derived concerning the behavior of the solutions of (2) as $t\to\infty$ with the help of suitably chosen functions $V(t,\boldsymbol{z})$ (so-called Lyapunov functions); see Antosiewicz (1958), Hahn (1963), Cesari (1959, § 7).

[1] The hypothesis that $V(t,z)$ has partial derivatives in t and z_ν everywhere [and thus also the hypothesis of Brauer and Sternberg (1958)] is not sufficient. Counterexample: $n=1$, $V(t,z)=r\sin 2\varphi$ with $r=\sqrt{t^2+z^2}$, $\varphi=\operatorname{arctg} z/t$; V_z and V_t exist everywhere and vanish at zero, but $V(t,t)=\sqrt{2}\,|t|$!!

If we consider the case where $f(t, \mathbf{0}) = \mathbf{0}$ and $\mathbf{u} = \mathbf{0}$ and where $\mathbf{v}$ is a solution of the differential equation $P\mathbf{v} = \mathbf{0}$ (in stability theory it is common to transform the differential equation so that the solution to be examined for stability is $\mathbf{u} = \mathbf{0}$), then the above theorem reads:

From

$$V_t(t, \mathbf{v}) + V_{\mathbf{z}}(t, \mathbf{v}) \cdot f(t, \mathbf{v}) \leqq \omega(t, V(t, \mathbf{v})), \qquad \varrho' > \omega(t, \varrho) \quad in \quad J_0$$

follows

$$V(t, \mathbf{v}) < \varrho(t) \quad in \quad J_0,$$

if this inequality holds for $t = 0$.

From this we can obtain the essential theorems on stability in the sense of Lyapunov without difficulty. *Lower* bounds for $V(t, \mathbf{v} - \mathbf{u})$ can be determined as in VI; they can be called upon for the proof of theorems on instability in the Lyapunov sense.

The introduction of a general distance $V(t, z)$ in estimation theorems based on differential inequalities goes back to a work of Conti (1956a). These ideas were deepened – in particular with respect to stability theory – by Brauer and Sternberg (1958), Antosiewicz (1960, 1962), Corduneanu (1959, 1960), Brauer (1961), Lakshmikantham (1961, 1962b), Chandra (1962, 1962a) among others. See also Wintner (1946a). In the last few years the literature on this subject has become voluminous.

VIII. Differential Equations in Banach Spaces. We refer back to notation introduced in 7 X. The class $Z(f)$ contains the functions $\varphi(t)$ which are continuous in J, differentiable in J_0, and for which $f(t, \varphi)$ is defined (φ, f are functions with values in a Banach space B). The inequality $D^- \|\varphi(t)\| \leqq \|\varphi'(t)\|$ which was proved in 7 II (δ) also holds under the present hypotheses.

Theorem I as well as the uniqueness theorem II hold without alteration for differential equations in a Banach space. The proofs can be taken over verbatim. The translation of the considerations of VI and VII to differential equations in a Banach space is also easily accomplished. Estimates and uniqueness theorems of this kind were proved by Hukuhara (1959). A stability theory from this point of view is given by Lakshmikantham (1962a). Ważewski (1960) and Olech (1960a) use the iteration procedure in connection with estimation theorems to prove existence and uniqueness theorems. See also the literature cited in 7 X.

IX. Bounds for Solutions, Global Existence. The following theorem is a variant of Theorem I. It is valid for differential equations in Banach spaces.

Let $\boldsymbol{u} \in Z(\boldsymbol{f}), \varrho \in Z(\omega)$, *where* ϱ *and* ω *are real-valued. If with a K-norm* $\|\cdot\|$

(α) $\|\boldsymbol{u}\|(0+) < \varrho(0+)$;

(β) $\boldsymbol{u}' = \boldsymbol{f}(t, \boldsymbol{u})$ *in* J_0;

(γ) $\varrho' > \omega(t, \varrho)$ *in* J_0;

(δ) $\|\boldsymbol{f}(t, \boldsymbol{u})\| \leqq \omega(t, \|\boldsymbol{u}\|)$ *for all t with* $\|\boldsymbol{u}(t)\| = \varrho(t)$, *then*

$$\|\boldsymbol{u}(t)\| < \varrho(t) \quad \text{in} \quad J_0 .$$

The proof is simple and similar to the one given in I. In the usual way the following corollary is derived:

If (β) and

(δ') $\|\boldsymbol{f}(t, \boldsymbol{z})\| \leqq \omega(t, \|\boldsymbol{z}\|)$

hold and if $\sigma(t)$ is the maximal solution of the initial value problem (ω continuous in $J \times \{z \geqq 0\}$)

$$\sigma' = \omega(t, \sigma) \quad \text{in } J, \quad \sigma(0) = \|\boldsymbol{u}(0)\| ,$$

then

$$\|\boldsymbol{u}(t)\| \leqq \sigma(t) \quad \text{in} \quad J.$$

This theorem can be used to calculate numbers T such that $[0, T]$ is an interval of existence for the solutions of the initial value problem (2). By way of example, consider the case $\omega(t, z) = \varphi(z)$, where $\varphi(z)$ is continuous and positive for positive z and furthermore

$$\int^{\infty} \frac{dz}{\varphi(z)} = \infty .$$

Then it is immediately clear from the behavior of the solutions of $\varrho' = \varphi(\varrho)$ that every solution $\boldsymbol{u}$ of (1) is bounded in bounded intervals. Therefore the solutions $\boldsymbol{u}$ of (1) exist in $[0, \infty)$ if f is defined and continuous in $[0, \infty) \times E^n$. This special result was proved by Wintner (1945, 1946b); cf. also Eisen (1966). The general statement given above contains results by Bihari (1956), Langenhop (1960), Lakshmikantham (1962b). For related works see Medvedev (1965), Bebernes, Fulks and Meisters (1966).

12. Systems of Differential Inequalities

The estimation theorem of the preceding section yields in many cases unsatisfactory error bounds. We can understand this easily by realizing that for an approximating function $\boldsymbol{v}(t)$ some components $v_\nu(t)$ may approximate the solution component $u_\nu(t)$ very well while other components $v_\nu(t)$ approximate the corresponding solution components $u_\nu(t)$ less well. But such differences are obliterated by the passage to the norm and do not appear in the resulting bound (they are also not given any consideration in the proof). We can correct this evil to a certain extent by weighting the individual components differently with a choice of suitable norms – a simple example is $\|z\| = c_1|z_1| + \cdots + c_n|z_n|$ with $c_\nu > 0$.

Now we shall develop a second method for obtaining bounds in the case of systems in which the single components are "individually" treated. It is characteristic of this method that we first translate the fundamental lemma 8 II itself to vector functions. Following this the theorems of Sections 8 and 9 are given an n-dimensional setting, in which the vectors are bounded by vectors and not, as in Section 11, by scalars via a norm. We proceeded similarly for systems of integral equations in Section 6.

I. Lemma. *Suppose the vector functions $\boldsymbol{\varphi}(t)$, $\boldsymbol{\psi}(t)$ are continuous in J_0. Suppose further that: if $\boldsymbol{\varphi} \leqq \boldsymbol{\psi}$, $\varphi_\nu = \psi_\nu$ for an index ν and a point $t_0 \in J_0$, then*

$$D_-\varphi_\nu(t_0) < D_-\psi_\nu(t_0) \quad \text{or} \quad D^-\varphi_\nu(t_0) < D^-\psi_\nu(t_0)$$

(thus $\varphi_\nu'(t_0) < \psi_\nu'(t_0)$, provided the derivatives exist).

Then we have precisely one of the following two cases:

(α) $\boldsymbol{\varphi} < \boldsymbol{\psi}$ in J_0;

(β) $\boldsymbol{\varphi}(0+) < \boldsymbol{\psi}(0+)$ does not hold, i.e., there are arbitrarily small $\bar{t} \in J_0$ satisfying

$$\varphi_\nu(\bar{t}) \geqq \psi_\nu(\bar{t}) \quad \text{for at least one index } \nu.$$

Let us assume for the proof that neither case (α) nor case (β) applies. This means that $\boldsymbol{\varphi}(0+) < \boldsymbol{\psi}(0+)$ and consequently that there is a $t_0 \in J_0$ satisfying

$$\boldsymbol{\varphi} < \boldsymbol{\psi} \quad \text{for} \quad 0 < t < t_0, \quad \boldsymbol{\varphi}(t_0) \leqq \boldsymbol{\psi}(t_0) \quad \text{and} \quad \varphi_\nu(t_0) = \psi_\nu(t_0)$$

for at least one index ν. From this we arrive at a contradiction if we deal with the functions φ_ν, ψ_ν exactly as we did with φ, ψ in the proof of 8 II.

As long as no assumptions on $\boldsymbol{f}$ are introduced, the transition from 8 V to systems is possible only in the following somewhat unwieldy form.

II. Theorem. *Suppose the vector functions $\boldsymbol{v}$, $\boldsymbol{w}$ are in $Z(\boldsymbol{f})$ and*

(α) $\boldsymbol{v}(0+) < \boldsymbol{w}(0+)$;

(β) $v_\nu' - f_\nu(t, \boldsymbol{z}) < w_\nu' - f_\nu(t, \boldsymbol{w})$ if $\boldsymbol{v}(t) \leqq \boldsymbol{z} \leqq \boldsymbol{w}(t)$, $z_\nu = v_\nu(t)$, $(t, \boldsymbol{z}) \in D(\boldsymbol{f})$ $(\nu = 1, \ldots, n)$; or

(β') $v_\nu' - f_\nu(t, \boldsymbol{v}) < w_\nu' - f_\nu(t, \boldsymbol{z})$ if $\boldsymbol{v}(t) \leqq \boldsymbol{z} \leqq \boldsymbol{w}(t)$, $z_\nu = w_\nu(t)$, $(t, \boldsymbol{z}) \in D(\boldsymbol{f})$ $(\nu = 1, \ldots, n)$.

Then

$$\boldsymbol{v}(t) < \boldsymbol{w}(t) \quad in \quad J_0 .$$

Furthermore, the assertion also holds for continuous functions $\boldsymbol{v}$, $\boldsymbol{w}$ with D_- or D^- instead of the derivative.

For if $\boldsymbol{v} \leqq \boldsymbol{w}$ and $v_\nu = w_\nu$ at the point t_0, then by (β) we also have $v'_\nu < w'_\nu$. Now the assertion follows from I.

An essential difference from 8 V is found in the sharpened hypothesis (β). Here the scope of applicability of II is strongly curtailed. For example, if the vector function $\boldsymbol{u}(t)$ is a solution of the initial value problem

$$\boldsymbol{u}' = \boldsymbol{f}(t, \boldsymbol{u}) \quad \text{in } J_0, \quad \boldsymbol{u}(0) = \boldsymbol{\eta}\,, \tag{1}$$

which is to be bounded from above by the vector function $\boldsymbol{w}(t)$, then if we have no further information about $\boldsymbol{u}$, we must require that $\boldsymbol{w}(0) > \boldsymbol{\eta}$ and for $\nu = 1, \ldots, n$

$$w'_\nu > f_\nu(t, \boldsymbol{z})$$

for $t \in J_0$ and all $\boldsymbol{z}$ with $z_\mu \leqq w_\mu(t)$ $(\mu \neq \nu)$, $z_\nu = w_\nu(t)$. Then according to the theorem just proved (with $\boldsymbol{u} = \boldsymbol{v}$) we have in fact $\boldsymbol{u}(t) < \boldsymbol{w}(t)$ in J. Similarly for bounds from below. Thus we have the following criterion for

III. Super- and Subfunctions for the initial value problem (1). If the functions $\boldsymbol{v}, \boldsymbol{w} \in Z(\boldsymbol{f})$ and if

(α) $\boldsymbol{v}(0+) < \boldsymbol{u}(0+)$ resp. $\boldsymbol{w}(0+) > \boldsymbol{u}(0+)$ for every solution of the initial value problem;

(β) $v'_\nu < f_\nu(t, \boldsymbol{z})$, if $\boldsymbol{z} \geqq \boldsymbol{v}(t)$, $z_\nu = v_\nu(t)$ and $(t, \boldsymbol{z}) \in D(\boldsymbol{f})$ or
$w'_\nu > f_\nu(t, \boldsymbol{z})$, if $\boldsymbol{z} \leqq \boldsymbol{w}(t)$, $z_\nu = w_\nu(t)$ and $(t, \boldsymbol{z}) \in D(\boldsymbol{f})$

then $\boldsymbol{v}$ is a subfunction, $\boldsymbol{w}$ a superfunction, i.e.,

$$\boldsymbol{v} < \boldsymbol{u} < \boldsymbol{w} \quad \text{in } J_0 \text{ for every solution } \boldsymbol{u} \text{ of the initial value problem.}$$

As has already been said, (β) is here a very strong requirement. We can make do with somewhat more modest hypotheses if we bound the solution *simultaneously* from above and below. Indeed we have, as was first proved by M. Müller (1927):

IV. Super- and Subfunctions *are provided by the functions* $\boldsymbol{w}(t)$, $\boldsymbol{v}(t) \in Z(\boldsymbol{f})$ *when they satisfy the two hypotheses*

(α) $\boldsymbol{v}(0+) < \boldsymbol{u}(0+) < \boldsymbol{w}(0+)$ *for every solution* $\boldsymbol{u}$;

(β) $v'_\nu < f_\nu(t, \boldsymbol{z})$ if $\boldsymbol{v}(t) \leqq \boldsymbol{z} \leqq \boldsymbol{w}(t)$, $v_\nu(t) = z_\nu$, $(t, \boldsymbol{z}) \in D(\boldsymbol{f})$,
$w'_\nu > f_\nu(t, \boldsymbol{z})$ if $\boldsymbol{v}(t) \leqq \boldsymbol{z} \leqq \boldsymbol{w}(t)$, $w_\nu(t) = z_\nu$, $(t, \boldsymbol{z}) \in D(\boldsymbol{f})$.

Again we have $\boldsymbol{v} < \boldsymbol{u} < \boldsymbol{w}$ *in* J_0.

Moreover, both criteria also hold for $\boldsymbol{v}, \boldsymbol{w} \in Z_c(\boldsymbol{f})$ with $D_- v_\nu$, $D_- w_\nu$ or $D^- v_\nu$, $D^- w_\nu$ in place of v'_ν, w'_ν.

For if $\boldsymbol{v} \leqq \boldsymbol{u} \leqq \boldsymbol{w}$ in an interval $0 < t \leqq t_0$, then, in the same interval, $\boldsymbol{v} < \boldsymbol{u} < \boldsymbol{w}$, which follows via two applications of II. The assertion is now obtained in the usual way.

The hypotheses (β) in II–IV are significantly simplified if $f_\nu(t, z)$ is monotone in several variables z_μ which are distinct from z_ν. Let us assume, say, that f_ν is monotone increasing [monotone decreasing] in z_λ $(\lambda \neq \nu)$. Then in the first line of (β) in II–IV, the range of variation of the component z_λ can be restricted to $z_\lambda = v_\lambda(t)$ $[z_\lambda = w_\lambda(t)]$, and in the second row of (β) to $z_\lambda = w_\lambda(t)$ $[z_\lambda = v_\lambda(t)]$.

These hypotheses become particularly simple if all components f_ν are monotone increasing in *all* variables z_μ with $\mu \neq \nu$; the expression "quasi-monotone increasing" was used in 6 II for this behavior. The systems with a right hand side $\boldsymbol{f}(t, \boldsymbol{z})$ which is quasi-monotone increasing in the variables $\boldsymbol{z}$ are of great theoretical significance. There is a series of properties of an ordinary differential equation which cannot be carried over to arbitrary systems, but only to systems with quasi-monotone increasing $\boldsymbol{f}$. These include so central a concept as the maximal and minimal solution as well as Theorems 8 V, VI.

V. Theorem. *If $\boldsymbol{f}(t, \boldsymbol{z})$ is quasi-monotone increasing in $\boldsymbol{z}$ and $\boldsymbol{v}, \boldsymbol{w} \in Z(\boldsymbol{f})$, then*

(α) $\boldsymbol{v}(0+) < \boldsymbol{w}(0+)$;

(β) $P\boldsymbol{v} < P\boldsymbol{w}$ *in* J_0

imply the inequality

$$\boldsymbol{v}(t) < \boldsymbol{w}(t) \quad \text{in} \quad J_0 .$$

Here, of course, $P\boldsymbol{v} = \boldsymbol{v}' - \boldsymbol{f}(t, \boldsymbol{v})$.

VI. Super- and Subfunctions. If the function $\boldsymbol{f}(t, \boldsymbol{z})$ is quasi-monotone increasing in $\boldsymbol{z}$, then $\boldsymbol{v} \in Z(\boldsymbol{f})$ is a subfunction, $\boldsymbol{w} \in Z(\boldsymbol{f})$ a superfunction, if

(α) $\boldsymbol{v}(0+) < \boldsymbol{u}(0+)$ resp. $\boldsymbol{u}(0+) < \boldsymbol{w}(0+)$ for every solution $\boldsymbol{u}$;

(β) $\boldsymbol{v}' < \boldsymbol{f}(t, \boldsymbol{v})$ resp. $\boldsymbol{w}' > \boldsymbol{f}(t, \boldsymbol{w})$ in J_0.

Every solution $\boldsymbol{u}$ of the initial value problem (1) then lies between $\boldsymbol{v}$ and $\boldsymbol{w}$:

$$\boldsymbol{v} < \boldsymbol{u} < \boldsymbol{w} \quad \text{in} \quad J_0 .$$

Both of these theorems, which are essentially due to M. Müller (1926, Satz 14) and Kamke (1932), follow immediately from II and III. They can be generalized to continuous functions in the usual way.

VII. Remarks on the Existence Problem ($\mathscr{F}^n, \mathscr{F}_0^n$). The results of 8 VII, VIII can be carried over to systems. If all components $f_\nu(t, z_1, \ldots, z_n)$ of $\boldsymbol{f}(t, \boldsymbol{z})$ are defined and continuous on a set $D(\boldsymbol{f})$ which is the intersection of an open set of $(n+1)$-dimensional $(t, \boldsymbol{z})$-space with the set $J \times E^n$, then $\boldsymbol{f} \in \mathscr{F}^n$; on the other hand if $\boldsymbol{f}$ is continuous only at points of

$D(f)$ with positive t, then $f \in \mathscr{F}_0^n$, provided that for every $\eta = (\eta_1, \dots, \eta_n)$ satisfying $(0, \eta) \in D(f)$ there exist a number $\delta > 0$ and a function $h(t) \in L(J)$ such that

$$|f_\nu(t, z)| \leqq h(t) \quad \text{for} \quad 0 < t < \delta, \quad |z_\mu - \eta_\mu| < \delta \quad (\mu, \nu = 1, \dots, n).$$

The initial value problem (1) has (at least) one solution if $f \in \mathscr{F}_0^n$ and $(0, \eta) \in D(f)$. Every solution can be continued to the right up to the boundary of $D(f)$. If $u \in Z(f)$ is a solution of the initial value problem which exists in J and is uniquely determined[1], then if $\delta = (\delta_1, \dots, \delta_n)$, $\varepsilon = (\varepsilon_1, \dots, \varepsilon_n)$ and the $|\delta_\nu|$, $|\varepsilon_\nu|$ are sufficiently small, all solutions of the initial value problem

$$\varphi' = f(t, \varphi) + \varepsilon, \quad \varphi(0) = \eta + \delta \tag{2}$$

exist in J, and as $\varepsilon \to 0$, $\delta \to 0$ they converge to $u(t)$ uniformly in J.

While all of this corresponds to the one-dimensional case 8 VII, VIII and is essentially proved in exactly the same way, the

VIII. Maximal and Minimal Solutions are among those concepts which can only be introduced for quasi-monotone increasing $f(t, z)$. As we said, they exist if f belongs to the class $\mathscr{F}_0^n$ and is quasi-monotone increasing in z and they can be continued to the right up to the boundary of $D(f)$. Every other solution u lies between the minimal solution u_* and the maximal solution u^*:

$$u_*(t) \leqq u(t) \leqq u^*(t)$$

(provided the respective solutions exist). The maximal [minimal] solution, if it exists in J, can be approximated uniformly in J by functions φ which are solutions of (2) with $\varepsilon > 0$, $\delta > 0$ [$\varepsilon < 0$, $\delta < 0$].

That these facts and their proofs can be taken over from 8 IX rests finally on the similarity between Theorems V and 8 V. The above results go back to Kamke (1932); a generalization is given by Perov (1965).

IX. Theorem. *If* $f(t, z) \in \mathscr{F}_0^n$ *is quasi-monotone increasing in* z *and further if* u^* [*resp.* u_*] $\in Z(f)$ *is the maximal* [*resp. minimal*] *solution of the initial value problem* (1) — *suppose it exists in* J — *and if for a function* v [*resp.* w] $\in Z(f)$

(α) $v(0) \leqq \eta$ [$w(0) \geqq \eta$];
(β) $v' \leqq f(t, v)$ *in* J_0 [$w' \geqq f(t, w)$ *in* J_0],

then

$$v \leqq u^* \; [w \geqq u_*] \quad \text{in} \quad J.$$

[1] See Footnote 2, p. 67.

The assertions also hold with the weaker assumption (β'), and then even for continuous functions $\boldsymbol{v}, \boldsymbol{w}$:

(β') $D_- v_\nu \leqq f_\nu(t, \boldsymbol{v})$ if $\boldsymbol{v}(t) \leqq \boldsymbol{u}^*(t) + \boldsymbol{\delta}$ and $u_\nu^*(t) < v_\nu(t) < u_\nu^*(t) + \delta_\nu$,
$D^- w_\nu \geqq f_\nu(t, \boldsymbol{w})$ if $\boldsymbol{w}(t) \geqq \boldsymbol{u}_*(t) - \boldsymbol{\delta}$ and $u_{*\nu}(t) - \delta_\nu < w_\nu(t) < u_{*\nu}(t)$,

where $\boldsymbol{\delta} > \boldsymbol{0}$.

This too is obtained just as in 8 X. Similar theorems were established by Szarski (1947, 1950, 1951). The sharpening 8 X (β'), which led Ważewski to the concept of "épiderme supérieur," is expressed for systems in (β') above. It was first given (in a less general form) by Mlak (1956) and further sharpened by Lasota (1959).

The fact that (β') has become more complicated here is due to the very nature of things. Indeed, the theorem is false if we simply write 8 X (β') instead of (β') and require $\boldsymbol{v}' \leqq \boldsymbol{f}(t, \boldsymbol{v})$ if $\boldsymbol{u}^*(t) < \boldsymbol{v}(t) < \boldsymbol{u}^*(t) + \boldsymbol{\delta}$ ($\boldsymbol{\delta} > \boldsymbol{0}$). Example: $\boldsymbol{f}(t, z) \equiv \boldsymbol{0}$, $\boldsymbol{u}^*(t) = \boldsymbol{\eta} = \boldsymbol{0}$, $\boldsymbol{v}(t) = (t, -1)$, $n = 2$.

X. Remarks. (α) Suppose the function $\boldsymbol{f}$ is quasi-monotone increasing in z and satisfies a uniqueness condition (14.2) with $\omega \in \mathscr{E}_5^n$. Further suppose $\boldsymbol{v}, \boldsymbol{w} \in Z(\boldsymbol{f})$. Then the Theorems V, VI can be given in more compact form, namely,

Va: *From* $\boldsymbol{v}(0) \leqq \boldsymbol{w}(0)$ *and* $\boldsymbol{P}\boldsymbol{v} \leqq \boldsymbol{P}\boldsymbol{w}$ *in* J_0 *follows* $\boldsymbol{v} \leqq \boldsymbol{w}$ *in* J.
VIa: $\boldsymbol{v}$ *is a subfunction if* $\boldsymbol{v}(0) \leqq \boldsymbol{\eta}$ *and* $\boldsymbol{v}' \leqq \boldsymbol{f}(t, \boldsymbol{v})$ *in* J_0;
$\boldsymbol{w}$ *is a superfunction if* $\boldsymbol{w}(0) \geqq \boldsymbol{\eta}$ *and* $\boldsymbol{w}' \geqq \boldsymbol{f}(t, \boldsymbol{w})$ *in* J_0.

Both correspond to 8 XI and are proved in a similar way.

(β) The situation is similar for Theorems II–IV. Thus if $\boldsymbol{f}$ satisfies a uniqueness condition (14.2) with $\omega \in \mathscr{E}_5^n$ (if $\boldsymbol{f}$ is continuous at the point $(0, \boldsymbol{\eta})$, then we can have $\omega \in \mathscr{E}_6^n$), then the criterion IV for super- and subfunctions takes the following form IVa:

IVa (α) $\boldsymbol{v}(0) \leqq \boldsymbol{\eta} \leqq \boldsymbol{w}(0)$;
(β) $v_\nu' \leqq f_\nu(t, z)$ *if* $\boldsymbol{v}(t) \leqq z \leqq \boldsymbol{w}(t)$ *and* $v_\nu(t) = z_\nu$,
$w_\nu' \geqq f_\nu(t, z)$ *if* $\boldsymbol{v}(t) \leqq z \leqq \boldsymbol{w}(t)$ *and* $w_\nu(t) = z_\nu$.

For the proof we show, as in 8 XI: if the functions $\boldsymbol{v}, \boldsymbol{w} \in Z(\boldsymbol{f})$ satisfy these conditions and if $\boldsymbol{\varrho}$ has the properties of 14 I, then the functions $\boldsymbol{v} - \boldsymbol{\varrho}$, $\boldsymbol{w} + \boldsymbol{\varrho}$ are sub- and superfunctions in the sense of IV.

(γ) It can happen that $\boldsymbol{f}$ is converted into a quasi-monotone increasing function by a simple change of sign. Example: If $n = 3$ and f_1 is monotone decreasing in z_2 and z_3, f_2 is monotone decreasing in z_1, increasing in z_3, f_3 is monotone decreasing in z_1, increasing in z_2, then the function $\boldsymbol{f}^*(t, z) = (f_1(t, z^*), -f_2(t, z^*), -f_3(t, z^*))$ with $z^* = (z_1, -z_2, -z_3)$ is quasi-monotone increasing in z. Further, we see that $\boldsymbol{u}' = \boldsymbol{f}(t, \boldsymbol{u})$ and $\boldsymbol{u}^* = (u_1, -u_2, -u_3)$ imply the equation $\boldsymbol{u}^{*\prime} = \boldsymbol{f}^*(t, \boldsymbol{u}^*)$. An enumeration of all the possible cases is given by Burton and Whyburn (1952).

XI. Differential Inequalities in Ordered Banach Spaces. Quasimonotonicity. The rest of this section is devoted to the study of differential inequalities in ordered Banach spaces. Since the main theorems of this section are connected with the concept of quasimonotonicity, we consider only Banach spaces B in which this concept can be defined in a natural way, i.e., in which it makes sense to speak of a component of an element of B. Thus B will be a space of real sequences or, more generally, of real-valued functions equipped with the natural ordering. The theory developed here is fundamental in our treatment of the line method for parabolic differential equations in Chapter IV.

Let A be a nonempty set (index set) and L the linear space of real-valued functions defined on A. For elements $\boldsymbol{z} \in L$ we use the index notation $\boldsymbol{z} = (z_\alpha)_{\alpha \in A}$ or simply $\boldsymbol{z} = (z_\alpha)$. Let $\boldsymbol{p} = (p_\alpha)$ be a fixed element of L such that $p_\alpha > 0$ for $\alpha \in A$ and B the Banach space of all elements $\boldsymbol{z}$ in L with finite norm

$$\|\boldsymbol{z}\| = \sup\{p_\alpha |z_\alpha| : \alpha \in A\}.$$

We set up the natural order relation in B defined by the "positive cone" $B_+ = \{\boldsymbol{z} \in B : z_\alpha \geqq 0 \text{ for } \alpha \in A\}$. Denote the interior of B_+ by B_+^0. Then $\boldsymbol{y} \leqq \boldsymbol{z}$ or $\boldsymbol{y} < \boldsymbol{z}$ is defined by $\boldsymbol{z} - \boldsymbol{y} \in B_+$ or $\boldsymbol{z} - \boldsymbol{y} \in B_+^0$ respectively; more explicitly,

$$\boldsymbol{y} \leqq \boldsymbol{z} \Leftrightarrow y_\alpha \leqq z_\alpha \qquad (\alpha \in A),$$

$$\boldsymbol{y} < \boldsymbol{z} \Leftrightarrow y_\alpha + \frac{\delta}{p_\alpha} \leqq z_\alpha \quad (\alpha \in A) \quad \text{for a positive } \delta.$$

Free use will be made of earlier definitions which are understood, as far as topology or ordering is involved, in the sense of the norm topology (in accordance with 7 X) and of the order relation just defined. For example, $\boldsymbol{v} \in Z$ means in the present context that the function $\boldsymbol{v}$ from J to B is continuous in J and differentiable in J_0 (in the sense of 7 X), $\boldsymbol{v}(0+) < \boldsymbol{w}(0+)$ means that $\boldsymbol{w}(t) - \boldsymbol{v}(t) \in B_+^0$ for small positive t (see 1 V).

The elements $\boldsymbol{e}$ and $\boldsymbol{e}^\beta (\beta \in A)$ of B are defined by

$$\boldsymbol{e} = \left(\frac{1}{p_\alpha}\right), \quad \boldsymbol{e}^\beta = (e_\alpha^\beta) \quad \text{where} \quad e_\beta^\beta = 1 \quad \text{and} \quad e_\alpha^\beta = 0 \quad \text{for} \quad \alpha \neq \beta.$$

In this notation $\boldsymbol{y} < \boldsymbol{z}$ if and only if $\boldsymbol{y} + \delta \boldsymbol{e} \leqq \boldsymbol{z}$ for a positive δ. The function $\boldsymbol{f}(t, \boldsymbol{z})$, defined on $D(\boldsymbol{f}) \subset J \times B$ and with values in B, is *quasimonotone increasing* in $\boldsymbol{z}$ if for $\alpha \in A$

$$f_\alpha(t, \boldsymbol{z}) \leqq f_\alpha(t, \bar{\boldsymbol{z}}) \quad \text{for} \quad z_\alpha = \bar{z}_\alpha, \quad z_\beta \leqq \bar{z}_\beta \quad (\alpha \neq \beta)$$

and $(t, \boldsymbol{z}), (t, \bar{\boldsymbol{z}}) \in D(\boldsymbol{f})$.

A modulus of continuity is a real-valued function $d(s)$ defined and continuous for all real s and such that $d(s)=d(-s)$, $d(0)=0$, $d(s)$ is monotone increasing for $s \geqq 0$. The following three propositions are immediate consequences of the definitions.

(α) The function $\boldsymbol{v}(t)$, defined in J, is continuous in J if and only if there is a modulus of continuity such that

$$p_\alpha |v_\alpha(t) - v_\alpha(\bar{t})| \leqq d(t-\bar{t}) \quad \text{for } \alpha \in A \text{ and } t, \bar{t} \in J .$$

(β) The function $\boldsymbol{v}(t)$ is differentiable at $t_0 \in J$, the value of the derivative being $\boldsymbol{v}'(t_0)$, if and only if a modulus of continuity d exists such that

$$p_\alpha \left| \frac{v_\alpha(t) - v_\alpha(t_0)}{t - t_0} - v'_\alpha(t_0) \right| \leqq d(t-t_0) \quad \text{for } t \in J \text{ and } \alpha \in A.$$

(γ) If $\boldsymbol{v}$ is continuous in J, then the real-valued function

$$\varphi(t) = \inf\{p_\alpha v_\alpha(t) : \alpha \in A\}$$

is also continuous in J (this follows easily from (α)).

XII. Monotonicity Theorem. *Let* $\boldsymbol{v}, \boldsymbol{w} \in Z(\boldsymbol{f})$, *where* $\boldsymbol{f}(t, \boldsymbol{z})$ *is quasimonotone increasing in* $\boldsymbol{z}$, *and*

(α) $\boldsymbol{v}(0+) < \boldsymbol{w}(0+)$;

(β) $\boldsymbol{P}\boldsymbol{v} < \boldsymbol{P}\boldsymbol{w}$ *in* J_0;

(γ) *for each* $(t_0, \boldsymbol{z}_0) \in D(\boldsymbol{f})$, $t_0 > 0$, *there exists a modulus of continuity* $d(s)$ *such that*

$$p_\alpha \{ f_\alpha(t_0, \boldsymbol{z}_0 + s\boldsymbol{e}^\alpha) - f_\alpha(t_0, \boldsymbol{z}_0)\} \geqq -d(p_\alpha s) \quad \textit{for} \quad \alpha \in A, s \geqq 0$$

and $(t_0, \boldsymbol{z}_0 + s\boldsymbol{e}^\alpha) \in D(\boldsymbol{f})$.

Then

$$\boldsymbol{v} < \boldsymbol{w} \quad \textit{in} \quad J_0 .$$

For the proof let $\boldsymbol{u} = \boldsymbol{w} - \boldsymbol{v}$ and

$$\varphi(t) = \inf\{p_\alpha u_\alpha(t) : \alpha \in A\} .$$

By (α) and XI (γ), $\varphi \in C(J)$ and $\varphi(0+) > 0$. Assume that the assertion of the theorem, $\varphi(t) > 0$ in J_0, is false and that $t_0 \in J_0$ is the first zero of φ. From the quasimonotonicity and (γ) we obtain, using the abbreviation $\boldsymbol{v}_0 = \boldsymbol{v}(t_0), \ldots$, since $\boldsymbol{v}_0 \leqq \boldsymbol{w}_0$,

$$\begin{aligned} f_\alpha(t_0, \boldsymbol{w}_0) - f_\alpha(t_0, \boldsymbol{v}_0) &\geqq f_\alpha(t_0, \boldsymbol{v}_0 + \boldsymbol{e}^\alpha u_\alpha(t_0)) - f_\alpha(t_0, \boldsymbol{v}_0) \\ &\geqq -p_\alpha^{-1} d(p_\alpha u_\alpha(t_0)) \end{aligned}$$

where d is the modulus of continuity corresponding to the point $(t_0, \boldsymbol{v}_0)$. By (β) there is a positive δ such that

$$p_\alpha u'_\alpha(t_0) \geqq \delta + p_\alpha \{f_\alpha(t_0, \boldsymbol{w}_0) - f_\alpha(t_0, \boldsymbol{v}_0)\} \geqq \delta - d(p_\alpha u_\alpha(t_0))$$

and hence a positive ε such that

$$p_\alpha u'_\alpha(t_0) > \delta/2 \quad \text{if} \quad p_\alpha u_\alpha(t_0) < \varepsilon \,. \tag{$*$}$$

By XI (β) the inequality

$$p_\alpha \left| \frac{u_\alpha(t_0) - u_\alpha(t_1)}{t_0 - t_1} - u'_\alpha(t_0) \right| < \frac{\delta}{2} \tag{$**$}$$

holds for a suitable t_1, $0 < t_1 < t_0$. Now, $p_\alpha u_\alpha(t_1) \geqq \varphi(t_1) > 0$. On the other hand, as a consequence of $\varphi(t_0) = 0$, there are indices α such that $p_\alpha u_\alpha(t_0) < \varepsilon$ and $< \varphi(t_1)$. For those α the difference quotient in ($**$) is negative and hence $p_\alpha u'_\alpha(t_0) < \delta/2$. But this contradicts the inequality ($*$). Q.E.D.

This proof is considerably more involved than the proof for finite A (Theorem V). The reason is easily pinpointed: for infinite A the assumption $\varphi(t_0) = 0$ does not imply $u_\alpha(t_0) = 0$ for some α. This difficulty necessitates another proof and an additional assumption (γ) on $\boldsymbol{f}$ which is, roughly speaking, a weak continuity requirement for f_α with regard to z_α. The following counterexample shows that the theorem is indeed false without the condition (γ).

XIII. Example. Let A be the set of natural numbers and $p_i = 1$ for all natural numbers i (i.e., B is the space l_∞ of bounded real sequences endowed with the maximum norm). Let (ε_i) be a sequence of numbers such that $1/2 < \varepsilon_{i+1} < \varepsilon_i < 1$ for all i and

$$f_i(t, \boldsymbol{z}) = \begin{cases} 4 z_i (t - \varepsilon_i)^{-3} & \text{for} \quad 0 \leqq t < \varepsilon_i \\ -3|z_i|^{2/3} + (t - \varepsilon_{i+1})^{-3} z_{i+1} & \text{for} \quad \varepsilon_i \leqq t < 1 \,. \end{cases}$$

For $\boldsymbol{v} = \boldsymbol{0}$, $\boldsymbol{w} = ((\varepsilon_i - t)^3)$ we have $\boldsymbol{P v} = \boldsymbol{0}$ and, after some calculation,

$$(\boldsymbol{P w})_i = w'_i - f_i(t, \boldsymbol{w}) \geqq 1 \quad \text{for} \quad 0 \leqq t \leqq 1 \quad \text{and all } i.$$

All assumptions of Theorem XII save (γ) are satisfied, but we have $v_i(t) = 0 > w_i(t)$ for $t > \varepsilon_i$.

XIV. Definition and Theorem. The transition from Theorem XII to a corresponding theorem with equality signs permitted is accomplished in the way outlined in X and 8 XI. The analog of the class $\mathscr{E}_5^n$ is a class $\mathscr{E}_5^B$ consisting of functions $\boldsymbol{\omega}(t, \boldsymbol{z})$ from $J_0 \times B_+$ to B such that for each $\boldsymbol{\varepsilon} > \boldsymbol{0}$ there is a $\boldsymbol{\delta} > \boldsymbol{0}$ and a function $\boldsymbol{\varrho} \in Z$ with

$$\boldsymbol{\varrho}' > \boldsymbol{\omega}(t, \boldsymbol{\varrho}) \quad \text{and} \quad \boldsymbol{\delta} \leqq \boldsymbol{\varrho} \leqq \boldsymbol{\varepsilon} \quad \text{in} \quad J_0 \,. \tag{3}$$

The condition on f is then given by

$$f(t,z)-f(t,\bar{z}) \leqq \omega(t, z-\bar{z}) \quad \text{for} \quad z \geqq \bar{z} \quad \text{and} \quad (t,z),(t,\bar{z}) \in D(f). \tag{4}$$

Theorem. *Suppose $\boldsymbol{v}, \boldsymbol{w} \in Z(f)$, where f is quasimonotone increasing in z and $(t, \boldsymbol{w}(t)+z) \in D(f)$ for $\mathbf{0} \leqq z \leqq \boldsymbol{\alpha}$ ($\boldsymbol{\alpha} > \mathbf{0}$). Then*

(α) $\boldsymbol{v}(0) \leqq \boldsymbol{w}(0)$;

(β) $P\boldsymbol{v} \leqq P\boldsymbol{w}$ *in* J_0;

(γ) f *satisfies condition* (4) *and* XII (γ);

imply

$$\boldsymbol{v} \leqq \boldsymbol{w} \quad \text{in} \quad J.$$

The proof by reduction to XII follows the usual pattern. If $\boldsymbol{\varrho}$ has the properties stated in (3), then $\boldsymbol{v}$ and $\bar{\boldsymbol{w}} = \boldsymbol{w} + \boldsymbol{\varrho}$ satisfy all assumptions of XII. Indeed $\bar{\boldsymbol{w}}(0) \geqq \boldsymbol{v}(0) + \boldsymbol{\delta}$ and

$$P\bar{\boldsymbol{w}} = \boldsymbol{w}' + \boldsymbol{\varrho}' - f(t, \boldsymbol{w}+\boldsymbol{\varrho}) > \boldsymbol{w}' + \omega(t, \boldsymbol{\varrho}) - f(t, \boldsymbol{w}+\boldsymbol{\varrho})$$

$$\geqq \boldsymbol{w}' - f(t, \boldsymbol{w}) = P\boldsymbol{w} \geqq P\boldsymbol{v}.$$

Therefore $\boldsymbol{v} \leqq \bar{\boldsymbol{w}} \leqq \boldsymbol{w} + \boldsymbol{\varepsilon}$ and hence $\boldsymbol{v} \leqq \boldsymbol{w}$.

Special cases of (γ) are:

(γ_1) f satisfies XII (γ) and

$$p_\alpha\{f_\alpha(t,z) - f_\alpha(t,\bar{z})\} \leqq \omega(t, \|z-\bar{z}\|) \quad \text{for} \quad z \geqq \bar{z}, \alpha \in A, \tag{5}$$

where $\omega(t,z) \in \mathscr{E}_5$.

(γ_2) f satisfies the condition

$$\|f(t,z) - f(t,\bar{z})\| \leqq \omega(t, \|z-\bar{z}\|)$$

where $\omega(t,z)$ has the properties of the Bompiani condition 10 IV (δ).

It is obvious that (γ_2) implies (γ_1) (including condition XII (γ)). Also (γ_1) implies (γ) since for a function $\omega(t,z)$ belonging to $\mathscr{E}_5$, the function $\boldsymbol{\omega}(t,z) = \boldsymbol{e}\,\omega(t, \|z\|)$ belongs to $\mathscr{E}_5^B$. For if we have $\varrho' > \omega(t,\varrho)$ and $\delta \leqq \varrho \leqq \varepsilon$ and put $\boldsymbol{\varrho} = \boldsymbol{e}\varrho$, $\boldsymbol{\delta} = \boldsymbol{e}\delta$, $\boldsymbol{\varepsilon} = \boldsymbol{v}\varepsilon$, then (3) holds.

In particular we have the following result which suffices for most applications:

If $f(t,z)$ is continuous and satisfies the condition (γ_2), then we have (a) *existence,* (b) *uniqueness, and* (c) *convergence of successive approximations for the initial value problem, and, finally,* (d) *the validity of the monotonicity theorems XII and XIV under the sole hypotheses* (α) (β). *In particular, this is true if f satisfies a Lipschitz condition.*

The material presented in XI–XIV is taken from Walter (1969a) where further examples and remarks are found. In this connection we mention the work of Mlak (1963) and Mlak and Olech (1963). These authors prove the existence of maximal solutions and an analog of Theorem IX for countably infinite systems. They do not work with the norm topology but with the weak topology in B. Lakshmikantham and Leela (1967) deal with the case, similar to X (γ), where other monotonicity properties of f are present.

13. Component-wise Bounds for Systems

Now that the necessary basis has been prepared, the estimates of Section 9 can be carried over to systems in sharpened form. The problem consists of finding bounds for the difference $\boldsymbol{v}(t)-\boldsymbol{u}(t)$ of an approximate solution $\boldsymbol{v}$ and an (no uniqueness is assumed!) exact solution of the initial value problem

$$\boldsymbol{u}' = \boldsymbol{f}(t, \boldsymbol{u}), \qquad \boldsymbol{u}(0) = \boldsymbol{\eta}\,. \tag{1}$$

Here we shall use only data which may be regarded as known, namely, the defect $\boldsymbol{P v} = \boldsymbol{v}' - \boldsymbol{f}(t, \boldsymbol{v})$ and the deviation from the initial value $\boldsymbol{v}(0) - \boldsymbol{\eta}$ as well as a bound of the function $\boldsymbol{f}(t, \boldsymbol{z})$. The available information about these quantities is now put to use in a way which is an improvement over Section 11. This means, for example, that the (known) defect $\boldsymbol{P v}$ is involved in the bound not only in an inequality $\|\boldsymbol{P v}\| \leqq \delta(t)$, but that for every component of the defect, bounds

$$|v_\nu' - f_\nu(t, \boldsymbol{v})| \leqq \delta_\nu(t) \qquad (\nu = 1, \ldots, n)$$

are given and also taken into account in obtaining the estimate. The same is true of the initial value and of the function $\boldsymbol{f}$; for the latter we use conditions of the form

$$f_\nu(t, z) - f_\nu(t, \bar{z}) \leqq \omega_\nu(t, |z - \bar{z}|)$$

(which are to be made more precise with respect to the permissible values of $z, \bar{z}$). The situation then corresponds to that of 9 III. We run into trouble in an attempt, similar to that of 9 I and II, to start out with even more general assumptions $-\bar{\delta}_\nu(t) \leqq v_\nu' - f_\nu(t, \boldsymbol{v}) \leqq \delta_\nu(t), \ldots$ and to separate the upper bound from the lower bound. Therefore we postpone this conversion to the end of the section and begin with 9 III.

Again, wherever possible, we shall use vector notation and in particular, recall that $|z|$ (in contrast to $\|\boldsymbol{z}\|$) is a *vector* with components $|z_\nu|$.

I. Theorem. *Suppose that for the vector functions* $\boldsymbol{u}, \boldsymbol{v} \in Z(\boldsymbol{f})$, $\boldsymbol{\varrho} \in Z(\boldsymbol{\omega})$, *where* $\boldsymbol{\omega}(t, \boldsymbol{z})$ *is defined for* $0 < t \leqq T$, $\boldsymbol{0} \leqq \boldsymbol{z} \leqq \boldsymbol{\varrho}(t)$ *and quasimonotone increasing in* $\boldsymbol{z}$, *and for* $\boldsymbol{\delta}(t)$, *we have*

(α) $|\boldsymbol{v}-\boldsymbol{u}|(0+) < \boldsymbol{\varrho}(0+)$;

(β) $\boldsymbol{u}' = \boldsymbol{f}(t, \boldsymbol{u})$ *and* $|\boldsymbol{P}\boldsymbol{v}| \equiv |\boldsymbol{v}' - \boldsymbol{f}(t, \boldsymbol{v})| \leqq \boldsymbol{\delta}(t)$ *in* J_0;

(γ) $\boldsymbol{\varrho}' > \boldsymbol{\omega}(t, \boldsymbol{\varrho}) + \boldsymbol{\delta}(t)$ *in* J_0;

(δ) $f_\nu(t, \boldsymbol{v}) - f_\nu(t, \boldsymbol{u}) \leqq \omega_\nu(t, |\boldsymbol{v}-\boldsymbol{u}|)$ *if* $|\boldsymbol{v}-\boldsymbol{u}| \leqq \boldsymbol{\varrho}$, $v_\nu - u_\nu = \varrho_\nu(t)$,
$f_\nu(t, \boldsymbol{u}) - f_\nu(t, \boldsymbol{v}) \leqq \omega_\nu(t, |\boldsymbol{v}-\boldsymbol{u}|)$ *if* $|\boldsymbol{v}-\boldsymbol{u}| \leqq \boldsymbol{\varrho}$, $u_\nu - v_\nu = \varrho_\nu(t)$
($\nu = 1, \ldots, n$).

Then

$$|\boldsymbol{v}(t) - \boldsymbol{u}(t)| < \boldsymbol{\varrho}(t) \quad in \quad J_0 .$$

II. Consequence. For two functions $\boldsymbol{u}, \boldsymbol{v} \in Z(\boldsymbol{f})$, a function $\boldsymbol{\delta}(t)$ which is continuous in J_0 and integrable over J, and an $\boldsymbol{\varepsilon} \geqq \boldsymbol{0}$, suppose

(α) $|\boldsymbol{v}(0) - \boldsymbol{u}(0)| \leqq \boldsymbol{\varepsilon}$;

(β) $\boldsymbol{u}' = \boldsymbol{f}(t, \boldsymbol{u}), |\boldsymbol{P}\boldsymbol{v}| \equiv |\boldsymbol{v}' - \boldsymbol{f}(t, \boldsymbol{v})| \leqq \boldsymbol{\delta}(t)$ in J_0.

Further suppose that the maximal solution $\boldsymbol{\sigma}(t) \in Z(\boldsymbol{\omega})$ of

(γ) $\boldsymbol{\sigma}' = \boldsymbol{\omega}(t, \boldsymbol{\sigma}) + \boldsymbol{\delta}(t), \boldsymbol{\sigma}(0) = \boldsymbol{\varepsilon}$

exists in J; of $\boldsymbol{\omega}(t, \boldsymbol{z})$ we assume that for a vector $\boldsymbol{\gamma} > \boldsymbol{0}$ this function is defined at least in the domain $0 \leqq t \leqq T, \boldsymbol{0} \leqq \boldsymbol{z} \leqq \boldsymbol{\sigma}(t) + \boldsymbol{\gamma}$ and is quasimonotone increasing in $\boldsymbol{z}$ as well as in the class $\mathscr{F}_0^n$.

From this and from

(δ) $f_\nu(t, \boldsymbol{v}) - f_\nu(t, \boldsymbol{u}) \leqq \omega_\nu(t, |\boldsymbol{v}-\boldsymbol{u}|)$ if $|\boldsymbol{v}-\boldsymbol{u}| \leqq \boldsymbol{\sigma} + \boldsymbol{\gamma}$,
$\sigma_\nu < v_\nu - u_\nu < \sigma_\nu + \gamma_\nu$,
$f_\nu(t, \boldsymbol{u}) - f_\nu(t, \boldsymbol{v}) \leqq \omega_\nu(t, |\boldsymbol{v}-\boldsymbol{u}|)$ if $|\boldsymbol{v}-\boldsymbol{u}| \leqq \boldsymbol{\sigma} + \boldsymbol{\gamma}$,
$\sigma_\nu < u_\nu - v_\nu < \sigma_\nu + \gamma_\nu$;

it follows that

$$|\boldsymbol{v}(t) - \boldsymbol{u}(t)| \leqq \boldsymbol{\sigma}(t) \quad \text{in} \quad J.$$

The proof of I goes through according to the usual scheme. Let us assume that $|\boldsymbol{v}-\boldsymbol{u}| \leqq \boldsymbol{\varrho}$ for $0 \leqq t \leqq T_1$. Then if t is a point in this interval and ν is an index with $v_\nu(t) - u_\nu(t) = \varrho_\nu(t)$, by I ($\beta$)–($\delta$)

$$\begin{aligned} v_\nu' - u_\nu' &\leqq \delta_\nu(t) + f_\nu(t, \boldsymbol{v}) - f_\nu(t, \boldsymbol{u}) \leqq \delta_\nu(t) + \omega_\nu(t, |\boldsymbol{v}-\boldsymbol{u}|) \\ &\leqq \delta_\nu(t) + \omega_\nu(t, \boldsymbol{\varrho}) < \varrho_\nu' \end{aligned}$$

at the point t (the quasi-monotonicity of $\boldsymbol{\omega}$ was used here). By Lemma 12 I applied to the interval $0 \leqq t \leqq T_1$ and $\boldsymbol{\varphi} = \boldsymbol{v} - \boldsymbol{u}$, $\boldsymbol{\psi} = \boldsymbol{\varrho}$, we have $\boldsymbol{v} - \boldsymbol{u} < \boldsymbol{\varrho}$ in $0 \leqq t \leqq T_1$. In the same way we show that $\boldsymbol{u} - \boldsymbol{v} < \boldsymbol{\varrho}$ in this interval as well. Thus we have found that from an inequality $|\boldsymbol{v}-\boldsymbol{u}| \leqq \boldsymbol{\varrho}$ in the closed interval $0 \leqq t \leqq T_1$ the sharper inequality $|\boldsymbol{v}-\boldsymbol{u}| < \boldsymbol{\varrho}$ follows in the same interval. Hence we obtain the assertion of I in a simple way.

The conclusion II is then easy to obtain by approximating the maximal solution $\boldsymbol{\sigma}(t)$ according to 12 VIII by superfunctions $\boldsymbol{\varrho}(t)$ which occur in Theorem I.

The hypothesis (δ) gives rise to several remarks. It differs from the corresponding hypotheses for *one* differential equation in Section 9 especially in that the solution $\boldsymbol{u}$ is involved, in $\boldsymbol{f}$ as well as in the "if" requirements. However, since $\boldsymbol{u}$ is in general unknown, the conditions of the requirements cannot be verified, and at first glance it seems as is for arbitrary $\boldsymbol{z}, \bar{\boldsymbol{z}}$ an inequality

$$|f_\nu(t, \boldsymbol{z}) - f_\nu(t, \bar{\boldsymbol{z}})| \leqq \omega_\nu(t, |\boldsymbol{z} - \bar{\boldsymbol{z}}|)$$

would be needed[1]. Upon closer examination we find, however, that each of the following two hypotheses (δ'), (δ'') is sufficient for the validity of (δ) in I and II (simply use "$v_\nu - u_\nu = \varrho_\nu$" or "$u_\nu - v_\nu = \varrho_\nu$" respectively in connection with the fact that $\varrho_\nu > 0$, $\sigma_\nu \geqq 0$):

$$(\delta') \quad f_\nu(t, \boldsymbol{v}) - f_\nu(t, \boldsymbol{v} - \boldsymbol{z}) \leqq \omega_\nu(t, |\boldsymbol{z}|) \text{ for } z_\nu \geqq 0, (t, \boldsymbol{v} - \boldsymbol{z}) \in D(\boldsymbol{f}),$$
$$f_\nu(t, \boldsymbol{v} + \boldsymbol{z}) - f_\nu(t, \boldsymbol{v}) \leqq \omega_\nu(t, |\boldsymbol{z}|) \text{ for } z_\nu \geqq 0, (t, \boldsymbol{v} + \boldsymbol{z}) \in D(\boldsymbol{f});$$
$$(\delta'') \quad f_\nu(t, \boldsymbol{z}) - f_\nu(t, \bar{\boldsymbol{z}}) \leqq \omega_\nu(t, |\boldsymbol{z} - \bar{\boldsymbol{z}}|) \text{ for } z_\nu \geqq \bar{z}_\nu, (t, \boldsymbol{z}), (t, \bar{\boldsymbol{z}}) \in D(\boldsymbol{f}).$$

Thus for systems we also do not need a bound on the absolute value of $\boldsymbol{f}(t, \boldsymbol{z}) - \boldsymbol{f}(t, \bar{\boldsymbol{z}})$ which holds for all $\boldsymbol{z}, \bar{\boldsymbol{z}}$, but rather a one-sided bound for certain $\boldsymbol{z}, \bar{\boldsymbol{z}}$ only. Therefore it is also possible, in some cases, to bound $\boldsymbol{f}$ by functions $\boldsymbol{\omega}$ which (for $\boldsymbol{z} \geqq \boldsymbol{0}$) take on negative values, and give monotone decreasing error bounds $\boldsymbol{\varrho}$ or $\boldsymbol{\sigma}$; compare this with the Examples III, IV. However, it should be noted that for *one* differential equation a bound for "one-half" of all the numbers $z, \bar{z}$ is required in (9.3), but for systems, according to (δ'') only a "2^nth" of all vectors $\boldsymbol{z}, \bar{\boldsymbol{z}}$ is excluded.

If we set $n = 1$ in both theorems, we obtain precisely 9 III. The generalizations mentioned in 9 IV (β) can also be carried out here. We note only the following in this regard. For the functions $\boldsymbol{u}, \boldsymbol{v}$ and $\boldsymbol{\varrho}$ involved in I and II the left-sided differentiability and the validity of the corresponding inequality with the left-sided derivative are sufficient; if $\boldsymbol{u}$ and $\boldsymbol{\varrho}$ are left-sided differentiable, then even the continuity of $\boldsymbol{v}$ suffices, where the Dini derivate $D^- \boldsymbol{v}$ or $D_- \boldsymbol{v}$ is to be substituted for $\boldsymbol{v}'$.

Theorems on error bounds and uniqueness for systems which are based on a certain estimate of $f_\nu(t, \boldsymbol{z}) - f_\nu(t, \bar{\boldsymbol{z}})$, for vectors $\boldsymbol{z}, \bar{\boldsymbol{z}}$ with $z_\nu \geqq \bar{z}_\nu$, were first given by M. Müller (1927)[2].

[1] Whether we put absolute values to the left of the $\leqq$ or omit them is of no import. We could exchange $\boldsymbol{z}$ and $\bar{\boldsymbol{z}}$, whereby the left hand side is multiplied by -1, but the right hand side remains the same.

[2] This work seems to be rather unknown according to a perusal of more recent works.

III. Example. Let us consider the initial value problem

$$u_1' = -2u_1 + g(t, u_2), \quad u_1(0) = \eta_1$$
$$u_2' = h(t, u_1) - 2u_2, \quad u_2(0) = \eta_2 .$$

Here suppose the functions g, h satisfy a Lipschitz condition with Lipschitz constant 1:

$$|g(t, z) - g(t, \bar{z})| \leqq |z - \bar{z}|, \quad |h(t, z) - h(t, \bar{z})| \leqq |z - \bar{z}| .$$

If v is a given approximate solution for which the following data are known,

$$|v_1' + 2v_1 - g(t, v_2)| \leqq \delta_1, \quad |v_1(0) - \eta_1| \leqq \varepsilon_1$$
$$|v_2' - h(t, v_1) + 2v_2| \leqq \delta_2, \quad |v_2(0) - \eta_2| \leqq \varepsilon_2$$

then from II with

$$\omega_1(t, z_1, z_2) = -2z_1 + z_2, \quad \omega_2(t, z_1, z_2) = z_1 - 2z_2$$

follows the inequality

$$|v_1 - u_1| \leqq \tfrac{1}{6}[3(\varepsilon_1 + \varepsilon_2 - \delta_1 - \delta_2)e^{-t} + (3\varepsilon_1 - 3\varepsilon_2 + \delta_2 - \delta_1)e^{-3t} + 4\delta_1 + 2\delta_2]$$
$$|v_2 - u_2| \leqq \tfrac{1}{6}[3(\varepsilon_1 + \varepsilon_2 - \delta_1 - \delta_2)e^{-t} + (3\varepsilon_2 - 3\varepsilon_1 + \delta_1 - \delta_2)e^{-3t} + 2\delta_1 + 4\delta_2].$$

On the other hand an estimate in connection with the requirement

$$|f(t, z) - f(t, \bar{z})| \leqq \omega(t, |z - \bar{z}|) \quad \text{for all} \quad z, \bar{z}$$

in this example would necessitate solving the system

$$\sigma_1' = 2\sigma_1 + \sigma_2 + \delta_1, \quad \sigma_2' = \sigma_1 + 2\sigma_2 + \delta_2$$

under the initial condition $\sigma_1(0) = \varepsilon_1, \sigma_2(0) = \varepsilon_2$ and would yield the significantly less favorable bounds

$$|v_1 - u_1| \leqq \tfrac{1}{6}[3(\varepsilon_1 - \varepsilon_2 + \delta_1 - \delta_2)e^{t} + (3\varepsilon_1 + 3\varepsilon_2 + \delta_1 + \delta_2)e^{3t} - 4\delta_1 + 2\delta_2]$$
$$|v_2 - u_2| \leqq \tfrac{1}{6}[3(\varepsilon_2 - \varepsilon_1 + \delta_2 - \delta_1)e^{t} + (3\varepsilon_1 + 3\varepsilon_2 + \delta_1 + \delta_2)e^{3t} + 2\delta_1 - 4\delta_2].$$

IV. The Lipschitz Condition. Suppose f satisfies the following Lipschitz condition

$$f_\nu(t, z) - f_\nu(t, \bar{z}) \leqq l_{\nu 1}(t)|z_1 - \bar{z}_1| + \cdots + l_{\nu n}(t)|z_n - \bar{z}_n| \quad \text{if} \quad z_\nu \geqq \bar{z}_\nu$$

and $(t, z), (t, \bar{z}) \in D(f)$ $(\nu = 1, \ldots, n)$. Here we let $l_{\mu\nu}$ be functions which are continuous in J_0 and integrable over J (instead of this the condition 9 VI (γ) will suffice) and $l_{\mu\nu}(t) \geqq 0$ for $\mu \neq \nu$. Then if $\boldsymbol{\sigma}(t)$ is the solution of

$$\sigma_\nu' = l_{\nu 1}(t)\sigma_1 + \cdots + l_{\nu n}(t)\sigma_n + \delta_\nu(t), \quad \sigma_\nu(0) = \varepsilon_\nu \quad (\nu = 1, \ldots, n),$$

the inequality

$$|\boldsymbol{v}(t)-\boldsymbol{u}(t)| \leqq \boldsymbol{\sigma}(t) \quad \text{in} \quad J$$

holds between the solution $\boldsymbol{u}$ of the initial value problem (1) and an approximate solution $\boldsymbol{v}(t)$ for which $|\boldsymbol{v}(0)-\boldsymbol{\eta}| \leqq \boldsymbol{\varepsilon}$, $|\boldsymbol{P}\boldsymbol{v}| \leqq \boldsymbol{\delta}(t)$.

For the special cases where the $l_{\mu\nu}$ and the δ_ν are constants, this was proved in a different way by Uhlmann (1957); see also Eltermann (1955) and Collatz (1955, pp. 108–109). Finally, it should be noted that this bound can also be derived from Theorem 6 VIII and thereby is also valid for $l_{\mu\nu}(t)$, $\delta_\nu(t) \in L(J)$ and $\boldsymbol{u}, \boldsymbol{v} \in Z_{ac}(\boldsymbol{f})$ (and indeed without an assumption about the sign of $l_{\nu\nu}$).

For the sake of completeness we now give the n-dimensional analog of 9 I.

V. Theorem. Suppose (for simplicity) that the functions $\boldsymbol{\omega}(t, \boldsymbol{z})$ and $\bar{\boldsymbol{\omega}}(t, \boldsymbol{z})$ are defined in the entire strip $J_0 \times E^n$ and are quasi-monotone increasing in $\boldsymbol{z}$. For the vector functions $\boldsymbol{u}, \boldsymbol{v} \in Z(\boldsymbol{f})$, $\boldsymbol{\varrho} \in Z(\boldsymbol{\omega})$, $\bar{\boldsymbol{\varrho}} \in Z(\bar{\boldsymbol{\omega}})$ and $\boldsymbol{\delta}(t)$, $\bar{\boldsymbol{\delta}}(t)$, suppose

(α) $-\bar{\boldsymbol{\varrho}}(0+) < (\boldsymbol{v}-\boldsymbol{u})(0+) < \boldsymbol{\varrho}(0+)$;
(β) $\boldsymbol{u}' = \boldsymbol{f}(t, \boldsymbol{u})$ and $-\bar{\boldsymbol{\delta}}(t) \leqq \boldsymbol{v}' - \boldsymbol{f}(t, \boldsymbol{v}) \leqq \boldsymbol{\delta}(t)$ in J_0;
(γ) $\boldsymbol{\varrho}' > \boldsymbol{\delta}(t) + \boldsymbol{\omega}(t, \boldsymbol{\varrho})$ and $\bar{\boldsymbol{\varrho}}' > \bar{\boldsymbol{\delta}}(t) + \bar{\boldsymbol{\omega}}(t, \bar{\boldsymbol{\varrho}})$ in J_0;
(δ) $f_\nu(t, \boldsymbol{v}) - f_\nu(t, \boldsymbol{v}-\boldsymbol{z}) \leqq \omega_\nu(t, \boldsymbol{\varrho})$ if $[-\bar{\boldsymbol{\varrho}}(t) \leqq] \boldsymbol{z} \leqq \boldsymbol{\varrho}(t)$, $z_\nu = \varrho_\nu(t)$,
$f_\nu(t, \boldsymbol{v}+\boldsymbol{z}) - f_\nu(t, \boldsymbol{v}) \leqq \bar{\omega}_\nu(t, \bar{\boldsymbol{\varrho}})$ if $[-\boldsymbol{\varrho}(t) \leqq] \boldsymbol{z} \leqq \bar{\boldsymbol{\varrho}}(t)$, $z_\nu = \bar{\varrho}_\nu(t)$

[and of course $(t, \boldsymbol{v}-\boldsymbol{z})$ or $(t, \boldsymbol{v}+\boldsymbol{z}) \in D(\boldsymbol{f})$].

Then

$$-\bar{\boldsymbol{\varrho}}(t) < \boldsymbol{v}(t) - \boldsymbol{u}(t) < \boldsymbol{\varrho}(t) \quad \text{in} \quad J_0 .$$

The bounding by $\boldsymbol{\varrho}$ is independent of the bounding by $\bar{\boldsymbol{\varrho}}$ if the expression in the square brackets in (δ) is removed. If bounding is done simultaneously by $\boldsymbol{\varrho}$ and $\bar{\boldsymbol{\varrho}}$, then the weaker hypothesis, where the square brackets in (δ) are taken into account, is sufficient.

This distinction of cases is completely analogous to that of 12 III and 12 IV. Let us prove the first inequality $-\bar{\boldsymbol{\varrho}} < \boldsymbol{v}-\boldsymbol{u}$, i.e., $\boldsymbol{u}-\boldsymbol{v} < \bar{\boldsymbol{\varrho}}$ independently of the other one. According to Lemma 12 I it suffices to show that $\boldsymbol{u}-\boldsymbol{v} \leqq \bar{\boldsymbol{\varrho}}$ and $u_\nu - v_\nu = \bar{\varrho}_\nu$ at the point t implies the inequality $u_\nu' - v_\nu' < \bar{\varrho}_\nu'$ at the same point t. This happens in the usual way with (β)–(δ):

$$u_\nu' - v_\nu' \leqq f_\nu(t, \boldsymbol{u}) - f_\nu(t, \boldsymbol{v}) + \bar{\delta}_\nu(t) \leqq \bar{\delta}_\nu(t) + \bar{\omega}_\nu(t, \bar{\boldsymbol{\varrho}}) < \bar{\varrho}_\nu' .$$

The second line of (δ) could be used because $\boldsymbol{u} = \boldsymbol{v}+\boldsymbol{z}$ and $\boldsymbol{u}-\boldsymbol{v} \leqq \bar{\boldsymbol{\varrho}}$, $u_\nu - v_\nu = \bar{\varrho}_\nu$ and thus $\boldsymbol{z} \leqq \boldsymbol{\varrho}$, $z_\nu = \bar{\varrho}_\nu$.

If we are faced with the case of simultaneous bounding by $\boldsymbol{\varrho}$ and $\bar{\boldsymbol{\varrho}}$, then a small additional consideration is involved at the beginning of the proof. We start with the largest interval $0 \leqq t \leqq T_1$ in which $-\bar{\boldsymbol{\varrho}} \leqq \boldsymbol{v}-\boldsymbol{u} \leqq \boldsymbol{\varrho}$. In this interval, as above, we then prove $-\bar{\boldsymbol{\varrho}} < \boldsymbol{v}-\boldsymbol{u}$ and then in the same way $\boldsymbol{v}-\boldsymbol{u} < \boldsymbol{\varrho}$. Hence we obtain $T_1 = T$ and the assertion. In using Lemma 12 I to prove $-\bar{\boldsymbol{\varrho}} < \boldsymbol{v}-\boldsymbol{u}$ for $0 \leqq t \leqq T_1$ we then have to show that $-\bar{\boldsymbol{\varrho}} \leqq \boldsymbol{v}-\boldsymbol{u} \leqq \boldsymbol{\varrho}$, $u_\nu - v_\nu = \bar{\varrho}_\nu$ (at the point t) implies the inequality $u_\nu' - v_\nu' < \bar{\varrho}_\nu'$ (at the same point). This happens as above, where (δ) is needed only in the form weakened by taking into account the square brackets.

The transition, analogous to Theorem 9 II, to the case where, instead of $\boldsymbol{\varrho}, \bar{\boldsymbol{\varrho}}$, bounds are obtained as maximal solutions $\boldsymbol{\sigma}, \bar{\boldsymbol{\sigma}}$ of the corresponding differential equations, is easy and is left to the reader.

The reason that this theorem has no great practical significance will be evident from a simple example. Let us assume that we are given a system in which the first of the n differential equations is $u_1' = -u_2 (f_1 = -z_2)$. For $\nu = 1$ the first line of (δ),

$$-v_2 + (v_2 - z_2) = -z_2 \leqq \omega_1(t, \varrho) \quad \text{for} \quad [-\bar{\varrho}_2(t) \leqq] z_2 \leqq \varrho_2(t),$$

can never be satisfied without taking account of the square brackets. [One could remedy this by restricting $D(f)$ but this would still yield a generally unfavorable ω_1.] Considering the case in square brackets, whether or not we can find a "reasonable" ω_1, depends on the size of ϱ_2 and $\bar{\varrho}_2$.

Theorem V is interesting in practice when $f(t, z)$ is quasi-monotone increasing in z. The condition (δ) then reads simply

$$f_\nu(t, \boldsymbol{v}) - f_\nu(t, \boldsymbol{v} - \boldsymbol{\varrho}) \leqq \omega_\nu(t, \boldsymbol{\varrho})$$
$$f_\nu(t, \boldsymbol{v} + \boldsymbol{\varrho}) - f_\nu(t, \boldsymbol{v}) \leqq \omega_\nu(t, \bar{\boldsymbol{\varrho}});$$

in other words: *If we have a system with right hand side which is quasi-monotone increasing in z, then we can read* 9 I *and moreover also* 9 II *"vectorially."*

14. Further Uniqueness Results for Systems

With the help of the results just obtained, we are now in a position to sharpen the uniqueness theory of Section 11 for the initial value problem

$$\boldsymbol{u}' = \boldsymbol{f}(t, \boldsymbol{u}) \quad \text{in} \quad J_0, \qquad \boldsymbol{u}(0) = \boldsymbol{\eta}. \tag{1}$$

We shall dwell briefly on this, since the way in which Theorem 13 I is used is transparent and completely analogous to the one-dimensional case handled in Section 10.

I. Definition ($\mathscr{E}_4^n - \mathscr{E}_7^n$). The function $\boldsymbol{\omega}(t, \boldsymbol{z})$ defined for $t \in J_0$, $\boldsymbol{z} \geqq \boldsymbol{0}$ and quasi-monotone increasing in $\boldsymbol{z}$ belongs to one of the classes $\mathscr{E}_4^n$ to $\mathscr{E}_7^n$ if it has the appropriate property given in 10 I resp. 10 V but with the change that now $\boldsymbol{\varepsilon} = (\varepsilon_1, \ldots, \varepsilon_n) > \boldsymbol{0}$, $\boldsymbol{\delta} = (\delta_1, \ldots, \delta_n) > \boldsymbol{0}$ and $\boldsymbol{\varrho}$ and $\boldsymbol{\omega}$ are vectors.

II. Uniqueness Theorem. *The initial value problem* (1) *has at most one solution* $\boldsymbol{u} \in Z(\boldsymbol{f})$, *and it depends continuously on the initial value [or on the initial value and the right hand side of the differential equation] if for* $\nu = 1, \ldots, n$ *the condition*

$$f_\nu(t, \boldsymbol{z}) - f_\nu(t, \bar{\boldsymbol{z}}) \leqq \omega_\nu(t, |\boldsymbol{z} - \bar{\boldsymbol{z}}|) \quad \text{for} \quad z_\nu \geqq \bar{z}_\nu \quad \text{and} \quad (t, \boldsymbol{z}), (t, \bar{\boldsymbol{z}}) \in D(\boldsymbol{f}) \tag{2}$$

with $\boldsymbol{\omega} \in \mathscr{E}_5^n$ [or $\boldsymbol{\omega} \in \mathscr{E}_4^n$] *holds. If the solutions also satisfy the differential equation for* $t = 0$, *then* $\boldsymbol{\omega} \in \mathscr{E}_6^n$ *or* $\mathscr{E}_7^n$ *is sufficient for uniqueness.*

The continuous dependence was defined in 11 II.

The proof is obtained without difficulty from 13 I by applying this theorem to a solution $\boldsymbol{u}$ and an approximate solution $\boldsymbol{v}$. The hypothesis

13 I (α) is satisfied in the cases $\boldsymbol{\omega} \in \mathscr{E}_4^n$, $\boldsymbol{\omega} \in \mathscr{E}_5^n$. In the other two cases $\boldsymbol{v}$ is a solution and we have to proceed as in 10 II and 10 VI.

The Remarks 10 III can be carried over without change to systems; they lead to classes $\mathscr{E}^n[o(\boldsymbol{g}(t))], \ldots$. We do not go into this any further since there are no difficulties.

III. Examples. Suppose the assumptions on $l(t)$ and $\psi(t)$ given at the beginning of 10 IV hold.

(α) If moreover $l(t) \in L(J)$, then the function $\boldsymbol{\omega}$ defined by

$$\omega_\nu(t, \boldsymbol{z}) = l(t)\,\psi(z_1 + \cdots + z_n) \qquad (\nu = 1, \ldots, n)$$

is of class $\mathscr{E}_4^n$. This criterion reduces to 10 IV (α) for $n = 1$.

Here we obtain suitable functions $\boldsymbol{\varrho}$ as in 10 IV (α), by determining $\varphi(t)$ as a solution of the differential equation $\varphi' = [l(t) + 1]\,\psi(n\varphi)$ with $\varphi(T) = \varepsilon > 0$ and setting $\varrho_\nu = \varphi$ for all ν. Then we have $\varphi > 0$ in J and

$$\varrho_\nu' = [l(t) + 1]\,\psi(\varrho_1 + \cdots + \varrho_n) > l(t)\;\psi(\varrho_1 + \cdots + \varrho_n) + \delta$$

for a positive δ.

(β) Let us now consider the more general form

$$\omega_\nu(t, \boldsymbol{z}) = l_{\nu 1}(t)\psi_{\nu 1}(z_1) + \cdots + l_{\nu n}(t)\psi_{\nu n}(z_n)$$

under the assumption that the $l_{\mu\nu} \geqq 0$ and the $\psi_{\mu\nu}$ have the properties of l and ψ in (α). If all $\psi_{\mu\nu}$ are equal, then this is trivially a uniqueness condition, since $\boldsymbol{\omega} \leqq \boldsymbol{\omega}^* \in \mathscr{E}_4^n$ if we set $\omega_\nu^*(t, \boldsymbol{z}) = l(t)\,\psi(z_1 + \cdots + z_n)$ with $\psi_{\mu\nu} = \psi$, $l(t) = n \max l_{\mu\nu}(t)$. It was stated in the German edition of this book that this is *not* true in the general case. But the hints given there for the construction of a counterexample lead to difficulties, so the question is open.

(γ) *Generalized Nagumo condition.* The two function $\boldsymbol{\omega}(t, \boldsymbol{z})$ defined by

$$\omega_\nu(t, \boldsymbol{z}) = \frac{1}{t}\,(a_{\nu 1} z_1 + \cdots + a_{\nu n} z_n)$$

and

$$\omega_\nu(t, \boldsymbol{z}) = \frac{1}{t}\,(a_{\nu 1} z_1 + \cdots + a_{\nu n} z_n) + l(t)(z_1 + \cdots + z_n)$$

are in the class $\mathscr{E}_6^n$ if the following holds:

The $n \times n$ matrix $A = (a_{\mu\nu})$ has only non-negative elements off the main diagonal, $a_{\mu\nu} \geqq 0$ for $\mu \neq \nu$, it is irreducible and the eigenvalue λ with the greatest real part (which is real according to the theorem of Frobenius) has the value $\lambda = 1$. For $l(t)$, 9 VI (γ) holds.

Approximate functions $\boldsymbol{\varrho}(t)$ can be constructed exactly as in 6 VI (β). The only thing new in this condition as compared to 6 VI (β) is that the sign of the diagonal elements $a_{\nu\nu}$ is now arbitrary.

IV. Supplementary Remarks, Conditions of the Krasnosel'skii-Krein Type. The numerous uniqueness criteria of the first two chapters should not be misconstrued as presenting a summary; they represent only a modest selection from the literature on the uniqueness problem. This problem, which is so simple at first glance and does not show its maliciousness until closer examination, seems to be especially enticing to a great many mathematicians. The wealth of ideas and theorems which have been and will be generated in this connection provides eloquent testimony for the (occasionally questioned by others) imagination of mathematicians.

(α) A new idea which has its beginnings in the work of Krasnosel'skii and Krein (1956), and was further pursued by Luxemburg (1958), Kooi (1958) and Brauer (1959, 1959a), should also be mentioned. Instead of *one* inequality, say $\|f(t, z)-f(t, \bar{z})\| \leqq \omega(t, \|z-\bar{z}\|)$, we use *two* inequalities

$$\|f(t, z)-f(t, \bar{z})\| \leqq \omega_i(t, \|z-\bar{z}\|) \quad (i=1,2) \tag{3}$$

to be satisfied simultaneously. Krasnosel'skii and Krein considered the case

$$\|f(t, z)-f(t, \bar{z})\| \leqq \begin{cases} C\|z-\bar{z}\|^{\alpha} \\ k\dfrac{\|z-\bar{z}\|}{t} \end{cases} \tag{4}$$

where $0<\alpha<1$, $k(1-\alpha)<1$. While it is not difficult to show by way of examples that one of the two inequalities (4) alone is not a uniqueness condition (if $k>1$), both together do imply uniqueness. If we define

$$\omega(t, z)=\min(Cz^{\alpha}, kz/t), \tag{5}$$

then the condition (4) is equivalent to

$$\|f(t, z)-f(t, \bar{z})\| \leqq \omega(t, \|z-\bar{z}\|), \tag{6}$$

i.e., to a familiar condition encountered in Section 11. An investigation of this function ω reveals the surprising fact:

The function ω given by (5) *belongs to $\mathscr{E}_4$; what is more, it belongs to the Bompiani class defined in* 10 IV (δ).

The proof is not difficult. It is easy to see, by putting $\omega(0, z)=0$, that ω is continuous in $J \times \{z \geqq 0\}$ and monotone increasing in z. According to 10 IV (δ) we have to show that a function σ which satisfies $\sigma'=\omega(t, \sigma)$ in J, $\sigma(T)>0$, has a value $\sigma(0)>0$. Let φ be the solution of $\varphi'=k\varphi/t$ in J_0, $\varphi(T)=\sigma(T)$. Since $\sigma' \leqq k\sigma/t$, we have $\varphi \leqq \sigma$ in J_0 by 8 X or 8 XI (note that inequality signs are reversed in the differential inequality since we are arguing "to the left"). Similarly, if $t_0 \in J_0$, $\psi'=C\psi^{\alpha}$ in $0 \leqq t \leqq t_0$, $\psi(t_0)=\varphi(t_0) \leqq \sigma(t_0)$, then $\psi \leqq \sigma$ for $0 \leqq t \leqq t_0$.

Now,

$$\varphi(t)=At^k, \qquad A=\sigma(T)T^{-k}>0.$$

A solution ψ of $\psi'=C\psi^{\alpha}$ is given by

$$\psi(t)=B(t+\varepsilon)^b, \quad \text{where} \quad b=1/(1-\alpha)>k, \quad B^{1-\alpha}=C(1-\alpha).$$

Since $b > k$, there is a $t_0 \in J_0$ such that $Bt_0^b < At_0^k$ and consequently a positive ε such that $\varphi(t_0) = \psi(t_0)$. This together with the first part of the proof gives the desired inequality $\sigma(0) \geqq \psi(0) = B\varepsilon^b > 0$.

The above considerations show that the condition of Krasnosel'skii and Krein is *not* a uniqueness condition of a new type. It is rather an example (to be sure, a new example) of a well-known general uniqueness criterion. The same is true for the generalizations of the Krasnosel'skii-Krein condition mentioned above. The two inequalities (3) are equivalent to the inequality (6) if one defines

$$\omega(t, z) = \min(\omega_1(t, z), \omega_2(t, z)).$$

In all cases considered in the literature this function is of the Kamke type described in 10 VII, *i.e.*, $\omega \in \mathscr{E}_7$. The proof is found in Walter (1964).

In connection with this we mention a result of Olech (1960): If $\boldsymbol{f}$ is continuous in $J \times D$, where D is a compact set in E^n, and if $\boldsymbol{f}$ satisfies a condition (6) with a function ω of the Kamke type (10 VII) then it also satisfies the same condition with a function ω of the Perron type (10 IV (δ)). In other words, for continuous $\boldsymbol{f}$, Kamke's condition is no more general than Perron's condition. For the proof we refer to Olech (1960) or to Walter (1964), where a number of other relations between uniqueness criteria are found.

(β) By means of a general distance function $V(t, z)$, another type of uniqueness theorem is obtained. Let $V(t, z)$ be a real-valued function of class $C^1(J \times E^n)$ (the conditions at the end of 11 VI actually suffice) which satisfies $V(t, \mathbf{0}) = 0$, $V(t, z) > 0$ for $z \neq \mathbf{0}$. The estimation theorem 11 VI immediately yields the following result:

If for two solutions $\boldsymbol{u}, \boldsymbol{v}$ of the initial value problem (1)

$$V_t(t, \boldsymbol{v} - \boldsymbol{u}) + V_z(t, \boldsymbol{v} - \boldsymbol{u}) \cdot [\boldsymbol{f}(t, \boldsymbol{v}) - \boldsymbol{f}(t, \boldsymbol{u})] \leqq \omega(t, V(t, \boldsymbol{v} - \boldsymbol{u})),$$

where $\omega \in \mathscr{E}_6$, then $\boldsymbol{u} = \boldsymbol{v}$.

Uniqueness theorems of this kind were given by Okamura (1942), Moyer (1966), George (1967); see also Heins (1963).

15. Differential Equations of Higher Order

The initial value problem for a differential equation of nth order ($n \geqq 1$)

$$u^{(n)} = f(t, u, u', \ldots, u^{(n-1)}) \tag{1}$$

$$u(0) = \eta_0, u'(0) = \eta_1, \ldots, u^{(n-1)}(0) = \eta_{n-1} \tag{2}$$

can be converted by setting

$$u(t) = u_1(t), u'(t) = u_2(t), \ldots, u^{(n-1)}(t) = u_n(t) \tag{3}$$

into an initial value problem for a system of differential equations of first order

$$\left.\begin{aligned} u_1' = u_2, u_2' = u_3, \ldots, u_{n-1}' = u_n \\ u_n' = f(t, u_1, \ldots, u_n) \end{aligned}\right\} \tag{4}$$

$$u_1(0) = \eta_0, \ldots, u_n(0) = \eta_{n-1} \tag{5}$$

in familiar fashion. Thus all the assertions for systems of differential equations of first order are also assertions for differential equations of nth order. In several places we naturally find peculiarities due to the special form of the system (4). Before treating these, we translate some earlier notation in accordance with the present circumstances.

I. Definition ($Z(f)$, *quasi-monotone increasing in* z, $f \in \mathcal{F}^n, \mathcal{F}_0^n$, *summation convention for* α, β). Suppose the function $f(t, z_1, \ldots, z_n) = f(t, z)$ is defined in a domain $D(f) \subset E^{n+1}$ (we note that the index for z_ν runs, as before, from 1 to n and hence the variable z_ν corresponds to the derivative $u^{(\nu-1)}$). If the function $\varphi(t)$ is $(n-1)$ times continuously differentiable in J, if $\varphi^{(n-1)}$ is differentiable in J_0 and all points $(t, \varphi(t), \ldots, \varphi^{(n-1)}(t))$ for $t \in J_0$ are in $D(f)$, then $\varphi \in Z(f)$. The function $f(t, z)$ is in the class $\mathcal{F}^n$ or $\mathcal{F}_0^n$ if the right hand side of (4), i.e., the vector $\mathbf{f}$ with $f_\nu = z_{\nu+1}$ $(\nu = 1, \ldots, n-1)$ and $f_n = f$ has this property. Likewise, $f(t, z)$ is quasi-monotone increasing in z if this holds for the right hand side of (4), i.e., if

$$f(t, z) \leqq f(t, \bar{z}) \quad \text{for} \quad z \leqq \bar{z}, \quad z_n = \bar{z}_n.$$

Since indices will frequently run from 0 to $n-2$ or $n-1$, we make the following conventions in this section: The index α always runs through the numbers 0 to $n-1$, the index β through the numbers 0 to $n-2$. In this notation, for example, the initial condition (2) reads simply $u^{(\alpha)}(0) = \eta_\alpha$.

II. Preliminary Remarks. Since we accomplish the transition from the differential equation (1) to the system (4) according to (3), we deal with the special vectors $\boldsymbol{\varphi} = (\varphi_1, \ldots, \varphi_n) = (\varphi, \varphi', \ldots, \varphi^{(n-1)})$ and $\boldsymbol{\psi} = (\psi_1, \ldots, \psi_n) = (\psi, \psi', \ldots, \psi^{(n-1)})$ "generated" by the functions $\varphi(t), \psi(t)$ which are $(n-1)$ times continuously differentiable in J. If for such vectors $\boldsymbol{\varphi}, \boldsymbol{\psi}$

$$\varphi^{(\beta)}(0) \leqq \psi^{(\beta)}(0), \quad \varphi^{(n-1)}(0+) < \psi^{(n-1)}(0+), \tag{6}$$

and if $\varphi^{(n-1)}(t) < \psi^{(n-1)}(t)$ say for $0 < t < t_0$, then we have $\boldsymbol{\varphi} < \boldsymbol{\psi}$ in this interval since all derivatives of order $< n-1$ can be represented as integrals of the $(n-1)$st derivative. Let us now consider Lemma 12 I, or more precisely its application to these special vectors $\boldsymbol{\varphi}, \boldsymbol{\psi}$. If (6) holds and we do *not* have the case 12 I (α) "$\boldsymbol{\varphi} < \boldsymbol{\psi}$ in J_0," then we have the equality sign first for the nth component, i.e., there then exists a t_0 with $\varphi < \psi$ in $0 < t < t_0$, $\varphi^{(n-1)}(t_0) = \psi^{(n-1)}(t_0)$.

This remark yields the following

Lemma. *Let φ, ψ be $(n-1)$ times continuously differentiable in J. Suppose that $\varphi^{(\beta)}(t_0) \leqq \psi^{(\beta)}(t_0)$, $\varphi^{(n-1)}(t_0) = \psi^{(n-1)}(t_0)$ $(t_0 \in J_0)$ implies*

$$D_-\varphi^{(n-1)}(t_0) < D_-\psi^{(n-1)}(t_0) \quad \text{or} \quad D^-\varphi^{(n-1)}(t_0) < D^-\psi^{(n-1)}(t_0)$$

(i.e., $\varphi^{(n)}(t_0) < \psi^{(n)}(t_0)$ if the derivatives exist).

Then we have precisely one of the following two cases:

(α) $\varphi^{(\alpha)} < \psi^{(\alpha)}$ in J_0;

(β) (6) does not hold.

This lemma is the analog of 12 I for the present case. Since the theorems of Section 12 were proved by going back to 12 I, in converting these theorems to the special system (4) we have to assume the corresponding differential inequality — that is, the hypothesis (β) of the corresponding theorem — only for the index $v = n$.

On the basis of this observation we obtain the following theorems automatically from those of Section 12.

III. Theorem. *Suppose the functions v, w are from $Z(f)$, and*

(α) $v^{(\beta)}(0) \leqq w^{(\beta)}(0)$, $\quad v^{(n-1)}(0+) < w^{(n-1)}(0+)$;

(β) $v^{(n)} - f(t, z_1, \ldots, z_n) < w^{(n)} - f(t, w, \ldots, w^{(n-1)})$ or

$\quad v^{(n)} - f(t, v, \ldots, v^{(n-1)}) < w^{(n)} - f(t, z_1, \ldots, z_n)$

at least for those points $t \in J_0$ and vectors $\mathbf{z}$ for which simultaneously

$$v^{(\beta)}(t) \leqq z_{\beta+1} \leqq w^{(\beta)}(t), \quad v^{(n-1)}(t) = z_n = w^{(n-1)}(t) \quad \text{and} \quad (t, \mathbf{z}) \in D(f).$$

Then

$$v^{(\alpha)}(t) < w^{(\alpha)}(t) \quad \text{in} \quad J_0.$$

IV. Super- and Subfunctions. We omit the conversion of 12 III and restrict ourselves to the simultaneous estimate from above and below.

Suppose v, w are two functions from $Z(f)$ which satisfy the hypotheses:

(α) $v^{(\beta)}(0) \leqq \eta_\beta \leqq w^{(\beta)}(0)$, $\quad v^{(n-1)}(0+) < u^{(n-1)}(0+) < w^{(n-1)}(0+)$

for every solution u of the initial value problem;

(β) $v^{(n)} < f(t, z_1, \ldots, z_{n-1}, v^{(n-1)})$ and $w^{(n)} > f(t, z_1, \ldots, z_{n-1}, w^{(n-1)})$,

if

$$v^{(\beta)}(t) \leqq z_{\beta+1} \leqq w^{(\beta)}(t)$$

and $(t, z_1, \ldots, z_{n-1}, v^{(n-1)})$ or $(t, z_1, \ldots, z_{n-1}, w^{(n-1)}) \in D(f)$.

Then we have

$$v^{(\alpha)} < u^{(\alpha)} < w^{(\alpha)} \quad \text{in } J_0 \text{ for every solution } u \text{ of the initial value problem.} \tag{7}$$

Of course the hypotheses (β) are again considerably simplified if f has monotonicity properties. As in Section 12 we restrict consideration to the quasi-monotone case.

V. Theorem. *If $f(t, z)$ is quasi-monotone increasing in z and $v, w \in Z(f)$, then from*

(α) $v^{(\beta)}(0) \leqq w^{(\beta)}(0)$, $\quad v^{(n-1)}(0+) < w^{(n-1)}(0+)$;

(β) $Pv < Pw$ *in* J_0

follow the inequalities

$$v^{(\alpha)}(t) < w^{(\alpha)}(t) \quad in \quad J_0 .$$

Here

$$Pv = v^{(n)} - f(t, v, \ldots, v^{(n-1)}) . \tag{8}$$

VI. Super- and Subfunctions. If $f(t, z)$ is quasi-monotone increasing in z, then $v \in Z(f)$ is a subfunction, $w \in Z(f)$ a superfunction, if

(α) $v^{(\beta)}(0) \leqq \eta_\beta \leqq w^{(\beta)}(0)$, $v^{(n-1)}(0+) < u^{(n-1)}(0+) < w^{(n-1)}(0+)$

for every solution u of the initial value problem;

(β) $v^{(n)} < f(t, v, \ldots, v^{(n-1)})$ or $w^{(n)} > f(t, w, \ldots, w^{(n-1)})$ in J_0.

Then (7) holds. The two bounds are independent of each other.

VII. Maximal and Minimal Solutions. According to 12 VII the initial value problem (1) (2) has a solution which can be continued to the right up to the boundary of $D(f)$ if $f \in \mathscr{F}_0^n$. If it exists in all of J and if it is uniquely determined, then the solutions of the initial value problems

$$\varphi^{(n)} = f(t, \varphi, \ldots, \varphi^{(n-1)}) + \varepsilon, \qquad \varphi^{(\alpha)}(0) = \eta_\alpha + \delta_\alpha , \tag{9}$$

as $\varepsilon \to 0$, $\delta_\alpha \to 0$, approach u uniformly in J (the derivatives up to order $n-1$ also converge uniformly).

A maximal solution u^* and a minimal solution u_* for the initial value problem (1) (2) exist, according to 12 VIII, if $f \in \mathscr{F}_0^n$ is quasi-monotone increasing in z. For every other solution u we have

$$u_*^{(\alpha)}(t) \leqq u^{(\alpha)}(t) \leqq u^{*(\alpha)}(t) .$$

The maximal solution can be approximated from above by superfunctions which are solutions of (9) with $\varepsilon > 0$, $\delta_\alpha > 0$.

VIII. Remarks. (α) As before, the hypotheses on the nth derivative can be weakened. It can be replaced everywhere by the left sided nth derivative. For the subfunctions even $D_- v^{(n-1)}$ instead of $v^{(n)}$ and for the superfunctions $D^- w^{(n-1)}$ instead of $w^{(n)}$ are permitted.

(β) As was already noted in 12 X (α), in Theorems III to VI equality signs are permitted in the appropriate places if f satisfies a uniqueness condition. We confine ourselves to formulating V for this case.

Suppose that in its domain of definition the function f satisfies the inequality (11) with a function $\omega \in \overline{\mathscr{E}}_5^n$ and that it is quasi-monotone increasing in z. Suppose the functions v, w are from the class $Z(f)$.

Then we have the theorem

Va: *$v^{(\alpha)}(0) \leqq w^{(\alpha)}(0)$ and $Pv \leqq Pw$ in J implies $v^{(\alpha)}(t) \leqq w^{(\alpha)}(t)$ in J.*

We come to the theorems of Sections 13 and 14.

IX. Theorem. Suppose $u, v \in Z(f)$ and $\delta(t)$ is a function defined in J. Suppose further that $\omega(t, z)$ is quasi-monotone increasing in z, $\varrho = (\varrho_1, \ldots, \varrho_n)$ is a vector function continuous in J and differentiable in J_0 and $(t, z) \in D(\omega)$ for $t \in J_0$, $0 \leqq z_\alpha \leqq \varrho_\alpha(t)$, $z_n = \varrho_n(t)$. Suppose

(α) $|v^{(\beta)}(0) - u^{(\beta)}(0)| \leqq \varrho_{\beta+1}(0)$, $\quad |v^{(n-1)} - u^{(n-1)}|(0+) < \varrho_n(0+)$;

(β) $u^{(n)} = f(t, u, \ldots, u^{(n-1)})$ and $|Pv| \equiv |v^{(n)} - f(t, v, \ldots, v^{(n-1)})| \leqq \delta(t)$ in J_0;

(γ) $\varrho'_\nu \geqq \varrho_{\nu+1} (1 \leqq \nu \leqq n-1)$ and $\varrho'_n > \omega(t, \varrho_1, \ldots, \varrho_n) + \delta(t)$ in J_0;

(δ) $f(t, z) - f(t, \bar{z}) \leqq \omega(t, |z - \bar{z}|)$ if $|z_{\beta+1} - \bar{z}_{\beta+1}| \leqq \varrho_{\beta+1}(t)$, $z_n - \bar{z}_n = \varrho_n(t)$ and $(t, z), (t, \bar{z}) \in D(f)$.

Then

$$|v^{(\alpha)}(t) - u^{(\alpha)}(t)| < \varrho_{\alpha+1}(t) \quad \text{in} \quad J_0 .$$

In particular, suppose that $\varrho(t)$ is a (scalar) function from $Z(\omega)$ and that

(α') $|v^{(\beta)}(0) - u^{(\beta)}(0)| \leqq \varrho^{(\beta)}(0)$, $\quad |v^{(n-1)} - u^{(n-1)}|(0+) < \varrho^{(n-1)}(0+)$;

(γ') $\varrho^{(n)} > \omega(t, \varrho, \ldots, \varrho^{(n-1)}) + \delta(t)$ in J_0

and that (β) and (δ) hold, where the "if" condition in (δ) now reads: $|z_{\beta+1} - \bar{z}_{\beta+1}| \leqq \varrho^{(\beta)}(t)$ and $z_n - \bar{z}_n = \varrho^{(n-1)}(t)$. Then

$$|v^{(\alpha)}(t) - u^{(\alpha)}(t)| < \varrho^{(\alpha)}(t) \quad \text{in} \quad J_0 .$$

The second part of the theorem represents a special case of the first. For if we put $\varrho = \varrho_1$ and use equalities in the first $n-1$ inequalities of (γ), then $\varrho_{\alpha+1} = \varrho^{(\alpha)}$ and $\varrho \in Z(\omega)$ (note that the higher derivatives of ϱ exist at zero).

The first part is reduced to 13 I by converting to systems via the transformation rule (3). It is *not* a special case of 13 I insofar as in the latter, in (γ), we have the $>$ symbol for all components; moreover (α) is changed.

First we show: If for a fixed ν $(1 \leqq \nu \leqq n-1)$

$$|v^{(\nu)} - u^{(\nu)}| < \varrho_{\nu+1} \quad \text{for} \quad 0 < t < t_0 , \tag{10}$$

then in the same interval we also have

$$|v^{(\nu-1)} - u^{(\nu-1)}| < \varrho_\nu .$$

For this we need the following fact: If the function $d(t)$ is continuous in J and differentiable in J_0 and if $d(0) \geqq 0$, $d'(t) > 0$ for $0 < t < t_0$, then $d(t) > 0$ for $0 > t > t_0$ (proof by the mean value theorem). Let us apply this now to the two functions $d = \varrho_\nu \pm (v^{(\nu-1)} - u^{(\nu-1)})$. By ($\alpha$) we have $d(0) \geqq 0$, and further by (10) and (γ)

$$d' = \varrho'_\nu \pm (v^{(\nu)} - u^{(\nu)}) \geqq \varrho_{\nu+1} \pm (v^{(\nu)} - u^{(\nu)}) > 0 \quad \text{for} \quad 0 < t < t_0,$$

i.e., $d(t) > 0$ as asserted. Hence it now follows that, first, (10) holds for $0 \leqq \nu \leqq n-1$ and second, if the assertion is false, we have the equality $|v^{(\nu)}(t) - u^{(\nu)}(t)| = \varrho_{\nu+1}(t)$ for the first time for $\nu = n-1$. On the basis of these remarks (which are, moreover, already contained in essence in II) we can now take over the proof of 13 I; there we now have $\nu = n$, hence $f_\nu = f, v'_\nu = v^{(n)}, \ldots$.

It is left to the reader as an exercise to find the other form of this theorem, corresponding to 13 II, in which we have instead of a differential inequality in (γ') the corresponding differential equation. A sufficient condition for the validity of (δ) is

$$f(t, z) - f(t, \bar{z}) \leqq \omega(t, |z - \bar{z}|) \quad \text{if} \quad z_n \geqq \bar{z}_n \quad \text{and} \quad (t, z), (t, \bar{z}) \in D(f). \quad (11)$$

A similar theorem was proved for linear ω by Uhlmann (1957a).

We had a good reason for giving Theorem IX in an asymmetric form first, involving simultaneously differential equations of nth order and systems of differential equations or differential inequalities of first order. Namely, in some cases it is easy to satisfy the condition (γ) while this is difficult for (γ'); see Example XII.

For questions of boundedness of solutions, asymptotic behavior, and global existence we have at our disposal Theorem VI on super- and subfunctions and the estimation theorem IX. Furthermore it is possible to adapt Theorem 11 IX to the present situation. For second order equations there is an abundant literature on these problems. For higher order equations reference is made to Malcharek (1965), Waltman (1965), Radziszewski (1966). Švec (1966, 1966a) (among others).

Now we turn to the uniqueness problem. The peculiarity in the foregoing proof, which is due to the difference in the hypotheses (α) and (γ) in IX and 13 I, forces us to make insignificant alterations in the definition of the classes $\mathscr{E}_i^n$.

X. Definition ($\bar{\mathscr{E}}_4^n - \bar{\mathscr{E}}_6^n$). The function $\omega(t, z)$ belongs to the class $\bar{\mathscr{E}}_i^n$ if it is defined for $t \in J_0$, $z \geqq \mathbf{0}$, is quasi-monotone increasing in z, and has the property:

For every $\varepsilon > 0$ there is a $\delta > 0$ and a vector function $\varrho(t) = (\varrho_1, \ldots, \varrho_n)$ continuous in J and differentiable in J_0 such that $\varrho'_\nu \geqq \varrho_{\nu+1}$ in J_0 for $\nu = 1, \ldots, n-1$ and

$\bar{\mathscr{E}}_4^n$: $\varrho'_n > \omega(t, \varrho_1, \ldots, \varrho_n) + \delta$ and $\delta \leqq \varrho_n(t) \leqq \varepsilon$ in J_0, $\varrho_\nu(0) \geqq \delta$ $(1 \leqq \nu \leqq n-1)$

$\bar{\mathscr{E}}_5^n$: $\varrho'_n > \omega(t, \varrho_1, \ldots, \varrho_n)$ and $\delta \leqq \varrho_n(t) \leqq \varepsilon$ in J_0, $\varrho_\nu(0) \geqq \delta$ $(1 \leqq \nu \leqq n-1)$

$\bar{\mathscr{E}}_6^n$: $\varrho'_n > \omega(t, \varrho_1, \ldots, \varrho_n)$ and $\delta t \leqq \varrho_n(t) \leqq \varepsilon$ in J_0, $\varrho_\nu(0) \geqq 0$ $(1 \leqq \nu \leqq n-1)$.

XI. Uniqueness Theorem. *Suppose the function $f(t,z)$ satisfies condition (11) with a function $\omega \in \bar{\mathscr{E}}_5^n$ [resp. $\omega \in \bar{\mathscr{E}}_4^n$]. Then the initial value problem (1) (2) has at most one solution $u \in Z(f)$ and it depends continuously on the initial values η_α [resp. on the initial values η_α and on the right hand side of the differential equation]. If f is continuous at the point $(0, \eta_0, \ldots, \eta_{n-1})$, then $\omega \in \bar{\mathscr{E}}_6^n$ is sufficient for uniqueness.*

This follows in familiar fashion from the previous theorem. Moreover, we can get along with the uniqueness theorem 14 II in most cases. For if $\boldsymbol{\omega}(t, z) = (\omega_1, \ldots, \omega_n)$ is in the class $\mathscr{E}_i^n$ and we have $\omega_\nu(t, z) \geqq z_{\nu+1}$ for $\nu = 1, \ldots, n-1$, then $\omega_n \in \bar{\mathscr{E}}_i^n$. In particular, the Lipschitz condition and Montel condition 14 III (α) are applicable [the latter when $\psi(z) \geqq z$]. In the following example we give two sharper criteria not contained in Section 14.

XII. Example *(generalized Nagumo condition)*. The function

$$\omega(t, z) = \sum_{\nu=1}^{n} \frac{\alpha_\nu z_\nu}{t^{n+1-\nu}} \quad \text{with} \quad \sum_{\nu=1}^{n} \frac{\alpha_\nu}{(n+1-\nu)!} \leqq 1 \quad \text{and} \quad \alpha_\nu \geqq 0 \text{ for } 1 \leqq \nu \leqq n-1$$

(the sign of α_n is arbitrary) belongs to $\bar{\mathscr{E}}_6^n$, and the same also holds for

$$\omega(t, z) = \sum_{\nu=1}^{n} \frac{\alpha_\nu z_\nu}{t^{n+1-\nu}} + l(t)(z_1 + \cdots + z_n),$$

provided $l(t) \geqq 0$ is continuous in J_0 and integrable over J.

To prove the second assertion we define functions $\varrho_1, \ldots, \varrho_n$ by

$$\varrho_\nu(t) = C \frac{t^{n+1-\nu}}{(n+1-\nu)!} e^{kL(t)}, \quad \text{where} \quad L(t) = \int_0^t l(\tau)\,d\tau,\ C > 0$$

and k is a positive constant still to be determined. As we see immediately, $\varrho'_\nu \geqq \varrho_{\nu+1}$ for $\nu = 1, \ldots, n-1$. The remaining inequality

$$\varrho'_n = C e^{kL}(1 + ktl) > C e^{kL} \sum_\nu \frac{\alpha_\nu}{(n+1-\nu)!} + C e^{kL} \sum_\nu \frac{t^{n+1-\nu}}{(n+1-\nu)!}$$

holds if

$$k > \sum_\nu \frac{t^{n-\nu}}{(n+1-\nu)!}, \quad \text{thus if} \quad k = 1 + \sum_\nu \frac{T^{n-\nu}}{(n+1-\nu)!}.$$

This example generalizes a criterion of Wintner (1956). A criterion similar to the one above was given by Kikodze (1966).

XIII. The Blasius Equation. (α) In flow over a flat plate the following initial value problem

$$u''' = uu'', \qquad u(0) = u'(0) = 0, \qquad u''(0) = 1 \tag{12}$$

arises [Blasius (1908)]. The equation is usually given in the form $u''' + uu'' = 0$, which results from ours if we replace t by $-t$. We want to compute the solution for positive t and in particular the value of the point T_a where the solution becomes infinite.

According to VI and VIII (β), conditions for subfunctions v and superfunctions w are

$$\left.\begin{array}{llll} v(0) \leqq 0, & v'(0) \leqq 0, & v''(0) \leqq 1, & v''' \leqq v v'' \\ w(0) \geqq 0, & w'(0) \geqq 0, & w''(0) \geqq 1, & w''' \geqq w w''. \end{array}\right\} \tag{13}$$

Here we must restrict consideration to comparison functions with nonnegative second derivatives, since $f(t, z_1, z_2, z_3) = z_1 z_3$ is monotone increasing in z_1 only for $z_3 \geqq 0$.

A solution of the differential equation is given by

$$\varphi(t) = \frac{3}{\alpha - t} \quad \text{for arbitrary } \alpha. \tag{14}$$

This function φ is a superfunction for the initial value problem (12) if $\varphi''(0) = 6/\alpha^3 \geqq 1$, and thus $\alpha \leqq \sqrt[3]{6}$. The simplest subfunction is $t^2/2$. Thus with almost no computation we obtain

$$v_1 = \frac{t^2}{2} < u(t) < w_1 = \frac{3}{\sqrt[3]{6} - t} \quad \text{and} \quad T_a \geqq \sqrt[3]{6} > 1.8\,.$$

Now we shall discuss two possibilities for obtaining more exact bounds.

(β) *A first "Ansatz"*. The power series expansion (only exponents $2 + 3k$ occur)

$$u(t) = \sum_{k=0}^{\infty} \frac{a_k}{3^k} t^{2+3k} \tag{15}$$

with

$$a_0 = \frac{1}{2}, \quad a_1 = \frac{1}{40}, \quad a_2 = \frac{11}{40 \cdot 112}, \quad a_3 = \frac{5}{22 \cdot 8 \cdot 112}, \ldots$$

yields usable subfunctions for small t ($a_k > 0$ follows immediately from the recursion formula for a_k, i.e., finite segments of the power series represent subfunctions). However, no upper bound for T_a can be obtained from this. In order to obtain one, we use the power series for small positive t and then the function φ defined in (14). Then the Ansatz is, for example,

$$v_2(t) = \begin{cases} \dfrac{1}{2} t^2 + \dfrac{1}{120} t^5 + \dfrac{11}{360 \cdot 112} t^8 & \text{for } 0 \leqq t \leqq t_0\,, \\[2ex] \dfrac{3}{\alpha - t} & \text{for } t_0 < t < \alpha\,. \end{cases}$$

Here we must assume the additional conditions[1]

$$v_2(t_0) \geqq \frac{3}{\alpha - t_0}, \quad v_2'(t_0) \geqq \frac{3}{(\alpha - t_0)^2}, \quad v_2''(t_0) \geqq \frac{6}{(\alpha - t_0)^3}\,. \tag{16}$$

If we choose, say, $t_0 = 2$, then $\alpha = 3.284$ is admissible.

---

[1] Here of course $v_2(t_0)$, $v_2'(t_0)$, ... denote the values which result from the *first* line in the definition of v_2.

The corresponding Ansatz for superfunctions

$$w_2(t)=\begin{cases}\frac{1}{2}t^2+\frac{1}{120}t^5+\beta t^8 & \text{for} \quad 0\leqq t\leqq t_0\\ \frac{3}{\alpha-t} & \text{for} \quad t_0<t<\alpha\end{cases}$$

requires more effort, since we have to determine the constant β such that the first line represents a superfunction for $0\leqq t\leqq t_0$ *before* verifying the additional conditions

$$w_2(t_0)\leqq\frac{3}{\alpha-t_0},\qquad w_2'(t_0)\leqq\frac{3}{(\alpha-t_0)^2},\qquad w_2''(t_0)\leqq\frac{6}{(\alpha-t_0)^3}. \tag{16'}$$

Thus we must have the inequality

$$\frac{1}{2}t^2+336\beta t^5\geqq\left(\frac{1}{2}t^2+\frac{1}{120}t^5+\beta t^8\right)\left(1+\frac{1}{6}t^3+56\beta t^6\right) \quad \text{for} \quad 0\leqq t\leqq t_0$$

which is equivalent to

$$336\beta\geqq\frac{11}{120}+t_0^3\left(29\beta+\frac{1}{720}\right)+\frac{19}{30}\beta t_0^6+56\beta^2t_0^9. \tag{17}$$

This cannot be satisfied for $t_0=2$, while for $t_0=1$ we obtain $\beta=3.038\cdot 10^{-4}$ and hence $\alpha=2.695$.

The function $f(t,z_1,z_2,z_3)=z_1z_3$ corresponding to the equation $u'''=uu''$ is monotone increasing in all variables z_ν (if we restrict consideration to $z_1\geqq 0$, $z_3\geqq 0$). In such a case it is always advantageous to go from the differential equation to the corresponding integral equation, which then contains a monotone increasing integral operator. For problem (12) the equivalent integral equation for $U(t)=u''(t)$ is

$$U(t)=1+KU=1+\int_0^t U(\tau)u(\tau)d\tau \quad \text{with} \quad u(t)=\int_0^t(t-\tau)U(\tau)d\tau. \tag{18}$$

The operator K (if its domain of definition is restricted to $U\geqq 0$) is monotone increasing in the sense of 1 IV. Subfunctions $V(t)$ and superfunctions $W(t)$ can thus be characterized by

$$V\leqq 1+KV,\qquad W\geqq 1+KW.$$

From this point of view, let us consider the above function w_2 or more precisely its second derivative $W_2(t)=1+\frac{1}{6}t^3+56\beta t^6$ in the interval $0\leqq t\leqq t_0$. The condition to be imposed on β

$$W_2\geqq 1+KW_2=1+\int_0^t\left(1+\frac{\tau^3}{6}+56\beta\tau^6\right)\left(\frac{\tau^2}{2}+\frac{\tau^5}{120}+\beta\tau^8\right)d\tau$$

for $0\leqq t\leqq t_0$ is equivalent to

$$336\beta\geqq\frac{11}{120}+\frac{2}{3}\left(29\beta+\frac{1}{720}\right)t_0^3+\frac{19}{60}\beta t_0^6+\frac{112}{5}\beta^2t_0^9 \tag{19}$$

and is hence weaker than (17). It follows that

$$\text{for} \quad t_0=1: \quad \beta=2.793\cdot 10^{-4}, \quad \alpha=2.695;$$
$$\text{for} \quad t_0=2: \quad \beta=6.448\cdot 10^{-4}, \quad \alpha=2.9496.$$

In this way, by piecing together a quadratic polynomial in t^3 and the solution φ, we can obtain super- and subfunctions whose difference in the interval $0 \leqq t \leqq 1$ is smaller than $6.5\cdot 10^{-6} t^8$ and in the interval $1 \leqq t \leqq 2$ is smaller than $3.72\cdot 10^{-4} t^8$. Further, from these bounds we obtain $2.949 < T_a < 3.284$.

(γ) *A second Ansatz.* If we seek a single closed expression for super- and subfunctions, then the Ansatz $v, w = P(t)/(1-\beta t)$ suggests itself (P a polynomial). To get the best possible agreement with the power series expansion (15), we change it to

$$v, w = \frac{t^2 P(t^3/3)}{1-\beta t^3/3} = \frac{t^2 P(s)}{1-\beta s} \quad \text{with} \quad s = \frac{t^3}{3}. \tag{20}$$

As is easily verified, we have a super- (sub-) function if

$$\left.\begin{aligned} &9s^2(A^3P''' + 3\beta A^2 P'' + 6\beta^2 A P' + 6\beta^3 P) \\ &\qquad + 36 s(A^3 P'' + 2\beta A^2 P' + 2\beta^2 A P) + 20(A^3 P' + A^2 \beta P) \\ &\underset{(\leqq)}{\geqq} P[2A^2 P + 18 s(A^2 P' + \beta A P) + 9 s^2 (A^2 P'' + 2\beta A P' + 2\beta^2 P)] \\ &\qquad\qquad \text{for} \quad 0 \leqq s < 1/\beta; \end{aligned}\right\} \tag{21}$$

here

$$A = 1 - \beta s \quad \text{and} \quad P(s) = \frac{1}{2} + b_1 s + \cdots + b_q s^q.$$

For $P(s) = \frac{1}{2}(q=0)$, for example, we obtain

$$\frac{1}{2}(20\beta - 1)A^2 + \frac{9}{2}(8\beta - 1)s\beta A + \frac{9}{2}(6\beta - 1)s^2\beta^2 \underset{(\leqq)}{\geqq} 0,$$

i.e., $\beta = \frac{1}{6}$ (resp. $\beta = \frac{1}{20}$). Hence

$$v_2 = \frac{t^2}{2} \cdot \frac{1}{1 - \frac{1}{60} t^3} < u(t) < \frac{t^2}{2} \cdot \frac{1}{1 - \frac{1}{18} t^3}$$

and

$$2.620 < \sqrt[3]{18} \leqq T_a \leqq \sqrt[3]{60} < 3.915.$$

For an Ansatz with more terms in P, we try for the best possible agreement with the expansion (15). This is achieved by

$$b_k = a_k - \beta a_{k-1} \qquad (k = 1, \ldots, q). \tag{22}$$

For example, for $q=1$, i.e., $P(s) = \frac{1}{2} + (\frac{1}{40} - \frac{\beta}{2})\, s$, we obtain the values $\beta = \sqrt{1/20}$ (subfunction) and $\beta = 0.10089$ (superfunction) and hence

$$3.098 < T_a < 3.203,$$

while for $q=3$, after some computation, $\beta = \frac{1}{4}\sqrt[4]{5/231}$ (subfunction) and $\beta = 9299/91\,000$ (superfunction), and thus

$$3.084 < T_a < 3.151.$$

Since the expansion (15) has only positive coefficients, T_a is equal to the radius of convergence of this series; it has been computed repeatedly. Oudard (1948) gave the bounds $2.884 < T_a < 3.203$ which were improved by Ostrowski (1948) to $3.1 < T_a < 3.18$. The sharper bounds $3.11 < T_a < 3.13$ given by Punnis (1956) require a more exact foundation, since a solution of an ordinary differential equation computed by the Runge-Kutta method was used in its derivation without a very precise discussion of error.

The solution was computed on the electronic computer Z23 at the Universität Karlsruhe with the help of the Runge-Kutta program available there. We obtained

for step size	$h=$	0.1	0.05	0.02	0.01
stop after	$t_a=$	3.3	3.2	3.16	3.14

("stop" means that starting from t_a, the Runge-Kutta procedure attains a value $u''(t_a+h) > 10^{38}$). If we assume that the Runge-Kutta method produces the solution in the interval $0 \leqq t \leqq t_0$ *exactly*, then at the point t_0 we can piece it together with the function $\varphi(t)$ according to the procedure described in (β), and by solving the additional conditions (16) and (16′) obtain two numbers $\alpha = \alpha_2$ and $\alpha = \alpha_1$. (Here $v(t_0), v'(t_0), \ldots, w''(t_0)$ are to be replaced by the corresponding values computed by the machine). Since the assumption is certainly not satisfied exactly, we hesitate to claim α_1 and α_2 as lower and upper bounds for T_a. Nevertheless the result seems worth communicating:

t_0	α_1	α_2	t_0	α_1	α_2
1	2.69	6.90	2.7	3.120	3.132
2	2.98	3.27	2.8	3.122	3.130
2.5	3.106	3.138	2.9	3.1256	3.1283
2.6	3.118	3.135	3.0	3.1271	3.1277

16. Supplement

In this section we take up some more problems which are directly amenable to treatment with our methods.

I. Complex Differential Equations. The theory carries over immediately to ordinary differential equations where all the quantities except t are complex. Such a system of n equations

$$\boldsymbol{u}'(t) = \boldsymbol{f}(t, \boldsymbol{u}(t)) \qquad (0 \leqq t \leqq T) \tag{1}$$

is equivalent to the real system of $2n$ equations ($\boldsymbol{u} = \boldsymbol{u}_1 + i\boldsymbol{u}_2, \boldsymbol{f} = \boldsymbol{f}_1 + i\boldsymbol{f}_2$, $i = \sqrt{-1}$)

$$\boldsymbol{u}_1' = \boldsymbol{f}_1(t, \boldsymbol{u}_1, \boldsymbol{u}_2), \qquad \boldsymbol{u}_2' = \boldsymbol{f}_2(t, \boldsymbol{u}_1, \boldsymbol{u}_2). \tag{2}$$

Hence the earlier theorems can be applied. There is, however, a second possibility which is better suited to some problems. It is based on the following

II. Estimation Theorem for Complex Systems. *Suppose the complex valued functions* $\boldsymbol{u}(t)$, $\boldsymbol{v}(t)$ *are from* $Z(\boldsymbol{f})$, *the scalar functions* $\varrho_i(t) \in Z(\omega_i)$, $i=1,2$ *and are positive in* J_0. *With the help of a constant positive definite Hermitian matrix*[1] C $(C=\bar{C}^T)$ *we define a scalar product* $(z, z') = \bar{z}^T C z'$ *and a norm* $\|z\| = \sqrt{(z, z)}$. *Suppose*

(α) $\varrho_1(0+) < \|\boldsymbol{v}-\boldsymbol{u}\|(0+) < \varrho_2(0+)$;

(β) $\boldsymbol{u}' = \boldsymbol{f}(t, \boldsymbol{u})$, $\|\boldsymbol{v}-\boldsymbol{u}\|\,\delta_1(t) \leqq \operatorname{Re}(\boldsymbol{v}-\boldsymbol{u}, P\boldsymbol{v}) \leqq \|\boldsymbol{v}-\boldsymbol{u}\|\,\delta_2(t)$;

(γ) $\varrho_1' < \omega_1(t, \varrho_1) + \delta_1(t)$, $\varrho_2' > \omega_2(t, \varrho_2) + \delta_2(t)$ *in* J_0;

(δ) $\|\boldsymbol{v}-\boldsymbol{u}\|\,\omega_1(t, \|\boldsymbol{v}-\boldsymbol{u}\|) \leqq \operatorname{Re}(\boldsymbol{v}-\boldsymbol{u}, \boldsymbol{f}(t,\boldsymbol{v}) - \boldsymbol{f}(t,\boldsymbol{u}))$
$\leqq \|\boldsymbol{v}-\boldsymbol{u}\|\,\omega_2(t, \|\boldsymbol{v}-\boldsymbol{u}\|)$,

where the inequality with ω_i *need hold only for those* t *with* $\|\boldsymbol{v}-\boldsymbol{u}\| = \varrho_i(t)$. *Then*

$$\varrho_1(t) < \|\boldsymbol{v}-\boldsymbol{u}\| < \varrho_2(t) \quad in \quad J_0 .$$

Here of course $P\boldsymbol{v} = \boldsymbol{v}' - \boldsymbol{f}(t, \boldsymbol{v})$. Moreover, the two bounds for $\|\boldsymbol{v}-\boldsymbol{u}\|$ are independent of each other.

For the proof we apply Lemma 8 II to the functions $\varphi = \|\boldsymbol{v}-\boldsymbol{u}\|^2$, $\psi = \varrho_2^2$. From $\varphi = \psi$, i.e., $\|\boldsymbol{v}-\boldsymbol{u}\| = \varrho_2$, it follows that

$$\begin{aligned}\varphi' &= 2\operatorname{Re}(\boldsymbol{v}-\boldsymbol{u}, \boldsymbol{v}'-\boldsymbol{u}') = 2\operatorname{Re}(\boldsymbol{v}-\boldsymbol{u}, P\boldsymbol{v} + \boldsymbol{f}(t,\boldsymbol{v}) - \boldsymbol{f}(t,\boldsymbol{u}))\\ &\leqq 2\|\boldsymbol{v}-\boldsymbol{u}\|\,\delta_2(t) + 2\|\boldsymbol{v}-\boldsymbol{u}\|\,\omega_2(t, \|\boldsymbol{v}-\boldsymbol{u}\|) = 2\varrho_2(\delta_2(t) + \omega_2(t, \varrho_2))\\ &< 2\varrho_2\varrho_2' = \psi' .\end{aligned}$$

Thus we have proved that ϱ_2 is an upper bound; the lower bound is treated quite similarly.

III. Remark on the Uniqueness Problem. The present theorem contains a series of uniqueness assertions which differ essentially from 11 II only in that the following condition

$$\operatorname{Re}(\boldsymbol{v}-\boldsymbol{u}, \boldsymbol{f}(t,\boldsymbol{v}) - \boldsymbol{f}(t,\boldsymbol{u})) \leqq \|\boldsymbol{v}-\boldsymbol{u}\|\,\omega(t, \|\boldsymbol{v}-\boldsymbol{u}\|) \tag{3}$$

or, without explicit consideration of the functions $\boldsymbol{u}$ and $\boldsymbol{v}$,

$$\operatorname{Re}(z-z', \boldsymbol{f}(t,z) - \boldsymbol{f}(t,z')) \leqq \|z-z'\|\,\omega(t, \|z-z'\|) \tag{3'}$$

replaces condition (11.3). The proof also corresponds completely to the transition from 11 I to 11II.

[1] For a complex matrix M, $\bar{M}$ denotes the complex conjugate and M^T the transposed matrix. In matrix products, vectors z are to be taken as column vectors; z^T is the corresponding row vector.

The special case where all the quantities in II and III are real and C is the identity matrix (which is related to, but not contained in previous theorems) leads to theorems which were first proved by Eltermann (1955); see also Collatz (1955, pp. 108–109). Ciliberto (1961c) proved uniqueness using conditions similar to (3') (but for real systems). Estimates for the complex equation (1) ($n=1$) are given by Deo and Lakshmikantham (1963).

IV. Remark on Stability Theory. Theorem II has important consequences in the theory of stability of ordinary differential equations. Conti (1956) proves a theorem which is contained in II and furnishes applications to stability problems. We now derive a result from II which relates to the linear equation

$$\boldsymbol{u}' = A(t)\boldsymbol{u} + \boldsymbol{b}(t) \tag{4}$$

and goes back to Ważewski (1948). In (4), $A(t)$ is a complex matrix, $\boldsymbol{u}$ and $\boldsymbol{b}$ are to be taken as (also complex) column vectors. If $\boldsymbol{u}(t)$ is a solution of (4) and $\boldsymbol{v}(t) \equiv \boldsymbol{0}$, then the hypotheses of II read

(α) $\varrho_1(0+) < \|\boldsymbol{u}\|(0+) < \varrho_2(0+)$;
(β) $\|\boldsymbol{u}\|\delta_1(t) \leqq \mathrm{Re}(\boldsymbol{u}, \boldsymbol{b}(t)) \leqq \|\boldsymbol{u}\|\delta_2(t)$ in J_0;
(γ) $\varrho_1' < \lambda_1(t)\varrho_1 + \delta_1(t)$ and $\varrho_2' > \lambda_2(t)\varrho_2 + \delta_2(t)$ in J_0;
(δ) $(\boldsymbol{u}, \boldsymbol{u})\lambda_1(t) \leqq \mathrm{Re}(\boldsymbol{u}, A(t)\boldsymbol{u}) \leqq (\boldsymbol{u}, \boldsymbol{u})\lambda_2(t)$ in J_0.

Then

$$\varrho_1(t) < \|\boldsymbol{u}(t)\| < \varrho_2(t) \quad \text{in} \quad J_0 .$$

If the functions $\delta_i(t)$, $\lambda_i(t)$ are continuous in J_0 and integrable over J (this hypothesis can be weakened), then after the usual passage to the differential equation in (γ) the bound is

$$e^{-\mu_1(t)}\left[\|\boldsymbol{u}(0)\| + \int_0^t \delta_1(\tau)e^{\mu_1(t)-\mu_1(\tau)}d\tau\right] \leqq \|\boldsymbol{u}(t)\|$$

$$\leqq e^{-\mu_2(t)}\left[\|\boldsymbol{u}(0)\| + \int_0^t \delta_2(\tau)e^{\mu_2(t)-\mu_2(\tau)}d\tau\right]$$

with

$$\mu_i(t) = \int_0^t \lambda_i(\tau)d\tau \qquad (i=1,2) .$$

If we now choose C as the identity matrix, then $(\boldsymbol{u}, \boldsymbol{u}) = |u_1|^2 + \cdots + |u_n|^2$ and we can interpret the functions $\lambda_1(t)$, $\lambda_2(t)$ in (δ) as lower and upper bounds for the (real!) eigenvalues of the matrix $\frac{1}{2}(A + \bar{A}^T)$. The bound was proved for this case by Ważewski (1948).

V. Implicit Differential Equations. We can also establish theorems on super- and subfunctions for implicit differential equations $F(t, u, u') = 0$

and for systems of such differential equations if they have the form $F_\nu(t, u_1, \ldots, u_n, du_\nu/dt)$ $(\nu = 1, \ldots, n)$. The same also holds for the problems treated in Chapter I, and thus for implicit integral equations; cf. Porath (1965). We can even go a step further and consider

VI. Integro-Differential Equations. The methods of proof of Chapters I and II are then used simultaneously. For example, the following theorem can be proved easily (all quantities are real again).

Suppose the function $F(t, z, p, q)$ is monotone increasing in p, monotone decreasing in q[1]. *Suppose further that K is a monotone increasing operator (in the sense of* 1 IV). *If $v, w \in Z_c(K)$ and*

(α) $v(0+) < w(0+)$;

(β) $F(t, v, v', Kv) < F(t, w, w', Kw)$ *in* J_0,

then

$$v < w \quad in \quad J_0.$$

The assumption that the assertion is false and $v < w$ for $0 < t \leqq t_0$, $v(t_0) = w(t_0)$, $v'(t_0) \geqq w'(t_0)$, together with $(Kv)(t_0) \leqq (Kw)(t_0)$ and both monotonicity properties of F leads to

$$F(t, v, v', Kv) \geqq F(t, w, w', Kw) \quad \text{for} \quad t = t_0$$

and thus to a contradiction of (β).

Thus we have a problem of monotonic type in the terminology of the introduction II. The application of this theorem to an integro-differential equation

$$F(t, u, u', Ku) = 0$$

is obvious. The extension to systems of the form

$$F_\nu(t, u_1, \ldots, u_n, \partial u_\nu/\partial t, K_\nu u_\nu) = 0 \qquad (\nu = 1, \ldots, n)$$

presents no difficulties ($F_\nu(t, z_1, \ldots, z_n, p, q)$ quasi-monotone decreasing in $z_1, \ldots, z_n$, monotone increasing in p, monotone decreasing in q, K_ν a monotone increasing operator).

Further investigations of integro-differential equations can be found in the work of Nickel (1961 a) (the above theorem is found there), Corduneanu (1964), Sugiyama (1965), Novruzov (1966).

VII. Solutions in the Sense of Carathéodory. Basically, all theorems on error bounds in this chapter can also be derived for solutions in the sense of Carathéodory with the methods of Chapter I. We gave only the most important of these theorems in Chapter I in order to avoid repetition. The essential alterations in the hypotheses in converting the theorems of this chapter to absolutely continuous functions can be described briefly

[1] Monotonicity is always understood in the weak sense.

as follows: The Condition 5 III (γ), or 6 VII (γ) for systems, appears as a new, more difficult hypothesis (for the super- and subfunctions with respect to f, for the theorems on error bounds with respect to ω). On the other hand, equality is permitted in the differential inequalities – that is, the hypothesis (β) for the super- and subfunctions, the hypothesis (γ) for the theorems on error bounds — and the validity of these inequalities is only required almost everywhere. What we have said refers particularly to differential equations of nth order, which were not considered in Chapter I. By a solution of the differential equation of nth order (1) in the sense of Carathéodory we mean a solution from the class $Z_{ac}(f)$ of functions which are $(n-1)$ times continuously differentiable in J and have an absolutely continuous $(n-1)$st derivative. If we have such an Equation (15.1) and f has the Property 6 VII (γ), then Theorems 15 IV–VI also hold for functions $v, w \in Z_{ac}(f)$; here we allow equality and an exceptional set of measure 0 in the hypotheses (β).

CHAPTER III

Volterra Integral Equations in Several Variables Hyperbolic Differential Equations

17. Monotone Operators

The primary goal of this chapter is to present an estimation and uniqueness theory for hyperbolic differential equations in two independent variables with a completeness similar to that for ordinary differential equations in the first two chapters. If it is approached with the methods of the first chapter, this problem leads to Volterra integral equations in two variables which can be handled largely as in the one-dimensional case; even consideration of such integral equations in an arbitrary number m of independent variables produces no new difficulties. A significant part of the present chapter is devoted to the development of this theory.

The differential approach carried out for ordinary differential equations in Chapter II can also be extended to the hyperbolic case. This method is given only a little space in Section 22, because (in contrast to the one-dimensional case) it is based on the same sharp monotonicity conditions as the first method[1].

I. Definition (x, G, ∂G, $\bar{G}$, R_v, G_v, $G(x)$, $\varphi < \bar{\varphi}$ *on* R_v^+). Let m be an integer and $G \subset E^m$ a bounded open point set (we shall always deal with *bounded* domains in this chapter). The set of boundary points of G is denoted by ∂G, and the closure by $\bar{G} = G + \partial G$. We do not use bold-face for points $x = (x_1, \ldots, x_m)$ (bold-face is still reserved for points in E^n; n is the number of equations for systems of operator and integral equations). Inequalities $x < \bar{x}, \ldots$ are defined as in 6 I[2]. We use $G(\bar{x})$ for the set of all points $x \in \bar{G}$ for which $x \leq \bar{x}$, R_v ("initial boundary") for the set of all $\bar{x}$ for which $G(\bar{x})$ consists only of the point $\bar{x}$, and G_v for the difference $\bar{G} - R_v$. We have $R_v \subset \partial G$ (Fig. 4).

[1] In this regard compare the Remarks 20 X and Theorem 22 VI.

[2] The notation $t = (t_1, \ldots, t_m)$ instead of x would make the analogy with Chapter I more evident. But since it is common practice to write hyperbolic differential equations in the form $u_{xy} = f$ and not $u_{st} = f$, we have chosen the above notation.

If, for example, G is the unit sphere, then R_v is the set of points of the surface of the sphere with $x_\mu \leqq 0$ ($\mu = 1, \ldots, m$); if $a, b \in E^m$, $a < b$ and G is the m-dimensional interval $a < x < b$, then R_v consists of the point a, while $G_v = [a, b] - \{a\}$. In the case $n = 1$ treated in Chapter I we have $G = (0, T)$, $R_v = \{0\}$, $G_v = (0, T] = J_0$.

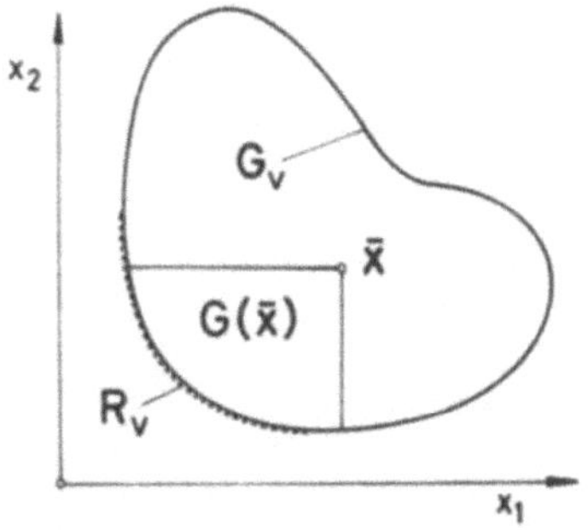

Fig. 4. The domain G (R_v is dotted)

In analogy to 1 V we write

$$\varphi < \overline{\varphi} \quad \text{on} \quad R_v^+$$

if the functions $\varphi, \overline{\varphi}$ are defined in G_v and if for every $\overline{x} \in R_v$ there is a neighborhood $U(\overline{x})$ such that

$$\varphi(x) < \overline{\varphi}(x) \quad \text{for} \quad x \in G_v \cdot U(\overline{x})\,.\text{[1]}$$

If the functions $\varphi, \overline{\varphi}$ are continuous on $\overline{G}$ and $\varphi < \overline{\varphi}$ on R_v, then $\varphi < \overline{\varphi}$ on R_v^+. But if, for example, G is the unit cube $0 < x_\mu < 1$ ($\mu = 1, \ldots, m$) and $\varphi = 0$, $\overline{\varphi}(x) = C(x_1 + x_2 + \cdots + x_m)$, $C > 0$, then we have $\varphi < \overline{\varphi}$ on R_v^+ but not on R_v.

II. Definition (*operator, monotone increasing operator,* $Z_c(K)$, $Z_c(k)$). Here we consider Volterra operators K of the following kind. An operator K associates a function $\psi = K\varphi$ defined in $\overline{G}$ with every function $\varphi(x)$ from its domain of definition $Z_c(K) \subset C(\overline{G})$; we write $\psi(x) = (K\varphi)(x)$. Furthermore, K has the property that $(K\varphi)(x_0) = (K\overline{\varphi})(x_0)$ if $x_0 \in \overline{G}$, $\varphi, \overline{\varphi} \in Z_c(K)$ and $\varphi(x) = \overline{\varphi}(x)$ for $x \in G(x_0)$. In analogy to 1 VIII (δ) it is natural to speak of "operators of Volterra type", for the value of $K\varphi$ at the point x is already completely determined by the values taken on by φ in $G(x)$.

[1] Thus it is not the set R_v^+, but the whole expression "$\varphi < \overline{\varphi}$ on R_v^+" which is defined here. However the usual interpretation "$\varphi(x) < \overline{\varphi}(x)$ for $x \in R_v^+$" is also possible; then R_v^+ would have to be defined as the intersection of G_v with a neighborhood of R_v.

The operator K is called a "*monotone increasing operator*" if it has the following property:

If $\varphi, \bar{\varphi} \in Z_c(K)$ and if for a point $x_0 \in G_v$ the inequality $\varphi(x) \leqq \bar{\varphi}(x)$ holds in $G(x_0)$, then

$$(K\varphi)(x_0) \leqq (K\bar{\varphi})(x_0).$$

The integral operator

$$(K\varphi)(x) = \int_{G(x)} k(x, \xi, \varphi(\xi))d\xi \tag{1}$$

provides an important example. If the "kernel" $k(x, \xi, z) = k(x_1, \ldots, x_m, \xi_1, \ldots, \xi_m, z)$ is defined on a set $D(k) \subset E^{2m+1}$, then $Z_c(k)$ consists of those continuous functions φ in $\bar{G}$ for which $(x, \xi, \varphi(\xi)) \in D(k)$ for $x, \xi \in \bar{G}, \xi \leqq x$ and $k(x, \xi, \varphi(\xi)) \in L(G(x))$ for every fixed x in $\bar{G}$ (briefly, for which the integral makes sense). If $k(x, \xi, z)$ is monotone increasing in z (compare with (1.2)), then the integral operator defined according to (1) is monotone increasing.

The following theorem is basic. It agrees almost verbatim with 1 VI in statement and proof.

III. Theorem. *For a monotone increasing operator K and two functions* $v, w \in Z_c(K)$, *suppose*

(α) $v < w$ *on* R_v^+;

(β) $v - Kv < w - Kw$ *in* G_v.

Then

$$v < w \quad in \quad G_v.$$

The hypothesis (α) can be omitted if $(K\varphi)(\bar{x}) = 0$ for $\bar{x} \in R_v$ and all $\varphi \in Z_c(K)$ and if (β) also holds on R_v (it then follows from (β)).

The proof goes through as in 1 VI. Let us assume that the assertion is false and that A is the set of those points from G_v at which $v \geqq w$. Let $s(x) = x_1 + \cdots + x_m$ and suppose s_0 is the lower bound of this function relative to A. If this lower bound is achieved at a point of A, i.e., if $s_0 = s(x_0)$, $x_0 \in A$, then in $G(x_0) - \{x_0\}$ we have $s(x) < s(x_0)$ and thus $v < w$. At the point x_0 by (β) and the monotonicity of K,

$$v = (v - Kv) + Kv < (w - Kw) + Kv \leqq w$$

which contradicts the assumption $x_0 \in A$. Thus the function $s(x)$ does not assume its infimum (relative to A) on A. Then there exists a sequence $x_1, x_2, \ldots$ of points from A with $s(x_k) \to s_0$ $(k \to \infty)$. If $\bar{x} \in \bar{G}$ is an accumulation point of this sequence, then $v(\bar{x}) \geqq w(\bar{x})$ because of the continuity of these functions, and furthermore $\bar{x} \notin A$ and thus $\bar{x} \in R_v$. But this leads to a contradiction of (α). Thus we have shown that the set A is empty and the assertion of the theorem is true.

IV. Definition and Theorem $(\boldsymbol{K}, Z_c(\boldsymbol{K}), Z_c(\boldsymbol{k}))$. Similarly to Section 6, for systems of integral equations we also deal with operators $\boldsymbol{K}$ which associate with a continuous vector function $\boldsymbol{\varphi}(x)=(\varphi_1(x), \ldots, \varphi_n(x))$ in $\bar{G}$ a vector function $\boldsymbol{K}\boldsymbol{\varphi}$ defined in $\bar{G}$, where $(\boldsymbol{K}\boldsymbol{\varphi})(x_0)$ depends only on the values which $\boldsymbol{\varphi}(x)$ takes on in $G(x_0)$. The meaning of $Z_c(\boldsymbol{K})$, $Z_c(\boldsymbol{k})$ and of "$\boldsymbol{\varphi}<\bar{\boldsymbol{\varphi}}$ on R_v^+" is self-evident. Likewise, monotone increasing operators are defined exactly as above in II: If $\boldsymbol{\varphi}, \bar{\boldsymbol{\varphi}} \in Z_c(\boldsymbol{K})$ and $\boldsymbol{\varphi}(x) \leqq \bar{\boldsymbol{\varphi}}(x)$ in $G(x_0)$, then we have $(\boldsymbol{K}\boldsymbol{\varphi})(x_0) \leqq (\boldsymbol{K}\bar{\boldsymbol{\varphi}})(x_0)$. We also speak of vector operators $\boldsymbol{K}$ in contrast to scalar operators K.

Theorem III *still holds without changes for vector operators* $\boldsymbol{K}$ *and vector functions* $\boldsymbol{v}, \boldsymbol{w}$.

In the proof A is now the set of points in G_v for which $\boldsymbol{v}<\boldsymbol{w}$ does *not* hold; everything else remains the same.

These theorems make it possible to characterize

V. Super- and Subfunctions for the equation

$$\boldsymbol{u}=\boldsymbol{g}+\boldsymbol{K}\boldsymbol{u} \tag{2}$$

by inequalities. A *solution* is a function $\boldsymbol{u} \in Z_c(\boldsymbol{K})$ which satisfies this equation in $\bar{G}$. The function $\boldsymbol{v} \in Z_c(\boldsymbol{K})$ is a subfunction [the function $\boldsymbol{w} \in Z_c(\boldsymbol{K})$ is a superfunction] with respect to Eq. (2) with monotone increasing $\boldsymbol{K}$ if

(α) $\boldsymbol{v}<\boldsymbol{u}$ $[\boldsymbol{u}<\boldsymbol{w}]$ *on* R_v^+ *for every solution* $\boldsymbol{u}$;
(β) $\boldsymbol{v}<\boldsymbol{g}+\boldsymbol{K}\boldsymbol{v}$ $[\boldsymbol{w}>\boldsymbol{g}+\boldsymbol{K}\boldsymbol{w}]$ *in* G_v.

If (β) also holds at the points of R_v and if $(\boldsymbol{K}\boldsymbol{\varphi})(x)=\mathbf{0}$ for $x \in R_v$ and $\boldsymbol{\varphi} \in Z_c(\boldsymbol{K})$, then (α) is superfluous since then we even have $\boldsymbol{v}<\boldsymbol{u}<\boldsymbol{w}$ on R_v. For every solution $\boldsymbol{u}$ of (2),

$$\boldsymbol{v}<\boldsymbol{u}<\boldsymbol{w} \quad \text{in} \quad G_v .$$

Of course, our assertion holds in particular for $n=1$.

Similarly to Section 4, we prove estimation theorems which hold for arbitrary, not necessarily monotone operators and which are based on a bound of the vector operator $\boldsymbol{K}$ by a vector operator $\boldsymbol{\Omega}$

$$|\boldsymbol{K}\boldsymbol{\varphi}-\boldsymbol{K}\bar{\boldsymbol{\varphi}}| \leqq \boldsymbol{\Omega}(|\boldsymbol{\varphi}-\bar{\boldsymbol{\varphi}}|) . \tag{3}$$

We go immediately to the n-dimensional case (note Definition 6 I).

VI. Theorem. *Suppose the condition* (3) *holds between the operator* $\boldsymbol{K}$ *and the monotone increasing operator* $\boldsymbol{\Omega}$, *for* $\boldsymbol{\varphi}, \bar{\boldsymbol{\varphi}} \in Z_c(\boldsymbol{K})$ *and* $|\boldsymbol{\varphi}-\bar{\boldsymbol{\varphi}}| \in Z_c(\boldsymbol{\Omega})$. *Suppose the functions* $\boldsymbol{u}, \boldsymbol{v}, \boldsymbol{\varrho}, \boldsymbol{d}, \boldsymbol{g}$ *are defined in* $\bar{G}$ *and* $\boldsymbol{u}, \boldsymbol{v} \in Z_c(\boldsymbol{K})$, $|\boldsymbol{v}-\boldsymbol{u}|, \boldsymbol{\varrho} \in Z_c(\boldsymbol{\Omega})$. *Then*

(α) $|\boldsymbol{v}-\boldsymbol{u}|<\boldsymbol{\varrho}$ *on* R_v^+;
(β) $\boldsymbol{u}=\boldsymbol{g}+\boldsymbol{K}\boldsymbol{u}$ *and* $|\boldsymbol{v}-\boldsymbol{g}-\boldsymbol{K}\boldsymbol{v}| \leqq \boldsymbol{d}$ in G_v;
(γ) $\boldsymbol{\varrho}>\boldsymbol{d}+\boldsymbol{\Omega}\boldsymbol{\varrho}$ *in* G_v

imply the inequality

$$|\boldsymbol{v}-\boldsymbol{u}|<\boldsymbol{\varrho} \quad in \quad G_v.$$

If the operators $\boldsymbol{K}$ and $\boldsymbol{\Omega}$ have the property that $(\boldsymbol{K}\boldsymbol{\varphi})(x)=(\boldsymbol{\Omega}\boldsymbol{\varphi})(x)=\mathbf{0}$ for $x\in R_v$ and $\boldsymbol{\varphi}\in Z_c(\boldsymbol{K})$ or $\boldsymbol{\varphi}\in Z_c(\boldsymbol{\Omega})$, and if ($\beta$) and ($\gamma$) also hold on R_v, then (α) is superfluous. We then have $|\boldsymbol{v}-\boldsymbol{u}|=|\boldsymbol{v}-\boldsymbol{g}|\leqq \boldsymbol{d}<\boldsymbol{\varrho}$ on R_v.

This assertion is proved exactly like the corresponding Theorem 4 I by reducing it to III or IV with the substitution of $|\boldsymbol{v}-\boldsymbol{u}|$, $\boldsymbol{\varrho}$, $\boldsymbol{\Omega}$ for $\boldsymbol{v}$, $\boldsymbol{w}$, $\boldsymbol{K}$.

In Section 7 we became acquainted with another kind of condition where the vector operator $\boldsymbol{K}$ is combined with a scalar operator Ω by

$$\|\boldsymbol{K}\boldsymbol{\varphi}-\boldsymbol{K}\overline{\boldsymbol{\varphi}}\|\leqq\Omega(\|\boldsymbol{\varphi}-\overline{\boldsymbol{\varphi}}\|). \tag{4}$$

To Theorem 7 III corresponds the

VII. Theorem. *Suppose $\boldsymbol{K}$ is a (vector) operator, Ω a monotone increasing (scalar) operator, $\|\cdot\|$ a K-norm, and the condition* (4) *holds if* $\boldsymbol{\varphi}, \overline{\boldsymbol{\varphi}}\in Z_c(\boldsymbol{K})$ and $\|\boldsymbol{\varphi}-\overline{\boldsymbol{\varphi}}\|\in Z_c(\Omega)$. *Suppose the functions $\boldsymbol{u}, \boldsymbol{v}, \varrho, d, \boldsymbol{g}$ are defined in $\overline{G}$, and $\boldsymbol{u}, \boldsymbol{v}\in Z_c(\boldsymbol{K})$, $\|\boldsymbol{v}-\boldsymbol{u}\|, \varrho\in Z_c(\Omega)$. Then*

(α) $\|\boldsymbol{v}-\boldsymbol{u}\|<\varrho$ *on* R_v^+;
(β) $\boldsymbol{u}=\boldsymbol{g}+\boldsymbol{K}\boldsymbol{u}$ *and* $\|\boldsymbol{v}-\boldsymbol{g}-\boldsymbol{K}\boldsymbol{v}\|\leqq d$ *in* G_v;
(γ) $\varrho>d+\Omega\varrho$ *in* G_v

imply the inequality

$$\|\boldsymbol{v}-\boldsymbol{u}\|<\varrho \quad in \quad G_v.$$

Under conditions similar to those given in connection with VI, (α) is not needed.

This theorem is proved like the previous one by reduction to III with $\|\boldsymbol{v}-\boldsymbol{u}\|$, ϱ, Ω instead of v, w, K.

The following uniqueness theorem corresponds in the one-dimensional case to the Theorems 4 III, 6 V and 7 IV, the class $\mathscr{D}^n$ to the class $\mathscr{E}^n$.

VIII. Definition ($\mathscr{D}^n$). The monotone increasing operator $\boldsymbol{\Omega}$ belongs to the class $\mathscr{D}^n$ if for every $\boldsymbol{\varepsilon}=(\varepsilon_1,\ldots,\varepsilon_n)>\mathbf{0}$ there exist a $\boldsymbol{\delta}=(\delta_1,\ldots,\delta_n)>\mathbf{0}$ and a function $\boldsymbol{\varrho}\in Z_c(\boldsymbol{\Omega})$ such that

$$\boldsymbol{\varrho}>\boldsymbol{\Omega}\boldsymbol{\varrho} \quad \text{in } G_v \text{ and } \quad \boldsymbol{\delta}<\boldsymbol{\varrho}<\boldsymbol{\varepsilon} \quad \text{in} \quad \overline{G}$$

and if, moreover, all continuous nonnegative functions $\boldsymbol{\varphi}=(\varphi_1,\ldots,\varphi_n)$ in $\overline{G}$ which vanish on R_v belong to $Z_c(\boldsymbol{\Omega})$.

IX. Uniqueness Theorem. *Suppose the operator $\boldsymbol{K}$ has the property that $(\boldsymbol{K}\boldsymbol{\varphi})(x)=\mathbf{0}$ for $x\in R_v$ and $\boldsymbol{\varphi}\in Z_c(\boldsymbol{K})$. Uniqueness holds for the operator eq.* (2) *if for all $\boldsymbol{\varphi}, \overline{\boldsymbol{\varphi}}\in Z_c(\boldsymbol{K})$ the condition* (3) *holds with $\boldsymbol{\Omega}\in\mathscr{D}^n$, or if for all $\boldsymbol{\varphi}, \overline{\boldsymbol{\varphi}}\in Z_c(\boldsymbol{K})$ the condition* (4) *holds with $\Omega\in\mathscr{D}^1$ and with a K-norm for which $\|z\|=\|-z\|=0$ is equivalent to $z=\mathbf{0}$.*

If for two arbitrary solutions $\boldsymbol{u}, \bar{\boldsymbol{u}}$ *of* (2) *the function* $\max(\boldsymbol{u}, \bar{\boldsymbol{u}})$ *is also a solution of* (2), *or if Eq.* (2) *has a maximal solution, then a one-sided condition is sufficient for uniqueness: There is an* $\boldsymbol{\Omega} \in \mathscr{D}^n$ *such that*

(α) *for all* $\boldsymbol{\varphi}, \bar{\boldsymbol{\varphi}} \in Z_c(\boldsymbol{K})$ *which satisfy the inequality* $\bar{\boldsymbol{\varphi}} \leqq \boldsymbol{\varphi}$ *in* $\bar{G}$,

$$\boldsymbol{K}\boldsymbol{\varphi} - \boldsymbol{K}\bar{\boldsymbol{\varphi}} \leqq \boldsymbol{\Omega}(\boldsymbol{\varphi} - \bar{\boldsymbol{\varphi}}). \tag{5}$$

This theorem can be reduced easily to VI or VII. Like the preceding theorems, it is found for the case $n=2$ and special domains in Walter (1961).

X. Super- and Subfunctions in the General Case. If the operator $\boldsymbol{K}$ is not monotone increasing, then V is false in general. However, for an arbitrary operator $\boldsymbol{K}$ we can *simultaneously* characterize a superfunction $\boldsymbol{w}$ and a subfunction $\boldsymbol{v}$ by the conditions

(α) $\boldsymbol{v} < \boldsymbol{u} < \boldsymbol{w}$ *on* R_v^+ *for every solution* $\boldsymbol{u}$;

(β) $\boldsymbol{v}(x) < \boldsymbol{g}(x) + (\boldsymbol{K}\boldsymbol{\varphi})(x) < \boldsymbol{w}(x)$ *for* $x \in G_v$ *and all functions* $\boldsymbol{\varphi} \in Z_c(\boldsymbol{K})$ *with* $\boldsymbol{v} \leqq \boldsymbol{\varphi} \leqq \boldsymbol{w}$ *in* $G(x)$

as in 4 VII.

We show, essentially as in III, that

$$\boldsymbol{v} < \boldsymbol{u} < \boldsymbol{w} \quad \text{in} \quad G_v$$

again follows. We choose a fixed solution $\boldsymbol{u}$ and define A as the set of points of G_v at which at least one of the inequalities $\boldsymbol{v} < \boldsymbol{u} < \boldsymbol{w}$ fails. Then if x_0 is determined as in the proof of III, from (β) with x_0 and $\boldsymbol{u}$ in place of x and $\boldsymbol{\varphi}$ we derive

$$\boldsymbol{v}(x_0) < \boldsymbol{g}(x_0) + (\boldsymbol{K}\boldsymbol{u})(x_0) = \boldsymbol{u}(x_0) < \boldsymbol{w}(x_0),$$

and thus a contradiction to $x_0 \in A$. As in III, the other possibility, that there exists no $x_0 \in A$ with $s_0 = s(x_0)$, leads to a contradiction of (α).

XI. Generalizations and Remarks. Theorems III to X form the foundation for all further discussions of this chapter. From them we can derive many important special results, and in particular all the known uniqueness criteria for hyperbolic differential equations in two variables (as far as they relate to the initial value problems considered by us).

(α) As for possible generalizations, we shall be satisfied with a reference to the Remarks 1 VIII, which are easily carried over to the present situation. In particular we can extend the theory to nonlinear equations $h(x, u, Ku) = 0$ as in 1 VIII (γ).

(β) Suppose $s(x) = x_1 + \cdots + x_m$, s_0 is the lower bound of $s(x)$ on $\bar{G}$ and R_0 is the set of all points $x \in \bar{G}$ with $s(x) = s_0$. We note that all theorems remain valid if we let R_v denote an arbitrary set containing R_0 where, as before, $G_v = \bar{G} - R_v$.

(γ) The important question in theory and practice of when equality signs are allowed in the inequalities of our theorems can be answered satisfactorily from the point of view of 1 IX. Since no difficulties and no new point of view relative to the one-dimensional case arise, we shall be satisfied with making the following remarks. Suppose $(\boldsymbol{K}\varphi)(x)=\boldsymbol{0}$ on R_v. Then we have a theorem

IIIa: *From* $\boldsymbol{v}\leqq \boldsymbol{g}+\boldsymbol{K}\boldsymbol{v}$, $\boldsymbol{u}^*=\boldsymbol{g}+\boldsymbol{K}\boldsymbol{u}^*$ *in* $\bar{G}$ *follows* $\boldsymbol{v}\leqq\boldsymbol{u}^*$ *in* $\bar{G}$,
if $\boldsymbol{u}^*$ is the maximal solution of Eq. (2) and can be approximated arbitrarily well by functions $\boldsymbol{w}$ which satisfy the inequality $\boldsymbol{w}>\boldsymbol{g}+\boldsymbol{K}\boldsymbol{w}$. Similarly for the minimal solution. In the following section we prove the existence of maximal and minimal solutions for a large class of integral equations.

We have a theorem

IIIb: *From* $\boldsymbol{v}\leqq \boldsymbol{g}+\boldsymbol{K}\boldsymbol{v}$, $\boldsymbol{w}\geqq\boldsymbol{g}+\boldsymbol{K}\boldsymbol{w}$ *in* $\bar{G}$ *follows* $\boldsymbol{v}\leqq\boldsymbol{w}$ *in* $\bar{G}$
if a uniqueness condition (3) holds with $\boldsymbol{\Omega}\in\mathscr{D}^n$.

The proofs of 1 IX remain valid, as do the remarks on the connection between a Theorem IIIb and the uniqueness problem for Eq. (2).

We have the analogous situation for VI and VII. Equality signs are permitted in VI (γ) and VII (γ) (we must then also have them in the statement) if $|\boldsymbol{\Omega}\varphi-\boldsymbol{\Omega}\bar{\varphi}|\leqq\boldsymbol{\Omega}^*(|\varphi-\bar{\varphi}|)$ with $\boldsymbol{\Omega}^*\in\mathscr{D}^n$ or $\|\boldsymbol{\Omega}\varphi-\boldsymbol{\Omega}\bar{\varphi}\|\leqq\Omega^*(\|\varphi-\bar{\varphi}\|)$ with $\Omega^*\in\mathscr{D}^1$.

(δ) For $n=1$ the two Theorems VI and VII are identical if we choose the norm $\|z\|=|z_1|$ in VII. However, Theorem VII is more general than Theorem VI in the case $n=1$. It can also be used for one-sided bounds, and the sign of ϱ is arbitrary; we might think of the K-norms $\|z\|=z_1$ or $\|z\|=-z_1$.

18. Existence Theorems

Now we consider a system of n integral equations

$$u_v(x)=g_v(x)+\int_{H_v(x)} k_v(x,\xi,u_1(\xi),\ldots,u_n(\xi))d\xi \quad \text{in} \quad \bar{G} \tag{1}$$

($v=1,\ldots,n$; $H_v(x)\subset G(x)$), where $H_v(x)$ is a measurable point set and the integral is an m-dimensional Lebesgue integral, and accordingly $d\xi$ is the volume element. An existence theorem for this problem is relatively easy to obtain with the methods of Section 2.

I. Definition ($\mathscr{H}$, $\mathscr{H}_0$). Suppose the scalar function $k(x,\xi,z)=k(x_1,\ldots,x_m,\xi_1,\ldots,\xi_m,z_1,\ldots,z_n)$ is defined for $x,\xi\in\bar{G}$, $\xi\leqq x$, $z\in E^n$, for fixed $(x,z)\in\bar{G}\times E^n$ is measurable in $\xi\in G(x)$, and for fixed $\xi\in\bar{G}$ is continuous in $(x,z)\in\{x\,|\,x\in\bar{G}, x\geqq\xi\}\times E^n$. Suppose further that for every constant $M>0$ there is a function $h(x)\in L(\overline{G})$ such that

$$|k(x,\xi,z)|\leqq h(\xi) \tag{2}$$

provided $x,\xi\in\overline{G}$, $z\in E^n$ and $\xi\leqq x$, $|z_\nu|<M$ ($\nu=1,\dots,n$). Then $k\in\mathscr{H}_0$. If it is possible to make do with one and the same function $h(\xi)$ for all $M>0$, then $k\in\mathscr{H}$.

II. Existence Theorem. *The system of integral equations* (1) *has at least one continuous solution* $\boldsymbol{u}(x)$ *in* $\overline{G}$ *if the functions* g_ν *are continuous in* $\overline{G}$ *and the functions* k_ν *belong to* $\mathscr{H}$ *for* $\nu=1,\dots,n$, *and if further the domains* $H_\nu(x)$ *are L-measurable and have the following property:*

(α) $|H_\nu(x)+H_\nu(\overline{x})-H_\nu(x)\cdot H_\nu(\overline{x})|\to 0$ *for* $x\to\overline{x}$,

uniformly in $\overline{x}\in\overline{G}$ ($|M|$ *denotes the m-dimensional L-measure of a set* $M\subset E^m$).

The condition (α) means that with "small" changes in x, $H_\nu(x)$ changes only "a little." We introduce some notation for this and the following proof only. Let $d(t)$, $\overline{d}(t)$, $d_\nu(t)$, ... denote moduli of continuity; these are functions which for real $t\geqq 0$ are continuous, nonnegative, monotone increasing, and vanish for $t=0$. We use $\|\cdot\|$ for the norm 7 I (δ); thus $\|x\|=|x_1|+\cdots+|x_m|$ for $x\in E^m$, $\|z\|=|z_1|+\cdots+|z_n|$ for $z\in E^n$. For $x\in E^m$ we write $d(x)$, ... instead of $d(\|x\|)$, ... for brevity (only in connection with moduli of continuity).

The proof, which is carried out largely as in Section 2, is based on Schauder's fixed point theorem. We consider the mapping $\boldsymbol{\varphi}\to\boldsymbol{g}+\boldsymbol{K}\boldsymbol{\varphi}$, where the components $(\boldsymbol{K}\boldsymbol{\varphi})_\nu$ are given by

$$(\boldsymbol{K}\boldsymbol{\varphi})_\nu(x)=\int_{H_\nu(x)} k_\nu(x,\xi,\boldsymbol{\varphi}(\xi))\,d\xi\,; \tag{3}$$

of course $\boldsymbol{g}=(g_1,\dots,g_n)$. Our task is to find a suitable compact and convex set $\Phi\subset C(\overline{G})$ and to show that the given mapping is continuous and takes Φ into itself. Then there is at least one fixed point in Φ, and this is a solution of (1).

Suppose $\boldsymbol{k}=(k_1,\dots,k_n)$,

$$M=\int_{\overline{G}} h(\xi)\,d\xi,\qquad M_1=nM+\max_{\overline{G}}\|\boldsymbol{g}\| \tag{4}$$

(suppose all the k_ν satisfy the inequality (2) with this $h(\xi)$) and let $\delta(\xi,\varepsilon)$ be the function

$$\delta(\xi,\varepsilon)=\sup\{|k_\nu(x,\xi,z)-k_\nu(\overline{x},\xi,\overline{z})|\,|\,x,\overline{x},\xi\in\overline{G},\ \xi\leqq x,\ \xi\leqq\overline{x},\ 1\leqq\nu\leqq n,$$
$$\|x-\overline{x}\|+\|z-\overline{z}\|\leqq\varepsilon,\ \|z\|<M_1\},$$

defined for $\varepsilon\geqq 0$ and $\xi\in\overline{G}$. This function has the same properties as the corresponding function $\delta(\tau,\varepsilon)$ of (2.4): it is nonnegative and $\leqq 2h(\xi)$, measurable in ξ, continuous and monotone decreasing in ε, and we have $\delta(\xi, 0) = 0$ in $\bar{G}$. Thus

$$\int_{\bar{G}} \delta(\xi, \varepsilon) d\xi \leqq d(\varepsilon) \tag{5}$$

for a suitable modulus of continuity $d(t)$.

If, when $d_0(t)$ is a (yet undetermined) modulus of continuity, we use Φ to denote the set of all functions $\boldsymbol{\varphi} \in C(\bar{G})$ for which

$$|\varphi_\nu(x) - g_\nu(x)| \leqq M \quad \text{and} \quad |\varphi_\nu(x) - \varphi_\nu(\bar{x})| \leqq d_0(x - \bar{x}) \quad \text{in} \quad \bar{G}$$

$(\nu = 1, \ldots, n)$, then the compactness and convexity of this set are seen at once. The continuity of the mapping (in the Banach space of continuous functions $\boldsymbol{\varphi}$ in $\bar{G}$ with norm $\max_{\bar{G}} \|\boldsymbol{\varphi}(x)\|$) is obtained from

$$\begin{aligned} \|(\boldsymbol{K}\boldsymbol{\varphi})(x) - (\boldsymbol{K}\bar{\boldsymbol{\varphi}})(x)\| &\leqq \int_{G(x)} \|\boldsymbol{k}(x, \xi, \boldsymbol{\varphi}) - \boldsymbol{k}(x, \xi, \bar{\boldsymbol{\varphi}})\| d\xi \\ &\leqq n \int_{\bar{G}} \delta(\xi, \max_{\bar{G}} \|\boldsymbol{\varphi} - \bar{\boldsymbol{\varphi}}\|) d\xi \end{aligned}$$

and (5). Let $e = (e_1, \ldots, e_m)$, $x, x + e \in \bar{G}$ and

$$A = A(x, e, \nu) = H_\nu(x) \cdot H_\nu(x + e), B = H_\nu(x) - A, C = H_\nu(x + e) - A\,.$$

By the hypothesis (α) and the continuity of g_ν, for suitable moduli of continuity $d_1(t)$, $d_2(t)$, $d_3(t)$ we have

$$|B| + |C| \leqq d_1(e), \qquad \int_{B+C} h(\xi) d\xi \leqq d_2(e), \qquad |g_\nu(x + e) - g_\nu(x)| \leqq d_3(e) \tag{6}$$

$(\nu = 1, \ldots, n)$, and thus

$$\begin{aligned} |(\boldsymbol{K}\boldsymbol{\varphi})_\nu(x) - (\boldsymbol{K}\boldsymbol{\varphi})_\nu(x + e)| &\leqq \int_A |k_\nu(x, \xi, \boldsymbol{\varphi}) - k_\nu(x + e, \xi, \boldsymbol{\varphi})| d\xi \\ &\quad + \int_B |k_\nu(x, \xi, \boldsymbol{\varphi})| d\xi + \int_C |k_\nu(x + e, \xi, \boldsymbol{\varphi})| d\xi \\ &\leqq \int_A \delta(\xi, \|e\|) d\xi + 2 d_2(e) \leqq d(e) + 2 d_2(e)\,. \end{aligned}$$

In conjunction with $|(\boldsymbol{K}\boldsymbol{\varphi})_\nu(x)| \leqq M$, this inequality shows that Φ is in fact mapped into itself if we let $d_0(t) = d(t) + 2d_2(t) + d_3(t)$. Thus the existence theorem is proved.

Now we proceed to the essentially more difficult case where integrals other than space integrals occur in (1), and we consider first an example of an integral equation without a continuous solution.

III. Example. Let $n=1$, $m=2$, $G=\{x|0<x_1<1,\ |x_2|<1\}$, $H_1(x)=\{\xi|0\leqq\xi_1\leqq x_1, \xi_2=x_2\}, g_1(x)=x_2$ and

$$k_1(x,\xi,z)=f(z)=\begin{cases}0 & \text{for } z\leqq 0\\ 2\sqrt{z} & \text{for } z>0.\end{cases}$$

The integral equation reads

$$u(x_1,x_2)=x_2+\int_0^{x_1} f(u(\xi_1,x_2))d\xi_1 .$$

If we set $u(t,x_2)=\varphi(t)$, where x_2 plays the role of a parameter, then $\varphi(t)$ is the solution of an initial value problem for an ordinary differential equation

$$\varphi'(t)=f(\varphi(t)), \qquad \varphi(0)=x_2 .$$

This initial value problem can be solved easily; we obtain

$$u(x_1,x_2)=\begin{cases}(x_1+\sqrt{x_2})^2 & \text{for } x_2>0\\ x_2 & \text{for } x_2<0,\end{cases}$$

while the values for $x_2=0$ are not uniquely determined. Thus the function u is discontinuous along the x_1-axis, i.e., the given integral equation has no continuous solution.

Similarly, from every *non-unique* initial value problem

$$\varphi'=f(t,\varphi), \qquad \varphi(0)=\alpha \tag{7}$$

we can derive a Volterra integral equation which is *unsolvable* in the domain of continuous functions; for example the equation

$$u(x_1,x_2)=\alpha+x_2(1+x_1)+\int_0^{x_1} f(\xi_1,u(\xi_1,x_2))d\xi_1 . \tag{8}$$

For constant $x_2>0$ the function $w(t)=u(t,x_2)$ is a superfunction relative to the initial value problem (7), i.e., we have $w'>f(t,w)$ and $w(0)>\alpha$; similarly $v(t)=u(t,x_2)$ for $x_2<0$ is a subfunction. By 8 VI, u thus has points of discontinuity on the x-axis (we started with the assumption that there are at least two distinct solutions of (7)). If, conversely, f is continuous and the initial value problems $\varphi'=f(t,\varphi)+\beta, \varphi(0)=\alpha$ (α,β real) are always uniquely solvable, then by 8 VIII the solutions depend continuously on α and β. Hence it follows, for example, that the integral equation (8) then has a continuous solution.

This example shows the close connection between the *existence problem* for a Volterra integral equation and the *uniqueness problem* for an associated ordinary differential equation. Similarly we can set up integral equations for $m=3$ and a domain of integration $H_1(x)$ in a plane, which are equivalent to hyperbolic differential equations which depend on a parameter; here also the existence problem for the integral equation is related to the uniqueness problem for the differential equation (more precisely, related to the problem of the continuous dependence on the initial values and on the right hand side of the differential equation). We shall return to this situation in 21 IV.

IV. General Case. By this we mean the system of Volterra integral equations

$$u_\nu(x)=g_\nu(x)+\int\limits_{H_\nu(x)} k_\nu(x,\xi,\boldsymbol{u}(\xi))(d\xi)_{p_\nu} \quad \text{in } \bar{G} \quad (\nu=1,\ldots,n). \tag{9}$$

Here p_ν is an integer, $1\leqq p_\nu\leqq m$, $H_\nu(x)$ is a subset of $G(x)$ in a p_ν-dimensional plane parallel to the axes [1], and $(d\xi)_{p_\nu}$ is the p_ν-dimensional volume element in this plane; suppose $H_\nu(x)$, taken as a point set of a p_ν- dimensional space, is L-measurable. By renumbering we can arrange that $p_\nu=m$ for $\nu=1,\ldots,n'$ and $p_\nu<m$ for $\nu=n'+1,\ldots,n$. The indices of the first kind are denoted by α, those of the second kind by β. Thus, below, α will run through the numbers $1,\ldots,n'$; β through the numbers $n'+1,\ldots,n$; and ν through the numbers $1,\ldots,n$.

The case where no β's are involved was solved in great generality in II. In view of the above example, it is clear that now an existence proof is possible only in a much more restricted class of kernels, and that, in particular, the continuity of the kernel alone is not sufficient. We assume:

(α) II (α) holds for the sets $H_\alpha(x)$. The set $H_\beta(x+e)$, $e\in E^m$, arises from $H_\beta(x)$ essentially by means of a parallel shift by e. More precisely: If for $D\subset E^m$ we let D_e denote the set $D_e=\{x\,|\,x-e\in D\}$ and define the sets A', B', C' by

$$A'=A'(x,e,\beta)=H_\beta(x)\cdot[H_\beta(x+e)_{-e}] \quad \text{and thus} \quad A'_e=H_\beta(x)_e\cdot H_\beta(x+e),$$

$$B'=H_\beta(x)-A', \qquad C'=H_\beta(x+e)-A'_e,$$

then

$$|B'|_{p_\beta}+|C'|_{p_\beta}\leqq d_4(e) \qquad (x,x+e\in\bar{G}) \tag{10}$$

($|\cdot|_p$ means the p-dimensional L-measure) for a suitable modulus of continuity $d_4(t)$.

(β) The functions $g_\nu(x)$ are continuous in $\bar{G}$, the kernels k_α belong to the class $\mathscr{H}$, and the kernels k_β are bounded on the point set given by $x,\xi\in\bar{G}$, $\xi\leqq x$, $z\in E^n$ and continuous in (x,ξ,z). Moreover, each of the

[1] By a p-dimensional plane parallel to the axes we mean the set of points x which can be represented in the form $x=x_0+\alpha_1\xi_1+\cdots+\alpha_p\xi_p$ (α_i real) with the help of $p+1$ points $x_0,\xi_1,\ldots,\xi_p\in E^m$ with all the ξ_i distinct, and where ξ_i has a component 1 and all other components zero.

kernels k_β satisfies a Lipschitz condition relative to all the variables z_β

$$\left.\begin{aligned}&|k_\beta(x,\xi,z_1,\ldots,z_n)-k_\beta(x,\xi,z_1,\ldots,z_{n'},\bar z_{n'+1},\ldots,\bar z_n)|\\&\qquad\leqq L[|z_{n'+1}-\bar z_{n'+1}|+\cdots+|z_n-\bar z_n|]\end{aligned}\right\}\tag{11}$$

(provided both arguments lie in the given point set).

Accordingly there is a constant M_2 such that

$$|g_\nu(x)|\leqq M_2,\qquad \int\limits_{H_\nu(x)}|k_\nu(x,\xi,z)|(d\xi)_{p_\nu}\leqq M_2,\qquad |k_\beta(x,\xi,z)|\leqq M_2 \tag{12}$$

and a modulus of continuity $d_5(t)$ such that for $x,\bar x,\xi,\bar\xi\in\bar G,\ \xi\leqq x,\ \bar\xi\leqq\bar x$ and $|z_\nu|,|\bar z_\nu|\leqq 2nM_2$

$$|k_\beta(x,\xi,z)-k_\beta(\bar x,\bar\xi,\bar z)|\leqq d_5(x-\bar x)+d_5(\xi-\bar\xi)+d_5(z-\bar z)\,. \tag{13}$$

Moreover, without loss of generality we can assume that G is contained in the cube $0<x_\mu<N$, $N\geqq 1$.

Now, again with the help of the Schauder fixed point theorem, we show that under the given assumptions Eq. (9) has at least one solution. Here we take over the notation used in the proof of II with the following changes: In the definition of $\delta(\xi,\varepsilon)$ we replace ν by α, $1\leqq\alpha\leqq n'$, and M_1 by the constant $2nM_2$; also the first two inequalities in (6) involve only the indices α.

We consider the Banach space of continuous functions $\boldsymbol{\varphi}(x)$ in $\bar G$ with norm[1] $\max\limits_{\bar G}\|\boldsymbol{\varphi}(x)\|$ and the subset Φ of functions $\boldsymbol{\varphi}$ for which $\max\limits_{\bar G}\|\boldsymbol{\varphi}(x)\|\leqq 2nM_2$ and

$$\left.\begin{aligned}&|\varphi_\alpha(x)-\varphi_\alpha(x+e)\|\leqq d(e)+2d_2(e)+d_3(e)=\bar d(e)\\&|\varphi_\beta(x)-\varphi_\beta(x+e)|\leqq e^{\gamma(x_1+\cdots+x_m)}d^*(e)\qquad(x,x+e\in\bar G)\end{aligned}\right\}. \tag{14}$$

The number $\gamma>0$ and the modulus of continuity $d^*(t)$ will be defined later. The mapping corresponding to Eq. (9) is again given by $\boldsymbol{\varphi}\to\boldsymbol{\psi}=\boldsymbol{g}+\boldsymbol{K}\boldsymbol{\varphi}$, $\boldsymbol{\varphi}\in\Phi$; however, the element of integration in the definition (3) is $(d\xi)_{p_\nu}$ instead of $d\xi$. The simple proof of the following facts is left to the reader: The mapping is continuous (proof for the α-components as in II, for the β-components with (13)); we have $\|\boldsymbol{\psi}\|\leqq 2nM_2$; Φ is convex and compact (the latter is trivial); we have $|\psi_\alpha(x)-\psi_\alpha(x+e)|\leqq\bar d(e)$ (proof as in II). Hence it remains only to show (and here is the difficulty of the proof) that

$$|\psi_\beta(x)-\psi_\beta(x+e)|\leqq e^{\gamma(x_1+\cdots+x_m)}d^*(e)\,; \tag{15}$$

more precisely, that γ and $d^*(t)$ can be determined so that (15) holds.

[1] Now, as before, $\|z\|=|z_1|+\cdots+|z_n|$.

From (10), (12), (13) (we write p for p_β) follows

$$|\psi_\beta(x)-\psi_\beta(x+e)| \leqq |g_\beta(x)-g_\beta(x+e)| + |(\boldsymbol{K}\varphi)_\beta(x)-(\boldsymbol{K}\varphi)_\beta(x+e)|$$

$$\leqq d_3(e) + \left|\int_{A'} k_\beta(x,\xi,\varphi(\xi))(d\xi)_p - \int_{A'_e} k_\beta(x+e,\xi,\varphi(\xi))(d\xi)_p\right|$$

$$+ \int_{B'} |k_\beta(x,\xi,\varphi)|\,(d\xi)_p + \int_{C'} |k_\beta(x+e,\xi,\varphi)|\,(d\xi)_p$$

$$\leqq d_3(e) + M_2 d_4(e) + \int_{A'} |k_\beta(x,\xi,\varphi(\xi)) - k_\beta(x+e,\xi+e,\varphi(\xi+e))|\,(d\xi)_p\,.$$

But now according to (11) and (13)

$$|k_\beta(x,\xi,\varphi(\xi)) - k_\beta(x+e,\xi+e,\varphi(\xi+e))|$$
$$\leqq 2d_5(e) + d_5(n'\bar{d}(e)) + L(n-n')d^*(e)\,e^{\gamma(\xi_1+\cdots+\xi_m)}$$

and (if, for example, $d\xi_1$ occurs in $(d\xi)_p$)

$$\int_{A'} e^{\gamma(\xi_1+\cdots+\xi_m)}(d\xi)_p \leqq e^{\gamma(x_2+\cdots+x_m)} N^{p-1}\int_0^{x_1} e^{\gamma\xi_1}\,d\xi_1 \leqq \frac{N^m}{\gamma}\,e^{\gamma(x_1+\cdots+x_m)},$$

and thus

$$|\psi_\beta(x)-\psi_\beta(x+e)| \leqq d_3(e) + M_2 d_4(e)$$
$$+ N^m\left[2d_5(e) + d_5(n\bar{d}(e)) + \frac{Ln}{\gamma}\,d^*(e)\,e^{\gamma(x_1+\cdots+x_m)}\right].$$

From this inequality we see that the requirement (15) can easily be satisfied; we let $\gamma = 2LnN^m$ and $d^*(t) = 2[d_3(t) + M_2 d_4 + 2N^m d_5(t) + N^m d_5(n\bar{d}(t))]$. With this choice of γ and $d^*(t)$ the set Φ is mapped into itself, which alone remained to be proved.

V. Existence Theorem. *The system of integral equations* (9) *under the hypotheses* IV (α) (β) *has at least one continuous solution in* $\bar{G}$.

VI. Remarks. (α) We note that the sharper hypotheses of V relate only to kernels k_β and that no Lipschitz condition is prescribed in the z_α. The case where all k_ν satisfy a Lipschitz condition in all variables $z_\varrho(\nu, \varrho = 1, \ldots, n)$ is significantly easier to handle by successive approximation; see VIII below. The type of equation studied by Kasatkina (1965, 1967) is a special case of (9) (in the second paper the assumptions are of Carathéodory type).

(β) The existence theorem also holds if the kernels k_α are only in $\mathscr{H}_0$ and the kernels k_β are not necessarily bounded. In general, then the solution exists not in all of $\bar{G}$, but only at those points of $\bar{G}$ which belong to a neighborhood of R_ν. This is shown in the usual manner by considering instead of $k_\nu(x,\xi,z)$ the kernels $k_\nu^C(x,\xi,z)=k_\nu(x,\xi,[z_1]_C,\ldots,[z_n]_C)$, where $[s]_C=\min(|s|,C)\operatorname{sgn}s$. For $C>0$ these kernels are in $\mathscr{H}$ for $\nu=\alpha$ and bounded for $\nu=\beta$.

In an important special case, however, the original statement of V *can be preserved; namely, if*

$$|k_\alpha(x,\xi,z)|\leqq h(\xi)(1+\|z\|),\qquad |k_\beta(x,\xi,z)|\leqq M(1+\|z\|) \tag{16}$$

or, more generally,

$$|k_\alpha(x,\xi,z)|\leqq h(\xi)\varphi(\|z\|),\qquad |k_\beta(x,\xi,z)|\leqq \varphi(\|z\|) \tag{16'}$$

hold, where $\varphi(s)$ is continuous and increasing in s for $s\geqq 0$,

$$\int^\infty \frac{ds}{\varphi(s)}=\infty\,,$$

and $h(\xi)\in L(\bar G)$.

To see this, we choose a continuous function w in $\bar G$ such that $\|\boldsymbol{u}\|<w$ for every solution $\boldsymbol{u}$, and instead of k_ν again consider the kernels k_ν^C with $C>w(x)$ which satisfy the hypotheses of V. We show how to obtain such an a priori bound w in the case (16′). The integral operator L defined for $u(x)\geqq 0$ by

$$(Lu)(x)=\sum_\alpha\int_{H_\alpha(x)} h(\xi)\varphi(u(\xi))d\xi+\sum_\beta\int_{H_\beta(x)}\varphi(u(\xi))(d\xi)_{p_\beta}$$

is monotone increasing in the sense of 17 II. Let $\|z\|=|z_1|+\cdots+|z_n|$, $\|\boldsymbol{g}(x)\|<C$ and $v(x)=\|\boldsymbol{u}(x)\|$, where $\boldsymbol{u}$ is a solution of (9). From (9) and (16′) we get $v<C+Lv$. We will construct a function w satisfying

$$w\geqq C+Lw\quad\text{in}\quad \bar G\,. \tag{17}$$

Then from 17 III the inequality $\|\boldsymbol{u}(x)\|=v(x)<w(x)$ follows (in the domain of existence of $\boldsymbol{u}$).

Let $s(x)=x_1+\cdots+x_n$. Without loss of generality we may suppose that G lies between the hyperplanes $s(x)=0$ and $s(x)=b>0$. We denote by $h_1(s_0)$ the $(n-1)$-dimensional integral of $h(\xi)$ over the region $\{\xi: s(\xi)=s_0\}\cap\bar G$. Clearly $h_1(s)\in L(0,b)$. For a function $w(s)$ of one variable we have

$$\int_{H_\alpha(x)} h(\xi)\varphi(w(s(\xi)))d\xi\leqq\int_0^{s(x)} h_1(s)\varphi(w(s))ds\,.$$

Furthermore

$$\int_{H_\beta(x)}\varphi(w(s(\xi)))(d\xi)_{p_\beta}\leqq b^n\int_0^{s(x)}\varphi(w(s))ds\,,$$

as is seen by performing only one of the p_β integrations. Combining these inequalities we get

$$Lw(s(x)) \leqq \int_0^{s(x)} h_2(s)\varphi(w(s))ds\,,$$

where $h_2 \in L(0, b)$ depends only on $h(\xi)$ and the size of G. Now let $w(s)$ be the solution of

$$w' = h_2(s)\varphi(w) \quad \text{in} \quad 0 \leqq s \leqq b, \qquad w(0) = C\,.$$

Since

$$w(s_0) = C + \int_0^{s_0} h_2(s)\varphi(w(s))ds\,,$$

it follows that the function $w(s(x))$ satisfies the inequality (17). The solution $w(s)$ of the above initial value problem exists and is bounded in $0 \leqq s \leqq b$ for each $b > 0$, i.e., (17) holds in $\bar{G}$.

This result generalizes one proved by Walter (1967).

For hyperbolic differential equations similar results have been given by Kisyński (1959a), Walter (1959a, p. 447–448), Shanahan (1962), Kisyński and Tym (1964), Aziz and Maloney (1965/66).

(γ) We note without proof that we can replace the Lipschitz condition (11) by weaker conditions which are similar to those given in the "hyperbolic case" by several authors; see Walter (1967).

(δ) In 2 VI we indicated how an existence theorem is proved for unbounded difference kernels. Similar reasoning can also be carried out for multidimensional difference kernels of the form $k_\nu(x, \xi, z) = k_\nu^*(x - \xi, z)$.

(ε) For ordinary differential equations $u' = f(t, u)$ with unbounded f a fundamental theorem states that every solution can be extended to the boundary of $D(f)$, the domain of definition of f. The same is true for one-dimensional Volterra integral equations. In the case $m > 1$ the situation is more intricate. First we need some definitions.

A subset C of $\bar{G}$ is called a V-set, if $x \in C$ implies $G(x) \subset C$. Obviously it makes sense to speak of a "solution of (9) in C" if C is a V-set. The notions "G-boundary of C", "C is G-open" refer to the relative topology in $\bar{G}$. The following theorem on the extension of solutions was proved by Walter (1967).

Suppose that IV (α) (β) *holds, but with* $k_\alpha \in \mathscr{H}_0$ *and* k_β *not necessarily bounded, and that for each* $x \in \bar{G}$ *the set* $H_\nu(x)$ *is contained in a* p_ν*-dimensional hyperplane parallel to the coordinate axes which contains the point* x. *Let* $\boldsymbol{u}^*$ *be a solution of* (9) *in a* G*-open* V*-set* C^* (*the existence is guaranteed by the remark in* (β)). *Then there exists a maximal extension of* $\boldsymbol{u}^*$, i.e., *a solution* $\boldsymbol{u}(x)$ *in a* G*-open* V*-set* $C \supset C^*$, *which is equal to* $\boldsymbol{u}^*$ *in* C^* *and which cannot be extended. For* x_0 *on the* G*-boundary of* C, $\boldsymbol{u}(x)$ *is unbounded in* $G(x_0) \cap C$.

VII. Maximal and Minimal Solutions *for Eq.* (9) *exist under the hypotheses of* V *if* $\boldsymbol{k}(x, \xi, z) = (k_1, \ldots, k_n)$ *is monotone increasing in* z. *If* $\boldsymbol{u}^*$ *is the maximal solution,* $\boldsymbol{u}_*$ *the minimal solution, then* $\boldsymbol{u}_* \leqq \boldsymbol{u} \leqq \boldsymbol{u}^*$ *for every solution* $\boldsymbol{u}$ *of* (9). *The minimal [maximal] solution can be approximated arbitrarily well by functions* $\boldsymbol{v}$ [$\boldsymbol{w}$] *which satisfy the corresponding inequality* $\boldsymbol{v} < \boldsymbol{g} + \boldsymbol{K}\boldsymbol{v}$ [$\boldsymbol{w} > \boldsymbol{g} + \boldsymbol{K}\boldsymbol{w}$].

The idea of the proof corresponds to that of 2 IV. If $\boldsymbol{u}(x; \boldsymbol{\varepsilon})$, $\boldsymbol{\varepsilon} \in E^n$, is a solution of (9) when g_ν is replaced by $g_\nu + \varepsilon_\nu$, then by 17 III we have $\boldsymbol{u}(x; \boldsymbol{\varepsilon}) < \boldsymbol{u}(x; \bar{\boldsymbol{\varepsilon}})$ for $\boldsymbol{\varepsilon} < \bar{\boldsymbol{\varepsilon}}$. Further, all these solutions (for, say, $\|\boldsymbol{\varepsilon}\| < 1$) are equicontinuous, as we conclude from the considerations given in the proof of V. For a sequence $\boldsymbol{\varepsilon} = \boldsymbol{\varepsilon}_n$ which monotonely increases or decreases to $\mathbf{0}$, the associated sequence $\boldsymbol{u}(x; \boldsymbol{\varepsilon}_n)$ is also monotone and thus uniformly convergent. The limit functions $\boldsymbol{u}_*$ or $\boldsymbol{u}^*$ are then solutions of (9) with the asserted properties.

In 2 V we learned of a second way to construct maximal and minimal solutions. This is also practical here. But it has little meaning for $m > 1$ because the property 2 V (β) is usually lost (though it holds, for example, for the example given in III).

VIII. Successive Approximation. Under the hypotheses of the existence theorems II and V we have in general neither uniqueness nor convergence of successive approximation. In what follows we state assumptions under which these properties are preserved.

Suppose $\boldsymbol{k}(x, \xi, z)$ satisfies a condition (with a norm $\|\cdot\|$)

$$\|\boldsymbol{k}(x, \xi, \boldsymbol{z}) - \boldsymbol{k}(x, \xi, \bar{\boldsymbol{z}})\| \leqq \omega(x, \xi, \|\boldsymbol{z} - \bar{\boldsymbol{z}}\|) \tag{18}$$

or

$$|k_\nu(x, \xi, \boldsymbol{z}) - k_\nu(x, \xi, \bar{\boldsymbol{z}})| \leqq \omega_\nu(x, \xi, |\boldsymbol{z} - \bar{\boldsymbol{z}}|) \quad (\nu = 1, \ldots, n). \tag{18'}$$

We introduce the assumptions:

(α) $\omega(x, \xi, z) \in \mathscr{H}_0$ is real-valued and monotone increasing in z;

(α') $\omega_\alpha(x, \xi, \boldsymbol{z}) \in \mathscr{H}$, $\omega_\beta(x, \xi, \boldsymbol{z})$ is bounded and continuous in $(x, \xi, \boldsymbol{z})$, $\boldsymbol{\omega}(x, \xi, \boldsymbol{z})$ is monotone increasing in $\boldsymbol{z}$;

(β) the equation $\sigma = \Omega\sigma$ has only one solution in $C(\bar{G})$, namely $\sigma = 0$;

(β') the equation $\boldsymbol{\sigma} = \boldsymbol{\Omega}\boldsymbol{\sigma}$ has only one solution which is bounded and such that σ_α is continuous, σ_β is upper semicontinuous in $\bar{G}$, namely $\boldsymbol{\sigma} = \mathbf{0}$.

Here Ω and $\boldsymbol{\Omega}$ are the integral operators defined in (19.17) and (19.15).

Then the following two theorems hold.

Under the assumptions of Theorem II *and* (18), (α) (β) *Eq.* (1) *has exactly one solution* $\boldsymbol{u}$. *The sequence* $(\boldsymbol{u}_n)$, *defined by* $\boldsymbol{u}_{n+1} = \boldsymbol{g} + \boldsymbol{K}\boldsymbol{u}_n$, *where* $\boldsymbol{u}_0 \in C(\bar{G})$ *is arbitrary, converges uniformly in* $\bar{G}$ *to* $\boldsymbol{u}$.

Under the assumptions IV $(\alpha)(\beta)$, *but without* (11), *and* (18'), $(\alpha')(\beta')$ *Eq.* (9) *has exactly one solution. Again we have uniform convergence of the sequence obtained by successive approximation.*

These theorems, in formulation and proof, are straightforward translations of the corresponding one-dimensional Theorem 7 XII. The assumptions $(\alpha)(\alpha')$ and $(\beta)(\beta')$ correspond to the assumptions (α) and (β) of 7 XII. The theorems remain true if we dispense with the boundedness assumptions on ω or $\boldsymbol{\omega}$ ($\mathcal{H}$ has to be replaced by $\mathcal{H}_0$) and add an assumption similar to 7 XII (γ). But this procedure does not produce more general theorems since we have boundedness assumptions on k; see the remarks at the end of 7 XII.

Both theorems apply, for instance, if an Osgood condition

$$\|\boldsymbol{k}(x,\xi,z)-\boldsymbol{k}(x,\xi,\bar{z})\| \leqq \psi(\|z-\bar{z}\|)$$

is satisfied, where ψ has the properties stated in 4 V (β). It is easily seen that in this case the functions $\omega(x,\xi,z)=\psi(z)$ resp. $\omega_\nu(x,\xi,z)=\psi(\|z\|)$ have the above properties.

The two convergence theorems are taken from Walter (1965). The case of a Lipschitz condition can also be handled most easily by an adaptation of the device developed at the end of 7 XII, i.e. by working with a weighted maximum norm; cf. Walter (1965). For hyperbolic differential equations the question of convergence of successive approximation has been treated by many authors, starting with the classical paper of Picard (1896). Conditions more general than a Lipschitz condition were first given by Walter (1959a, 1960) and Kisyński (1960). Conditions of the Krasnosel'skii-Krein type were investigated by Palczewski (1963/64), Palczewski and Pawelski (1963/64), Wong (1965, 1966). Walter (1970b) showed that these conditions are in fact special cases of the condition (18), $(\alpha)(\beta)$ given above; compare 14 IV (α) for the analogous situation in the one-dimensional case. — See also Bielecki (1956), Shanahan (1960), Kisyński and Tym (1964), Pelczar (1964).

19. Estimates for Integral Equations

In the following sections of this chapter we shall particularize the results obtained so far to Volterra integral equations and hyperbolic differential equations. First we discuss integral equations.

I. Super- and Subfunctions $(Z_c(\boldsymbol{k}))$. We consider a Volterra integral equation

$$u(x)=g(x)+\int_{H(x)} k(x,\xi,u(\xi))\,d\xi \quad \text{in} \quad \bar{G} \quad (H(x)\subset G(x)) \tag{1}$$

and more generally a system of Volterra integral equations

$$u_\nu(x)=g_\nu(x)+\int_{H_\nu(x)} k_\nu(x,\xi,\boldsymbol{u}(\xi))(d\xi)_{p_\nu} \qquad (\nu=1,\ldots,n;\, H_\nu(x)\subset G(x)) \quad (2)$$

in the sense of 18 IV [1].

Those continuous functions $\boldsymbol{\varphi}(x)$ in $\bar{G}$ for which $k_\nu(x,\xi,\boldsymbol{\varphi}(\xi))$ for $x\in\bar{G}$ is defined on the set $H_\nu(x)$ and is integrable over this set, belong to the class $Z_c(\boldsymbol{k})$[2]. For the integral operator $\boldsymbol{K}$ corresponding to Eq. (2) we have $(\boldsymbol{K}\varphi)(x)=\boldsymbol{0}$ on R_v. Further, $\boldsymbol{K}$ is a monotone increasing operator if $\boldsymbol{k}(x,\xi,\boldsymbol{z})=(k_1,\ldots,k_n)$ is monotone increasing in $\boldsymbol{z}$ (see the definition in 6 II). Thus according to 17 V the function $\boldsymbol{v}$ or $\boldsymbol{w}\in Z_c(\boldsymbol{k})$ respectively is a sub- or superfunction respectively for monotone increasing $\boldsymbol{k}$ if 17 V (β), i.e.,

$$\left.\begin{aligned} v_\nu(x)&<g_\nu(x)+\int_{H_\nu(x)} k_\nu(x,\xi,\boldsymbol{v})(d\xi)_{p_\nu}\\ w_\nu(x)&>g_\nu(x)+\int_{H_\nu(x)} k_\nu(x,\xi,\boldsymbol{w})(d\xi)_{p_\nu}\end{aligned}\right\} \qquad (\beta)$$

holds in $\bar{G}$. By 17 XI (γ) equality signs are allowed here if $\boldsymbol{K}$ satisfies a uniqueness condition (see the Examples VI).

II. The Linear Integral Equation

$$u(x)=g(x)+\int_{H(x)} u(\xi)h(\xi)d\xi \qquad (3)$$

is considered under the assumption that $g(x)$ is continuous in $\bar{G}$, $h(x)$ is integrable over $\bar{G}$ and $H(x)\subset G(x)$ has the property 18 II (α).

(α) The solution (which exists by 18 II) can be represented as a Neumann series

$$u=\sum_{i=0}^{\infty} H^i g \text{ with } H\varphi=\int_{H(x)} h(\xi)\varphi(\xi)d\xi,\ H^0\varphi=\varphi,\ H^{i+1}\varphi=H(H^i\varphi). \qquad (4)$$

We carry out the proof in the usual way by showing that H, considered as a linear transformation of a suitable Banach space, has a norm <1. For this we choose the space of continuous functions in $\bar{G}$ with the norm

$$\|\varphi\|=\max_{\bar{G}}|\varphi(x)|\,e^{-2M(x)},\quad M(x)=\int_{\xi\leqq x}|h(\xi)|d\xi\,,$$

[1] We can also allow more general systems of equations, such as those with Stieltjes integrals and with domains $H_\nu(x)$ which do not lie in a plane parallel to the axes.

[2] This class also depends on the $H_\nu(x)$, which is not expressed explicitly in the notation.

where we set $h=0$ outside of $\bar{G}$. Now we have

$$|(H\varphi)(x)|e^{-2M(x)} \leqq e^{-2M(x)} \int_{H(x)} |h(\xi)\varphi(\xi)| e^{2M(\xi)-2M(\xi)} d\xi$$
$$\leqq \|\varphi\| e^{-2M(x)} \int_{\xi \leqq x} |h(\xi)| e^{2M(\xi)} d\xi \leqq \tfrac{1}{2} \|\varphi\|,$$

the latter because the function $z(x)=e^{2M(x)}$ has the properties

$$z_{x_1 x_2 \dots x_\mu} \geqq 0 \quad (\mu=1,\dots,m-1), \qquad z_{x_1 \dots x_m} \geqq 2|h(x)|z(x),$$
$$2 \int_{\xi \leqq x} |h(\xi)| z(\xi) d\xi \leqq \int_{-a}^{x_1} \cdots \int_{-a}^{x_m} z_{x_1 \dots x_m}(\xi) d\xi_1 \dots d\xi_m \leqq z(x)$$

(with a so large that G lies in the domain $x_\mu > -a$) [1]. Accordingly $\|H\| = \max_{\|\varphi\|=1} \|H\varphi\| \leqq \frac{1}{2}$. Incidentally, uniqueness is immediately obtained form this (it also follows from later theorems). Thus if w is the difference of two solutions of (3), then $w=Hw$ and thus $\|w\| = \|Hw\| \leqq \|w\|/2$, i.e., $w=0$.

(β) With the help of a "resolvent kernel" we can write the Neumann series (4) as an integral. We shall indicate briefly the derivation for the case $H(x)=G(x)$. Thus suppose $h(x)=0$ outside of $\bar{G}$ and $H(x)=G(x)$ for all x. Then for $i \geqq 0$ (proof by induction)

$$(H^i\varphi)(x) = \int_{G(x)} h(\xi) h_i(x,\xi) \varphi(\xi) d\xi \tag{5}$$

with

$$h_1(x,\xi)=1, \quad h_{i+1}(x,\xi) = \int_{\xi \leqq \eta \leqq x} h(\eta) h_i(\eta,\xi) d\eta \tag{6}$$

(note that we can integrate over all $\xi \leqq x$ instead of over $G(x)$). Now if we set (assuming convergence)

$$h^*(x,\xi) = \sum_{i=1}^{\infty} h_i(x,\xi), \tag{7}$$

then by (4), we obtain, for the solution u in the case $G(x)=H(x)$,

$$u(x) = g(x) + \int_{G(x)} g(\xi) h(\xi) h^*(x,\xi) d\xi \quad \text{in } \bar{G}. \tag{4'}$$

If $\bar{h}(x)$ is a nonnegative integrable function, if $|h(x)| \leqq \bar{h}(x)$, and if the operator $\bar{H}$ and the functions $\bar{h}_i(x,\xi)$ are defined corresponding to the above, then we can show without difficulty (by induction) that

$$\left.\begin{aligned} &|h_i(x,\xi)| \leqq \bar{h}_i(x,\xi) \leqq \bar{H}^{i-1}\psi \\ &\text{with} \quad \psi(x) \equiv 1 \end{aligned}\right\}. \tag{8}$$

[1] It follows from Fubini's theorem (or by approximating h by continuous functions) that this reasoning, which is transparent for continuous h, also works in the case $h \in L(\bar{G})$.

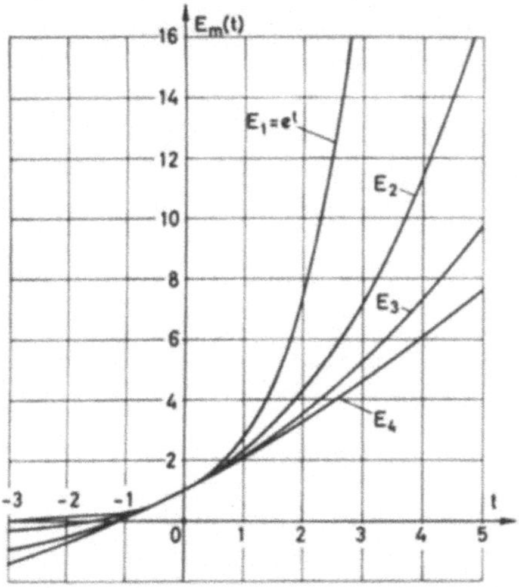

Fig. 5. The functions $E_m(t)$ for $m = 1, 2, 3, 4$

If, in particular, we let $\bar{h} = |h|$, then $\|\bar{H}\| \leqq \frac{1}{2}$, $\|\bar{H}^{i-1}\| \leqq 2^{1-i}$ and hence the series (7) is absolutely and uniformly convergent.

In particular, for a product

$$\left.\begin{aligned} &h(x) = l_1(x_1) \dots l_m(x_m)\,, \\ &l_\mu(t) \in L(a_\mu \leqq t \leqq b_\mu)\,, \end{aligned}\right\} \tag{9}$$

we obtain

$$h_{i+1}(x, \xi) = \frac{1}{(i!)^m} \prod_{\mu=1}^{m} [L_\mu(x_\mu) - L_\mu(\xi_\mu)]^i \qquad (i \geqq 0)\,,$$

where

$$L_\mu(t) = \int_{a_\mu}^{t} l_\mu(s)\,ds\,.$$

Thus

$$h^*(x, \xi) = E_m([L_1(x_1) - L_1(\xi_1)] \dots [L_m(x_m) - L_m(\xi_m)]) \tag{10}$$

holds for $a \leqq \xi \leqq x \leqq b$, $a = (a_1, \dots, a_m)$, $b = (b_1, \dots, b_m)$. Here we have used the notation [1]

$$E_m(t) = \sum_{i=0}^{\infty} \frac{t^i}{(i!)^m}\,, \quad \text{in particular,} \quad E_1(t) = e^t,\ E_2(t) = I_0(2\sqrt{t})\,; \tag{11}$$

see Fig. 5.

III. Gronwall's Inequality for Several Variables. *Suppose* $v(x)$, $g(x) \in C(\bar{G})$, $0 \leqq h(x) \in L(\bar{G})$, *and*

$$v(x) \leqq g(x) + \int_{G(x)} h(\xi) v(\xi)\,d\xi \quad \textit{in} \quad \bar{G}\,. \tag{12}$$

[1] I_0 is the modified Bessel function of order 0.

Then if we set $h(x)=0$ *outside of* $\bar{G}$ *and* $h^*(x,\xi)$ *is defined according to* (6), (7),

$$v(x) \leqq g(x) + \int_{G(x)} g(\xi)h(\xi)h^*(x,\xi)d\xi \quad in \quad \bar{G}. \tag{13}$$

If in addition we assume $v(x) \geqq 0$, *then* (13) *also follows from a bound*

$$v(x) \leqq g(x) + \int_{H(x)} h(\xi)v(\xi)d\xi \quad in \quad \bar{G} \quad with \quad H(x) \subset G(x). \tag{12'}$$

If $g(x)=A$ *and* $h(x)=L$ *are constant and if* G *lies in* $\{x>0\}$, *then we have the bound*

$$v(x) \leqq A E_m(Lx_1 \dots x_m). \tag{13'}$$

First we see that for $v \geqq 0$ (12′) implies the inequality (12), so that we must concern ourselves only with the assertion proposed in connection with (12). But this follows at once from (4′) and 17 III applied to the operator H and the functions v and $u=w$. Finally we obtain the bound (13′) if according to (10) we substitute the function $E_m(L(x_1-\xi_1) \dots L(x_m-\xi_m))$ for h^* in (13) and integrate this infinite series term by term over $0 \leqq \xi \leqq x$. (Note: h^* in general is not exactly this function since we set $h=0$ outside $\bar{G}$; however, if we continue h outside $\bar{G}$ by $L>0$ rather than by 0, then h^* is increased.)

We remark that the assertion relative to (12′) is false without the hypothesis $v \geqq 0$. Counterexample: $m=1$, $G(x)=[0,x]$, $H(x)=\emptyset$, $g(x)=-1$, $v(x)=-1$; from (13) we get $-1 \leqq -e^x$!

IV. Estimation Theorems. Although no difficulties arise in applying the estimation theorems VI and VII of Section 17 to integral equations, several remarks will be of use. If we have a system (2), then in order to apply 17 VI we need a bound for the (vector) kernel $\boldsymbol{k}$ by a (vector) kernel $\boldsymbol{\omega}$ which is monotone increasing in $\boldsymbol{z}$

$$|\boldsymbol{k}(x,\xi,\boldsymbol{z}) - \boldsymbol{k}(x,\xi,\bar{\boldsymbol{z}})| \leqq \boldsymbol{\omega}(x,\xi,|\boldsymbol{z}-\bar{\boldsymbol{z}}|), \tag{14}$$

and we consider the integral operator

$$(\boldsymbol{\Omega}\boldsymbol{\varphi})_\nu = \int_{H'_\nu(x)} \omega_\nu(x,\xi,\boldsymbol{\varphi})(d\xi)_{p_\nu} \quad (\nu=1,\dots,n), \tag{15}$$

generated by $\boldsymbol{\omega}$, where $H_\nu(x) \subset H'_\nu(x) \subset G(x)$ (and H'_ν is p_ν-dimensional). The condition (17.3) then follows from (14) and (15).

The use of K-norms and Theorem 17 VII is convenient if all $H_\nu(x)$ are of the same dimension p ($1 \leqq p \leqq m$). Under this assumption, (17.4) is a consequence of

$$\|\boldsymbol{k}(x,\xi,\boldsymbol{z}) - \boldsymbol{k}(x,\xi,\bar{\boldsymbol{z}})\| \leqq \omega(x,\xi,\|\boldsymbol{z}-\bar{\boldsymbol{z}}\|), \tag{16}$$

if we set

$$(\Omega\varphi)(x) = \int_{H(x)} \omega(x, \xi, \varphi)(d\xi)_p \text{ with } H_1(x) + \cdots + H_n(x) \subset H(x) \subset G(x) \quad (17)$$

($H(x)$ is p-dimensional). Of course the kernel $\omega(x, \xi, z)$ in (16) must be monotone increasing in z.

If on the other hand not all the p_ν are equal, then difficulties arise in Theorem 17 VII. They make it desirable to have available a theorem which, in a certain sense, is half-way between 17 VI and 17 VII, where certain components (namely those of the same dimensionality) are grouped together. Such a possibility has already been indicated, at the beginning of Section 7. Since the general formulation is rather awkward, we restrict ourselves to setting up such a theorem for systems of hyperbolic differential equations in Section 21.

For the

V. Uniqueness Criteria we also content ourselves with a few remarks. Here we cannot just define a relationship $\boldsymbol{\omega}(x, \xi, \boldsymbol{z}) \in \mathscr{D}^n$ as in 4 II, since an operator $\boldsymbol{\Omega}$ is determined only by the simultaneous assignment of the kernel $\boldsymbol{\omega}$ and the domains $H_\nu(x)$ and it may very well happen that a bound (14) or (16) is sufficient for uniqueness for certain domains H_ν and not for other domains H_ν for one and the same right hand side $\boldsymbol{\omega}$. Therefore we shall use the notation $\boldsymbol{\omega} \in \mathscr{D}^n$ only when *every* integral operator $\boldsymbol{\Omega}$ generated by $\boldsymbol{\omega}$ belongs to the class $\mathscr{D}^n$ in the sense of 17 VIII (i.e., for arbitrary choice of G and $H_\nu(x)$). We shall also dispense with the introduction of a class $\mathscr{D}_1^n$, analogous to the class $\mathscr{E}_1^n$ in 6 V, which provides for the case where the difference between two arbitrary solutions vanishes of a higher order on R_v (we shall go into this case for hyperbolic differential equations).

VI. Examples. (α). *Lipschitz Condition.* The function $\boldsymbol{\omega}$ defined by

$$\omega_\nu(x, \xi, \boldsymbol{z}) = L(z_1 + \cdots + z_n) \qquad (L > 0) \tag{18}$$

is in $\mathscr{D}^n$. *Thus the system* (2) *with G and $H_\nu(x)$ arbitrary has at most one solution if* **k** *satisfies a Lipschitz condition.*

In this case it is easy to obtain an explicit error bound for approximate solutions. Suppose $\boldsymbol{u}$ is a solution of (2), $\boldsymbol{v}$ is an approximate solution, and we have (d constant)

$$\left|v_\nu(x) - g_\nu(x) - \int_{H_\nu(x)} k_\nu(x, \xi, \boldsymbol{v})(d\xi)_{p_\nu}\right| \leqq d \qquad (\nu = 1, \ldots, n). \tag{19}$$

Furthermore, for the sake of simple notation we assume that the domain G lies entirely in $\{x > 0\}$. Then we claim that the inequality

$$|v_\nu(x) - u_\nu(x)| \leqq d e^{Ln(x_1 + \cdots + x_m)} \quad \text{in } \bar{G} \qquad (\nu = 1, \ldots, n), \tag{20}$$

holds if in addition $L \geqq 1/n$.

In the proof we assume for simplicity that in the νth Eq. (2) we integrate with respect to the variables $\xi_1, \xi_2, \ldots, \xi_{p_\nu}$. If we denote the right hand side of (20) by $\varphi(x)$ and write p instead of p_ν, then, since φ is monotone increasing in all variables,

$$\int_{H_\nu(x)} \varphi(\xi)(d\xi)_p \leqq \int_0^{x_1} \cdots \int_0^{x_p} d e^{Ln(\xi_1+\cdots+\xi_p+x_{p+1}+\cdots+x_m)} d\xi_1 \ldots d\xi_p$$

$$\leqq d(Ln)^{-p} e^{Ln(x_{p+1}+\cdots+x_m)} \prod_{\varrho=1}^{p} (e^{Lnx_\varrho}-1) \leqq \frac{d}{Ln}\,[e^{Ln(x_1+\cdots+x_m)}-1]$$

(and we obviously obtain the same result if we integrate with respect to different ξ_ν). Thus if we set $\varrho_\nu(x)=\varphi(x)$ for all ν, by (18) we have

$$\int_{H_\nu(x)} \omega_\nu(x,\xi,\varrho)(d\xi)_p = \int_{H_\nu(x)} Ln\varphi(\xi)(d\xi)_p \leqq \varphi(x)-d\,.$$

Now all the hypotheses of 17 VI are satisfied [1]. Of course (20) immediately implies the above uniqueness statement as well as a theorem on the continuous dependence of the solution on $\boldsymbol{k}$ and $\boldsymbol{g}$.

(β) *Generalized Lipschitz Condition.* If we are dealing with m-dimensional integrals ($p_\nu = m$) in (2) for all ν, then we can also allow a generalized Lipschitz condition in which there is an integrable function $l(\xi)$ in place of the Lipschitz constant L as uniqueness condition.

This will again be obtained as a special case of an estimate for which we now refer to 17 VII in contrast to (α). Accordingly, suppose the Lipschitz condition

$$\|\boldsymbol{k}(x,\xi,z)-\boldsymbol{k}(x,\xi,\bar{z})\| \leqq l(\xi)\,\|z-\bar{z}\| \tag{21}$$

holds for a K-norm and a nonnegative function $l(\xi)\in L(G)$ (provided both arguments lie in $D(\boldsymbol{k})$). Suppose the function $\boldsymbol{u}$ is a solution of (2), and that for $\boldsymbol{v}\in Z(\boldsymbol{k})$ we have

$$\|\boldsymbol{v}-\boldsymbol{g}-\boldsymbol{K}\boldsymbol{v}\| \leqq d(x) \quad \text{in} \quad \bar{G} \tag{22}$$

with a function $d(x)$ continuous in G ($\boldsymbol{K}$ is the operator defined in (18.3)).

Then we have the bound

$$\|\boldsymbol{v}-\boldsymbol{u}\| \leqq d(x) + \int_{G(x)} d(\xi)l(\xi)l^*(x,\xi)d\xi \quad \text{in} \quad \bar{G}\,, \tag{23}$$

where $l^*(x,\xi)$ is defined with the help of $l(\xi)$ exactly like $h^*(x,\xi)$ with $h(\xi)$ in (6) (7).

In particular for constant $l(\xi)=L$

$$\|\boldsymbol{v}-\boldsymbol{u}\| \leqq d(x) + L\int_{G(x)} d(\xi)E_m(L(x_1-\xi_1)\ldots(x_m-\xi_m))d\xi\,. \tag{23'}$$

If moreover $d(x)=d$ is constant and G lies in the domain $\{x>0\}$, then we obtain the particularly simple formula

$$\|\boldsymbol{v}-\boldsymbol{u}\| \leqq dE_m(Lx_1\ldots x_m)\,. \tag{23''}$$

[1] The $<$ symbol in 17 VI (γ) is obtained in the usual way, e.g., by beginning with $d+\varepsilon$ instead of d and then letting $\varepsilon \to 0$.

Formula (23) follows directly from 17 VII and II, in particular (4′). The special cases (23′) and (23″) are then obtained from (10)(11) by elementary computations (term by term integration of the series (11) over $0\leqq\xi\leqq x$).

Finally, we want to give formula (23″) a form convenient for applications. If $d(x)$ is a function determined according to (22), then we have

$$\|\boldsymbol{v}-\boldsymbol{u}\|\leqq E_m(Lx_1\ldots x_m)\max_{\xi\in G(x)} d(\xi). \tag{23‴}$$

This is clear since, for fixed x, we are allowed to consider the domain $G(x)$ instead of $\bar{G}$ and to use (23‴) on it.

(γ) *Osgood's Condition.* If $\psi(z)$ has the properties noted in 4 V (β), then the function $\boldsymbol{\omega}$ with the components

$$\omega_\nu(x,\xi,\boldsymbol{z})=\psi(z_1+\cdots+z_n)$$

is likewise in $\mathscr{D}^n$.

For the proof we assume that G is contained in the cube $0<x_\mu<N$, $N\geqq 1$. The ordinary differential equation

$$\varphi'(t)=\alpha\psi(n\varphi) \quad \text{with} \quad \alpha=N^m$$

has arbitrarily small positive and monotone increasing solutions $\varphi(t)$ in the interval $0\leqq t\leqq mN$. If we set $\varrho_\nu(x)=\varphi(x_1+\cdots+x_n)$ for all ν and assume, say, that integration is with respect to ξ_1 in the νth Eq. (2), then by the monotonicity of φ and ψ

$$\int\limits_{H_\nu(x)}\omega_\nu(x,\xi,\boldsymbol{\varrho})(d\xi)_{p_\nu}=\int\limits_{H_\nu(x)}\psi(n\varphi)(d\xi)_{p_\nu}\leqq N^{p_\nu-1}\int\limits_0^{x_1}\psi(n\varphi(\xi_1+x_2+\cdots+x_n))d\xi_1$$
$$<\frac{N^m}{\alpha}\varphi(x_1+\cdots+x_n)=\varrho_\nu(x),$$

which is all that was left to show.

VII. Remarks. In evaluating the integral in (4′) or (23), we can sometimes make use of the fact that (almost everywhere)

$$\frac{\partial^m}{\partial\xi_1\ldots\partial\xi_m}h^*(x,\xi)=(-1)^m h(\xi)h^*(x,\xi).$$

Then — assuming appropriate differentiability properties for $d(\xi)$ — it is possible to put the integral into a more suitable form for computations by integrating by parts. This transformation was carried out in (1.4) in the case $m=1$.

The proof of the above formula is trivial in the special case (10); in the general case we have to go back to the definition of h^* and differentiate the integrals in (6).

20. The Hyperbolic Differential Equation $u_{xy}=f(x,y,u)$

If $I=[a,b]=\{x\,|\,a\leqq x\leqq b\}$, $a=(a_1,\ldots,a_m)\leqq b=(b_1,\ldots,b_m)$, is a closed m-dimensional interval and $\varphi(x)$ is a given function, then by the "interval function" $\varphi(I)$ we mean the function

$$\varphi(I)=\sum_c(-1)^{\nu(c)}\varphi(c),$$

where $c \in E^m$ runs through all corner points of I and $\nu(c)$ is the number of components a_μ occuring in c (see Kamke (1956, p. 42)). For $m=1$ $\varphi(I)=\varphi(b)-\varphi(a)$, for $m=2$: $\varphi(I)=\varphi(b_1,b_2)+\varphi(a_1,a_2)-\varphi(a_1,b_2)-\varphi(b_1,a_2)$. This concept plays a fundamental role in the theory of the Lebesgue-Stieltjes integral; we mention it here only because it lets us depict clearly the relationship between ordinary and hyperbolic differential equations. Using the notation $D=\partial^m/\partial x_1 \ldots \partial x_m$ and $|I|=(b_1-a_1)\ldots(b_m-a_m)$, we have (with suitable regularity hypotheses) $D\varphi(x)=\lim \varphi(I)/|I|$ if I is shrunk to a point x (i.e., $x \in I$, diameter of $I \to 0$). In this notation the fundamental theorem of differential and integral calculus reads ($m=1$)

$$\int_I D\varphi(x)dx=\varphi(I). \tag{1}$$

In this form the theorem is still valid for arbitrary $m \geqq 1$. For example, for $m=2$ (we write (x, y) instead of (x_1, x_2)),

$$\int_I \varphi_{xy}(x, y)dx\,dy=\varphi(I)=\varphi(b_1, b_2)+\varphi(a_1, a_2)-\varphi(a_1, b_2)-\varphi(a_2, b_1), \tag{2}$$

which is easily verified.

Now from this point of view the derivative $\varphi_{xy}(x, y)$ is the natural generalization of the derivative $\varphi'(t)$ and hence, too, the hyperbolic differential equation

$$u_{xy}=f(x, y, u) \quad \text{for} \quad u=u(x, y) \tag{3}$$

is the two-dimensional analog of the ordinary differential equation

$$u'=f(t, u) \quad \text{for} \quad u=u(t). \tag{4}$$

This close relationship can be traced in the entire theory. However, there are some essential differences, namely in estimation and uniqueness problems; we will discuss these later.

I. The Three Initial Value Problems. (*Problem* (a), (b), (c), Z^*, $Z^*(f)$.) Of the various initial value problems for Eq. (3), we shall consider three in detail here and designate them as (a), (b), (c). Let R be the rectangle $0 \leqq x \leqq a$, $0 \leqq y \leqq b$ $(a, b>0)$. In problem (a), the *characteristic initial value problem*, the domain $\bar{G}$ coincides with R (Fig. 6). Here the values of the solution u of (3) are prescribed on both characteristics $x=0$ and $y=0$. The initial condition is then

$$u(x, 0)=\sigma(x) \quad \text{for} \quad 0 \leqq x \leqq a, \qquad u(0, y)=\tau(y) \quad \text{for} \quad 0 \leqq y \leqq b, \tag{5}$$

where $\sigma(x)$ and $\tau(y)$ are given functions defined in $[0, a]$ and $[0, b]$ respectively, and $\sigma(0)=\tau(0)$ [1].

[1] The exact regularity hypotheses will be given later.

Now suppose C is a curve in R which connects the two corner points $(0, b)$ and $(a, 0)$ and is, moreover, strictly monotone. Suppose it is defined by a function $y=\bar{y}(x)$ with $\bar{y}(0)=b, \bar{y}(a)=0$ which is continuous in an interval $[0, a]$ and strictly monotone decreasing. Then the function $\bar{y}(x)$ has an inverse function $x=\bar{x}(y)$ which is also continuous and monotone. The domain for the *Cauchy problem* or problem (b) is the curvilinear

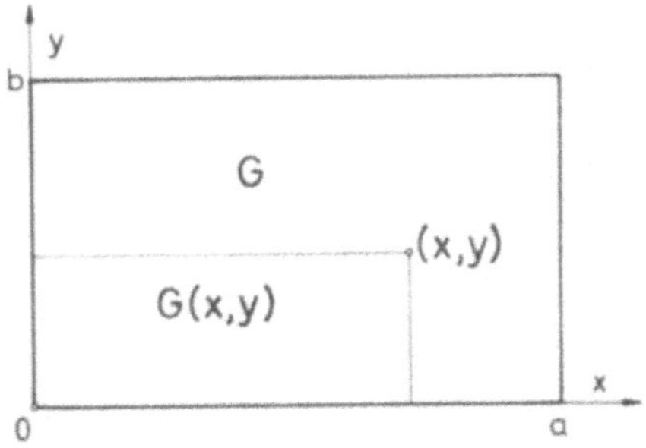

Fig. 6. The domain for the characteristic initial value problem (a)

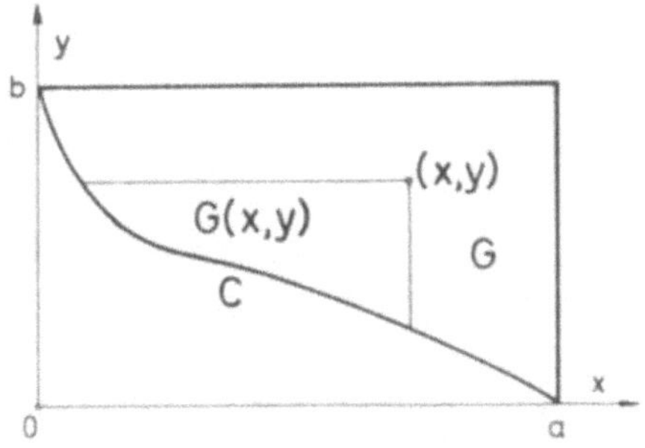

Fig. 7. The domain for the Cauchy problem (b)

triangle lying above C (Fig. 7), $\bar{G}=\{(x,y)\,|\,0\leqq x\leqq a, \bar{y}(x)\leqq y\leqq b\}$ $=\{(x,y)\,|\,0\leqq y\leqq b, \bar{x}(y)\leqq x\leqq a\}$. The values of u, u_x and u_y are prescribed on C:

$$u(x,\bar{y}(x))=z(x),\ u_x(x,\bar{y}(x))=p(x),\ u_y(x,\bar{y}(x))=q(x) \quad \text{for} \quad 0\leqq x\leqq a\,. \quad (6)$$

However the functions $z(x), p(x), q(x)$ cannot be chosen arbitrarily. Rather they are subject to a restriction called a "*consistency condition*" which we formulate as follows: There is a continuously differentiable function $\psi(x,y)$ in $\bar{G}$ such that $z(x), p(x), q(x)$ coincide with the values of ψ, ψ_x, ψ_y on C (i.e., by replacing y by $\bar{y}(x)$). If $\bar{y}(x)$ is continuously differentiable, then this condition becomes identical with the usual consistency condition: In the interval $[0, a]$, $z(x)$ is continuously differentiable, $p(x)$ and $q(x)$ are continuous, and we have

$$z'(x) = p(x) + \bar{y}'(x)q(x)\,. \tag{7}$$

Finally, we consider another problem, (c), formulated by Goursat (1927, p. 120). Here $\bar{G}$ is the part of R to the right of a curve C; the curve C is defined by a function $x = \bar{x}(y)$ continuous in $[0, b]$ and strictly monotone increasing, or by its inverse function $y = \bar{y}(x)$, $0 \leqq x \leqq \bar{x}(b)$, and we have $\bar{G} = \{(x, y) | 0 \leqq y \leqq b,\ \bar{x}(y) \leqq x \leqq a\}$, where $\bar{x}(0) = 0$, $\bar{x}(b) \leqq a$ (Fig. 8).

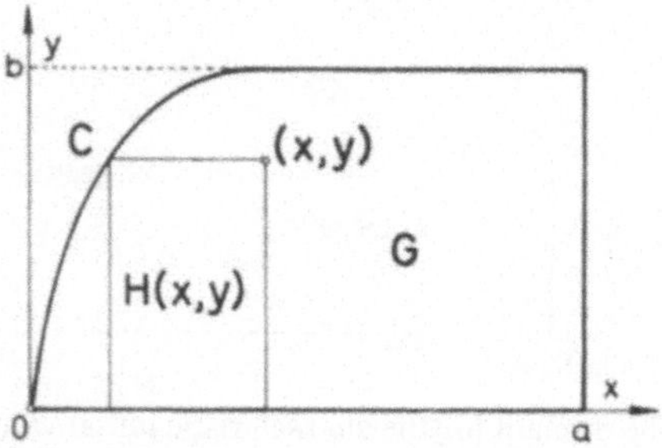

Fig. 8. The domain for the Goursat problem (c)

The functional values

$$u(x, 0) = \sigma(x) \quad \text{for} \quad 0 \leqq x \leqq a, \qquad u(\bar{x}(y), y) = \tau(y) \quad \text{for} \quad 0 \leqq y \leqq b \tag{8}$$

are prescribed on the curve C and on the characteristic $y = 0$, and again σ and τ are given functions, and $\sigma(0) = \tau(0)$ [1].

For each of the three problems, the set $G(x, y)$ is (as in 17 I) the set of points $(\xi, \eta) \in \bar{G}$ for which $\xi \leqq x, \eta \leqq y$. Moreover, we need the set $H(x, y)$, which is equal to $G(x, y)$ for problems (a), (b), and is

$$H(x, y) = \{(\xi, \eta) | 0 \leqq \eta \leqq y, \bar{x}(y) \leqq \xi \leqq x\}$$

for problem (c). By $C(x, y)$ we mean the part of the curve C in $G(x, y)$. Two-dimensional integrals are denoted by double integral signs; one-dimensional and line integrals by a single integral sign.

Z^* is the class of functions $\varphi(x, y)$ which are continuous in $\bar{G}$ together with their derivatives $\varphi_x, \varphi_y, \varphi_{xy}$ [2]. The class $Z_c(f)$ was defined in 19 I[3]. If φ belongs to this class and to Z^*, then φ belongs to the class $Z^*(f)$.

[1] The exact regularity hypotheses will be given later.

[2] Here at those boundary points where we can differentiate only on one side, derivative means one-sided derivative. The class Z^* can also be defined as follows: $\varphi \in Z^*$ means that the derivatives $\varphi_x, \varphi_y, \varphi_{xy}$ exist in G and are uniformly continuous there.

[3] Of course we mean the class $Z_c(k)$ with $k(x, y, \xi, \eta, z) = f(\xi, \eta, z)$.

II. Transformation to an Integral Equation. Every function $\varphi \in Z^*$ allows a representation

$$\varphi(x,y)=\varphi^0(x,y)+\iint\limits_{H(x,y)} \varphi_{xy}(\xi,\eta)\,d\xi\,d\eta\,, \tag{9}$$

where

$$\varphi^0(x,y)=\varphi(x,0)+\varphi(0,y)-\varphi(0,0)\,, \tag{10a}$$

$$\left.\begin{aligned}\varphi^0(x,y)&=\varphi(x,\bar{y}(x))+\int\limits_{\bar{y}(x)}^{y}\varphi_y(\bar{x}(\eta),\eta)\,d\eta\\&=\varphi(\bar{x}(y),y)+\int\limits_{\bar{x}(y)}^{x}\varphi_x(\xi,\bar{y}(\xi))\,d\xi\,,\end{aligned}\right\} \tag{10b}$$

$$\varphi^0(x,y)=\varphi(x,0)+\varphi(\bar{x}(y),y)-\varphi(\bar{x}(y),0)\,. \tag{10c}$$

This formula is only another way of writing (2) for the cases (a) and (c), while in case (b) it is obtained from

$$\begin{aligned}\iint\limits_{H(x,y)}\varphi_{xy}\,d\xi\,d\eta&=\int\limits_{\bar{y}(x)}^{y}\int\limits_{\bar{x}(\eta)}^{x}\varphi_{xy}\,d\xi\,d\eta=\int\limits_{\bar{y}(x)}^{y}[\varphi_y(x,\eta)-\varphi_y(\bar{x}(\eta),\eta)]\,d\eta\\&=\varphi(x,y)-\varphi(x,\bar{y}(x))-\int\limits_{\bar{y}(x)}^{y}\varphi_y(\bar{x}(\eta),\eta)\,d\eta\end{aligned}$$

(here if we change the order of integration, then we obtain the expression in the second line in (10b)). By (10), φ^0 is of the form

$$\varphi^0(x,y)=\sigma_1(x)+\tau_1(y) \tag{11}$$

with continuous functions σ_1, τ_1. In cases (a) and (b) these functions are also continuously differentiable, i.e., $\varphi^0 \in Z^*$ and $\varphi^0_{xy}=0$; the same is true of (c) if $\bar{x}(y)$ is continuously differentiable. These assertions are trivial for (a) and (c), and for (b) we obtain $\varphi^0_y(x,y)=\varphi_y(\bar{x}(y),y)$ from the first line, $\varphi^0_x(x,y)=\varphi_x(x,\bar{y}(x))$ from the second line in (10b), and thus $\varphi^0 \in Z^*$ and $\varphi^0_{xy}=0$, whence (11) follows.

On the basis of Eq. (9), each of the three initial value problems can be converted into a Volterra integral equation

$$u(x,y)=g(x,y)+\iint\limits_{H(x,y)} f(\xi,\eta,u(\xi,\eta))\,d\xi\,d\eta\,. \tag{12}$$

The function $g(x,y)$ is uniquely determined by the prescribed initial values according to

$$g(x,y)=\sigma(x)+\tau(y)-\sigma(0)\,, \tag{13a}$$

$$g(x,y)=z(x)+\int\limits_{\bar{y}(x)}^{y} q(\bar{x}(\eta))\,d\eta=z(\bar{x}(y))+\int\limits_{\bar{x}(y)}^{x} p(\xi)\,d\xi\,, \tag{13b}$$

$$g(x,y)=\sigma(x)+\tau(y)-\sigma(\bar{x}(y))\,. \tag{13c}$$

The two expressions in (13b) coincide since by hypothesis there is a continuously differentiable function ψ in G so that ψ, ψ_x and ψ_y coincide with z, p and q on C, and since for every such function ψ we have the identity

$$\psi(x, \bar{y}(x)) - \psi(\bar{x}(y), y) = \int_{C(x,y)} (\psi_x d\xi + \psi_y d\eta)$$

$$= \int_{\bar{x}(y)}^{x} \psi_x(\xi, \bar{y}(\xi)) d\xi - \int_{\bar{y}(x)}^{y} \psi_y(\bar{x}(\eta), \eta) d\eta$$

(the first integral is a line integral where $C(x, y)$ is oriented "from upper left to lower right"; it is well known and easy to see that this trivial formula for continuously differentiable $\bar{y}(x)$ also holds for continuous $\bar{y}(x)$).

From these discussions we directly obtain the following connection between the initial value problems (a), (b), (c) and the associated integral equation (12).

If the function $u \in Z^*(f)$ satisfies the differential equation (3) *in* $\bar{G}$ *and if we set* $u^0 = g$ [1], *then g is continuous and u is a solution of the integral equation* (12); *here g is given by* (13) *if u satisfies the initial conditions* (5) *or* (6) *or* (8).

Conservely, if the functions $\sigma(x)$ *and* $\tau(y)$ *in case* (a), *the functions* $\sigma(x)$, $\tau(y)$ *and* $\bar{x}(y)$ *in case* (c) *are continuously differentiable* (*in case* (b) *the assumptions made in connection with* (6) *are sufficient*), *if g is defined by* (13) *and u is a continuous solution of the integral equation* (12) *in* $\bar{G}$ *where f is a continuous function in all three variables, then the function u belongs to* $Z^*(f)$, *it satisfies the differential equation* (3) *in* $\bar{G}$ *and the initial conditions* (5) *or* (6) *or* (8).

In our approach the integral equation stands in the foreground. This has the advantage that we do not have to restrict ourselves to solutions of class Z^* and therefore require only continuity of the functions $\bar{y}(x), \sigma(x), \tau(y)$. Now the function $f(x, y, z)$ is also completely arbitrary; according to 19 I, $Z_c(f)$ is the class of those continuous functions in $\bar{G}$ for which $f(x, y, \varphi(x, y)) \in L(\bar{G})$. A *solution* is a function $u \in Z_c(f)$ satisfying Eq. (12) in $\bar{G}$.

By 18 II and 18 VI we have the following

III. Existence Theorem. *The integral equation* (12) *has at least one solution if g is continuous in* $\bar{G}$ *and* $f \in \mathcal{H}$ [2]. *If* $f \in \mathcal{H}_0$, *then in general the solution will exist only in a neighborhood of the origin* (*case* (a) (c)) *or the curve C* (*case* (b)). *However, it exists in the entire domain* $\bar{G}$ *if we have*

[1] u^0 is defined according to (9) (10).

[2] Of course, by 18 I, this means that $k(x_1, x_2, \xi_1, \xi_2, z) = f(\xi_1, \xi_2, z)$ is in $\mathcal{H}$.

a bound

$$|f(x,y,z)| \leqq h(x,y)(1+|z|)$$

or more generally

$$|f(x,y,z)| \leqq h(x,y)\varphi(|z|),$$

where φ *has the properties specified in* 18 VI (β) *and* $h \in L(\bar{G})$.

This theorem, which goes back to Picard (1896) for the case where f satisfies a Lipschitz condition in z, was proved for problem (a) by Guglielmino (1958); see also Pulvirenti (1960a). According to 18 VII we know that a maximal and a minimal solution exist if f is monotone increasing in z. A counter-example constructed by Walter (1967a) shows that under the single assumption of continuity of $f(x,y,z)$ problems (a) and (b) have no maximal and minimal solution (contrary to statements made by Zwirner (1951) and Santoro (1959)).

Very simple examples show that the initial value problem does not have unique solutions in general if we assume $f \in \mathscr{H}$ or even assume the continuity of f. Thus the characteristic initial value problem for the equation $u_{xy}=4\sqrt{u}$ with $u(x,0)=u(0,y)=0$ has the solutions $u=0$ and $u=x^2y^2$.

IV. Super- and Subfunctions. *Suppose the function* f *is monotone increasing in* z. *Then if* $v, w \in Z_c(f)$,

$$v < g + \iint\limits_{H(x,y)} f(\xi,\eta,v)\,d\xi\,d\eta, \qquad w > g + \iint\limits_{H(x,y)} f(\xi,\eta,w)\,d\xi\,d\eta \quad \text{in} \quad \bar{G}$$

and if u *is a solution of* (12), *we have*

$$v < u < w \quad \text{in} \quad \bar{G}.$$

In particular, if the functions $u, v, w \in Z^*(f)$ *and if*

(α) $v^0(x,y) < u^0(x,y) < w^0(x,y)$ *in* $\bar{G}$,

(β) $v_{xy} \leqq f(x,y,v)$, $u_{xy}=f(x,y,u)$, $w_{xy} \geqq f(x,y,w)$ *in* $\bar{G}$,

then the above inequality again holds[1].

The second part of the assertion reduces to the first part with the aid of (9), and the first part is a special case of 19 I.

Note: The inequalities (α) must hold in all of $\bar{G}$. If we have a problem (a) with given functions σ, τ and if $\sigma(x) < w(x,0)$ and $\tau(y) < w(0,y)$, this does *not* imply $u^0 < w^0$ in $\bar{G}$. Example: $\sigma=\tau=0$, $a=b=1$, $w(x,0)=\frac{3}{2}-x$, $w(0,y)=\frac{3}{2}-y$; we have $u^0=0$, $w^0=\frac{3}{2}-x-y$, and thus $u^0(1,1) > w^0(1,1)$.

V. Theorem[1]. *Suppose the function* $\omega(x,y,z)$ *is defined for* $z \geqq 0$ *and monotone increasing in* z. *Suppose that for the functions* $u, v \in Z^*(f)$,

[1] The theorems of this section always hold for all three problems.

$\varrho \in Z^*(\omega)$, $\delta \in L(\bar{G})$, we have

(α) $|v^0 - u^0| < \varrho^0$ *in* $\bar{G}$;

(β) $u_{xy} = f(x, y, u)$ *and* $|v_{xy} - f(x, y, v)| \leqq \delta(x, y)$ *in* $\bar{G}$;

(γ) $\varrho_{xy} \geqq \delta(x, y) + \omega(x, y, \varrho)$ *in* $\bar{G}$;

(δ) $|f(x, y, z) - f(x, y, \bar{z})| \leqq \omega(x, y, |z - \bar{z}|)$ *for* $(x, y, z), (x, y, \bar{z}) \in D(f)$.

Then

$$|v - u| < \varrho \quad in \quad \bar{G}.$$

This too is just a special case of the corresponding Theorem 19 IV or 17 VI. Indeed, if we set $g = u^0$ and

$$d(x, y) = |v^0 - g| + \iint_{H(x,y)} \delta(\xi, \eta) d\xi d\eta,$$

then all the hypotheses of 17 VI are satisfied.

We shall deal with

VI. Uniqueness Problems in detail in the next section. According to 19 V, VI we know that a Lipschitz condition or an Osgood condition, i.e., a bound V (δ) with $\omega(x, y, z) = l(x, y)z$, $l(x, y) \in L(\bar{G})$ or $\omega(x, y, z) = \psi(z)$ respectively (ψ having the properties given in 4 V (β)) is sufficient for uniqueness. Moreover, by referring to Theorem V and the considerations given in 19 VI, it is easy to show that the solution depends continuously on the initial values and on the right hand side f of the differential equation (3) in both cases. Here continuous dependence is defined in the usual way: for $\varepsilon > 0$ there is a $\delta > 0$ such that $|v - u| < \varepsilon$ if u is the solution of the initial value problem, $|v_{xy} - f(x, y, v)| \leqq \delta$, $|v^0 - g| \leqq \delta$ in $\bar{G}$.

Finally it should be noted that a one-sided bound

$$f(x, y, z) - f(x, y, \bar{z}) \leqq \omega(x, y, z - \bar{z}) \quad \text{for} \quad z \geqq \bar{z}$$

suffices instead of V (δ) provided the existence of a maximal solution is assured (see the remark in III).

VII. The Lipschitz Condition. Suppose the function $f(x, y, z)$ satisfies the Lipschitz condition

$$|f(x, y, z) - f(x, y, \bar{z})| \leqq h(x) l(y) |z - \bar{z}|, \tag{14}$$

where h and l are integrable functions with integrals

$$H(x) = \int_0^x h(\xi) d\xi \quad \text{and} \quad L(y) = \int_0^y l(\eta) d\eta.$$

If $v \in Z_c(f)$,

$$\left| v - g - \iint_{H(x,y)} f(\xi, \eta, v) d\xi d\eta \right| \leqq d(x, y) \quad \text{in} \quad \bar{G} \tag{15}$$

with a continuous function $d(x, y)$, and furthermore, if u is a solution of (12), then we have

$$|v-u| \leqq E_2(H(x)L(y)) \cdot \max_{G(x,y)} d(\xi, \eta). \tag{16}$$

The function $E_2(t)$ was defined in (19.11).

Often instead of (15) we have a bound

$$|v_{xy}-f(x, y, v)| \leqq \delta(x, y) \quad \text{and} \quad |v^0-u^0|=|v^0-g| \leqq \varepsilon(x, y) \text{ in } \bar{G}; \tag{17}$$

then (16) also holds with

$$d(x, y)=\varepsilon(x, y)+\iint_{H(x,y)} \delta(\xi, \eta)\, d\xi\, d\eta.$$

If δ and ε are constant here, then from (16)

$$\left.\begin{aligned} |v-u| &\leqq E_2(H(x)L(y))(\varepsilon+\delta \cdot \max_{G(x,y)} |H(\xi, \eta)|) \\ &\leqq E_2(H(x)L(y))(\varepsilon+\delta|G(x, y)|). \end{aligned}\right\} \tag{18}$$

These formulas follow from (19.23) and hold for all three problems.

If both functions h and l are constant, say $h(x)l(y)=L$, then besides (18) we have the bound

$$|v-u| \leqq E_2(Lxy)\left(\varepsilon+\frac{\delta}{L}\right)-\frac{\delta}{L}. \tag{19}$$

For problem (a) this yields somewhat smaller bounds than (18) and can be proved easily with V. The right hand side of (19), denoted by ϱ, satisfies the equations

$$\varrho_{xy}=L\left(\varepsilon+\frac{\delta}{L}\right) E_2(Lxy)=L\varrho+\delta,$$

and in all three cases we have $\varrho^0 \geqq \varepsilon$, which follows from (10) by simple computations.

VIII. A Special Cauchy Problem. Many vibration problems lead to a Cauchy problem in which the curve C is a 45° line, say $x+y=0$ (it is obviously of no importance for our theorems that it does not go through the first quadrant). They arise mostly in a coordinate system which has been rotated 45°. If we convert the expression

$$P\varphi=\varphi_{xy}-f(x, y, \varphi) \tag{20}$$

accordingly to the new variables (t, s) by

$$x+y=t, \quad x-y=s, \quad \text{thus} \quad x=\tfrac{1}{2}(s+t), \quad y=\tfrac{1}{2}(t-s)$$

(physically t is a time and s a space variable), then, using the notation $\bar{\varphi}(t,s)=\varphi\left(\frac{t+s}{2},\frac{t-s}{2}\right)$, we have

$$P\varphi=\bar{\varphi}_{tt}-\bar{\varphi}_{ss}-f\left(\frac{t+s}{2},\frac{t-s}{2},\bar{\varphi}\right)=\bar{\varphi}_{tt}-\bar{\varphi}_{ss}-\bar{f}(t,s,\bar{\varphi})=\bar{P}\bar{\varphi}\,. \tag{21}$$

Suppose then that $\bar{G}$ is the triangle defined by the inequalities $x+y\geqq 0, x\leqq a, y\leqq b$, that u is a solution of a problem (b) in $\bar{G}$ and $v\in Z^*(f)$, where f satisfies the condition (14) with the Lipschitz constant $L=h(x)l(y)$. Suppose further that $|Pv|\leqq\delta$ in $\bar{G}$, and that on the curve C: $x+y=0$ we have

$$|v-u|\leqq\varepsilon,\qquad \alpha|v_x-u_x|+(1-\alpha)|v_y-u_y|\leqq\varepsilon_1\,, \tag{22}$$

where α is a suitable constant, $0\leqq\alpha\leqq 1$ [1]. Then in $\bar{G}$

$$|v-u|\leqq Ae^{\sqrt{L}(x+y)}+Be^{-\sqrt{L}(x+y)}-\frac{\delta}{L} \tag{23}$$

with

$$A=\frac{1}{2}\left(\varepsilon+\frac{\delta}{L}+\frac{\varepsilon_1}{\sqrt{L}}\right),\qquad B=\frac{1}{2}\left(\varepsilon+\frac{\delta}{L}-\frac{\varepsilon_1}{\sqrt{L}}\right).$$

In the (t,s) coordinate system the hypothesis reads

$$\bar{P}\bar{u}=0\quad\textit{and}\quad|\bar{P}\bar{v}|\leqq\delta\quad\textit{in}\quad\bar{G}$$

$$|\bar{v}-\bar{u}|\leqq\varepsilon\quad\textit{and}\quad|\bar{v}_t-\bar{u}_t|\leqq\varepsilon_1\quad\textit{for}\quad t=0$$

and the assertion is

$$|\bar{v}-\bar{u}|\leqq Ae^{\sqrt{L}t}+Be^{-\sqrt{L}t}-\frac{\delta}{L}\,. \tag{23'}$$

For $L=0$ we should interpret this as the limiting value for $L\to 0$, and thus the formula

$$|\bar{v}-\bar{u}|\leqq\varepsilon+\varepsilon_1 t+\frac{\delta}{2}t^2\,.$$

This bound follows easily from V. For if we denote the right hand side of (23) by $\varrho(x,y)=\varphi(x+y)$, then $\varphi(t)$ is just the solution of the initial value problem

$$\varphi''=L\varphi+\delta,\qquad\varphi(0)=\varepsilon,\qquad\varphi'(0)=\varepsilon_1\,,$$

i.e., we have $\varrho_{xy}=L\varrho+\delta$ and $\varrho^0=\varepsilon+\varepsilon_1(x+y)$. Furthermore, as can be computed from (10b), we have $|u^0-v^0|\leqq\varepsilon+\varepsilon_1(x+y)=\varrho^0(x,y)$ in $\bar{G}$.

[1] Note that (22) must hold only on C.

IX. Remarks. (α) The theorems of this section can be extended without difficulty to systems of differential equations

$$\boldsymbol{u}_{xy}=\boldsymbol{f}(x,y,\boldsymbol{u})$$

in which $\boldsymbol{u}=(u_1,\ldots,u_n)$, $\boldsymbol{f}=(f_1,\ldots,f_n)$ with $f_\nu=f_\nu(x,y,z_1,\ldots,z_n)$ and $\boldsymbol{\sigma}=(\sigma_1,\ldots,\sigma_n),\ldots$. Each of the three problems is equivalent to an integral equation (12) (with bold-face for u, g, f) for which the existence Theorem III holds. The statements of IV hold without change. Theorem V can be generalized in two ways: we bound either with a vector function $\boldsymbol{\omega}$ (ω, ϱ, δ bold-face, absolute value signs remain) or we go over to scalar functions ω, δ, ϱ with the help of a K-norm in place of the absolute value. If we choose the latter possibility, then *the bounds* (16), (18), (19), (23) *remain valid if we replace the quantities* $f, z, \bar{z}, u, v, g$ *in* (14) *to* (23) *by the corresponding vectors and the absolute values by K-norms.*

(β) We have distinguished a direction, "from lower left to upper right," in the three problems. Every problem is equivalent to three others which are obtained from it by a reflection in the x-axis, the y-axis, and the x- and y- axis respectively. For example if we have a Cauchy problem in which $\bar{G}$ is that part of the rectangle R lying *under* the curve C (here as before C should be monotone decreasing), then we have to introduce new variables $\bar{x}=-x$, $\bar{y}=-y$; if on the other hand C is a monotone increasing curve going through the rectangle R from the origin to the point (a,b) and $\bar{G}$ is the upper portion of the rectangle, then the transformation is $\bar{x}=-x$, $\bar{y}=y$, etc.

(γ) We note two more of the initial value problems for hyperbolic differential equations. Kisyński (1957) formulates an initial value problem which contains problems (a) and (b) as special cases. Here (Fig. 9), function values are given on

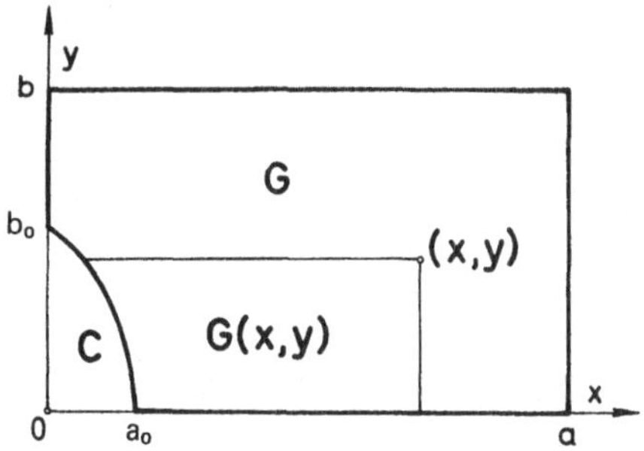

Fig. 9. The domain for a Kisyński problem

the characteristic segments $[b_0, b]$ and $[a_0, a]$; function values and the values of the first derivatives are given on the curve C which connects the points $(0, b_0)$, $(a_0, 0)$. The conversion into an integral equation (12) and the extension of the above results to this case present no difficulties; see Kisyński (1960). However, it is a different matter with the following Goursat problem (Fig. 10), in which $\bar{G}$ is the part of R lying between the two curves C_1, C_2 which start at the origin and are monotone increasing. Function values are prescribed on C_1, C_2. Here too we can give an equivalent integral equation under suitable hypotheses (for more detail see Kisyński (1957)). However, what is new is that the region of integration $H(x, y)$ is a sum of (finitely or infinitely many) rectangles $H^k(x, y)$ over which we integrate alternately with positive and negative signs:

$$u(x, y) = g(x, y) + \sum_k (-1)^k \iint_{H^k(x,y)} f(\xi, \eta, u)\, d\xi\, d\eta\,. \tag{12'}$$

It seems to be impossible to convert IV for this problem. On the other hand it is easy to set up an estimate similar to V. If we let K denote the integral operator corresponding to Eq. (12′), then $|f(x, y, z) - f(x, y, \bar{z})| \leqq \omega(x, y, |z - \bar{z}|)$ implies the bound

$$|K\varphi - K\bar{\varphi}| \leqq \iint_{H'(x,y)} \omega(\xi, \eta, |\varphi - \bar{\varphi}|)\, d\xi\, d\eta \quad \text{with} \quad \bigcup_k H^k(x, y) = H'(x, y)\,.$$

This is just the inequality (17.3) if we define Ω by

$$\Omega\varphi = \iint_{H'(x,y)} \omega(\xi, \eta, \varphi)\, d\xi\, d\eta\,.$$

Then Theorem 17 VI is applicable; in particular, we have uniqueness by 19 VI when f satisfies a Lipschitz or an Osgood condition.

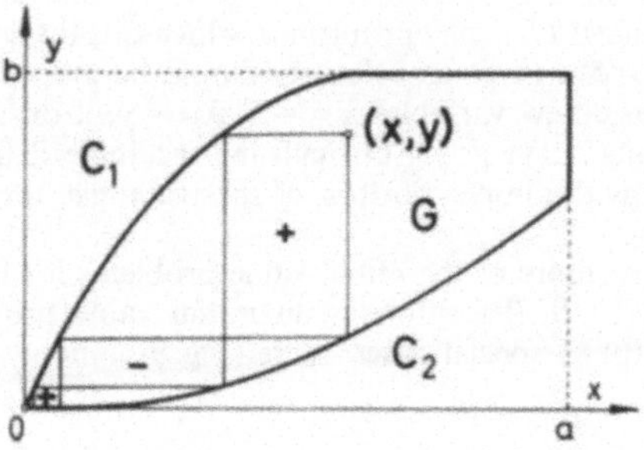

Fig. 10. The domain for a Goursat problem

(δ) Equality signs are permitted in Theorem IV if f satisfies a uniqueness condition, and also in V (α) if ω satisfies a uniqueness condition, on the basis of the general remark 17 XI (γ). The other possibility discussed in 17 XI (γ) for passing to theorems with equality signs in all inequalities makes use of maximal and minimal solutions. It is easily accomplished for IV and V.

X. Remark on the Monotonicity Hypothesis. The theorems on super- and subfunctions are based on the essential hypothesis of monotonicity of the kernels. In Section 5 we saw that we can rid ourselves of this restriction for ordinary differential equations. The analogous question for hyperbolic differential equations is fundamental: For example, does assertion IV still hold if f is not monotone increasing in z? As the following example shows, this question is to be answered in the negative; and this is true even if IV (β) is sharpened and if f satisfies a Lipschitz condition. Thus monotonicity is not a superfluous hypothesis here. This difference from the case $m=1$ is already evident for the function $E_2(t)$. While $E_1(t)=e^t$ by (19.11), and is thus positive for negative t, $E_2(t)$ can take on negative values. And now for our example, which uses precisely this property of E_2! We consider the problem (a)

$$u_{xy}=-u \quad \text{for} \quad x,y \geqq 0, \qquad u(x,0)=u(0,y)=0\,.$$

The solution is $u=0$. Now we construct a function $w(x,y)$ with the properties

$$w_{xy}=-w+\delta\ (\delta>0), \qquad w(x,0)=\sigma(x), \qquad w(0,y)=\tau(y)\,,$$

where $\sigma(0)=\tau(0)>0$, $\sigma'(x)>0$, $\tau'(y)>0$ and $\delta>0$. By (19.4′) and 19 VII,

$$w(x,y)=d(x,y)+\int_0^x\int_0^y d(\xi,\eta)\frac{\partial^2}{\partial\xi\,\partial\eta}E_2(-(x-\xi)(y-\eta))\,d\xi\,d\eta$$

with

$$d(x,y)=\sigma(x)+\tau(y)-\sigma(0)+\delta xy\,.$$

After two integrations by parts this becomes

$$w(x,y)=\sigma(0)E_2(-xy)+\int_0^x \sigma'(\xi)E_2(-(x-\xi)y)\,d\xi$$
$$+\int_0^y \tau'(\eta)E_2(-x(y-\eta))\,d\eta+\delta(1-E_2(-xy))\,.$$

Since $|E_2(t)|\leqq 1$ for $t\leqq 0$ and $E_2(t)<-0.1$ for $-1.8>t>-6.8$, we can easily find a point $(x_0,y_0)>(0,0)$, a $\delta>0$ and two functions σ,τ with the given properties such that $w(x_0,y_0)<0$. Therefore this function is not a superfunction even though $w_{xy}>-w$, $w^0>0$, $w_x^0>0$, $w_y^0>0$.

Let us return briefly to the beginning of this section where we pointed out the analogy between the ordinary differential equation (3) and the hyperbolic differential equation (4). The above example shows that this analogy breaks down when we come to monotonicity theorems. For ordinary differential equations we have

(a) $v(0)<w(0)$, $v'-f(t,v)<w'-f(t,w)$ implies $v<w$,

without any conditions on f (8 V), while for hyperbolic differential equations

(a′) $v^0<w^0$ and $v_{xy}-f(x,y,v)<w_{xy}-f(x,y,w)$ implies $v<w$

holds only if f is increasing in the last variable (IV).

Under these circumstances it is not surprising that for hyperbolic differential equations, in contrast to ordinary differential equations, maximal solutions exist only if f possesses this monotonicity property (compare the remarks in III and the counter-example quoted there). As a consequence, monotonicity theorems given by Zwirner (1951) (1951a), which are similar to (a′), but with v and w as maximal solutions of two different hyperbolic differential equations, also require monotonicity assumptions on f.

21. The Differential Equation $u_{xy}=f(x, y, u, u_x, u_y)$

For the general nonlinear hyperbolic differential equation

$$u_{xy}=f(x, y, u, u_x, u_y) \tag{1}$$

we shall restrict ourselves to the first two of the problems formulated in 20 I and deal with the different situation in the Goursat problem only briefly in XII (α). In order to avoid being too lengthy, we make the following restrictive

I. Assumptions. Suppose that the functions $\sigma(x)$, $\tau(y)$ in problem (a) and the functions $z(x)$, $\bar{y}(x)$ in problem (b) are continuously differentiable in $[0, a]$ and $[0, b]$ respectively; further suppose that $p(x)$ and $q(x)$ in problem (b) are continuous and (20.7) holds.

Now the function $g(x, y)$ defined by (20.13) is in the class Z^*. The integral equation

$$u(x, y)=g(x, y)+\iint\limits_{H(x, y)} f(\xi, \eta, u(\xi, \eta), u_x(\xi, \eta), u_y(\xi, \eta))\,d\xi\, d\eta \tag{2}$$

corresponding to the initial value problem differs from (20.12) in one essential aspect: We are dealing with an integro-differential equation. However, we can eliminate the derivatives u_x, u_y in the usual way by introducing two new functions u_1 and u_2; we ensure that $u_1=u_x$ and $u_2=u_y$ by writing two new equations. Thus we obtain the system

$$\left.\begin{aligned} u(x, y)&=g(x, y)+\iint\limits_{H(x, y)} f(\xi, \eta, u(\xi, \eta), u_1(\xi, \eta), u_2(\xi, \eta))\,d\xi\, d\eta\\ u_1(x, y)&=g_x(x, y)+\int_{y_0}^{y} f(x, \eta, u(x, \eta), u_1(x, \eta), u_2(x, \eta))\,d\eta\\ u_2(x, y)&=g_y(x, y)+\int_{x_0}^{x} f(\xi, y, u(\xi, y), u_1(\xi, y), u_2(\xi, y))\,d\xi\,, \end{aligned}\right\} \tag{3}$$

where

$$\text{(a)}\quad x_0=y_0=0 \qquad \text{(b)}\quad x_0=\bar{x}(y), \quad y_0=\bar{y}(x)\,.$$

The expressions on the right hand side of (3) in the second or third line are, as is easily seen, the partial derivatives of the expression in the first line (if the integrand is continuous), i.e., (3) is equivalent to (2).

II. Definition ($Z^*(f)$, *solution*). We define $(u, u_1, u_2) \in Z_c(f)$ according to 19 I (relative to the system (3))[1]. The notation $u \in Z^*(f)$ means that $u \in Z^*$ and the triple $(u, u_x, u_y) \in Z_c(f)$. By a *solution* of Eq. (1) or of an initial value problem we always mean a solution from the class $Z^*(f)$.

If $u \in Z^*(f)$ is a solution of Eq. (1) and if we set $u^0 = g$, then (u, u_x, u_y) is a solution of (3); $g(x, y)$ is given by (20.13) if u is a solution of an initial value problem. Conversely: If g is determined from (20.13), $(u, u_1, u_2) \in Z_c(f)$ is a solution of the integral equation (3) and $f(x, y, z, p, q)$ is continuous, then $u_1 = u_x$ and $u_2 = u_y$ and u is a solution of the initial value problem from the class $Z^*(f)$. According to Section 18 we have the following

III. Existence Theorem. *Suppose the function $g(x, y)$ is continuously differentiable in G and the function $f(x, y, z, p, q)$ is bounded and continuous for $(x, y, z, p, q) \in \bar{G} \times E^3$. Suppose further that f satisfies a Lipschitz condition in p and q*

$$|f(x, y, z, p, q) - f(x, y, z, \bar{p}, \bar{q})| \leqq L(|p - \bar{p}| + |q - \bar{q}|). \tag{4}$$

Then there exists at least one solution of the integral equation (3). *A bound $|f(x, y, z, p, q)| \leqq M(1 + |z| + |p| + |q|)$ suffices in place of the boundedness. Furthermore, a maximal and a minimal solution exist if f is monotone increasing in the last three variables.*

Significant portions of the results presented in this section, including the above existence theorem (in somewhat more general form) can be found in two works of Satō (1941, 1945) which were apparently little known until recently; see also the book (in Japanese) of Hukuhara and Satō (1957)[2]. From the extensive literature which deals with generalizations and extensions of the existence theorem to other initial value problems, we note the work of Leehey (1950), Alexiewicz and Orlicz (1956), Szmydt (1956, 1956a, 1957, 1958a), Diaz (1958), Walter (1959a, 1960), Conlan (1959), Santagati (1959), Zitarosa (1960), Ciliberto (1955, 1956, 1956a, 1961), Conti (1953, 1958), Guglielmino (1959, 1959a, 1960) and Arnese (1962), most of which contain rich bibliographies. Hartman and Wintner (1952) published an example of a continuous function $f(x, y, z, p, q)$ for which there is *no* solution of Eq. (1) in the class Z^*. The example is of fundamental significance; it shows that the existence theorem of Peano — see 8 VII — can be extended to Eq. (20.3) but not

---

[1] We have a slight inconsistency here. To be more precise, we should write $Z_c(\mathbf{f})$ with $\mathbf{f} = (f, f, f)$.

[2] Prof. Satō kindly sent me a mimeographed French translation of the relevant Chapter 6 of this book.

to (21.1). We shall not present this relatively complicated example but instead construct

IV. An Unsolvable Initial Value Problem. We refer back to Example 18 III but write x, y instead of x_1, x_2. For the domain G given there (it obviously makes no difference that it does not lie in the first quadrant) and the function $f(z)$ defined there, we pose the following problem (a)

$$u_{xy}=f(u_y) \quad \text{in} \quad \bar{G}, \qquad u(x,-1)=\sigma(x), \qquad u(0,y)=\tfrac{1}{2}y^2$$

where σ can be an arbitrary continuously differentiable function. It is easy to see that the y-derivative of every solution indeed satisfies the integral equation of 18 III. By integrating the solution given there we obtain

$$u(x,y)=\begin{cases}\sigma(x)+\tfrac{1}{2}(y^2-1) & \text{for} \quad y\leqq 0,\\ \sigma(x)+\tfrac{1}{6}(6x^2y+8xy\sqrt{y}+3y^2-3) & \text{for} \quad y>0.\end{cases}$$

However this function u is not in the class Z^*; its y-derivative has a jump on the x-axis.

In constructing this example we made use of the fact that if u is a solution of Eq. (1), the derivative $u_y(x, y)=w(x)$ satisfies the ordinary differential equation

$$w'(x)=f(x, y, u(x, y), u_x(x, y), w) \qquad (y \text{ a parameter})$$

along a characteristic $y=\text{const}$. Likewise, for fixed x, $u_x(x, y)=v(y)$ satisfies

$$v'(y)=f(x, y, u(x, y), v, u_y(x, y)) \qquad (x \text{ a parameter}).$$

If u_y is continuous this means that the solutions $w(x)$ of the first differential equation depend continuously on the parameter y. Now the two concepts of "uniqueness" and "continuous dependence on parameters" are most closely related for ordinary differential equations; see Kamke (1945; p. 145) or Hartman (1964; p. 94) (note also the "recipe" given in 18 III for constructing unsolvable problems which indeed starts with non-unique initial value problems for ordinary differential equations and can be extended to the present case). From this viewpoint a uniqueness condition for the two given ordinary differential equations seems like a natural hypothesis for an existence theorem. In fact, (4) is a condition of this sort. Instead of this Lipschitz condition, weaker conditions suffice for the existence of a solution (see the literature cited in III).

V. Super- and Subfunctions. Let $f(x, y, z, p, q)$ be monotone increasing in z, p, and q. If (u, u_1, u_2), (v, v_1, v_2), $(w, w_1, w_2)\in Z_c(f)$ and if we have an equality sign, a $<$ or a $>$ in all three lines of (3) when we put in (u, u_1, u_2),

(v, v_1, v_2), (w, w_1, w_2), then

$$(v, v_1, v_2) < (u, u_1, u_2) < (w, w_1, w_2) \quad \text{in} \quad \bar{G}.$$

It is often more convenient to work with the *differential* equation. *Then the hypotheses are* $u, v, w \in Z^*(f)$,

(α) (a) $v(0,0) < u(0,0) < w(0,0)$,
$v_x < u_x < w_x$ *for* $y=0$, $v_y < u_y < w_y$ *for* $x=0$,
(α) (b) $(v, v_x, v_y) < (u, u_x, u_y) < (w, w_x, w_y)$ *on* C;
(β) $v_{xy} \leqq f(x, y, v, v_x, v_y)$, $u_{xy} = f(x, y, u, u_x, u_y)$,
$w_{xy} \geqq f(x, y, w, w_x, w_y)$ *in* $\bar{G}$,

and the conclusion is

$$(v, v_x, v_y) < (u, u_x, u_y) < (w, w_x, w_y) \quad \text{in} \quad \bar{G}.$$

If v is a subfunction in the sense of (α) (β), then (v, v_x, v_y) is a subfunction for the integral equation (3), as we can easily establish by integration. However, the converse is not true in general. Thus there is an essential difference between working with (α) (β) and thus with the differential equation and working with the integral equation (3); with (3) we sometimes obtain better bounds, as for instance in our numerical example XIV. On the other hand, the checking of the three inequalities in (3) is often more troublesome. We can reduce this work by making use of the following easily verified fact: If $v \in Z^*(f)$ satisfies the inequalities (α) and if the second and third lines of (3) hold for $v_1 = v_x$, $v_2 = v_y$ with the $<$ symbol, then this implies the first line of (3) with the $<$ symbol.

Moreover, we note that the equality sign is allowed in all of these inequalities if f satisfies a uniqueness condition, e.g., a Lipschitz condition; see 17 XI (γ). The corresponding modification when ω satisfies a Lipschitz condition will work in the following

VI. Theorem. *Suppose that the function* $\omega(x, y, u, p, q)$ *is defined for* $z, p, q \geqq 0$ *and monotone increasing in* z, p *and* q *and that the function* $\delta(x, y)$ *is continuous in* $\bar{G}$. *Suppose that for the functions* $u, v \in Z^*(f)$, $\varrho \in Z^*(\omega)$ *we have*

(α) (a) $|v-u| < \varrho$ *at the point* $(0,0)$,
$|v_x - u_x| < \varrho_x$ *for* $y=0$, $|v_y - u_y| < \varrho_y$ *for* $x=0$,
(α) (b) $|v-u| < \varrho$, $|v_x - u_x| < \varrho_x$, $|v_y - u_y| < \varrho_y$ *on* C;
(β) $u_{xy} = f(x, y, u, u_x, u_y)$ *and* $|v_{xy} - f(x, y, v, v_x, v_y)| \leqq \delta(x, y)$ *in* $\bar{G}$;
(γ) $\varrho_{xy} \geqq \delta(x, y) + \omega(x, y, \varrho, \varrho_x, \varrho_y)$ *in* $\bar{G}$;
(δ) $|f(x, y, z, p, q) - f(x, y, \bar{z}, \bar{p}, \bar{q})| \leqq \omega(x, y, |z-\bar{z}|, |p-\bar{p}|, |q-\bar{q}|)$ *in* $D(f)$.

Then

$$|v-u| < \varrho, \ |v_x - u_x| < \varrho_x, \ |v_y - u_y| < \varrho_y \quad \text{in} \quad \bar{G}.$$

We briefly indicate the proof, i.e., the reduction to 17 VI, for the problem (a). In 17 VI we now have $\boldsymbol{u}=(u, u_x, u_y)$, $\boldsymbol{v}=(v, v_x, v_y)$, $\boldsymbol{\varrho}=(\varrho, \varrho_x, \varrho_y)$, $\boldsymbol{g}=(g, g_x, g_y)$ with $g=u^0=u(x,0)+u(0,y)-u(0,0)$, and further the three components of $\boldsymbol{K u}$ are identical with the three integrals in (3). Since $R_v=\{(0,0)\}$, 17 VI (α) is true. From

$$\left|v-g-\int_0^x\int_0^y f(v,\ldots)d\xi\,d\eta\right| = \left|v^0-u^0+\int_0^x\int_0^y (v_{xy}-f(v,\ldots))d\xi\,d\eta\right|$$

$$\leqq |v^0-u^0|+\int_0^x\int_0^y \delta(\xi,\eta)d\xi\,d\eta$$

$$\left|v_x-g_x-\int_0^y f(v,\ldots)d\eta\right| = \left|v_x(x,0)-u_x(x,0)+\int_0^y (v_{xy}-f(v,\ldots))d\eta\right|$$

$$\leqq |v_x(x,0)-u_x(x,0)|+\int_0^y \delta(x,\eta)d\eta$$

and a third similar formula the validity of 17 VI (β) follows if we denote the right hand sides of these inequalities separately by d, d_1, d_2 and combine them in a vector $\boldsymbol{d}=(d, d_1, d_2)$. Then, for example, the second of the three inequalities 17 VI (γ) reads

$$\varrho_x > |v_x(x,0)-u_x(x,0)|+\int_0^y (\delta(x,\eta)+\omega(\varrho,\ldots))d\eta\,.$$

It is obtained from

$$\varrho_x(x,y)=\varrho_x(x,0)+\int_0^y \varrho_{xy}(x,\eta)d\eta$$

together with (γ) and (α) (a). The other two equations of 17 VI (γ) are derived similarly.

VII. Examples. (α) Let u be a solution, v an approximate solution of a problem (a) with a right hand side f satisfying a Lipschitz condition

$$|f(x,y,z,p,q)-f(x,y,\bar{z},\bar{p},\bar{q})| \leqq L|z-\bar{z}|+L_1|p-\bar{p}|+L_2|q-\bar{q}| \quad (5)$$

and suppose ($L, L_1, L_2, \varepsilon, \delta$ constant)

$$|v_{xy}-f(x,y,v,v_x,v_y)| \leqq \delta e^{L_1x+L_2y} \quad \text{in } \bar{G}$$
$$|v(0,0)-\sigma(0)| \leqq \varepsilon$$
$$|v_x(x,0)-\sigma'(x)| \leqq \varepsilon L_2 e^{L_2x} \quad \text{for } 0 \leqq x \leqq a$$
$$|v_y(0,y)-\tau'(y)| \leqq \varepsilon L_1 e^{L_1y} \quad \text{for } 0 \leqq y \leqq b\,.$$

Then we have

$$|v-u| \leqq e^{L_1y+L_2x}\left\{E_2(Mxy)\left(\varepsilon+\frac{\delta}{M}\right)-\frac{\delta}{M}\right\} \text{ with } M=L+L_1L_2\,. \quad (6)$$

As is easily verified, this is a special case of VI. The function on the right hand side of (6), denoted by ϱ, satisfies the differential equation $\varrho_{xy}=L\varrho+L_1\varrho_x+L_2\varrho_y+\delta$.

We remark that (6) also holds for a Cauchy problem if the initial conditions VI (α)(b) hold (with the $\leqq$ symbol). The latter is true, for example, when

$$|v-u|\leqq\varepsilon, \quad |v_x-u_x|\leqq L_2\varepsilon, \quad |v_y-u_y|\leqq L_1\varepsilon \quad \text{on} \quad C.$$

(β) We consider the Cauchy problem already handled in 20 VIII but where now $f=f(x, y, z, p, q)$. If for suitable constants $L, L_1, L_2, \delta, \varepsilon, \eta$ the inequalities

$$|v_{xy}-f(x, y, v, v_x, v_y)|\leqq\delta \quad \text{in} \quad \bar{G}$$

$$|v-u|\leqq\varepsilon, |v_x-u_x|\leqq\eta, |v_y-u_y|\leqq\eta \quad \text{for} \quad x+y=0$$

and (5) hold, then

$$|v-u|\leqq A e^{\lambda(x+y)}+B e^{\mu(x+y)}-\frac{\delta}{L} \tag{7}$$

with

$$2\lambda=L_1+L_2+W, \quad 2\mu=L_1+L_2-W, \quad W=\sqrt{(L_1+L_2)^2+4L}$$

$$AW=\eta-\mu\left(\varepsilon+\frac{\delta}{L}\right), \quad BW=\lambda\left(\varepsilon+\frac{\delta}{L}\right)-\eta.$$

This too is a special case of VI. If we make the substitution $\varrho(x, y)=\varphi(x+y)$ for the bound, then we are led to an ordinary differential equation for $\varphi(t)$

$$\varphi''=L\varphi+(L_1+L_2)\varphi'+\delta \quad \text{with} \quad \varphi(0)=\varepsilon, \quad \varphi'(0)=\eta$$

and in solving it we are led to exactly the above bound.

Now as for the uniqueness problem, the earlier criteria 19 VI (α) and (γ) are at our disposal (the first is also contained in the previous example). However, on the basis of VI, we can prove essentially sharper results in this direction and settle the uniqueness problem in great generality and with a completeness similar to that for ordinary differential equations.

VIII. Definition ($G_0, \mathscr{D}^*, \mathscr{D}_1^*$). Let G_0 in problem (a) be the set of points in $\bar{G}$ with $xy>0$, and in problem (b) the set $\bar{G}-C$. The function $\omega(x, y, z, p, q)$ is in the class $\mathscr{D}^*$ or $\mathscr{D}_1^*$ if it is defined for $(x, y)\in G_0$ and $z\geqq 0, p\geqq 0, q\geqq 0$, is monotone increasing in z, p and q, and if for every $\varepsilon>0$ there is a $\delta>0$ and a function ϱ, continuous in G_0 and equipped with continuous derivatives $\varrho_x, \varrho_y, \varrho_{xy}$, for which

$$\mathscr{D}^*: \varrho_{xy} \geqq \omega(x, y, \varrho, \varrho_x, \varrho_y) + \delta \quad \text{and} \quad \delta \leqq \varrho \leqq \varepsilon, \delta \leqq \varrho_x, \delta \leqq \varrho_y \quad \text{in} \quad G_0,$$

$$\mathscr{D}_1^*: \varrho_{xy} \geqq \omega(x, y, \varrho, \varrho_x, \varrho_y) \quad \text{and} \quad \delta|G(x, y)| \leqq \varrho \leqq \varepsilon, \delta(y - y_0) \leqq \varrho_x, \delta(x - x_0) \leqq \varrho_y \quad \text{in} \quad G_0.$$

Here $|G(x, y)|$ is the measure of $G(x, y)$ (and thus $= xy$ in the characteristic problem), $y_0 = y_0(x)$ and $x_0 = x_0(y)$ are defined just after (3).

IX. Uniqueness Theorem. *For the function f and a function $\omega \in \mathscr{D}_1^*$ suppose that*

$$|f(x, y, z, p, q) - f(x, y, \bar{z}, \bar{p}, \bar{q})| \leqq \omega(x, y, |z - \bar{z}|, |p - \bar{p}|, |q - \bar{q}|) \tag{8}$$

for $(x, y) \in G_0$ and (x, y, z, p, q), $(x, y, \bar{z}, \bar{p}, \bar{q}) \in D(f)$. Then every Cauchy problem for Eq. (1) has at most one solution from the class $Z^*(f)$. The same conclusion holds for a characteristic initial value problem if, moreover, both of the initial value problems for ordinary differential equations

$$\left.\begin{aligned} \frac{dv}{dy} &= f(0, y, \tau(y), v, \tau'(y)), \qquad v(0) = \sigma'(0) \qquad (0 \leqq y \leqq b) \\ \frac{dw}{dx} &= f(x, 0, \sigma(x), \sigma'(x), w), \qquad w(0) = \tau'(0) \qquad (0 \leqq x \leqq a) \end{aligned}\right\} \tag{9}$$

have at most one solution $v(y)$ or $w(x)$.

If (8) *holds with a function $\omega \in \mathscr{D}^*$, then the additional hypothesis imposed with respect to* (9) *is superfluous, and the (unique) solution depends continuously on the initial value and on the right hand side of the differential equation.*

The latter means: For $\varepsilon > 0$ there is a $\delta > 0$ such that $|v - u| < \varepsilon$ if VI (α) (β) hold when $\varrho, \varrho_x, \varrho_y, \delta(x, y)$ are replaced everywhere by the constant δ. — For a similar (somewhat weaker) result see Santoro (1963).

We need some additional considerations to reduce the first assertion to VI. In order to be able to handle both problems simultaneously, we agree to take the curve C in problem (a) as the part of the boundary of $\bar{G}$ where $xy = 0$. If u and v are two solutions of an initial value problem, then they coincide on C together with their derivatives of first order; this is trivial in the Cauchy problem and follows from the uniqueness of the two initial value problems (9) in the characteristic problem[1]. Then by (1) we also have $u_{xy} = v_{xy}$ on C. If we let G_α $(\alpha > 0)$ denote the set of points of $\bar{G}$ which are at a distance $> \alpha$ from C, then for every

[1] This is not true in general without the hypothesis regarding (9); see Example XI (α).

positive δ there is a positive α such that

$$|v_{xy}-u_{xy}|<\delta \quad \text{in} \quad \bar{G}-G_\alpha .$$

Hence, by integration we obtain

$$|v_x-u_x|<\delta(y-y_0),\ |v_y-u_y|<\delta(x-x_0),\ |v-u|<\delta|G(x, y)| \quad \text{in} \quad \bar{G}-G_\alpha .$$

Now if for given $\varepsilon>0$ we determine a δ and a function ϱ according to VIII, then for sufficiently small positive α we can apply VI to u, v, ϱ and $\bar{G}_\alpha$ instead of $\bar{G}$; the condition VI (α) is satisfied according to the inequalities just set up and those given in VIII. It follows that $|v-u| <\varrho<\varepsilon$, i.e., $u=v$. The second assertion is proved similarly; the additional considerations above are thereby simplified.

X. Examples. In the following let α, β, γ be three nonnegative functions defined in G_0 whose sum $\alpha+\beta+\gamma=1$, let $l(x, y)$ and $l_1(t)$ be two continuous nonnegative functions in G_0 and $(0, a+b]$ respectively whose integrals

$$L(x, y)=\iint_{G(x, y)} l(\xi, \eta)\, d\xi\, d\eta, \qquad L_1(t)=\int_0^t l_1(s)\, ds$$

are finite in G_0 and $(0, a+b]$ respectively, and finally let $\psi(z)$ be an "Osgood function" equipped with the properties of 4 V (β).

(α) *Generalized Lipschitz Condition.* We have

$$\omega=l(x, y) z+l_1(y) p+l_1(x) q \in \mathscr{D}^* .$$

Here we can set

$$\varrho(x, y)=A \exp [L(x, y)+2 L_1(x)+2 L_1(y)+x+y] \qquad (A>0) .$$

As is easily verified, ϱ, ϱ_x and $\varrho_y \geqq A>0$,

$$\varrho_{xy}=l \varrho+[L_x+2 l_1(x)+1]\,[L_y+2 l_1(y)+1] \varrho$$

and

$$\omega(x, y, \varrho, \varrho_x, \varrho_y)=l \varrho+l_1(y)\,[L_x+2 l_1(x)+1] \varrho+l_1(x)\,[L_y+2 l_1(y)+1] \varrho ,$$

and thus indeed $\varrho_{xy} \geqq \omega(x, y, \varrho, \varrho_x, \varrho_y)+\delta$ and hence $\omega \in \mathscr{D}^*$.

(β) The example

$$\omega=\alpha(x, y) \frac{z}{|G(x, y)|}+\beta(x, y) \frac{p}{y-y_0}+\gamma(x, y) \frac{q}{x-x_0} \in \mathscr{D}_1^*$$

can be considered an *extension of the Nagumo condition* 4 V (γ) to hyperbolic differential equations. Here the determination of suitable functions ϱ is especially simple:

$$\varrho(x, y)=A|G(x, y)| \qquad (A>0)$$

satisfies all the requirements of VIII.

This criterion was found by Walter (1959) and independently by Shanahan (1960). The former work contains an example which shows that this theorem, like the corresponding Nagumo theorem 4 V (γ) for ordinary differential equations, is "best" in the following sense: If α, β, γ are three nonnegative constants with

$\alpha+\beta+\gamma>1$, then we can give a characteristic initial value problem with infinitely many solutions in which f is continuous and we have a bound (8) with $\omega=\frac{\alpha z}{xy}+\frac{\beta p}{y}+\frac{\gamma q}{x}$. See also Diaz and Walter (1960).

(γ) As in 4 V (δ) we can also add the Lipschitz and Nagumo condition:

$$\omega=z\left[\frac{\alpha}{|G(x,y)|}+l(x,y)\right]+p\left[\frac{\beta}{y-y_0}+l_1(y-y_0)\right]+q\left[\frac{\gamma}{x-x_0}+l_1(x-x_0)\right]\in\mathscr{D}_1^*,$$

if in addition $\alpha(x,y)\geqq\delta>0$.

Here we can set

$$\varrho(x,y)=A|G(x,y)|\exp\left[L(x,y)+BL_1(x-x_0)+BL_1(y-y_0)+x+y\right]\qquad (A>0)$$

with $B\geqq 2$ and $\geqq 1/\delta$, as a longer but elementary computation shows.

(δ) Now a preliminary remark. For given ω and $M>0$ we let ω_M denote the function

$$\omega_M(x,y,z,p,q)=\omega(x,y,\min(z,M),\min(p,M),\min(q,M)).$$

Requiring only "$\omega_M\in\mathscr{D}^*$ for every positive M" instead of "$\omega\in\mathscr{D}^*$" in Theorem IX obviously suffices for uniqueness. For it is a matter of showing that for two arbitrary but fixed solutions u, v we have $|v-u|<\varepsilon$. However, here we can assume that the domain of definition of f is a *bounded* point set and therefore $\omega_M=\omega$ for all arguments occuring in (8) if only we choose M sufficiently large.

The following example makes use of this remark. Suppose the function $\omega(x,y,z,p,q)$ is continuous in $\bar{G}\times\{z\geqq 0, p\geqq 0, q\geqq 0\}$ and monotone increasing in z, p, q; further suppose that $\omega(x,y,0,0,0)=0$. Suppose that for every $(\bar{x},\bar{y})\in G_0$ the following holds: There is only one function $\sigma(x,y)$ which is nonnegative and continuous in $G(\bar{x},\bar{y})$, has nonnegative derivatives σ_x, σ_y almost everywhere in $G(\bar{x},\bar{y})$ and satisfies the integral equation

$$\sigma(x,y)=\iint_{G(x,y)}\omega(\xi,\eta,\sigma,\sigma_x,\sigma_y)\,d\xi\,d\eta\quad\text{in}\quad G(\bar{x},\bar{y}),\tag{10}$$

namely, the trivial solution $\sigma\equiv 0$. Then ω is a uniqueness condition; more precisely, $\omega_M\in\mathscr{D}^*$ for every $M>0$.

For the proof we let

$$\left.\begin{aligned}\varrho^0(x,y)&=1+x+y+\iint_{G(x,y)}\omega(\xi,\eta,M,M,M)\,d\xi\,d\eta\\ \varrho^{n+1}(x,y)&=\frac{1+x+y}{n+1}+\iint_{G(x,y)}\omega_M(\xi,\eta,\varrho^n,\varrho^n_x,\varrho^n_y)\,d\xi\,d\eta\end{aligned}\right\}\tag{11}$$

for given $M>0$ $(n=0,1,2,\ldots)$. By induction we can show easily that $\varrho^n\in Z^*$ and that each of the three sequences of functions $\varrho^n, \varrho^n_x, \varrho^n_y$ is monotone decreasing. If we denote their (bounded and nonnegative) limiting values by $\varphi(x,y), \varphi_1(x,y), \varphi_2(x,y)$, then for $n\to\infty$ (11) implies

$$\varphi(x,y)=\iint_{G(x,y)}\omega_M(\xi,\eta,\varphi,\varphi_1,\varphi_2)\,d\xi\,d\eta\quad\text{in}\quad\bar{G}.\tag{12}$$

Similarly, by a passage to the limit,

$$\varrho^n(x,y)=\frac{1+y+x_0}{n}+\int_{x_0}^{x}\varrho^n_x(\xi,y)\,d\xi$$

implies

$$\varphi(x, y)=\int_{x_0}^{x} \varphi_1(\xi, y)d\xi$$

as well as a corresponding equation for φ_2. Thus φ is continuous in $\bar{G}$ and $\varphi_1=\varphi_x$, $\varphi_2=\varphi_y$ almost everywhere in $\bar{G}$. We want to show that $\varphi\equiv 0$ in $\bar{G}$. If it has already been shown that $\varphi\equiv 0$ for the points of $\bar{G}$ with $x+y\leqq t$ for fixed t, then φ, φ_x and $\varphi_y\leqq M$ for the points of $\bar{G}$ with $x+y\leqq t+\alpha$ if α is a sufficiently small positive number (these three functions can actually be represented as integrals with *bounded* integrands ω_M). Therefore $\varphi\equiv 0$ for $x+y\leqq t+\alpha$ because of the hypothesis on σ and because of (12). Hence we conclude in familiar fashion that φ vanishes identically in $\bar{G}$. Thus the functions ϱ^n converge monotonically and therefore uniformly to 0. Hence they have all the properties which are needed to verify that $\omega_M\in\mathscr{D}^*$.

The example (δ) is identical to a uniqueness theorem of Shanahan (1960); additional examples can be found in Walter (1960a). Further investigations of uniqueness problems are contained in work of Kisyński (1957), Szmydt (1956a, 1958), Walter (1959, 1960), Ciliberto (1961a, 1961b), Lakshmikantham (1962). See also the literature quoted in 18 VIII.

XI. Remarks on the Uniqueness Theorem. For the characteristic initial value problem, besides the bound (8) which — to be sure — need hold only in G_0, we need the uniqueness of (9). We cannot get by without the latter, as the following example shows.

Suppose we are given problem (a) with

$$f(x, y, z, p, q)=\begin{cases} p/y & \text{for} \quad y>0 \\ 2q/x & \text{for} \quad y=0, x>0 \\ 0 & \text{for} \quad x=y=0 \end{cases}$$

and $\sigma(x)=\tau(y)=0$. Although a Nagumo condition X (β) holds in G_0 with $\alpha=\gamma=0$, $\beta=1$, there are infinitely many solutions $u=Cx^2y$.

However, if f is continuous and both of the following limiting values

$$\omega_1(x, q)=\lim_{y\to 0}\omega(x, y, 0, 0, q), \qquad \omega_2(y, p)=\lim_{x\to 0}\omega(x, y, 0, p, 0)$$

exist and if the functions ω_1, ω_2 are in one of the classes $\mathscr{E}, \ldots, \mathscr{E}_7$, then the uniqueness of (9) is a superfluous requirement. In these circumstances (8) implies for $y\to 0$

$$|f(x, 0, \sigma(x), \sigma'(x), q)-f(x, 0, \sigma(x), \sigma'(x), \bar{q})|\leqq\omega_1(x, |q-\bar{q}|),$$

and this bound is sufficient for the uniqueness of the second initial value problem in (9). We have the corresponding result for the first of these problems.

In particular we can disregard (9) in the examples X (α) (β) (γ) if f is continuous.

(β) For ordinary differential equations, uniqueness "in the large" follows from uniqueness "in the small"; see the remark at the end of Section 10. The answer to the question of whether this is also true for hyperbolic differential equations is not trivial. It can be answered affirmatively under certain restrictions; see Walter (1960a, p. 276).

XII. Remarks. (α) *The Goursat problem.* Let us assume that the functions $\sigma(x)$, $\tau(y)$, $\bar{x}(y)$ are continuously differentiable in the problem (c) (rewritten in more detail in 20 I) for Eq. (1). An equivalent system of integral equations is

$$\left.\begin{aligned} u(x,y) &= g(x,y) + \iint\limits_{H(x,y)} f(\xi,\eta,u,u_1,u_2)\,d\xi\,d\eta \\ u_1(x,y) &= g_x(x,y) + \int\limits_0^y f(x,\eta,u,u_1,u_2)\,d\eta \\ u_2(x,y) &= g_y(x,y) + \int\limits_{\bar{x}(y)}^x f(\xi,y,u,u_1,u_2)\,d\xi \\ &\quad - \bar{x}'(y)\int\limits_0^y f(\bar{x}(y),\eta,u,u_1,u_2)\,d\eta \end{aligned}\right\} \qquad (3')$$

with $g(x,y) = \sigma(x) + \tau(y) - \sigma(\bar{x}(y))$. This system does not have the "normal form" (18.9) but can easily be so transformed. If we replace u by u_1, u_1 by u_2 and u_2 by $u_3 + u_4$ then we obtain the system equivalent to (3′) for a 4-tuple (u_1, u_2, u_3, u_4)

$$\left.\begin{aligned} u_1(x,y) &= g(x,y) + \iint\limits_{H(x,y)} f(\xi,\eta,u_1,u_2,u_3+u_4)\,d\xi\,d\eta \\ u_2(x,y) &= g_x(x,y) + \int\limits_0^y f(x,\eta,u_1,u_2,u_3+u_4)\,d\eta \\ u_3(x,y) &= g_y(x,y) + \int\limits_{\bar{x}(y)}^x f(\xi,y,u_1,u_2,u_3+u_4)\,d\xi \\ u_4(x,y) &= -\bar{x}'(y)\int\limits_0^y f(\bar{x}(y),\eta,u_1,u_2,u_3+u_4)\,d\eta\,, \end{aligned}\right\} \qquad (3'')$$

to which the theory of Sections 18, 19 can be applied. By 18 V the existence theorem III holds for the system (3″) and we have uniqueness if f satisfies a special Lipschitz condition 19 VI (α) or an Osgood condition 19 VI (γ). It is impossible to carry over Theorems V and VI because there is a minus sign in the last equation of (3″) and thus, roughly speaking, the monotonicity of f no longer implies the monotonicity of the corresponding integral operator. On the other hand we can subject (3″) to the estimation theorem 17 VI without further ado. In 17 VI we now have $\boldsymbol{u} = (u_1, u_2, u_3, u_4)$, $\boldsymbol{v} = (v_1, v_2, v_3, v_4)$, $\boldsymbol{g} = (g, g_x, g_y, 0)$, $\boldsymbol{K}$ is the integral operator given by the four integrals in (3″) (the integrand in the last integral is $-\bar{x}'(y)f$) and, if a bound (8) holds, Ω is the integral operator which is formed when we replace f by ω in the first three integrals of (3″) and $-\bar{x}'(y)f$ by

$\bar{x}'(y)\omega$ (without the minus sign!) in the fourth integral. If for example we have a special Lipschitz condition $\omega(x, y, z, p, q)=Lz+L_1 p+L_2 q$ and if the function $\boldsymbol{d}=(d_1, d_2, d_3, d_4)$ is constant in 17 VI (β), then the substitution $\varrho_\nu = A_\nu e^{B(x+y)}$ $(\nu=1, 2, 3, 4)$ leads to the four conditions for the positive numbers A_ν and B

$$A_1 \geqq d_1 + C/B^2, \quad A_2 \geqq d_2 + C/B, \quad A_3 \geqq d_3 + C/B, \quad A_4 \geqq d_4 + \bar{x}'(y) C/B$$

with

$$C = LA_1 + L_1 A_2 + L_2(A_3 + A_4).$$

If these are satisfied, it leads to the bound

$$|v_\nu - u_\nu| \leqq A_\nu e^{B(x+y)} \qquad (\nu = 1, 2, 3, 4).$$

This contains a theorem on continuous dependence of the solution on the initial values and on the right hand side of the differential equation and the uniqueness theorem already mentioned.

(β) Similar reasoning is needed in the more general Goursat problem given in 20 IX (γ). However no difficulties arise in the Kysiński problem mentioned in the same place; we can proceed exactly as we did for the problems (a) (b).

(γ) *Systems of differential equations.* The conversion of the theorems to systems of differential equations

$$\boldsymbol{u}_{xy} = \boldsymbol{f}(x, y, \boldsymbol{u}, \boldsymbol{u}_x, \boldsymbol{u}_y) \tag{1'}$$

— see 20 IX (α) for the notation — presents no new problem. If $\boldsymbol{f}$ is monotone increasing in $\boldsymbol{z}, \boldsymbol{p}, \boldsymbol{q}$ then V holds. In Theorem VI if $\boldsymbol{f}, \boldsymbol{z}, \boldsymbol{p}, \boldsymbol{q}, \boldsymbol{u}$ and $\boldsymbol{v}$ are vectors, then we have to use double bars instead of absolute value signs. In particular we have the bounds VII and Theorem IX:

Uniqueness is obtained as a result of a condition

$$\|\boldsymbol{f}(x, y, \boldsymbol{z}, \boldsymbol{p}, \boldsymbol{q}) - \boldsymbol{f}(x, y, \bar{\boldsymbol{z}}, \bar{\boldsymbol{p}}, \bar{\boldsymbol{q}})\| \leqq \omega(x, y, \|\boldsymbol{z}-\bar{\boldsymbol{z}}\|, \|\boldsymbol{p}-\bar{\boldsymbol{p}}\|, \|\boldsymbol{q}-\bar{\boldsymbol{q}}\|) \tag{8'}$$

with $\omega \in \mathcal{D}_1^*$ and with a K-norm for which $\|-\boldsymbol{z}\| = \|\boldsymbol{z}\| = 0$ if and only if $\boldsymbol{z} = \boldsymbol{0}$ (in the case of problem (a) *the hypothesis made for* (9) *is added). If in* (8') $\|\cdot\|$ *is a norm and $\omega \in \mathcal{D}^*$, then the solution depends continuously on the initial value and on the right hand side of the differential equation (the hypothesis relative to* (9) *is then superfluous).*

(δ) Certain initial value problems for differential equations of the form

$$\frac{\partial^{r+s} u}{\partial x^r \partial y^s} = f\left(x, y, u, \frac{\partial u}{\partial x}, \ldots, \frac{\partial^{\mu+\nu} u}{\partial x^\mu \partial y^\nu}, \ldots, \frac{\partial^{r+s-1} u}{\partial x^r \partial y^{s-1}}\right)$$

$(0 \leqq \mu \leqq r, 0 \leqq \nu \leqq s, \mu+\nu < r+s)$ also lead to Volterra integral equations. For the characteristic initial value problem we seek a solution in the rectangle $0 \leqq x \leqq a, 0 \leqq y \leqq b$ which takes on prescribed values

$$\frac{\partial^\mu u(x,0)}{\partial x^\mu} = \sigma_\mu(x) \quad \text{for} \quad 0 \leqq x \leqq a, \qquad \frac{\partial^\nu u(0,y)}{\partial y^\nu} = \tau_\nu(y) \quad \text{for} \quad 0 \leqq y \leqq b$$

($\mu = 0, 1, \ldots, r-1$; $\nu = 0, 1, \ldots, s-1$). The problem can be formulated as an integral equation. With the help of the theorems of Sections 17–19 we can answer questions of existence, uniqueness and stability (relative to changes in the right hand side of the differential equation or the initial values) in great generality.

Initial value problems for this differential equation (also with boundary conditions of a different kind) were treated by Volpato (1952/53), Kovač (1965), de Lucia (1967).

(ε) The situation is similar for the equation

$$u_{xyz} = f(x, y, z, u, u_x, u_y, u_z, u_{xy}, u_{xz}, u_{yz}) \quad \text{for} \quad u = u(x, y, z)$$

investigated by Conlan (1962), Palczewski (1963), Wu (1963), Frasca (1966), and its n-th order analog, treated by Glick (1963), Conlan and Diaz (1963), Chu and Diaz (1967).

XIII. Super- and Subfunctions in the General Case. If the initial value problem under consideration is not of monotonic type, i.e., if V does not hold, then we can obtain super- and subfunctions by 17 X.

If we have problem (a) *and if for two functions* $v, w \in Z^*(f)$

(α) $v(0,0) < \sigma(0) < w(0,0)$,
$v_x(x,0) < \sigma'(x) < w_x(x,0)$, $v_y(0,y) < \tau'(y) < w_y(0,y)$;

(β) $v_{xy} \leqq f(x, y, z, p, q) \leqq w_{xy}$ *for all* z, q, p *with* $v \leqq z \leqq w, v_x \leqq p \leqq w_x,$
$v_y \leqq q \leqq w_y$;

then we have

$$(v, v_x, v_y) < (u, u_x, u_y) < (w, w_x, w_y) \quad in \quad \bar{G}$$

for every solution $u \in Z^*(f)$.

For the system (3) associated with problem (a), the hypotheses (α) and (β) of 17 X hold. For (α) this is immediately clear because $R_v = \{0, 0\}$; for the system of integral inequalities 17 X (β) the second inequality, for example, now reads

$$v_x(x,y) < \sigma'(x) + \int_0^y f(x, \eta, \varphi, \varphi_1, \varphi_2)\,d\eta < w_x(x,y)$$

for all $(\varphi, \varphi_1, \varphi_2)$ with $(v, v_x, v_y) \leqq (\varphi, \varphi_1, \varphi_2) \leqq (w, w_x, w_y)$ in $G(x,y)$. It follows from

$$v_x(x,y) = v_x(x,0) + \int_0^y v_{xy}(x,\eta)\,d\eta$$

and (α) (β).

With the help of this estimate we can, for example, show quite easily that the existence theorem III remains valid if f is not bounded but satisfies an inequality

$$|f(x, y, z, p, q)| \leqq M(1 + |z| + |p| + |q|).$$

Indeed if $M > 1$ is chosen so large that $|\sigma|, |\sigma'|, |\tau|, |\tau'| < M$, then the functions $-v(x, y) = w(x, y) = M e^{M(3x+3y+xy)}$ satisfy all the inequalities of (α) (β).

These considerations show that we also obtain existence theorems "in the large" for unbounded f if there are two functions v, w which satisfy the above inequalities (α) (β). Systematic use of this fact is made in the book by Hukuhara and Satō (1957). The extension to the Cauchy problem presents no difficulties.

XIV. Numerical Example. Suppose we are given the nonlinear characteristic initial value problem

$$u_{xy} = 1 + u(u_x + u_y) \quad \text{with} \quad u(x, 0) = u(0, y) = 0. \tag{13}$$

The solution is uniquely determined since the right hand side $f = 1 + z(p+q)$ satisfies a Lipschitz condition in z, p and q if we assume, say, that $|z|, |p|, |q| < M$. Since furthermore f is monotone increasing in z, p, and q for $z, p, q \geqq 0$, we can determine super- and subfunctions according to V (provided they are nonnegative together with their first derivatives); here the equality sign is permitted in all the inequalities of V. The simplest Ansatz $v, w = Cxy$ satisfies the initial condition exactly and the inequality V (β) reads

$$P \equiv C - 1 - C^2 xy(x+y) \underset{(\leqq)}{\geqq} 0. \tag{14}$$

Thus for $C = 1$ we have a subfunction $v = xy$. Since the polynomial $C - 1 - \alpha C^2$ possesses the zero $C_0 = (1 - \sqrt{1 - 4\alpha})/2\alpha$, we have, according to (14), $P \geqq 0$ for $C = C_0$ and $0 \leqq xy(x+y) \leqq \alpha$, i.e., $w = C_0 xy$ is a superfunction in this domain. For every point (x, y) we choose the best α, i.e., we set $\alpha = xy(x+y)$ and thus obtain the bounds

$$v = xy \leqq u \leqq w = \frac{1}{2(x+y)}\left(1 - \sqrt{1 - 4xy(x+y)}\right) = xy + x^2 y^2 (x+y) + \cdots. \tag{15}$$

If we determine the superfunction $w = Cxy$ not from the *differential* equation but — as presented at the beginning of V — from the corresponding *integral* equation, then instead of (14) we are led to three integral inequalities of which the last two are

$$w_x = Cy \geqq \int_0^y [1 + C^2 x\eta(x+\eta)]\,d\eta = y + C^2 xy^2\left(\frac{x}{2} + \frac{y}{3}\right),$$

$$w_y = Cx \geqq \int_0^x [1 + C^2 \xi y(\xi + y)]\,d\xi = x + C^2 x^2 y\left(\frac{x}{3} + \frac{y}{2}\right).$$

In this way we obtain a sufficient condition which is better than (14) (then the first of the three integral inequalities also holds)

$$C-1-\tfrac{1}{2}C^2xy(x+y)\geqq 0 \tag{16}$$

and hence the better superfunction

$$w_1=\frac{1}{x+y}\left(1-\sqrt{1-2xy(x+y)}\right)=xy+\frac{1}{2}x^2y^2(x+y)+\cdots. \tag{17}$$

A second obvious Ansatz $v, w=\varphi(xy)$ leads to the ordinary differential inequality for $\varphi(s)$ $(s=xy)$

$$P\equiv\varphi'+s\varphi''-1-(x+y)\varphi\varphi' \underset{(\leqq)}{\geqq} 0.$$

In our search for subfunctions we first replace the troublesome term $x+y$ by $2\sqrt{s}$ $(\leqq x+y)$. A simple solution of the resulting inequality is $\varphi(s)=s+0.32\,s^{5/2}$. If (x, y) belongs to the square $0\leqq x, y\leqq a$, then for a superfunction we can replace $x+y$ by the factor $a+\dfrac{s}{a}$ $(\geqq x+y)$. For example for $a=\frac{1}{2}$, $\varphi(s)=s+0.32\,s^2$ is a solution of this inequality. Thus we have

$$v_1=xy+0.32(xy)^{5/2}\leqq u\leqq w_2=xy+0.32(xy)^2$$

(the second inequality holds for $0\leqq x, y\leqq\frac{1}{2}$).

The difficulty in this Ansatz is obviously that we have to bound $x+y$ by an expression in s. If we want to adapt the Ansatz better to the given differential equation, then (this is made evident by the form of the bounds in (15) and (17) and by the power series expansion $u=xy+\frac{1}{6}x^2y^2(x+y)+\ldots$) we will have to combine the quantities xy and $x+y$. An attempt with $v, w=xy\psi(xy(x+y))$ yields as a condition for $\psi(r)$ with $r=xy(x+y)$

$$P=\psi+5r\psi'+2r^2\psi''-1-r\psi^2-r^2\psi\psi'+(xy)^3(\psi''-2\psi\psi') \underset{(\leqq)}{\geqq} 0. \tag{18}$$

Using a quadratic function $\psi(r)=1+\alpha r+\beta r^2$, a subfunction is determined by $\alpha=\frac{1}{6}$, $\beta=\frac{1}{30}$, a superfunction by $\alpha=\frac{1}{6}$, $\beta=0.0424$, the latter for $0\leqq x, y\leqq\frac{1}{2}$. Both of these bounds

$$v_2=xy+\tfrac{1}{6}x^2y^2(x+y)+\tfrac{1}{30}x^3y^3(x+y)^2\leqq u\leqq$$
$$w_3=xy+\tfrac{1}{6}x^2y^2(x+y)+0.0424\,x^3y^3(x+y)^2$$

are quite usable in the square $0\leqq x, y\leqq\frac{1}{2}$: at the worst point $(\frac{1}{2},\frac{1}{2})$ we have

$$v=0.25<v_1=0.26<v_2=0.26093<u<w_3$$
$$=0.26108<w_2=0.27<w_1=0.293<w=0.5\,.$$

The characteristic difficulties which arise in determining superfunctions for larger squares and which are caused by the rapid growth of the solution become apparent in all three cases. In order to find a bound for the domain of existence of the solution, we try an Ansatz $v, w=xy\psi(r)$ with $\psi(r)=c/(c-r)$. From (18) we then have

$$c^2(6-c)-3cr+r^2-2c(c-1)\frac{(xy)^3}{r} \underset{(\leqq)}{\geqq} 0 \quad \text{for} \quad 0\leqq r<c\,.$$

It follows from $4(xy)^3 \leqq r^2$ that for $c=3$ we have a superfunction, for $c=6$ a subfunction:

$$v_3 = \frac{6xy}{6-xy(x+y)} \leqq u \leqq \frac{3xy}{3-xy(x+y)} = w_4 .$$

Thus the solution surely exists in a square with $a < \sqrt[3]{1.5} \approx 1.145$, but not in the square with $a = \sqrt[3]{3} \approx 1.442$.

Similarly, we can achieve significantly better bounds with somewhat greater numerical effort.

22. Supplements. The Local Method of Proof

Our considerations so far concerning the differential equation

$$u_{xy} = f(x, y, u, u_x, u_y) \tag{1}$$

have been based on the (narrow) concept of solution $u \in Z^*$ usually found in textbooks. Now we define several

I. Other Concepts of Solution (Z_1^*, Z_2^*, Z_3^*). The function $\varphi(x, y)$ belongs to the class Z_1^* if it is continuous in $\bar{G}$ together with its derivatives φ_x, φ_y and if, moreover, φ_x is absolutely continuous in y for every fixed x, φ_y is absolutely continuous in x for every fixed y and $\varphi_{xy} \in L(G)$; φ belongs to Z_3^* if $\varphi \in Z_1^*$ and if, moreover, the derivatives $\varphi_{xy}, \varphi_{yx}$ exist in $\bar{G}$ and are equal; it belongs to Z_2^* if it is absolutely continuous as a function of two variables in every (two-dimensional) interval contained in $\bar{G}$. Roughly speaking, the latter means that Eq. (20.2) holds for every $I \subset \bar{G}$ (see Carathéodory (1918, p. 653)). Diaz (1960) bases a further concept of solution, which is also very general, on his uniqueness theorem.

II. The Class Z_1^*. All the theorems of Section 21 can be immediately extended to this class, along with their proofs. Obviously, it suffices in 21 V that hypothesis (β) hold a.e. in $\bar{G}$. A corresponding remark is true for the three differential equations or differential inequalities in 21 VI (β) (γ). A modification is needed only in the uniqueness theorem 21 IX. Since we can no longer make the sharp assertions as in the proof of 21 IX about the behavior of the difference between solutions near the initial curve, the first part of 21 IX becomes untenable. *However, as before, for $\omega \in \mathscr{D}^*$ we have uniqueness and continuous dependence.*

III. The Class Z_2^*. An existence and uniqueness theory can also be built up in this class — as in Section 5, it is natural to speak of *solutions in the sense of Carathéodory*. Nevertheless, since the partial derivatives of first order are not in general continuous, the methods of proof have to be modified in various places. For the existence we need compactness criteria in L_1, and for uniqueness we make appropriate use of a procedure related to successive approximation in a way similar to that of 21 X (δ). This program was carried out by Walter (1959b). There have been several recent investigations for hyperbolic differential equations and more general Volterra integral equations under Carathéodory hypotheses, e.g. Arnese (1963), Santagati (1963, 1965, 1967), Fiorenza (1965, Pulvirenti (1965), Fedele (1966), Kasatkina (1967), Deimling (1969, 1970), Reichert (1969).

The class Z_3^* leads us to the real subject matter of this section. In Chapter I our treatment of ordinary differential equations started with the *integral* equation, in Chapter II with the *differential* equation. For hyperbolic differential equations we have, up to now, worked only with *integral* equation methods. This gap will now be closed. We shall show that a local method of proof oriented to the *differential* equation is possible and leads to similar results.

IV. Theorem. If (in one of the problems (a) (b)) the functions $v, w \in Z_3^*$, if $f(x, y, z, p, q)$ is monotone increasing in z, p, q and if

(α) (a) $v(0,0) \leqq w(0,0)$, $v_x < w_x$ for $y=0$, $v_y < w_y$ for $x=0$;

(α) (b) $v \leqq w$, $v_x < w_x$, $v_y < w_y$ on C;

(β) $Pv < Pw$ in $\bar{G}$ with $P\varphi = \varphi_{xy} - f(x, y, \varphi, \varphi_x, \varphi_y)$,

then

$$v < w, \quad v_x < w_x, \quad v_y < w_y \quad \text{in } G_0\,[1].$$

One should compare this theorem and the subsequent proof with 8 V. We consider problem (a). It suffices to show that $v_x < w_x$ and $v_y < w_y$ in $\bar{G}$. Let us assume that this is false and that t_0 is the largest number such that both of these inequalities hold for $x + y < t_0$. Then there is a point (x_0, y_0) with $x_0 + y_0 = t_0$ at which one of the two inequalities fails and we have, say, $v_y(x_0, y_0) = w_y(x_0, y_0)$. By ($\alpha$) (a) we have $x_0 > 0$. The two functions $\varphi(x) = v_y(x, y_0)$, $\psi(x) = w_y(x, y_0)$ then have the properties $\varphi < \psi$ for $0 < x < x_0$, $\varphi(x_0) = \psi(x_0)$ and

$$\begin{aligned}\varphi'(x_0) &= (Pv)(x_0, y_0) + f(x_0, y_0, v, v_x, v_y)\\ &< (Pw)(x_0, y_0) + f(x_0, y_0, w, w_x, w_y) = \psi'(x_0),\end{aligned}$$

which contain a contradiction according to Lemma 8 II.

The proof can be adapted easily to the Cauchy problem.

V. Consequence. From IV we can deduce theorems on super- and subfunctions as well as estimation and uniqueness theorems. For this we proceed exactly as in Sections 8 and 9 for ordinary differential equations. As a result we obtain 21 V, VI, IX in an insignificantly modified form.

[1] G_0 is defined in 21 VIII.

This indicates a situation which is perhaps surprising compared with the case $m=1$. Although the two approaches are so different, we achieve approximately the same goal. It was quite different in Chapters I and II. The monotonicity assumption in Section 4 played a dominant role for integral equations. On the other hand, we did without it in Sections 8 and 9 in the second chapter. We know from the discussion in 20 X that the monotonicity in z is indispensable. The question comes up as to whether the monotonicity in p and q is necessary in IV. It can be answered in the negative for at least one special class of differential equations, namely, those independent of u_y

$$u_{xy}=f(x,y,u,u_x)\,. \tag{2}$$

VI. Theorem. *Suppose the functions v, w are from the class Z_3^*, the function $f(x,y,z,p)$, independent of q, is monotone increasing in z, and suppose*

(α) (a) $v \leqq w$ *for* $x=0$, $v_x < w_x$ *for* $y=0$,
(α) (b) $v \leqq w$ *and* $v_x < w_x$ *on* C,
(α) (c) $v \leqq w$ *on* C, $v_x < w_x$ *for* $y=0$,
(β) $Pv < Pv$ *with* $P\varphi = \varphi_{xy} - f(x,y,\varphi,\varphi_x)$ *in* $\bar{G}$.

Then

$$v<w \quad \text{and} \quad v_x<w_x \quad \text{in} \quad G_0\,.$$

This theorem does not require monotonicity in p, and it also holds for the Goursat problem (c). For the proof we assume that we do not have $v_x < w_x$ in $\bar{G}$. Then there exist a point (x_0, y_0) and a number t_0 with the properties $x_0 + y_0 = t_0$, $v_x(x_0,y_0) = w_x(x_0,y_0)$, $v_x < w_x$ for $x+y<t_0$. Hence, as in the proof of IV, for the two functions $\varphi(y) = v_x(x_0, y)$, $\psi(y) = w_x(x_0,y)$ we obtain $\varphi < \psi$ for $y < y_0$, $\varphi(y_0) = \psi(y_0)$, $\varphi'(y_0) < \psi'(y_0)$ and thus a contradiction to Lemma 8 II.

Of course, an analogous theorem also holds if the variable p is missing in f.[1] With the help of VI we easily prove the following

VII. Theorem. *Suppose the function $\omega(x,y,z,p)$ is defined for $z \geqq 0, p \geqq 0$ and is monotone increasing in z. Suppose that for $u, v, \varrho \in Z_3^*$ we have*

(α) (a) $|v-u| \leqq \varrho$ *for* $x=0$, $|v_x - u_x| < \varrho_x$ *for* $y=0$,
(α) (b) $|v-u| \leqq \varrho$ *and* $|v_x - u_x| < \varrho_x$ *on* C,
(α) (c) $|v-u| \leqq \varrho$ *on* C, $|v_x - u_x| < \varrho_x$ *for* $y=0$;

[1] Then the initial condition in the Goursat problem is $v \leqq w$ on C, $v \leqq w$ and $v_y < w_y$ on $y=0$. It should be noted that the values of u_y can also be considered as known for given initial values on $y=0$ since they can be obtained from an ordinary differential equation (21.13); the initial value $u_y(0,0)$ likewise follows from the given data.

(β) $u_{xy}=f(x,y,u,u_x)$, $|v_{xy}-f(x,y,v,v_x)|\leqq\delta(x,y)$ *in* $\bar{G}$;
(γ) $\varrho_{xy}>\delta(x,y)+\omega(x,y,\varrho,\varrho_x)$ *in* $\bar{G}$;
(δ) $f(x,y,z,p)-f(x,y,\bar{z},\bar{p})\leqq\omega(x,y,|z-\bar{z}|,p-\bar{p})$ *for* $p\geqq\bar{p}$.

Then

$$|v-u|<\varrho \quad \textit{and} \quad |v_x-u_x|<\varrho_x \quad \textit{in} \quad G_0 .$$

See 9 I for the proof, which proceeds like the preceding one and whose details we leave to the reader. Likewise, we shall not carry out the specialization to uniqueness assertions. Zwirner (1952) was the first to note the special role taken on by equations of the form (2) and to prove the theorem VI and a uniqueness theorem for problem (a) which are easy to derive from VII [1].

VIII. General Hyperbolic Equations. The theory for Eq. (1) has been developed to some extent for the more general equations

$$-h^2(x,y)u_{xx}+u_{yy}=F(x,y,u,u_x,u_y) \tag{3}$$

and

$$a(x,y)u_{xx}+2b(x,y)u_{xy}+c(x,y)u_{yy}=F(x,y,u,u_x,u_y) \tag{4}$$

where (condition for hyperbolicity)

$$ac-b^2<0 .$$

We mention the works of Protter (1951, 1958), Agmon, Nirenberg and Protter (1953), Weinberger (1956), Gloistehn (1963), Sather (1966), Payne and Sather (1967), their methods having much in common with the theory developed here. Many of these results are presented in Chapter 4 of the book by Protter and Weinberger (1967). The questions discussed in this chapter and various related problems are also treated in the book by Cinquini-Cibrario and Cinquini (1964).

---

[1] In the uniqueness theorem Zwirner does not proceed via an estimation theorem, but derives it directly from VI and therefore has to make the additional assumption that f is monotone increasing in z.

CHAPTER IV

Parabolic Differential Equations

23. Notation

For parabolic differential equations the domain G is an open set of points $(t, x)=(t, x_1, \ldots, x_m)$ with $x=(x_1, \ldots, x_m) \in E^m$, $m \geqq 1$, in the $m+1$-dimensional space E^{m+1}. This notation expresses the fact that the independent variables fall into two essentially different groups. The scalar variable t is also called the "time" variable, the variable $x \in E^m$ the "space" variable. We note that another n-dimensional space will occur later for parabolic systems. In order to distinguish these two spaces clearly in the notation, we shall (as in the preceding chapter) not use bold-face for the points $x \in E^m$, but will use it for n-dimensional vector-functions $\boldsymbol{\varphi}=(\varphi_1, \ldots, \varphi_n)$.

In the statement of physical problems which lead to parabolic differential equations, the domain is often a hypercylinder, i.e., G is equal to the topological product of an open domain D of the x-space E^m and a t-interval, say $0<t<T$:

$$G=(0, T) \times D .$$

We denote the boundary and the closed hull of D or G (with respect to E^m or E^{m+1}) respectively by ∂D and $\bar{D}$ or ∂G and $\bar{G}$. We partition the boundary ∂G into three subsets $\partial_0 G, \partial_1 G, \partial_2 G$ which we call the lower surface, the lateral surface and the upper surface; we have

$$\partial_0 G=\{0\} \times \bar{D}, \quad \partial_1 G=(0, T] \times \partial D, \quad \partial_2 G=\{T\} \times D$$

(we note that the "lower edge" $\{0\} \times \partial D$ is taken with the lower surface, and in contrast the "upper edge" $\{T\} \times \partial D$ with the lateral surface). This partitioning of the boundary is suggested by the various initial value problems. Usually, function values ("initial values") are prescribed on $\partial_0 G$, function values or relations between function values and their normal derivatives ("boundary conditions") on $\partial_1 G$, while $\partial_2 G$ is considered with the interior of the domain in the sense that the differential equation must also be satisfied on $\partial_2 G$. The "interior" is $J_0 \times D$ (see 1 I).

We use an analogous partitioning of the boundary points for an arbitrary, not necessarily cylindrical, domain.

I. Definition ($U, U_-, G, \bar{G}, \partial G, \partial_0 G, \partial_1 G, \partial_2 G, G_p, R_p$). By a neighborhood, or more precisely a δ-neighborhood ($\delta > 0$) of the point $(\bar{t}, \bar{x})$ we mean the set $U = U^\delta(\bar{t}, \bar{x})$ of points (t, x) for which

$$(t-\bar{t})^2 + (x_1 - \bar{x}_1)^2 + \cdots + (x_m - \bar{x}_m)^2 < \delta^2 .$$

The *lower half-neighborhood* U_- is that subset of U for whose points we have $t < \bar{t}$. Thus, according to this definition U_- does not contain the point $(\bar{t}, \bar{x})$[1].

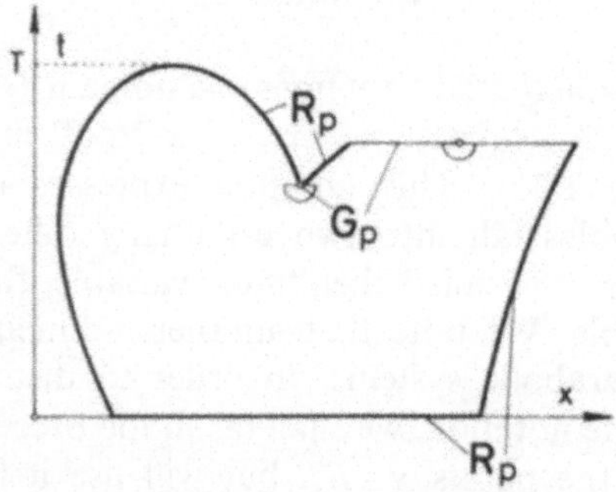

Fig. 11. The domain for parabolic differential equations

We use G to denote an open connected set of points $(t, x) = (t, x_1, \ldots, x_m) \in E^{m+1}$ ($m \geqq 1$); ∂G is its boundary and $\bar{G} = G + \partial G$ its closure. We assume that $\bar{G}$ lies completely between two hyperplanes $t = \text{const}$, i.e., that the projection of G on the t-axis fills out a bounded interval, namely, the interval $J = [0, T]$ (the latter can always be achieved by a parallel shift of $\bar{G}$). Again we divide the boundary into three pairwise disjoint subsets $\partial_0 G, \partial_1 G, \partial_2 G$. All boundary points for which there is a lower half-neighborhood U_- lying entirely outside of G belong to $\partial_0 G$; those boundary points for which a lower half-neighborhood lies entirely in G form $\partial_2 G$; the remaining boundary points are combined in the set $\partial_1 G = \partial G - \partial_0 G - \partial_2 G$. We further set

$$G_p = G + \partial_2 G ,$$
$$R_p = \partial_0 G + \partial_1 G = \partial G - \partial_2 G .$$

We make some remarks. We easily convince ourselves that the two definitions of $\partial_0 G, \partial_1 G, \partial_2 G$ coincide in the special case of the hyper-

[1] Thus the system of lower half-neighborhoods U_- does not define a topology in E^{m+1}. We could remove this inconvenience without difficulty by adding the point $(\bar{t}, \bar{x})$ to U_-. Topological problems related to parabolic differential equations are investigated by Kamke (1959a).

cylinder described above. The remark made there, that a differential equation and an initial condition are always prescribed in the problems of this chapter where the differential equation refers to G_p and the initial condition to R_p, also holds in the general case; the subcript p indicates that for G_p and R_p we are concerned with an "interior" and "boundary" relative to a *parabolic* problem[1].

The theory can be immediately extended to the case where the projection of G on the t-axis fills out the interval $0<t<\infty$. For this we need only apply the appropriate theorems to that part of G between the two hyperplanes $t=0$ and $t=T$ and then let $T\to\infty$ (see the corresponding remark 8 XII (γ) for ordinary differential equations). The restriction that G be bounded in the negative t-direction is, however, essential. Nevertheless, as we shall see in Section 28, there are also certain "problems without initial conditions" for which, for example, $G=\{-\infty<t<\infty\}\times D$, and which are amenable to treatment within the framework of the theory under consideration.

II. Definition *(φ_x, φ_{xx}, $\mathcal{P}$, defect P, $r\leqq\bar r$ for matrices, $D(f)=M\times M_r$).* If $\varphi(t,x)$ is a given function equipped with appropriate differentiability properties, then φ_x denotes the m-dimensional vector of the partial derivatives of first order, φ_{xx} denotes the $m\times m$ matrix of the partial derivatives of second order with respect to the x_μ,

$$\varphi_x=\left(\frac{\partial\varphi}{\partial x_\mu}\right)\quad(1\leqq\mu\leqq m),\qquad \varphi_{xx}=\left(\frac{\partial^2\varphi}{\partial x_\lambda\partial x_\mu}\right)\quad(1\leqq\lambda\leqq m;\,1\leqq\mu\leqq m).$$

To simplify the notation, we define the inequality $r\leqq\bar r$ for symmetric $m\times m$ matrices $r=(r_{\lambda\mu})$, $\bar r=(\bar r_{\lambda\mu})$

$$r\leqq\bar r\quad\text{if and only if}\quad\sum_{\lambda,\mu=1}^{m}(\bar r_{\lambda\mu}-r_{\lambda\mu})\alpha_\lambda\alpha_\mu\geqq 0\quad\text{for arbitrary }\alpha_\mu,\tag{1}$$

i.e., if and only if the matrix $\bar r-r$ is positive semi-definite. For $m=1$ this is the ordinary $\leqq$ concept.

Now suppose we have a nonlinear parabolic differential equation

$$u_t=f(t,x,u,u_x,u_{xx})\tag{2}$$

or, written out in detail,

$$\frac{\partial u}{\partial t}=f\left(t,x_1,\dots,x_m,u,\frac{\partial u}{\partial x_1},\dots,\frac{\partial u}{\partial x_m},\frac{\partial^2 u}{\partial x_1^2},\frac{\partial^2 u}{\partial x_1\partial x_2},\dots,\frac{\partial^2 u}{\partial x_m^2}\right).$$

[1] We could allow R_p to properly contain $\partial_0 G+\partial_1 G$ and thus contain points of $\partial_2 G$. But since our theorems assert that an inequality which is assumed to hold on R_p is also true in the interior $\bar G-R_p$, we are led to use the smallest possible R_p.

Suppose that the function $f(t, x, z, p, r)$, where $x=(x_1, \ldots, x_m)$, $p=(p_1, \ldots, p_m) \in E^m$ and $r=(r_{\lambda\mu})$ is an $m \times m$-matrix, is defined on a set $D(f)=M \times M_r$ where M is a set of the $(1+m+1+m)$-dimensional (t, x, z, p)-space and M_r is a set of the m^2-dimensional r-space [1]. Suppose further — this is the essential hypothesis which defines the parabolic character of Eq. (2) — that f is *monotone increasing in* r, i.e., that we have

$$f(t, x, z, p, r) \leqq f(t, x, z, p, \bar{r}) \quad \text{for} \quad r \leqq \bar{r}, \tag{3}$$

provided $(t, x, z, p) \in M$ and $r, \bar{r} \in M_r$. If f satisfies the given conditions, then f belongs to the class $\mathscr{P}$, or briefly, $f \in \mathscr{P}$.

The *defect* $P\varphi$ of a function φ is defined by

$$P\varphi = \varphi_t - f(t, x, \varphi, \varphi_x, \varphi_{xx}). \tag{4}$$

We write $(P\varphi)(\bar{t}, \bar{x})$ for the value of the defect at the point $(\bar{t}, \bar{x})$.

We now give two sufficient conditions for $f \in \mathscr{P}$.

(α) If f is quasilinear

$$f(t, x, z, p, r) = g(t, x, z, p) + \sum_{\lambda, \mu=1}^{m} k_{\lambda\mu}(t, x, z, p) r_{\lambda\mu}$$

and if the matrix $k=(k_{\lambda\mu})$ is symmetric and positive semi-definite at every point, then $f \in \mathscr{P}$.

For when $r \leqq \bar{r}$ we have

$$f(t, x, z, p, \bar{r}) - f(t, x, z, p, r) = \sum_{\lambda, \mu} k_{\lambda\mu} s_{\lambda\mu} \quad \text{with} \quad s = \bar{r} - r \geqq 0$$

(in the last inequality 0 is the null matrix, $s \geqq 0$ means that s is positive semi-definite in accordance with (1)). The sum occurring in the last formula is equal to the sum of the diagonal elements of the product matrix ks and thus equal to the trace $\mathrm{tr}(ks)$ [2]. Now if c is an orthogonal matrix which transforms k (for fixed t, x, z, p) to diagonal form, $c^{-1}kc = d$ ($d_{\mu\mu} \geqq 0$, $d_{\lambda\mu} = 0$ otherwise), then, since such a transformation leaves the

[1] It should be noted that we can get along with more modest hypotheses on $D(f)$. Actually we only need: $(t, x, z, p, r), (t, x, \bar{z}, p, \bar{r}) \in D(f)$ implies $(t, x, z, p, \bar{r}) \in D(f)$. The extension of such a hypothesis to systems, however, becomes unwieldy and therefore we shall not use it here. The hypothesis that the projection of $D(f)$ onto the (t, x)-plane covers the set G_p (see Szarski (1955, Theorème 2.1,1°) is *not* sufficient. Let us consider the following counterexample to 25 III: $m=1$, $f=2z+r$, but where $D(f)$ contains only the points (t, x, z, p, r) with $z+r=0$, $G_p=(0, T] \times (0, \pi)$. For the two solutions $u=0$, $v=e^t \sin x$ we have $|v-u| \leqq 1$ on R_p. Furthermore $f(t, x, z, p, r) - f(t, x, \bar{z}, p, r) = 0$ provided both arguments lie in $D(f)$; thus we can set $\omega = 0$. We would obtain $|v-u| \leqq 1$!

[2] By the trace $\mathrm{tr}(c)$ of an $m \times m$ matrix $c=(c_{\lambda\mu})$ we mean the sum $\mathrm{tr}(c) = c_{11} + c_{22} + \cdots + c_{mm}$.

trace invariant, we have

$$\sum_{\lambda,\mu} k_{\lambda\mu} s_{\lambda\mu} = \mathrm{tr}(ks) = \mathrm{tr}(c^{-1}kcc^{-1}sc)$$
$$= \mathrm{tr}(dc^{-1}sc) = \sum_{\mu} d_{\mu\mu}(c^{-1}sc)_{\mu\mu} \geqq 0,$$

since also $c^{-1}sc \geqq 0$ and since the diagonal elements of a matrix $\geqq 0$ are themselves $\geqq 0$.

(β) If f is continuously differentiable with respect to the $r_{\lambda\mu}$, M_r is a convex set and the matrix $(\partial f/\partial r_{\lambda\mu})$ is positive semi-definite, then $f \in \mathscr{P}$. For by the mean value theorem we have

$$f(t,x,z,p,\bar{r}) - f(t,x,z,p,r) = \sum_{\lambda,\mu} \frac{\partial f}{\partial r_{\lambda\mu}}(t,x,z,p,r+\alpha(\bar{r}-r))(\bar{r}_{\lambda\mu}-r_{\lambda\mu})$$

with $0<\alpha<1$, whence follows the inequality (3) as in (α).

III. Definition $(Z, Z_0, Z(f), Z_0(f))$. The function $\varphi(t,x)$ belongs to the class Z or Z_0 if it is defined and continuous in $\bar{G}$ or G_p respectively and if in G_p it has the derivative φ_t as well as continuous derivatives φ_x and φ_{xx}. If, moreover, (for given $f \in \mathscr{P}$) $(t,x,\varphi,\varphi_x,\varphi_{xx}) \in D(f)$ for $(t,x) \in G_p$, then $\varphi \in Z(f)$ or $Z_0(f)$ respectively.

We remark, however, that the differentiability requirements can be made considerably weaker without affecting the validity of the following theorems. The following two properties (α) (β) suffice:

(α) the backwards t-derivative[1]

$$\varphi_t^-(t,x) = \lim_{h\to+0} \frac{\varphi(t)-\varphi(t-h)}{h}$$

exists in G_p;

(β) the derivatives φ_x and φ_{xx} exist in G_p; φ_x is continuous, the matrix φ_{xx} is symmetric (i.e., the value of a mixed derivative is independent of the order of differentiation), and

$$\varphi(t,x+h) = \varphi(t,x) + \sum_{\mu=1}^{m} h_\mu \varphi_{x_\mu}(t,x)$$
$$+ \tfrac{1}{2} \sum_{\lambda,\mu=1}^{m} h_\lambda h_\mu \varphi_{x_\lambda x_\mu}(t,x) + o(h_1^2 + \cdots + h_m^2)$$

for $h=(h_1,\ldots,h_m)\to 0$[2].

---

[1] We then have u_t^- in (2) instead of u_t; moreover, this substitution must occur at the points of $\partial_2 G$ in every case.

[2] We might also say that φ possesses a total derivative (or a total differential) of second order (with respect to x).

We can go even further if we have a parabolic differential equation without mixed derivatives — an example is $u_t = g(t, x, u, u_x)\,\Delta u + h(t, x, u, u_x)$. In this case ($\beta$) can be replaced by

(β') the derivatives φ_x and $\varphi_{x_\mu x_\mu}$ ($\mu = 1, \ldots, m$) exist in G_p; φ_x is continuous in G_p.

In particular, for $m = 1$ the existence (without continuity) of φ_{xx} always suffices.

What we do need in the proofs (besides the trivial requirement that the derivatives occurring in the differential equation exist) is the following: *If the difference $\psi - \varphi$ of two functions $\varphi, \psi \in Z_0$ has a minimum relative to x at the point $(t_0, x_0) \in G_p$ ($t = t_0$ fixed), then*

(γ) $\varphi_x = \psi_x$ *and*[1] $\varphi_{xx} \leqq \psi_{xx}$ *at the point* (t_0, x_0)*; or*

(γ') $\varphi_x = \psi_x$ *and* $\varphi_{x_\mu x_\mu} \leqq \psi_{x_\mu x_\mu}$ ($\mu = 1, \ldots, m$) *at the point* (t_0, x_0).

However, under the hypothesis (β) or (β'), (γ) or (γ') is true. (It is known that $\varphi' = 0$, $\varphi'' \geqq 0$ for a function of *one* variable $\varphi(s)$ at a minimum point, and indeed this follows simply from the existence of the second derivative).

IV. Definition ($\varphi < \psi$, $\varphi \leqq \psi$, $\varphi = \psi$ *on* R_p^+ *and on* R_∞). Even quite simple physical problems lead to problems with discontinuous initial and boundary values. In order to be able to handle them too, we have introduced the class Z_0. Now it becomes necessary to extend the statements "$\varphi < \psi$ on R_p", ..., which at first sight are only meaningful in the class Z, to the class Z_0.

In problems with unbounded domains, conditions at infinity are involved in addition to the initial and boundary conditions on R_p. However, these can be of a very different type. In many heat conduction problems we have quite vague requirements at infinity, as for example: The solution must be bounded, or as $x \to \infty$ must not grow faster than a certain function. But there are also cases, for example in boundary layer theory, where the solution must approach a given function at infinity. A uniform mathematical description of all these possibilities, which introduces a set R_∞ of "ideal" or "infinitely distant" boundary points on which additional conditions are then imposed, is very natural, but leads to difficulties. Therefore, we by-pass the definition of R_∞ and instead define the statements "$\varphi < \psi$ on R_∞", ...

For two functions φ, ψ defined in G_p, we denote:

$$\varphi < \psi \text{ on } R_p^+ \text{ or on } R_\infty\text{:}\quad \limsup_{k \to \infty} [\psi(t_k, x_{(k)}) - \varphi(t_k, x_{(k)})] > 0;$$

$$\varphi \leqq \psi \text{ on } R_p^+ \text{ or on } R_\infty\text{:}\quad \limsup_{k \to \infty} [\psi(t_k, x_{(k)}) - \varphi(t_k, x_{(k)})] \geqq 0;$$

$$\varphi = \psi \text{ on } R_p^+ \text{ or on } R_\infty\text{:}\quad \lim_{k \to \infty} [\psi(t_k, x_{(k)}) - \varphi(t_k, x_{(k)})] = 0,$$

[1] The meaning of the following inequality was defined in (1).

and indeed for every sequence of points $(t_k, x_{(k)}) \in G_p$ for which the t_k form a monotone decreasing sequence of numbers and for which

$$R_p^+:\ (t_k, x_{(k)}) \to (\bar{t}, \bar{x}) \in R_p; \qquad R_\infty:\ \|x_{(k)}\|_e \to \infty \qquad (k \to \infty).$$

Here is an example which corresponds to the situation for the boundary layer equation in Section 30. If the functions $\varphi(t, x)$, $\psi(t, x)$ are continuous in $\bar{G} = [0, T] \times [0, \infty)$ $(m = 1)$ and converge, as $x \to \infty$, uniformly in $t \in [0, T]$ to continuous functions $\Phi(t)$, $\Psi(t)$, then the three statements "$\varphi < \psi$ on R_∞", ... are identical to the three relations "$\Phi(t) < \Psi(t)$ for $0 \leqq t \leqq T$", ...

If the functions φ, ψ are continuous in $\bar{G}$, then one would expect that the statement "$\varphi < \psi$ on R_p^+" is identical to $\varphi < \psi$ on R_p, and that the same holds for the other two concepts. This is indeed true in all practical cases; more exactly, it is true if to each point $(\bar{t}, \bar{x}) \in R_p$ there corresponds a sequence of points $(t_k, x_{(k)}) \in G_p$ such that $(t_k, x_{(k)}) \to (\bar{t}, \bar{x})$ and $t_k \geqq \bar{t}$.

24. The Nagumo-Westphal Lemma

Our approach corresponds to that of Chapter II. We again start out with a lemma in which we do not yet speak of differential equations and from which all subsequent monotonicity and estimation theorems follow relatively easily. We recall the convention by which the projection of $\bar{G}$ on the t-axis fills the interval $J = [0, T]$.

I. Lemma. *Suppose that for two functions $\varphi, \psi \in Z_0$ we have: If $\varphi = \psi$, $\varphi_x = \psi_x$, $\varphi_{xx} \leqq \psi_{xx}$ at a point in G_p, then $\varphi_t < \psi_t$ at this point. Then we have precisely one of the following cases:*

(α) *$\varphi < \psi$ in G_p;*

(β) *There exists a maximal $\bar{t}$ $(0 \leqq \bar{t} < T)$ such that*

$$\varphi < \psi \quad \textit{for all points in } G_p \textit{ with } \quad t \leqq \bar{t};$$

thus there is a sequence of points

$$(t_k, x_{(k)}) \in G_p,\ t_k > \bar{t}, \quad \textit{with} \quad \varphi(t_k, x_{(k)}) \geqq \psi(t_k, x_{(k)}) \qquad (k = 1, 2, \ldots)$$

such that, as $k \to \infty$,

$$\textit{either} \quad (t_k, x_{(k)}) \to (\bar{t}, \bar{x}) \in R_p \quad \textit{or} \quad t_k \to \bar{t},\ \|x_{(k)}\|_e \to \infty.$$

The essence of assertion (β) is that the inequality $\varphi < \psi$ also holds for $t = \bar{t}$ and that consequently the $(t_k, x_{(k)})$ always approach a boundary point or infinity.

For the proof let $\bar{t}$ be the largest number with the property that $\varphi<\psi$ on the set $G_p\cdot\{(t,x)|t<\bar{t}\}$. Then because of continuity $\varphi\leqq\psi$ on $H=G_p\cdot\{(t,x)|t=\bar{t}\}$ (the set H can be empty, for example, if $\bar{t}=0$). Now we prove that $\varphi<\psi$ on H. For if $\varphi=\psi$ for a point $(\bar{t},\bar{x})\in H$, then the difference $\psi-\varphi$ would vanish at this point and take on a minimum relative to x (for fixed $t=\bar{t}$) at this point. Hence by 23 III (γ) at this point we would have

$$\varphi=\psi, \qquad \varphi_x=\psi_x, \qquad \varphi_{xx}\leqq\psi_{xx},$$

and thus by hypothesis $\varphi_t<\psi_t$. But on the other hand $\varphi(t,\bar{x})<\psi(t,\bar{x})$ for $t<\bar{t}$ and $\varphi(\bar{t},\bar{x})=\psi(\bar{t},\bar{x})$, whence (with the same reasoning as in 8 II) we obtain $\varphi_t^-(\bar{t},\bar{x})\geqq\psi_t^-(\bar{t},\bar{x})$. This contradiction shows that in fact $\varphi<\psi$ on H.

If $\bar{t}=T$, then these considerations show that the inequality $\varphi<\psi$ also holds for the points of G_p which lie on $t=T$; we then have case (α). If $\bar{t}<T$, then by the definition of $\bar{t}$ there is surely a sequence of points $(t_k,x_{(k)})\in G_p$ with the property that $\varphi\geqq\psi$ at these points and that the t_k strictly decrease to $\bar{t}$. If the $x_{(k)}$ form an unbounded set in E^m, then we can assume — by going to a subsequence if necessary — that $\|x_{(k)}\|_e\to\infty$. If they form a bounded set then we can assume that they converge, say, to the point $\bar{x}$. Then $(t_k,x_{(k)})\to(\bar{t},\bar{x})$ and the assumption $(\bar{t},\bar{x})\in G_p$ leads immediately to a contradiction of the fact that $\varphi<\psi$ on H, proved at the beginning. Thus the lemma has been completely proved.

The proof was carried out in such a way that the following generalization was proved along with it.

II. Generalization. Suppose that the functions φ,ψ are continuous in G_p and that for all points $(t,x)\in G_p$ at which $\varphi=\psi$ they satisfy the two differentiability conditions 23 III (α) (β). Suppose further that if $\varphi=\psi$, $\varphi_x=\psi_x$, $\varphi_{xx}\leqq\psi_{xx}$ at a point in G_p then $\varphi_t^-<\psi_t^-$ at this point. Then the assertion of I holds.

It should be noted that we can get along without the differentiability in t; instead of $\varphi_t^-<\psi_t^-$, we then have to require

$$D_-\varphi<D_-\psi \quad \text{or} \quad D^-\varphi<D^-\psi$$

(here D_-, D^- are Dini derivates in the t direction).

With regard to differential equations without mixed partial derivatives we now give a

III. Corollary. Suppose the functions φ,ψ are continuous in G_p. Suppose that at the points of G_p at which $\varphi=\psi$, the derivatives φ_t^-, ψ_t^- exist, the derivatives φ_x, ψ_x exist and are continuous, and the pure derivatives of second order $\varphi_{x_\mu x_\mu}$, $\psi_{x_\mu x_\mu}$ ($\mu=1,\ldots,m$) exist. Suppose further that if $\varphi = \psi$, $\varphi_x = \psi_x$, $\varphi_{x_\mu x_\mu} \leqq \psi_{x_\mu x_\mu}$ ($\mu = 1, \ldots, m$) at a point in G_p, then $\varphi_t^- < \psi_t^-$ at this point. Then the assertion of I remains valid.

In the above proof of Lemma I we now have to appeal to 23 III (γ').

The following central lemma goes back to Nagumo (1939). It remained relatively unknown (the work is written in Japanese) until it was rediscovered by Westphal (1949) and was used by a series of mathematicians in investigating parabolic differential equations; see also Nagumo-Simoda (1951). On the basis of our preliminaries, we can formulate it in such a way that it also applies to problems with discontinuous boundary values and unbounded domains.

IV. Nagumo-Westphal Lemma. *Suppose $f \in \mathscr{P}$ and $v, w \in Z_0(f)$. Further suppose*

(α) $v < w$ *on* R_p^+ *and on* R_∞;

(β) $Pv < Pw$ *in* G_p.

Then

$$v < w \quad \text{in} \quad G_p.$$

The proof is obtained directly from I. If $v = w$, $v_x = w_x$, $v_{xx} \leqq w_{xx}$ at a point in G_p, then, by the monotonicity assumption (23.3), at this point

$$\begin{aligned} 0 < Pw - Pv &= w_t - v_t + f(t, x, v, v_x, v_{xx}) - f(t, x, w, w_x, w_{xx}) \\ &\leqq w_t - v_t + f(t, x, w, w_x, w_{xx}) - f(t, x, w, w_x, w_{xx}) = w_t - v_t . \end{aligned}$$

Thus Lemma I (of course with v, w instead of φ, ψ) is applicable. But since I (β) is incompatible with the hypothesis (α), I (α) holds, as asserted.

V. Remark. If f has the special form $f(t,x,z,p,r) = f(t,x,z,p,r_{11},r_{22},\ldots,r_{mm})$ then, as III shows, the existence of the pure derivatives of second order $v_{x_\mu x_\mu}$, $w_{x_\mu x_\mu}$ ($\mu = 1, \ldots, m$) is sufficient for the validity of the lemma.

Now we give two variants of the lemma. The first concerns functions f which satisfy a uniqueness condition. Interestingly, this uniqueness condition is of the same kind as that which came up in Section 10 for ordinary differential equations. See also Walter (1960a), Kaplan (1963).

VI. Corollary. Suppose that for $f \in \mathscr{P}$ and the functions $v, w \in Z_0(f)$ there exists an $\omega \in \mathscr{E}_5$ such that

$$f(t, x, w + z, w_x, w_{xx}) - f(t, x, w, w_x, w_{xx}) \leqq \omega(t, z) \quad \text{for} \quad z > 0 \tag{1}$$

[and $(t, x, w + z, w_x, w_{xx}) \in D(f)$]. Further suppose

(α) $v \leqq w$ on R_p^+ and on R_∞;

(β) $Pv \leqq Pw$ in G_p.

Then

$$v \leqq w \quad \text{in} \quad G_p.$$

For the proof suppose $\varrho(t)$ is a function with the properties given in 10 I. It obviously suffices to show that $v < w + \varrho$ in G_p. We reduce the assertion to I, now with $\varphi = v$, $\psi = w + \varrho$. As in the proof of III, $\varphi = \psi$, $\varphi_x = \psi_x$, $\varphi_{xx} \leqq \psi_{xx}$, i.e., $v = w + \varrho$, $v_x = w_x$, $v_{xx} \leqq w_{xx}$ with the help of (1)[1] implies

$$\begin{aligned} 0 &\leqq Pw - Pv = w_t - v_t + f(t, x, v, v_x, v_{xx}) - f(t, x, w, w_x, w_{xx}) \\ &\leqq w_t - v_t + f(t, x, w + \varrho, w_x, w_{xx}) - f(t, x, w, w_x, w_{xx}) \\ &\leqq w_t - v_t + \omega(t, \varrho) < w_t - v_t + \varrho' = \psi_t - \varphi_t \,. \end{aligned}$$

Thus Lemma I is applicable. By (α) together with the fact that $\varrho(t) \geqq \delta > 0$, we cannot have I ($\beta$). Thus I ($\alpha$) holds, i.e., we have precisely $v < w + \varrho$.

We note that the hypothesis (1) can be replaced by

$$f(t, x, v, v_x, v_{xx}) - f(t, x, v - z, v_x, v_{xx}) \leqq \omega(t, z) \quad \text{for} \quad z > 0 \,. \tag{1'}$$

A sufficient condition for (1) or (1') and hence for the validity of the corollary is

(γ) w, w_x, w_{xx} (or v, v_x, v_{xx}) are bounded in G_p; $f_z(t, x, z, p, r)$ exists and is bounded above in every set $G_p \times C$, where C is a bounded set in (z, p, r)-space.

In particular we can apply VI to two solutions of Eq. (1) whose boundary values coincide, i.e., a general uniqueness theorem is hidden in VI. We shall come back to this later.

For the sake of completeness we now give the

VII. Corollary. Suppose the function $f \in \mathscr{P}$ is monotone increasing in z and the domain G is bounded . For two functions $v, w \in Z(f)$ suppose

(α) $v < w$ on R_p;
(β) $Pv \leqq Pw$ in G_p.

Then we have $v < w$ and also

$$w(t, x) - v(t, x) \geqq \min\,[w(\bar{t}, \bar{x}) - v(\bar{t}, \bar{x}) \,|\, (\bar{t}, \bar{x}) \in R_p, \bar{t} \leqq t] \,. \tag{2}$$

If for the purposes of the proof we denote the expression on the right hand side of the last inequality by $h(t)$, then for a fixed $t_0 \in J_0$ we can choose a positive ε such that $\varepsilon(1 + t_0) < h(t_0)$. The function

$$\bar{w}(t, x) = w(t, x) - h(t_0) + \varepsilon(1 + t)$$

is $< w$ for $0 \leqq t \leqq t_0$. We want to apply Lemma I to the functions $\varphi = v$, $\psi = \bar{w}$ and the part of G lying between $t = 0$ and $t = t_0$; for this we need merely verify that $v = \bar{w}(< w)$, $v_x = \bar{w}_x (= w_x)$, $v_{xx} \leqq \bar{w}_{xx} (= w_{xx})$ implies the inequality

$$\begin{aligned} \bar{w}_t - v_t > w_t - v_t &= Pw - Pv + f(t, x, w, w_x, w_{xx}) - f(t, x, v, v_x, v_{xx}) \\ &\geqq f(t, x, w, v_x, v_{xx}) - f(t, x, v, v_x, v_{xx}) \geqq 0 \end{aligned}$$

[1] The hypothesis 23 II on $D(f)$ is used.

which has indeed just been done. Since $v < \overline{w}$ in the part of R_p lying under t_0, I (β) is eliminated and we obtain the assertion $v < \overline{w}$ in the part of G_p under t_0 and thus in particular on the hyperplane $t = t_0$. But the assertion now follows at the point t_0 if we let $\varepsilon \to 0$. Since t_0 was arbitrary, VII is proved.

25. The First Boundary Value Problem

Up to now Theorems 8 II and 8 V have been extended from ordinary to parabolic differential equations. In 24 I we needed a new method of proof which goes back to Nagumo (which, by hindsight, arises quite naturally from that of 8 II). There are no great difficulties to be overcome in applying the results to boundary value problems; rather we can proceed via the path familiar from Chapter II.

I. Definition (*first boundary value problem, limit functions* η_*, η^*, u_*, u^*). The first boundary value problem[1] consists of the following: for a given domain G lying as usual between $t=0$ and $t=T$, a given function $f(t, x, z, p, r) \in \mathscr{P}$ and a given function $\eta(t, x)$ defined and continuous on R_p, find a function $u \in Z(f)$ for which

$$u_t = f(t, x, u, u_x, u_{xx}) \quad \text{in} \quad G_p \tag{1}$$

and $u = \eta$ on R_p; every such function is called a solution. If G_p is unbounded, conditions at infinity may be added. They are specially stated in each case.

In order to take care of discontinuous boundary values also, we formulate the problem in somewhat greater generality in accord with Perron (1923) and Sternberg (1929). If $\eta(t, x)$ is an arbitrary function defined on R_p, then we define its *upper limit function* η^* on R_p by

$$\eta^*(\bar{t}, \bar{x}) = \lim_{\delta \to +0} \{\sup [\eta(t, x) \mid (t, x) \in R_p \cdot U^\delta(\bar{t}, \bar{x})]\} \quad \text{for} \quad (\bar{t}, \bar{x}) \in R_p .$$

For the lower limit function η_* we simply replace "sup" by "inf". If the function $u(t, x)$ is defined in G_p, then we also define an *upper limit function* u^* on R_p by

$$u^*(\bar{t}, \bar{x}) = \lim_{\delta \to +0} \{\sup [u(t, x) \mid (t, x) \in G_p \cdot U^\delta(\bar{t}, \bar{x})]\} \quad \text{for} \quad (\bar{t}, \bar{x}) \in R_p$$

and the corresponding lower limit function u_*.

By a solution of the first boundary value problem for arbitrary given $\eta(t, x)$ we mean a function $u \in Z_0(f)$ which satisfies the differential equation (1) in G_p and the boundary condition

$$\eta_* \leqq u_* \leqq u^* \leqq \eta^* \quad \text{on} \quad R_p .$$

[1] Some authors prefer the name initial value problem or initial-boundary value problem.

This second definition is consistent with the first, i.e., if u is a solution in the latter sense and η is continuous on R_p, then u is a solution in the former sense when we define $u=\eta$ on R_p.

II. Super- and Subfunctions for a given first boundary value problem are provided by the functions $w, v \in Z(f)$ if they satisfy the conditions

(α) *$v \leqq \eta$ or $w \geqq \eta$ on R_p;*

(β) *$Pv \leqq 0$ or $Pw \geqq 0$ in G_p,*

and if the following hypotheses are satisfied: G is bounded, η is continuous, and $f \in \mathscr{P}$ has the property (24.1′) or (24.1) required in 24 VI. For the (now uniquely determined) solution u we have

$$v \leqq u \leqq w \quad \text{in} \quad \bar{G}.$$

This follows immediately from 24 VI. In the general case we have to require considerably more according to 24 IV, namely,

(α') *$v^* < \eta_*, \eta^* < w_*$ on R_p,*
$v < u < w$ on R_∞ for every solution u;

(β') *$Pv < Pw$ in G_p.*

Then we have $v < u < w$ in G_p for every solution u.

We gave the main theorem of Section 9 in three different forms. Here we restrict consideration to the conversion of 9 I and 9 III and leave the other to the reader. An essential role is played by a condition

$$f(t, x, z, p, r) - f(t, x, \bar{z}, p, r) \leqq \omega(t, z-\bar{z}) \quad \text{for} \quad z \geqq \bar{z}, \tag{2}$$

similar to the one involved in 24 VI. To understand the proof it is helpful to note that here ω is regarded as a special case of a function which depends on t, x, z, p, r and later $\varrho(t)$ is regarded as a special case of a function which depends on t and x. In this sense the ordinary differential equation $\varrho' = \omega(t, \varrho)$ is a special parabolic differential equation. We obtain different classes $Z(\omega)$ depending on whether we take ω as a function of (t, x) or (t, x, z, p, r); however in 5 I or 23 III the relationship $\varrho(t) \in Z(\omega)$ has the same meaning.

III. Theorem. *For the functions $u(t, x), v(t, x) \in Z_0(f), \delta(t), \bar{\delta}(t), \omega(t, z), \bar{\omega}(t, z)$ and $\varrho(t) \in Z(\omega), \bar{\varrho}(t) \in Z(\bar{\omega})$ suppose*

(α) *$-\bar{\varrho} < v - u < \varrho$ on R_p^+ and on R_∞;*

(β) *$u_t = f(t, x, u, u_x, u_{xx})$ and $-\bar{\delta}(t) \leqq v_t - f(t, x, v, v_x, v_{xx}) \leqq \delta(t)$ in G_p;*

(γ) *$\varrho' > \omega(t, \varrho) + \delta(t)$ and $\bar{\varrho}' > \bar{\omega}(t, \bar{\varrho}) + \bar{\delta}(t)$ in J_0.*

Suppose further that $f \in \mathscr{P}$ satisfies the conditions

(δ) *$f(t, x, v, v_x, v_{xx}) - f(t, x, v-\varrho, v_x, v_{xx}) \leqq \omega(t, \varrho)$,*
$f(t, x, v+\bar{\varrho}, v_x, v_{xx}) - f(t, x, v, v_x, v_{xx}) \leqq \bar{\omega}(t, \bar{\varrho})$

(provided the arguments lie in $D(f)$).

Then

$$-\bar{\varrho}(t) < v(t, x) - u(t, x) < \varrho(t) \quad \text{in} \quad G_p.$$

The generalizations given in connection with the corresponding Theorem 9 I remain valid; in particular the upper and lower bounds are independent of each other.

For the proof of the inequality $v-u<\varrho$ we use 23 I with $\varphi=v-\varrho$, $\psi=u$. Then it suffices to show that $v_t-\varrho'<u_t$ for all points $(t,x)\in G_p$ with $v-\varrho=u$, $v_x=u_x$, $v_{xx}\leqq u_{xx}$. In fact according to $(\beta)-(\delta)$ we have for these points

$$\begin{aligned} v_t-u_t &\leqq \delta(t)+f(t,x,v,v_x,v_{xx})-f(t,x,u,u_x,u_{xx}) \\ &\leqq \delta(t)+f(t,x,v,v_x,v_{xx})-f(t,x,v-\varrho,v_x,v_{xx})\leqq \delta(t)+\omega(t,\varrho)<\varrho', \end{aligned}$$

whereby this part of the assertion has been proved. We proceed quite similarly for the second part.

IV. Consequence. *Suppose that for $u(t,x)$, $v(t,x)\in Z_0(f)$, $\varrho(t)\in Z(\omega)$ we have*

(α) $|v-u|<\varrho$ *on* R_p^+ *and* R_∞;

(β) $u_t=f(t,x,u,u_x,u_{xx})$ *and* $|v_t-f(t,x,v,v_x,v_{xx})|\leqq\delta(t)$ *in* G_p;

(γ) $\varrho'(t)>\omega(t,\varrho)+\delta(t)$ *in* J_0;

(δ) $f(t,x,z+\varrho,v_x,v_{xx})-f(t,x,z,v_x,v_{xx})\leqq\omega(t,\varrho)$,

provided $(t,x,z+\varrho,v_x,v_{xx}),(t,x,z,v_x,v_{xx})\in D(f)$. *Then*

$$|v(t,x)-u(t,x)|<\varrho(t) \quad in \quad G_p.$$

In III and IV we found it worthwhile to formulate the central hypothesis (δ) very precisely and to assume only what was really necessary. The reasons for this were discussed in Section 9; they are even more compelling here. Thus it is crucial here in the application to nonlinear differential equations that we do not have the inequality (2) (indeed this cannot be satisfied even for very simple nonlinear equations) but that we are allowed to substitute for p and r the derivatives of the approximate solution v which is considered as known. We also note that the Generalizations and Remarks 9 IV (α)–(δ) all carry over to the parabolic case.

Estimation theorems of this kind have been given by Westphal (1949), Szarski (1955, 1959) (for systems), Collatz (1956), Nickel (1959), and Cinquini (1964) (for systems and with Carathéodory hypotheses with respect to t). Kolodner and Pederson (1966) develop an effective method for constructing sub- and superfunctions.

V. Uniqueness Theorem. *Suppose that the function $f(t,x,z,p,r)\in\mathscr{P}$ satisfies the condition (2) if $(t,x,z,p,r),(t,x,\bar z,p,r)\in D(f)$ and $z\geqq\bar z$, where $\omega(t,z)\in\mathscr{E}_5$ resp. $\mathscr{E}_4$. Then, if G is bounded and η is continuous on R_p, the first boundary value problem has at most one solution of the class $Z(f)$, and this solution depends continuously on the boundary values resp. on the boundary values and the right hand side of the differential equation.*

This assertion holds also for unbounded domains if boundary values at infinity are prescribed in such a way that the relationship "$u=\bar{u}$ on R_∞" holds for any two solutions u and $\bar{u}$ [1].

By the continuous dependence of the solution on the boundary values [on the boundary values and the right hand side of the differential equation] we mean — as for the ordinary differential equations in 10 II — the statement: for every $\varepsilon>0$ there is a $\delta>0$ such that if u is the solution of the boundary value problem and $v\in Z(f)$, then $Pv=0$ in G_p, $|v-u|<\delta$ on R_p and on R_∞ [or $|Pv|<\delta$ in G_p, $|v-u|<\delta$ on R_p and on R_∞] imply the inequality $|v-u|<\varepsilon$ in $\bar{G}$.

Indeed for a given $\varepsilon>0$ there is, according to 10 I, a $\delta>0$ and a function $\varrho(t)$ such that $\delta\leqq\varrho\leqq\varepsilon$ and $\varrho'>\omega(t,\varrho)$ $[+\delta]$ in J_0. Now $|v-u|<\varrho$ follows from Theorem IV, and hence the assertion.

VI. Supplements. The present uniqueness theorem — due (in somewhat more specialized form) to Giuliano [2] (1952) — yields very general results when applied to linear problems. However in the above form it breaks down even for relatively simple nonlinear problems. If we have, say, the simplest case of the nonlinear heat equation in *one* space dimension

$$u_t=k(u)u_{xx}, \tag{3}$$

then the uniqueness condition reads

$$[k(z)-k(\bar{z})]r\leqq\omega(t,z-\bar{z}) \quad \text{for} \quad z\geqq\bar{z}, \tag{4}$$

and this obviously cannot be satisfied. However, as we shall show now, we can arrive at uniqueness results in many nonlinear cases by a more careful wording of the condition (2). As before, we are now concerned with the first boundary value problem for continuous boundary values on R_p and (if G is unbounded) boundary conditions at infinity such that $u=\bar{u}$ on R_∞ for any two solutions $u,\bar{u}$.

(α) *It is sufficient for uniqueness that for every (fixed) solution $\bar{u}$ we can find a number $\alpha>0$ and a function $\omega\in\mathscr{E}_5$ such that*

$$f(t,x,\bar{u}+z,\bar{u}_x,\bar{u}_{xx})-f(t,x,\bar{u},\bar{u}_x,\bar{u}_{xx})\leqq\omega(t,z), \quad \textit{provided} \quad 0<z<\alpha$$

and provided $(t,x,\bar{u}+z,\bar{u}_x,\bar{u}_{xx})\in D(f)$.

Then, if $u\in Z(f)$ is another solution and $\bar{\varrho}(t)$ is a function for which $\bar{\varrho}'>\omega(t,\bar{\varrho})$ and $0<\bar{\varrho}<\alpha$, III (with $v=\bar{u}$) implies the one-sided bound $-\bar{\varrho}<\bar{u}-u$. But since there are arbitrarily small functions $\bar{\varrho}$, $\bar{u}-u\geqq0$.

[1] An example of this is furnished by the boundary layer equation in Section 30; see also VII.

[2] As noted by Szarski in his review in Zentralblatt für Mathematik **47** (1953), pp. 92–93, the theorem of Giuliano should be corrected as follows: in the hypothesis b$'_1$), $\lim_{t\to0}\vartheta(t)=\lim_{t\to0}\vartheta'(t)=0$ should be replaced by $\lim_{t\to0}\vartheta(t)=0$.

The same reasoning can be used when u and $\bar{u}$ are interchanged. Hence $\bar{u}-u\leqq 0$ and thus $u=\bar{u}$.

(β) Suppose that for at least *one* solution $\bar{u}$ it is possible to determine an $\alpha>0$ and a function $\omega\in\mathscr{E}_5$ so that both inequalities

$$\left.\begin{array}{l} f(t,x,\bar{u},\bar{u}_x,\bar{u}_{xx})-f(t,x,\bar{u}-z,\bar{u}_x,\bar{u}_{xx}) \\ f(t,x,\bar{u}+z,\bar{u}_x,\bar{u}_{xx})-f(t,x,\bar{u},\bar{u}_x,\bar{u}_{xx}) \end{array}\right\}\leqq\omega(t,z)\quad\text{for}\quad 0<z<\alpha$$

hold (provided the arguments involved lie in $D(f)$). Then we also have uniqueness.

If u is another solution, we find as in (α) that $|\bar{u}-u|<\varrho$. Hence the assertion follows. *Moreover in both cases the continuous dependence on the initial values and, if $\omega\in\mathscr{E}_4$, on the right hand side are preserved.*

In Sections 29 and 30 we shall show in various examples how this sharpened version of V is used. In the above example (3) according to (α) we have to bound the difference

$$[k(\bar{u}+z)-k(\bar{u})]\bar{u}_{xx},$$

instead of (4), for small positive z. As long as we have no knowledge of $\bar{u}_{xx}$ we have gained nothing by this. However the situation is completely different when we know that, for example, $\bar{u}_{xx}$ is bounded. Then a Lipschitz condition on $k(z)$ is sufficient for uniqueness. See 27 IX and 29 VII.

The variant (β) can be of use if *one* solution is explicitly known and its uniqueness is to be verified; or in a situation where we know on the basis of an existence theorem that at least one solution exists with (e.g.) bounded second derivatives. Then the uniqueness of this solution can be verified not only in the class of all solutions with (for example) bounded second derivatives, but in the much broader class $Z(f)$.

If it is impossible to arrive at a result in this way, then as a last recourse we restrict the class of solutions and consider, say, only solutions with bounded derivatives u_{xx}. Thus under certain circumstances we obtain at least a weaker statement, say, "there is only one solution with bounded second derivatives."

This was just a crude sketch of the reasoning. By making use of the fact that we have to bound the f-difference only from one side, we can sharpen some of the results; see Sections 29 and 30.

VII. Remark on Unbounded Domains. The uniqueness theorem comprises unbounded domains if appropriate conditions are imposed at infinity. Several examples will clarify of what type these can be.

(α) In addition to the continuous function $\eta(t,x)$ on R_p, suppose we are given a continuous function $\Phi(t)$ in $[0,T]$ and we require that every solution converge uniformly to $\Phi(t)$ as $\|x\|_e\to\infty$. More precisely: For every $\varepsilon > 0$ there is a $K > 0$ such that

$$|u(t,x) - \Phi(t)| < \varepsilon \quad \text{for all} \quad (t,x) \in G_p \quad \text{with} \quad \|x\|_e > K$$

(K may depend on u!) [1].

(β) Let $\bar{G} = [0, T] \times E^m$. Then $\partial_1 G$ is empty. Suppose the function $\Phi(t, \alpha)$ is defined when t varies in $[0, T]$ and α varies on the m-dimensional unit sphere (thus $\alpha \in E^m$, $\|\alpha\|_e = 1$). As boundary condition at infinity we require the uniform convergence of $u(t, r\alpha)$ to $\Phi(t, \alpha)$ as $r = \|x\|_e \to \infty$, and thus

$$\left| u(t,x) - \Phi\left(t, \frac{x}{\|x\|_e}\right) \right| < \varepsilon \quad \text{for} \quad \|x\|_e > K = K(\varepsilon).$$

(γ) Let $m = 1$, $\bar{G} = [0, T] \times E^1$. As $x \to \infty$ or $\to -\infty$ suppose u approaches the (given) continuous functions $\Phi_+(t)$ or $\Phi_-(t)$ uniformly. We can interpret this as a special case of (β).

In all three cases we have "$u = \bar{u}$ on R_∞ for any two solutions $u, \bar{u}$ of the boundary value problem," as the uniqueness theorem requires.

(δ) For unbounded domains there is also the possibility of a transition to a bounded domain by a suitable transformation of variables (for example, reciprocal radii, or $\bar{x} = x/(1 + \|x\|_e^2)$, or $\bar{x} = \arctan x (m = 1), \ldots$). Here the differential equation is also transformed. Such transformations were used by Collatz (1958) relative to error bounds and also by Nickel (1958).

Until now we have used only bounds which are themselves constant in a plane $t = \text{const}$. The extension of III to the general case of a bound ϱ depending on t and x causes no difficulties. It just entails a more complicated hypothesis (δ).

VIII. Theorem. *Suppose the functions $f(t,x,z,p,r)$, $\omega(t,x,z,p,r)$, $\bar{\omega}(t,x,z,p,r)$ belong to $\mathscr{P}$. For the functions $u(t,x), v(t,x) \in Z_0(f)$, $\varrho(t,x) \in Z_0(\omega)$, $\bar{\varrho}(t,x) \in Z_0(\bar{\omega})$, $\delta(t,x)$ and $\bar{\delta}(t,x)$, suppose*

(α) $-\bar{\varrho} < v - u < \varrho$ *on* R_p^+ *and on* R_∞;

(β) $u_t = f(t,x,u,u_x,u_{xx})$ *and* $-\bar{\delta}(t,x) \leqq v_t - f(t,x,v,v_x,v_{xx}) \leqq \delta(t,x)$ *in* G_p;

(γ) $\varrho_t > \delta(t,x) + \omega(t,x,\varrho,\varrho_x,\varrho_{xx})$ *and* $\bar{\varrho}_t > \bar{\delta}(t,x) + \bar{\omega}(t,x,\bar{\varrho},\bar{\varrho}_x,\bar{\varrho}_{xx})$ *in* G_p;

(δ) $f(t,x,v,v_x,v_{xx}) - f(t,x,v-\varrho,v_x-\varrho_x,v_{xx}-\varrho_{xx}) \leqq \omega(t,x,\varrho,\varrho_x,\varrho_{xx})$
$f(t,x,v+\bar{\varrho},v_x+\bar{\varrho}_x,v_{xx}+\bar{\varrho}_{xx}) - f(t,x,v,v_x,v_{xx}) \leqq \bar{\omega}(t,x,\bar{\varrho},\bar{\varrho}_x,\bar{\varrho}_{xx})$

in G_p (we require that, with $D(f) = M \times M_r$, the points $v_{xx} + \bar{\varrho}_{xx}$ and $v_{xx} - \varrho_{xx}$ belong to M_r).

Then

$$-\bar{\varrho}(t,x) < v(t,x) - u(t,x) < \varrho(t,x) \quad \textit{in} \quad G_p.$$

[1] $\eta(t,x)$ and $\Phi(t)$ must satisfy an appropriate compatibility condition

$$|\eta(t,x) - \Phi(t)| < \varepsilon \quad \text{for} \quad (t,x) \in R_p, \ \|x\|_e > K.$$

We make use of the method of proof of III and, as before, have to show that: If $v-u=\varrho$, $v_x-u_x=\varrho_x$, $v_{xx}-u_{xx}\leqq\varrho_{xx}$ at a point of G_p, then $v_t-u_t<\varrho_t$ at this point. But this follows from the inequalities

$$\begin{aligned} v_t-u_t &\leqq \delta(t,x)+f(t,x,v,v_x,v_{xx})-f(t,x,u,u_x,u_{xx}) \\ &\leqq \delta(t,x)+f(t,x,v,v_x,v_{xx})-f(t,x,v-\varrho,v_x-\varrho_x,v_{xx}-\varrho_{xx}) \\ &\leqq \delta(t,x)+\omega(t,x,\varrho,\varrho_x,\varrho_{xx})<\varrho_t\,. \end{aligned}$$

We prove the other inequality $-\bar{\varrho}<v-u$ correspondingly.

IX. Special Case. *Suppose the function f is linear in the second derivatives,*

$$f(t,x,z,p,r)=g(t,x,z,p)+\sum_{\lambda,\mu=1}^{m} k_{\lambda\mu}(t,x)r_{\lambda\mu} \tag{4}$$

with a positive semi-definite matrix $(k_{\lambda\mu})$ at every point of G_p. Then if $u,v\in Z_0(f)$ and $\varrho\in Z_0$, with $\varrho(t,x)\geqq\alpha>0$ in G_p, and if

(α) $|v-u|\leqq\varrho$ *on* R_p^+ *and on* R_∞;

(β) $u_t=f(t,x,u,u_x,u_{xx})$ *and* $|v_t-f(t,x,v,v_x,v_{xx})|\leqq\delta(t,x)$ *in* G_p;

(γ) $\varrho_t\geqq\delta(t,x)+a(t,x)\varrho+\sum_\mu b_\mu(t,x)|\varrho_{x_\mu}|+\sum_{\lambda,\mu}k_{\lambda\mu}(t,x)\varrho_{x_\lambda x_\mu}$ *in* G_p

(the functions a, b_μ are defined in G_p), and if $g(t,x,z,p)$ satisfies a one-sided Lipschitz condition relative to z;

(δ) $g(t,x,z,p)-g(t,x,\bar{z},\bar{p})\leqq a(t,x)(z-\bar{z})+\sum_\mu b_\mu(t,x)|p_\mu-\bar{p}_\mu|$ *for* $z\geqq\bar{z}$, *in its domain of definition,*

then we have the bound

$$|v-u|\leqq\varrho \quad in \quad G_p\,.$$

It is easy to establish that the four hypotheses (α) through (δ) yield the corresponding hypotheses of VIII if we set $\varrho=\bar{\varrho}$, $\delta(t,x)=\bar{\delta}(t,x)$ and

$$\omega(t,x,z,p,r)=\bar{\omega}=az+\sum b_\mu|p_\mu|+\sum k_{\lambda\mu}r_{\lambda\mu}\,.$$

However, since we have allowed the equality sign in (α) and (γ), we need a small additional consideration: instead of ϱ we have to consider the function $\varrho^*(t,x)=\varrho(t,x)(1+\varepsilon+\varepsilon t)$ and then let $\varepsilon\to+0$. Now (α) and (γ) hold with the > symbol for ϱ^*; here we need the positivity of ϱ.

According to (γ) we can determine bounds ϱ in simple cases as solutions of a linear parabolic differential equation. First we take several solutions or a family of solutions and start with bounds in the form of a linear combination of such solutions. The free parameters are chosen in such a way as to conform as well as possible to the boundary conditions.

In the following

X. Example this procedure is especially easy to see. Let $m=2$ and $G_p = J_0 \times D$, where D is contained in the circle $x_1^2 + x_2^2 < R^2$. Suppose the differential equation is

$$u_t = \Delta u + g(t, x_1, x_2, u) \tag{5}$$

with

$$g(t, x_1, x_2, z) - g(t, x_1, x_2, \bar{z}) \leqq L(z - \bar{z}) \quad \text{for} \quad z \geqq \bar{z}.$$

Let us assume that u and v are two solutions of this differential equation whose initial values differ by at most ε for $t=0$ while for $t>0$ their boundary values coincide. The inequality IX (γ), which now reads

$$\varrho_t \geqq L\varrho + \Delta \varrho,$$

has, for example, the solutions

$$\varrho = e^{(L-\beta^2)t} J_0(\beta r) \quad \text{with} \quad r^2 = x_1^2 + x_2^2,$$

where $J_0(s)$ is the Bessel function of order 0. Thus we arrive at a bound of the form

$$|v - u| \leqq C e^{(L-\beta^2)t} J_0(\beta r).$$

If j_0 is the first positive zero of $J_0(s)$, then we must have $\beta R < j_0$. Thus the final form of the bound is

$$|v - u| \leqq \varepsilon e^{(L-\beta^2)t} \cdot \frac{J_0(\beta r)}{J_0(\beta R)};$$

here every positive $\beta < j_0/R$ is allowed. The question of the "best" β cannot be answered unambiguously. If we examine the behavior for large t we will choose β near j_0/R; if we are looking at the point $r=0$ at time $t=1$, then the problem consists of determining the minimum of $e^{-\beta^2}/J_0(\beta R)$ as a function of β. Of course we could also work with a linear combination of two or more functions ϱ of the above form.

XI. Lipschitz Bound. Suppose

$$Pu = 0, \quad -\bar{\delta}(t) \leqq Pv \leqq \delta(t) \quad \text{in} \quad G_p$$

for two functions $u, v \in Z_0(f)$. Suppose the function $f \in \mathscr{P}$ satisfies a condition

$$f(t, x, z, v_x, v_{xx}) - f(t, x, \bar{z}, v_x, v_{xx}) \leqq l(t)(z - \bar{z}) \quad \text{for} \quad z \geqq \bar{z} \tag{6}$$

[provided both arguments lie in $D(f)$]. Suppose the functions $\delta, \bar{\delta}, l$ are continuous in J_0 and integrable over J,[1] and

$$L(t) = \int_0^t l(\tau)\, d\tau.$$

Then we have the bound

$$-e^{L(t)} \left\{ \bar{\varepsilon} + \int_0^t \bar{\delta}(\tau) e^{-L(\tau)} d\tau \right\} \leqq v - u \leqq e^{L(t)} \left\{ \varepsilon + \int_0^t \delta(\tau) e^{-L(\tau)} d\tau \right\}$$

[1] Hypothesis 9 VI (γ) suffices in place of this.

in G_p if the constants $\varepsilon, \bar{\varepsilon}$ are determined such that these inequalities hold on R_p^+ and on R_∞ and that the left hand side is $\leqq 0$ while the right hand side is $\geqq 0$.

We should especially note that the functions $\delta, \bar{\delta}, l$ may also take on negative values. Since only a one-sided Lipschitz bound is required, we can determine a negative Lipschitz "constant" $l(t)$ in some cases.

A special case of this estimate was given by Westphal (1949) (but with a bound on the absolute value of f).

XII. Remarks. (α) As we have previously noted, we can also prove Theorems III and VIII by a reduction to II; $v + \bar{\varrho}$ is indeed a super- and $v - \varrho$ a subfunction (relative to a suitable initial value problem for Eq. (1)). The proof given here is no more difficult and has the advantage that it carries over to systems; see the corresponding remark in the introduction VIII.

(β) A general existence proof for the heat equation was carried out by Sternberg (1929). Existence problems for quasilinear differential equations have been investigated by Ciliberto (1956b, 1956c), Pini (1957), Agaev and Mamazov (1958), Friedman (1958, 1960, 1960a), Kružkov (1966), Fateeva (1967), among others. The two monographs by Friedman (1964) and Ladyzhenskaja, Solonnikov and Ural'ceva (1968) furnish a comprehensive existence theory. Mlak (1960) gives conditions for the existence of maximal and minimal solutions (the critical remark of Friedman in Mathematical Reviews, Vol. 22, A 6930, is incorrect). At the end of this chapter we shall treat existence problems with the line method.

XIII. Difference Methods for Parabolic Differential Equations. The most familiar procedure for numerical solution of a first boundary value problem consists of replacing the derivatives involved by finite differences. This can be done in quite different ways.

If only the derivatives in x are replaced by differences, then we obtain a system of *ordinary* differential equations. This is the "Line Method" treated in Sections 35 and 36.

If we replace all the derivatives in (1) by finite differences, then we obtain a system of pure difference equations (different systems are obtained depending on whether we choose forward, backward, ... differences in the t-direction). The problems associated with this have been thoroughly investigated for special f and in particular for the simple heat equation. Krawczyk (1963) dealt with the difference equations corresponding to the general eq. (1) for arbitrary domains. Among other things he was able to extend the Nagumo-Westphal lemma to these difference equations and to prove a series of important results (maximum-minimum principle for the solutions of the difference equation, estimation theorems, convergence of the solutions of the difference equation to the solution of the differential equation if the mesh width converges to 0). The work of Krawczyk deals in particular with the boundary layer equation (30.6′).

26. The Maximum-Minimum Principle

The simplest numerical statements about solutions of the differential equation

$$u_t = f(t, x, u, u_x, u_{xx}) \tag{1}$$

are obtained by comparison with constants. They lead to the well-known maximum-minimum principle.

I. Definition ($M(\varphi)$, $m(\varphi)$, $H(t)$, $H(t, x)$, $H^*(t, x)$, *maximum principle, strong maximum principle*, ...). For a function $\varphi(t, x)$ defined in G_p, $M(\varphi)$ is the infimum of all numbers A for which $\varphi < A$ on R_p^+ and on R_∞. The *maximum principle* holds for Eq. (1) if $u \leqq M(u)$ in G_p for every solution $u \in Z_0(f)$ of Eq. (1). If in particular $u \in Z(f)$, G is bounded and R_p is closed, this obviously means that u takes on its maximum (also!) on R_p.

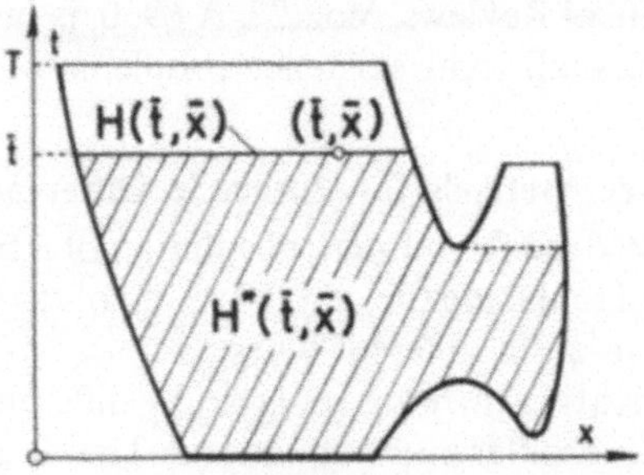

Fig. 12

Now suppose $(\bar{t}, \bar{x})$ is a point of G_p, $H(\bar{t})$ is the intersection of G_p with the hyperplane $t = \bar{t}$, $H(\bar{t}, \bar{x})$ is that connected component of $H(\bar{t})$ which contains the point $(\bar{t}, \bar{x})$, and $H^*(\bar{t}, \bar{x})$ is the set of all points of G_p which can be connected with $(\bar{t}, \bar{x})$ by a piecewise linear curve which is monotone increasing in the t-direction (Fig. 12). If for example G_p is the hypercylinder $J_0 \times D$ and if $D \subset R^m$ is connected, then $H(\bar{t}) = H(\bar{t}, \bar{x}) = \{\bar{t}\} \times D$ and $H^*(\bar{t}, \bar{x}) = (0, \bar{t}] \times D$. The *strong maximum principle* holds if the maximum principle holds and if every solution $u \in Z_0(f)$ of (1) has the property: If $u(\bar{t}, \bar{x}) = M(u)$ for a point $(\bar{t}, \bar{x}) \in G_p$, then $u \equiv M(u)$ in $H^*(\bar{t}, \bar{x})$.

We define $m(\varphi)$ and the minimum principle in the same way. If both principles hold, then we say that the maximum-minimum principle holds. Finally we recall the notation $D(f) = M \times M_r$.

II. Theorem (Maximum-Minimum Principle). *Let $f \in \mathscr{P}$ and $0 \in M_r$. Then we have the*

maximum principle	*if*	$f(t, x, z, 0, 0) \leqq 0$,
minimum principle	*if*	$f(t, x, z, 0, 0) \geqq 0$,
maximum-minimum principle	*if*	$f(t, x, z, 0, 0) = 0$

for $(t, x, z, 0, 0) \in D(f)$.[1]

This theorem has long been known in special cases and was stated in general by Nickel (1958). A similar boundary maximum theorem is given by Collatz (1956, Theorem 4) (in this theorem one has to set $g \equiv 0$).

Theorem II is contained in the following

III. Theorem. Let $f \in \mathscr{P}$, $v \in Z_0(f)$ *and* $0 \in M_r$. *Then*

$$v \leqq M(v) \text{ in } G_p \text{ if } Pv \leqq 0 \text{ and } f(t, x, z, 0, 0) \leqq 0 \quad \text{for } M(v) < z < M(v) + \varepsilon,$$

$$v \geqq m(v) \text{ in } G_p \text{ if } Pv \geqq 0 \text{ and } f(t, x, z, 0, 0) \geqq 0 \quad \text{for } m(v) - \varepsilon < z < m(v)$$

for some $\varepsilon > 0$ *(as long as the given arguments lie in* $D(f)$*).*

It suffices to prove the first part of the assertion. Thus suppose $v < A$ on R_p^+ and on R_∞. We now have to derive the inequality $v \leqq A$ in G_p. For this we use 24 I with $\varphi = v$, $\psi = A + \delta t$, $\delta > 0$; A and δ are chosen so that $A + \delta T < M(v) + \varepsilon$. From $v = \psi$, $v_x = \psi_x = 0$, $v_{xx} \leqq \psi_{xx} = 0$ it follows that

$$v_t = Pv + f(t, x, v, v_x, v_{xx}) \leqq f(t, x, v, 0, 0) \leqq 0 < \delta = \psi_t\,.$$

Thus Lemma 24 I can be applied and yields $v < \psi$ in G_p. Hence, since δ is arbitrarily small and A can be chosen arbitrarily close to $M(v)$, $v \leqq M(v)$ in G_p.

Under additional hypotheses we can sharpen the assertion of this theorem considerably, as was shown by Nirenberg (1953).

[1] In this theorem, as well as later, 0 occurs with three different meanings, namely, as a number, as the origin of E^m and as the null matrix. The context always indicates which is meant.

IV. Theorem (Strong Maximum-Minimum Principle). *Suppose the function $f \in \mathscr{P}$ is of the form*

$$f(t, x, z, p, r) = g(t, x, z, p) + \sum_{\lambda,\mu=1}^{m} k_{\lambda\mu}(t, x) r_{\lambda\mu},$$

where g is defined in a domain M and $k_{\lambda\mu}$ is defined in G_p. For every $(t_0, x_0) \in G_p$ suppose there is a lower half-neighborhood $U_- = U_-(t_0, x_0)$ and two positive numbers α, β such that

(α) $g(t,x,z,p) \leqq \alpha \sum_{\mu=1}^{m} |p_\mu| \left[\geqq -\alpha \sum_{\mu=1}^{m} |p_\mu|\right]$ *for* $(t, x) \in U_-, (t,x,z,p) \in M$;

(β) $k_{\mu\mu}(t, x) \leqq \alpha$ *in* U_-;

(γ) $\sum_{\lambda,\mu=1}^{m} k_{\lambda\mu}(t, x) c_\lambda c_\mu \geqq \beta \sum_{\mu=1}^{m} c_\mu^2$ *for arbitrary* c_μ *and* $(t, x) \in U_-$.

Then if $v \in Z_0(f)$ and $Pv \leqq 0$ $[Pv \geqq 0]$ in G_p, we have $v \leqq M(v)$ $[v \geqq m(v)]$ in G_p and, if $v = M(v)$ $[v = m(v)]$ at a point $(\bar{t}, \bar{x}) \in G_p$, $v \equiv M(v)$ $[v \equiv m(v)]$ in $H^*(\bar{t}, \bar{x})$.

First we concern ourselves with the statement in square brackets. Since III can be applied, $v \geqq m(v)$. We assume that the assertion is false and that there exists in G_p a point with $v = m(v)$ and a piecewise linear curve starting from it which is monotone decreasing in the coordinate t and which ends at a point with $v > m(v)$. Suppose (t', x') is the last point on this curve at which we still have $v = m(v)$. All the following considerations refer to the half-neighborhood U_- associated with this point (t', x') by hypothesis.

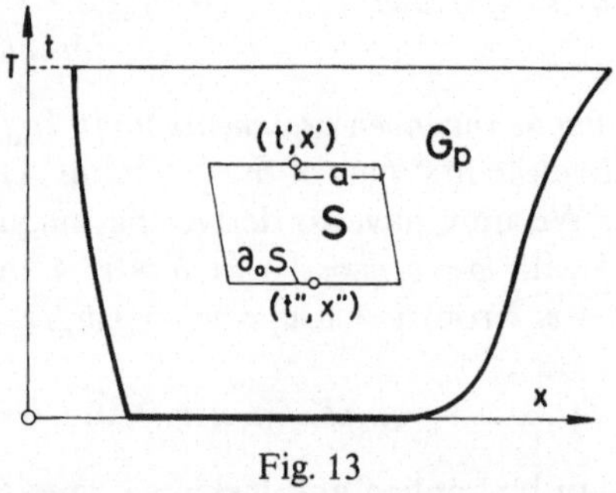

Fig. 13

In U_- there is a point (t'', x'') with $v > m(v)$ and $t'' < t'$. Now we consider an oblique circular cylinder S which has the points (t'', x''), (t', x') as midpoint of the lower and upper surfaces respectively and has radius a (Fig. 13); analytically

$$S = \{(t, x) \,|\, t'' < t < t', \|x - x'' - (t - t'')\xi\|_e < a\} \quad \text{with} \quad \xi = \frac{x' - x''}{t' - t''};$$

here and below we write $x^2 = x_1^2 + \cdots + x_m^2$, and thus $\|x\|_e = \sqrt{x^2}$. Moreover, let a be so small that

$$v > m(v) \quad \text{on} \quad \partial_0 S$$

(the sets $\partial_0 S, S_p, \overline{S}, \ldots$ are defined according to 23 I). Now we consider the function

$$w(t,x) = (a^2 - (x - t\xi)^2)^2 e^{-Bt} = A^2 e^{-Bt}, B > 0$$

for $0 < t \leqq t' - t''$, $A = a^2 - (x - t\xi)^2 > 0$. First we show that for all sufficiently large B

$$w_t < \sum_{\lambda,\mu} k_{\lambda\mu}(t'' + t, x'' + x) w_{x_\lambda x_\mu} - \alpha \sum_\mu |w_{x_\mu}|.$$

As we can easily verify, this relationship is equivalent to

$$\begin{aligned} 8 \sum_{\lambda,\mu} k_{\lambda\mu}(x_\lambda - t\xi_\lambda)(x_\mu - t\xi_\mu) - 4A \sum_\mu k_{\mu\mu} - 4\alpha A \sum_\mu |x_\mu - t\xi_\mu| \\ + BA^2 - 4A \sum_\mu \xi_\mu (x_\mu - t\xi_\mu) \\ \geqq 8\beta(x - t\xi)^2 + BA^2 - 4A\Big(m\alpha + m\alpha a + a \sum_\mu |\xi_\mu|\Big) > 0 \end{aligned}$$

(at the points where A is small, the first summand predominates in the last line and at the remaining points the second summand predominates provided we choose B sufficiently large). Now we apply Lemma 24 I to S_p and the two functions

$$\varphi = m(v) - \varepsilon + \delta w(t - t'', x - x'') \qquad (\delta, \varepsilon > 0)$$

and $\psi = v$. This is allowed since $\varphi = v$, $\varphi_x = v_x$, $\varphi_{xx} \leqq v_{xx}$ implies, according to the inequality just proved and with the help of (α) (β),

$$\begin{aligned} \varphi_t &< -\alpha \sum |\varphi_{x_\mu}| + \sum k_{\lambda\mu} \varphi_{x_\lambda x_\mu} \leqq g(t, x, \varphi, \varphi_x) + \sum k_{\lambda\mu} \varphi_{x_\lambda x_\mu} \\ &\leqq g(t, x, v, v_x) + \sum k_{\lambda\mu} v_{x_\lambda x_\mu} = f(t, x, v, v_x, v_{xx}) \leqq v_t = \psi_t \end{aligned}$$

(it follows from a well-known theorem of matrix theory that the double sum does not become smaller when φ_{xx} is replaced by v_{xx}; see the inequalities in 23 II (α)). Further, by a suitable choice of $\delta > 0$ we can ensure that

$$m(v) + \delta w \leqq v \quad \text{and thus} \quad \varphi < v = \psi$$

on the (parabolic) boundary of S_p. This is easily possible since w is distinct from zero only on $\partial_0 S$, and there $m(v) < v$. Thus $\varphi < v$ in S_p follows from 24 I; since ε is arbitrary, we then have $m(v) + \delta w \leqq v$ and in particular $m(v) < v$ in S_p. But this result contradicts the above assumptions.

We deal correspondingly with the functions $\varphi = v$, $\psi = M(v) + \varepsilon - \delta w$ in the other part of the theorem.

V. Corollary. The statements of Theorem IV can be carried over immediately to quasilinear differential equations

$$f = g(t, x, z, p) + \sum_{\lambda, \mu = 1}^{m} k_{\lambda\mu}(t, x, z, p) r_{\lambda\mu}.$$

If $v \in Z_0(f)$, if g satisfies the hypothesis IV (α), and if $k^*_{\lambda\mu}(t, x) = k_{\lambda\mu}(t, x, v, v_x)$ satisfies the hypotheses IV (β) (γ), then the assertion of IV holds.

We see this immediately if we apply IV to the function $f^*(t, x, z, p, r) = g(t, x, z, p) + \sum k^*_{\lambda\mu}(t, x) r_{\lambda\mu}$.

The theorem can also be carried over to nonlinear equations if f is continuously differentiable in the $r_{\lambda\mu}$. By the mean value theorem we have

$$f(t, x, v, v_x, v_{xx}) = f(t, x, v, v_x, 0) + \sum_{\lambda, \mu = 1}^{m} \frac{\partial f}{\partial r_{\lambda\mu}}(t, x, v, v_x, \vartheta v_{xx}) v_{x_\lambda x_\mu},$$

where $0 < \vartheta = \vartheta(t, x) < 1$. In this case it suffices for $g(t, x, z, p) = f(t, x, z, p, 0)$ to have the property IV (α) and, moreover, for $k^*_{\lambda\mu}(t, x) = \partial f(t, x, v, v_x, \vartheta v_{xx}) / \partial r_{\lambda\mu}$ to have the properties IV (β) (γ).

In particular the strong maximum principle [strong minimum principle] holds for Eq. (1) under these hypotheses.

Further investigation in connection with the maximum principle for parabolic and elliptic differential equations, based on the classical work of Hopf (1927), have been made by Hopf (1952), Oleinik (1952), Výborný (1958), Friedman (1958a, 1961), Pucci (1957, 1958), Kusano (1963), Velte (1963), Nitsche (1966), for weak solutions by Aronson (1965), Kadlec and Výborný (1965), Ikeda (1967), Aronson and Serrin (1967, 1967a), among others.

The theorems proved so far provide an answer to the question of when a constant function is a super- or subfunction. Now we investigate other functions and their usefulness as super- or subfunctions. The choice of "good" approximate functions depends very much on the particular type of differential equation and the boundary conditions. Thus general prescriptions cannot be given. The following forms are simple and natural; in some cases they will yield usable bounds. For the sake of brevity we always assume $Pu = 0$ (as in II above). The extension to the case $Pu \leqq 0$ for upper bounds, $Pu \geqq 0$ for lower bounds (this corresponds to the conversion from II to III) is left to the reader.

In the following let a, A and $b = (b_1, \ldots, b_m)$, $B = (B_1, \ldots, B_m)$ be constants. We use bx to denote the scalar product

$$bx = b_1 x_1 + \cdots + b_m x_m.$$

VI. Theorem. *Let $u \in Z_0(f)$, $Pu=0$, $0 \in M_r$ and*

$$f(t,x,z,p,0)=0$$

provided $(t,x,z,p,0) \in D(f)$. Then the estimates

$$a+bx \leqq u \leqq A+Bx$$

hold in G_p if they hold on R_p^+ and on R_∞.

The proof is very simple. We merely apply 24 I once to $\varphi=a+bx-\varepsilon(1+t)$, $\psi=u$ and once to $\varphi=u$, $\psi=A+Bx+\varepsilon(1+t)$.

This theorem, given by Nickel (1958), has a simple geometric interpretation. The equations $z=a+bx$, $z=A+Bx$ represent two hyperplanes in the $(m+1)$-dimensional (x,z)-space. Now if we associate with given boundary values η a set $H=\{(x,z)\,|\,z=\eta_*(t,x)$ or $z=\eta^*(t,x)$ for some $t \in J\}$ in the (x,z)-space, we can say: If G is bounded and H lies between the given hyperplanes, then the surface $z=u(t,x)$, taken as a function of x with t held fixed, lies between these hyperplanes for every $t \in J$. If we carry out this reasoning for every hyperplane which lies completely on one side of the set H, we are led to the following conclusion:

If G is bounded, then for every fixed $t \in J$ the surface $z=u(t,x)$ lies in the convex hull of H.

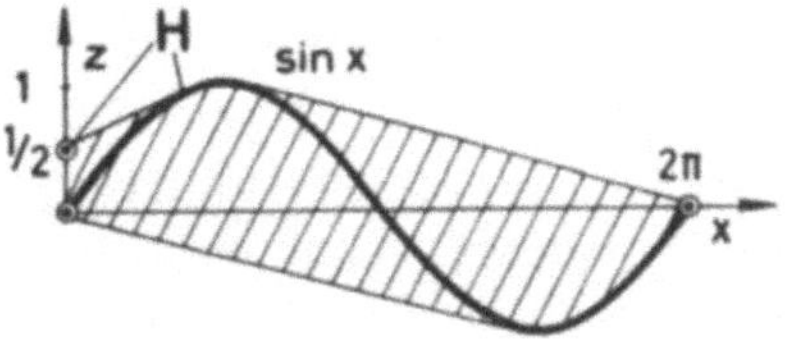

Fig. 14. The set H and its convex hull (hatched)

This situation is illustrated in Fig. 14 for the example of a heat conduction problem (with discontinuous boundary values): $m=1$, $G=\{0<t<\infty\}\times\{0<x<2\pi\}$, $u(0,x)=\sin x$, $u(t,0)=1/2$, $u(t,2\pi)=0$, $f(t,x,z,p,r)=kr$. The temperature remains in the shaded region for all positive times. Here $k \geqq 0$ can depend in an arbitrary fashion on t, x, z, p.

VII. Further Forms. In the following we assume that $u \in Z_0(f)$ is a solution of the differential equation (1). The application to the first boundary value problem is obvious.

(α) *Bounds which depend only on t.* For the functions $a(t), A(t) \in Z(f)$ suppose

$$a(t)<u(t,x)<A(t) \quad \text{on} \quad R_p^+ \text{ and } R_\infty, \tag{2}$$

$$a'(t)<f(t,x,a(t),0,0), A'(t)>f(t,x,A(t),0,0) \quad \text{in} \quad G_p.$$

Then the inequalities (2) hold in G_p.

We can obtain such bounds as solutions of the ordinary differential equations

$$a' = g(t, a), \qquad A' = G(t, A),$$

where g, G are determined so that

$$g(t, z) < f(t, x, z, 0, 0) < G(t, z).$$

(β) *Bounds which are linear in x.* The simplest case was already discussed in VI. Now let $a = a(t)$, $b = b(t) = (b_1(t), \ldots, b_m(t))$ and let A, B be functions depending on t with $a + bx$, $A + Bx \in Z(f)$. Then the three conditions

$$\begin{aligned} a(t) + b(t)x < u(t, x) < A(t) + B(t)x \quad &\text{on} \quad R_p^+ \text{ and } R_\infty \qquad (3)\\ a' + b'x < f(t, x, a + bx, b, 0) \quad &\text{in} \quad G_p\\ A' + B'x > f(t, x, A + Bx, B, 0) \quad &\text{in} \quad G_p \end{aligned}$$

are sufficient for the inequalities (3) to hold in G_p.

(γ) *Product form.* Suppose the functions $a(t)$, $A(t)$, $c(x)$, $C(x)$ are such that ac, $AC \in Z_0(f)$ and

$$\begin{aligned} a(t)c(x) < u(t, x) < A(t)C(x) \quad &\text{on} \quad R_p^+ \text{ and } R_\infty \qquad (4)\\ a'(t)c(x) < f(t, x, ac, ac_x, ac_{xx}) \quad &\text{in} \quad G_p\\ A'(t)C(x) > f(t, x, AC, AC_x, AC_{xx}) \quad &\text{in} \quad G_p\,. \end{aligned}$$

Then (4) is valid in G_p.

These are just special cases of 24 IV. In most practical cases, to be precise, if we have a uniqueness condition as in 24 VI, we can allow equality signs in the inequalities. The form (γ) is especially useful if f is linear and thus the principle of superposition holds. Let us consider this case in somewhat more detail.

VIII. The Linear Differential Equation. Let

$$f(t, x, z, p, r) = g(x)z + \sum_{\mu=1}^{m} h_\mu(x)p_\mu + \sum_{\lambda,\mu=1}^{m} k_{\lambda\mu}(x)r_{\lambda\mu} \qquad (5)$$

or briefly

$$f = gz + hp + kr$$

with a symmetric and positive semi-definite matrix k. We use $c(x; \alpha)$ to denote a solution of the ordinary or elliptic differential equation

$$(g - \alpha)c + hc_x + kc_{xx} = 0 \quad (\alpha \text{ constant}).$$

Then, as is easily verified, the function

$$u(t, x) = e^{\alpha t}c(x; \alpha)$$

is a solution of $Pu = 0$, and the same is also true for sums of such functions.

If we know a finite or infinite sequence of such solutions $c(x;\alpha_k)$ and if the a_k, A_k are determined so that (for a given solution $u\in Z_0(f)$ of (5))

$$\sum_k a_k e^{\alpha_k t} c(x;\alpha_k) \leqq u(t,x) \leqq \sum_k A_k e^{\alpha_k t} c(x;\alpha_k) \quad \text{on} \quad R_p^+ \text{ and } R_\infty,$$

then the solution u also lies between the two bounds throughout G_p provided $g(x)$ is bounded above. In this case f satisfies a one-sided Lipschitz condition; the assertion then follows from 25 II. Of course in the case of infinitely many α_k, convergence hypotheses are to be made such that both series represent solutions of (5) of the class $Z_0(f)$.

27. The Shape of Profiles

This section contains the most important of the results of Nickel (1958, 1962) (only the consequence IX is new).

Our considerations are based on a parabolic differential equation in *one* space dimension

$$u_t = f(t, x, u, u_x, u_{xx}) \qquad (m=1)\,. \tag{1}$$

Our aim is to make qualitative statements about the behavior of solutions with respect to x and thus along a characteristic $t=\text{const}$ (in connection with this we also speak of "solution profiles," "temperature profiles" for heat conduction problems, etc.), such as the monotonicity or the number of relative maxima of a solution profile[1].

I. Notation. (A, B, C, D, R_α, *α-point, extremum*, $A_0(\alpha), A_p(\alpha), M_0, M_p$). In this section $m=1$ and the domain G is a rectangle $0<t<T, 0<x<a$ with vertices A, B, C, D (in the positive direction from the origin). Let $R_0=\overline{DC}$ be the closed upper side of G. Then $R_p=\partial_0 G+\partial_1 G=\overline{DA}+\overline{AB}+\overline{BC}$ and $R_0=\partial_2 G+\{C, D\}$; thus the points C, D belong to R_0 as well as to R_p; see Fig. 15.

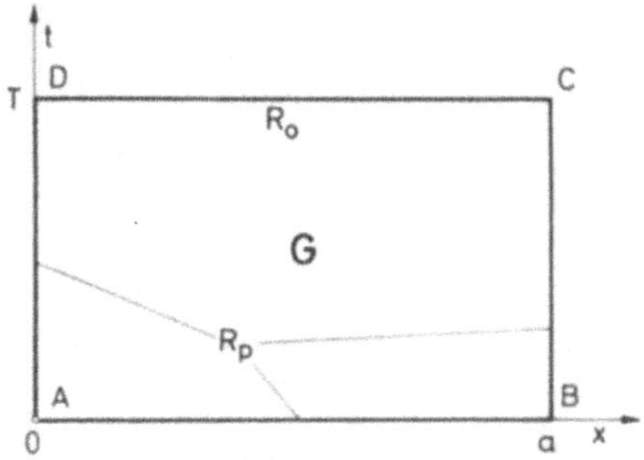

Fig. 15.

[1] Nickel used the term „Gestaltaussagen" for statements of this type.

Suppose the function $\varphi(s)$ is continuous in the interval $s_0 \leqq s \leqq s_1$; $\bar{s}$ is an α-point (α real) of φ if $\varphi(\bar{s})=\alpha$. If $\varphi(\bar{s})=\alpha$ and $a \leqq s \leqq b$ is the largest interval containing the point $\bar{s}$ in which $\varphi(s)\equiv\alpha$, the number $A(\alpha)$ of α-points counts this interval as a single α-point. There is a (relative) maximum at the point $\bar{s}$ if for the largest interval $a \leqq s \leqq$ b containing the point $\bar{s}$ in which $\varphi(s)\equiv\varphi(\bar{s})$ there exist two numbers $\bar{a}<a, b<\bar{b}$ such that $\varphi(s)\leqq\varphi(\bar{s})$ in $(\bar{a},\bar{b})\cdot[s_0, s_1]$. Such an interval of constancy $a \leqq s \leqq b$ is again counted as a single maximum in the number M of maxima. A corresponding definition holds for the minimum. If the function u is continuous in $\bar{G}$, then $A_0(\alpha)$ and M_0 or $A_p(\alpha)$ and M_p denote respectively the number of α-points and maxima of u taken as a function defined on R_0 or R_p respectively; here we should think of R_p in the direction of D over A, B toward C as an interval (or u as a function of arc length along R_p starting at D).

II. Lemma. Suppose $u \in Z(f)$ is a solution of Eq. (1) and $0 \in M_r$ [1]. Suppose that if $(t, x, z, 0, 0) \in D(f)$ the function $f \in \mathscr{P}$ satisfies the equation (condition for the maximum-minimum principle)

$$f(t, x, z, 0, 0)=0\,. \tag{2}$$

If α is a real number, $(T, \bar{x}) \in R_0$ and $u(T, \bar{x})>\alpha$, we consider the largest connected point set $H \subset \bar{G}$ containing the point $(T, \bar{x})$ in which $u>\alpha$ (thus, the set of those points of $\bar{G}$ which can be connected with $(T, \bar{x})$ by a piecewise linear curve in $\bar{G}$ on which $u>\alpha$) and the sets $j_0=H\cdot R_0$, $j_p=H\cdot R_p$.

Then we have

(α) j_0 is an interval [namely the largest interval in R_0 containing the point $(T, \bar{x})$ in which $u>\alpha$];

(β) j_p is not empty (it is a union of intervals).

(γ) If, as above , for the two numbers α', α'' and two points (T, x'), $(T, x'') \in R_0$ with $u(T, x')>\alpha'$, $u(T, x'')>\alpha''$ we form the corresponding sets $H', H'', j_0', \ldots$ then H' and H'' are disjoint provided j_0' and j_0'' are disjoint.

The corresponding assertions with the $<$ symbol are also true.

The assertion (β) is a simple consequence of the maximum principle 26 II applied to H. If j_p were empty we would have $u=\alpha$ on the parabolic boundary $\partial_0 H+\partial_1 H$ of H and thus $u=\alpha$ in H in contradiction to the hypothesis $u(T, \bar{x})>\alpha$. For the proof of (α) we take an arbitrary point $(T, x^*) \in j_0$ and show that $u>\alpha$ at the points of R_0 lying between $(T, \bar{x})$ and (T, x^*). If the piecewise linear curve P connecting these two points and on which $u>\alpha$ remains in R_0, it is trivial; if it extends (partially) outside R_0, then we consider the domain G^* bounded by the piecewise linear curve P and R_0. Since on the piecewise linear curve we have $u>\alpha$

[1] By 23 II, $D(f)=M \times M_r$.

and thus $u \geqq \alpha + \varepsilon$ for some $\varepsilon > 0$, we have (again by 26 II) $u \geqq \alpha + \varepsilon$ in $\bar{G}^*$, which also proves this assertion.

We proceed quite analogously for (γ). If, say, $\alpha' \leqq \alpha''$ and $H' \cdot H'' \neq \emptyset$, then there is a piecewise linear curve P in $H' + H''$ which connects (T, x') with (T, x''). On P, and thus by 26 II in the domain enclosed by P and R_0, we have $u \geqq \alpha' + \varepsilon$ $(\varepsilon > 0)$, i.e., j'_0 and j''_0 have points in common.

III. Theorem. *Suppose that for* $(t, x, z, 0, 0) \in D(f)$, $f \in \mathscr{P}$ *satisfies Eq.* (2) *and* $0 \in M_r$. *Then for the number of* α*-points of a solution* $u \in Z(f)$ *of Eq.* (1) *on* R_0 *or* R_p *we have*

$$A_0(\alpha) \leqq A_p(\alpha) .$$

In the following proof we associate with every α-point on R_0 an α-point on R_p. If $u \equiv \alpha$ on R_0, then we choose, say, the point D as α-point on R_p. Then $A_0(\alpha) = 1 \leqq A_p(\alpha)$. Otherwise the set of points of R_0 at which $u \neq \alpha$ is a nonempty union of open (for boundary intervals, half-open) pairwise disjoint intervals j_0^k. If for each of these intervals (starting from a point $(T, x^k) \in j_0^k$) we form the associated sets H^k, j_p^k as in II, then by II (α) (γ) the H^k are pairwise disjoint (this is trivial if in one set $u > \alpha$ and in the other $u < \alpha$). Now there is at least one α-point of u between two sets j_p^k on R_p. If $A_0(\alpha)$ is infinite, so is $A_p(\alpha)$. If $A_0(\alpha)$ is finite, then every α-point in the interior of R_0 lies between two intervals j_0^k; with it we associate an α-point between the two corresponding intervals j_p^k on R_p. Finally if C or D is an α-point, we can associate it with itself since it also lies on R_p. This reasoning leads directly to the proof of the above inequality.

IV. Consequence. *Under the hypotheses of* III, *if the function* u *is monotone increasing or decreasing on* R_p *(in the orientation* $DABC$*), then it has the same behavior on* R_0.

For $A_p(\alpha) = 1$ implies $A_0(\alpha) \leqq 1$. Freud (1957) proved IV for linear differential equations.

Next we prove a theorem about the number of (relative) extrema of u.

V. Theorem. *Under the hypotheses of* III *we have*

$$M_0 \leqq M_p, \quad m_0 \leqq m_p$$

for the number of maxima M_0, M_p *and for the number of minima* m_0, m_p *of* u *on* R_0 *or* R_p.

If the number of extrema on R_0 *is finite and if they are assumed (enumerated from left to right) at the points* (T, x_k), $k = 1, \ldots, k_0$, *then there are* k_0 *points* $(t_k, \bar{x}_k)$ *on* R_p, *where the enumerations correspond (from* D *over* A, B *towards* C*), with the properties: If* u *has a maximum* [*minimum*] *relative to* R_0 *at the point* (T, x_k), *then* u *also has a maximum* [*minimum*] *relative to* R_p *at the point* $(t_k, \bar{x}_k)$, *and*

$$u(T, x_k) \leqq u(t_k, \bar{x}_k) \qquad [u(T, x_k) \geqq u(t_k, \bar{x}_k)] .$$

First we prove the second part. If a maximum [minimum] corresponds to the index k, let H^k be the largest connected point set containing the point (T, x_k) for whose points $u > u(T, x_k) - \varepsilon$ $[u < u(T, x_k) + \varepsilon]$. The corresponding sets j_0^k are then obviously disjoint if $\varepsilon > 0$ is sufficiently small, and the same holds for the j_p^k on R_p. Because of the way in which the j_p^k are defined, the supremum [infimum] of u relative to j_p^k is assumed at a point $(t_k, \bar{x}_k) \in j_p^k$, and there u has a maximum [minimum] relative to R_p. The function value at this point is $\geqq u(T, x_k)$ $[\leqq u(T, x_k)]$; for in our reasoning we can let $\varepsilon \to +0$, and the j_p^k contract.

Hence not only has the second part been proved, but also the first part for finite M_0, m_0. But if $M_0 = \infty$ and hence also $m_0 = \infty$, then we can carry out the above reasoning for every finite number of extreme points on R_0 and obtain $M_p = m_p = \infty$. Indeed these considerations are feasible as soon as we have k_0 points (T, x_k) with $x_{k-1} < x_k$ on R_0 such that the differences $u(T, x_k) - u(T, x_{k-1})$ are alternately positive and negative. If $2\varepsilon < \min |u(T, x_k) - u(T, x_{k-1})|$, then the corresponding sets j_0^k are disjoint. In the way indicated above we find k_0 consecutive points $(t_k, \bar{x}_k)$ on R_p, where u has a maximum or minimum relative to R_p and $u(T, x_k) \leqq$ or $\geqq u(t_k, \bar{x}_k)$.

From this we obtain an interesting

VI. Consequence[1]. *If the hypotheses of* III *hold and if* V_0 *resp.* V_p *is the total variation of* u *on* R_0 *resp.* R_p, *then*

$$V_0 \leqq V_p .$$

Indeed if u has only finitely many extrema on R_0, then we can write V_0 as the sum of the differences of maximal and minimal values, and the assertion follows from the inequalities of Theorem V. In the general case we can approximate V_0 arbitrarily well by finite sums $\sum |u(T, x_k) - u(t, x_{k-1})|$ of the type considered at the end of the proof of V, whence the assertion follows.

All the theorems up to now allow an important and obvious

VII. Generalization. We assume that for $0 \leqq x \leqq a$ there exists a twice continuously differentiable function $U(x)$ such that $U''(x) \in M_r$ $(0 < x < a)$ and

$$f(t, x, z, U', U'') = 0$$

[provided the argument lies in $D(f)$]. Then the previous theorems can be applied to the new function

$$u^*(t, x) = u(t, x) - U(x) .$$

[1] I owe this to an oral communication from Dr. Brakhage, Karlsruhe.

If $u \in Z(f)$ is a solution of $Pu=0$ and we define f^* by

$$f^*(t, x, z, p, r) = f(t, x, U+z, U'+p, U''+r),$$

then $u^* \in Z(f^*)$, $u_t^* = f^*(t, x, u^*, u_x^*, u_{xx}^*)$ and f^* satisfies the hypotheses of III.

In particular we note the

VIII. Special Case corresponding to a linear function $U(x)$. Let $u \in Z(f)$, $0 \in M_r$, $Pu=0$ in G_p, and suppose that instead of (2) we have

$$f(t, x, z, p, 0) = 0 \tag{3}$$

for $(t, x, z, p, 0) \in D(f)$. Then for every number β, the number of α-points on R_0 as well as the number of maxima and minima of the function $u(t, x) - \beta x$ on R_0, is at most as large as on R_p.

Hence we also immediately obtain a noteworthy

IX. Consequence. *If under the hypotheses of* VIII *the functions $u(t, 0)$ and $u(t, a)$ are monotone decreasing [increasing] functions of t (in the direction of increasing t) and if the function $u(0, x)$ is convex*[1] *from above [below] in $[0, a]$, then for fixed t, $u(t, x)$ is, as a function of x, also convex from above [below] in $[0, a]$, i.e.,*

$$u_{xx} \leqq 0 \quad [u_{xx} \geqq 0] \quad \text{in} \quad G_p.$$

Let us consider the assertion which is not in square brackets. With our hypotheses, for arbitrary numbers α, β the number of α-points of the function $u(t, x) - \beta x$ on the boundary R_p is at most two (this situation is depicted in Fig. 16, where the two sides AD and BC are

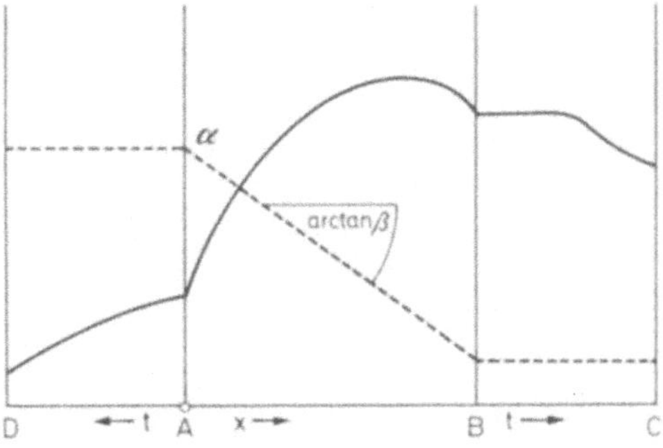

Fig. 16. The boundary values and the function $\alpha + \beta x$ (dotted). The two sides $\overline{AD}$ and $\overline{BC}$ are folded down

[1] By convexity we here mean weak convexity. The function $\varphi(x)$ is convex from above in $[0, a]$ if for two arbitrary points x_1, x_2 $(0 \leqq x_1 < x_2 \leqq a)$ the inequality $u(x) \geqq l(x)$ holds in the interval $x_1 < x < x_2$, where $l(x)$ is the linear function with $l(x_i) = \varphi(x_i)$ $(i = 1, 2)$. In particular a linear function is convex from above as well as from below.

"folded down"; the number of α-points is equal to the number of points of intersection of the extended u-curve with the dotted line which is given between A and B by $\alpha+\beta x$ and is constant between AD and BC). According to VIII the number of points of intersection of the function $u(T,x)$ with the straight line $\alpha+\beta x$ is thus also at most two, and the same also holds of course for $u(t,x)$ with fixed $t\in J_0$. But since a function is convex if and only if it meets every straight line in at most two points [1], $u(t,x)$ (for a fixed t) is either convex from above or convex from below. Thus we have only to show that the latter cannot be the case. Now we assume that $u(\bar{t},x)$ $(0<\bar{t}\leqq T)$ is convex from below but not linear in $[0,a]$ and from this we shall derive a contradiction. Because of the continuity of u, $u(t,x)$ (for fixed t) is also convex from below and not linear for all t in a neighborhood of $\bar{t}$ (the set of all t with this property is open) [2]. If t' is the smallest number such that for fixed t, $t'<t\leqq\bar{t}$, $u(t,x)$ is convex from below and not linear, then $u(t',x)$ is linear in x. Then $u_{xx}\geqq 0$ for those points of G_p with $t'\leqq t\leqq\bar{t}$ and thus by (3) and the monotonicity of f we have $u_t\geqq 0$ and therefore $u(t,x)\geqq u(t',x)$. But this is a contradiction to the above assumption that $u(t,0)$ and $u(t,a)$ are monotone decreasing and u lies below the straight line connecting the two points $(0,u(t,0))$, $(a,u(t,a))$.

X. Remarks. (α) Since an extremum of $u-\beta x$ at an interior point of R_0 is simultaneously a β-point of u_x, one might try to bound their number also. Here several difficulties arise. The most convenient approach to bounding the number of β-points seems to be a passage to a differential equation for u_x, which is possible in many cases, including the simplest heat equation. If the new differential equation for u_x satisfies the hypothesis (2), then III can be brought into play. However this approach is practicable only when the values of u_x on R_p are known, i.e., for the so-called boundary value problems of second kind. For the first boundary value problem, however, we do not know u_x on $\overline{DA}$ and $\overline{BC}$. As Nickel (1962) has shown, we can circumvent these difficulties and draw conclusions about the number of β-points from a knowledge of the boundary values alone. We shall not go into any more detail about this.

(β) The theorems of this section are not dependent on the rectangular form of G and can be extended to domains G which are defined by the inequalities

$$0<t<T,\quad \varphi(t)<x<\psi(t)$$

[1] Here again an interval in which the function and the straight line are identical is to be considered as a point.

[2] This conclusion is true because we have already shown that for fixed t the function u is convex from above or convex from below.

(φ and ψ are continuous in J, $\varphi<\psi$ in J_0). In IX the hypothesis has to be modified appropriately [the boundary values must be such that for arbitrary constants α, β the function $u(t,x)-\alpha-\beta x$ has at most two zeros on R_p].

(γ) Theorem IX is false if we define convexity in the strong sense and thus do not count the linear function with the convex functions. Indeed it can happen that the stationary state (linear solution profile) is achieved in finite time. Example: $u_t=k(t)u_{xx}$ with $k(t)=2/(1-t)$ for $0\leqq t<1$, $k(t)\geqq 0$ (arbitrary) for $t\geqq 1$, $a=\pi$, initial values $\sin x$, boundary values 0, solution $u(t,x)=(1-t)^2\sin x$ for $0\leqq t\leqq 1$, $u(t,x)=0$ for $t\geqq 1$.

(δ) Extension of the results of this section to the case of several space dimensions encounters considerable difficulties of a predominantly topological nature. See the counterexample given by Nickel (1962). For other results concerning the shape of profiles and the number of α-points for solutions of linear equations, see Ivanov (1965), Godunova and Levin (1966), Landis (1966).

28. Infinite Domains, Discontinuous Initial Values, Problems Without Initial Values

In this section we deal with uniqueness problems for a differential equation

$$u_t=g(t,x,u,u_x)+\sum_{\lambda,\mu=1}^{m}k_{\lambda\mu}(t,x)u_{x_\lambda x_\mu} \tag{1}$$

which is linear in the second derivatives; of course we assume that the matrix $(k_{\lambda\mu})$ is positive semi-definite at every point of G_p. The general uniqueness theorem 25 V is unsatisfactory in two respects: it cannot be applied to discontinuous initial values and, if the behavior at infinity is not precisely prescribed, it cannot be applied to infinite domains. However the essential estimation theorems of Sections 24 and 25 were formulated so generally that even such problems can be handled without difficulties. This will be done below, but with no attempt to present a complete theory. Rather it will become evident that far-reaching conclusions can be drawn even from very simple comparison functions. We shall use the notation

$$x^2=x_1^2+\cdots+x_m^2,\qquad \|x\|_e=\sqrt{x^2}\,.$$

I. Infinite Domain. Here the relationships in the problem of infinitely long (heat) conductors

$$u_t=u_{xx}\qquad (m=1) \tag{2}$$

have been thoroughly investigated and known for a long time. In this problem $\bar{G}=[0, T]\times E^1$ and we seek a solution $u\in Z$ with $u(0, x)=\eta(x)$ for given continuous initial values $\eta(x)$. It has exactly one solution if as an additional requirement we impose a growth condition (for a fixed $K>0$)

$$|u(t, x)| \leqq K e^{Kx^2} \quad \text{in} \quad \bar{G} \tag{3}$$

acting to a certain degree as a boundary condition on the infinitely distant boundary (of course the initial values themselves have to satisfy such a growth condition). Theorems of this kind have been proved by Holmgren (1924) and Picone (1929) among others. A famous example constructed by Tychonoff (1935) shows that (3) is not only a sufficient condition but is also in a certain sense a necessary condition for uniqueness; a very simple example was given by Rosenbloom and Widder (1958). Sharpened versions of these results and extensions to more general differential equations and other boundary conditions at infinity can be found in Widder (1944), Cooper (1950), Birkhoff and Kotik (1954), Friedman (1959b), Aronson (1962), Smirnova (1963, 1966), Nicolescu and Foiaş (1965, 1965a, 1966), Kalašnikov (1966), Aronson and Besala (1966, 1967), Shapiro (1966); we take special note of the works of Krzyżański (1945, 1948, 1962, 1962a), Besala and Krzyżański (1962). Very little is known in this direction for quasilinear equations. We shall now prove several results for the (mildly nonlinear) differential equation (1), using essentially Krzyżański's method.

We can expect uniqueness for a first boundary value problem with unbounded domain G only if we impose certain restrictions on the behavior of the solution for large x. Here we take a condition of the form

$$|u(t, x) - \bar{\eta}(x)| \leqq \Phi(x) \quad \text{for large } x; \tag{4}$$

$\bar{\eta}(x)$ and $\Phi(x)$ are fixed given functions. The phrase "for large x" means: there is a constant K such that the relationship under discussion holds for all $(t, x)\in\bar{G}$ with $\|x\|_e > K$. The inclusion of the function $\bar{\eta}(x)$ in the growth condition gives us greater generality in subsequent assertions. If we have a problem for the domain $\bar{G}=[0, T]\times E^m$ with given initial values $\eta(x)$, then if the growth condition is $|u(t, x)|\leqq\Phi(x)$ we have to require that $|\eta(x)|\leqq\Phi(x)$. However in (4) for example we can set $\bar{\eta}(x)=\eta(x)$ i.e., we do not need to impose any bound on the initial values η. For linear equations this remark is of no consequence and the introduction of $\bar{\eta}(x)$ is superfluous since we can always go over to initial values 0.

Now we specialize Theorem 25 IX appropriately.

II. Uniqueness Theorem. Suppose G is an unbounded domain lying between $t=0$ and $t=T$ on whose parabolic boundary R_p a continuous function $\eta(t, x)$ is defined, and suppose further that $\bar{\eta}(x)$ and $\Phi(x)$ are two functions defined in E^m. Suppose the function $g(t, x, z, p)$ satisfies a condition

$$\left.\begin{aligned} &g(t, x, z, p) - g(t, x, \bar{z}, \bar{p}) \\ &\quad \leqq a(t, x)(z - \bar{z}) + \sum_\mu b_\mu(t, x)|p_\mu - \bar{p}_\mu| \quad \text{for} \quad z \geqq \bar{z}. \end{aligned}\right\} \tag{5}$$

For these functions a, b_μ (we assume only that they are defined in G_p) suppose there exists a function $\varphi(t, x) \in Z$ which satisfies the inequality

$$\varphi_t > a(t, x)\varphi + \sum_\mu b_\mu(t, x)|\varphi_{x_\mu}| + \sum_{\lambda, \mu} k_{\lambda\mu}(t, x)\varphi_{x_\lambda x_\mu} \quad \text{in} \quad G_p \tag{6}$$

as well as the two relations

$$\varphi > 0 \quad \text{in } \bar{G}, \qquad \varphi(t, x)/\Phi(x) \to \infty \qquad \text{for } \|x\|_e \to \infty \text{ uniformly in } t. \tag{7}$$

Then the first boundary value problem for Eq. (1) has at most one solution in the class $Z(f)$ satisfying condition (4).

This theorem can be reduced easily to 25 IX. If u and v are two solutions, then by (4) we have $|v - u| \leqq 2\Phi(x)$ for large x. With the function $\varrho(t, x) = \varepsilon\varphi(t, x)$ $(\varepsilon > 0)$ all the hypotheses of 25 IX are satisfied. The assertion is then obtained for $\varepsilon \to +0$.

The problem now consists of finding suitable functions $\varphi(t,x)$ for which (6), (7) holds. We make an attempt (inspired by the cited work of Krzyżański) with the function

$$\varphi(t, x) = \alpha(t)e^{\beta(t)x^2}.$$

Then (6) becomes

$$\frac{\alpha'}{\alpha} + \beta' x^2 > a + 2|\beta| \sum_\mu b_\mu |x_\mu| + 2\beta \sum_\mu k_{\mu\mu} + 4\beta^2 \sum_{\lambda, \mu} k_{\lambda\mu} x_\lambda x_\mu \tag{6'}$$

or, if we choose

$$\alpha(t) = e^{At}, \ \beta(t) = B + Ct \qquad (A, B, C \text{ positive constants}),$$

$$\left.\begin{aligned} A + Cx^2 > a &+ 2(B + Ct)\sum_\mu b_\mu |x_\mu| + 2(B + Ct)\sum_\mu k_{\mu\mu} \\ &+ 4(B + Ct)^2 \sum_{\lambda, \mu} k_{\lambda\mu} x_\lambda x_\mu. \end{aligned}\right\} \tag{6''}$$

Now suppose that for a constant $K > 0$

$$a(t, x) \leqq K(1 + x^2), \ |b_\mu(t, x)| \leqq K(1 + |x_1| + \cdots + |x_m|), \ k_{\lambda\mu}(t, x) \leqq K \tag{8}$$

and

$$\Phi(x) = e^{Kx^2} \tag{9}$$

(if we have different constants in the different places, then we can always take the largest of them). If we restrict t to the domain $0 \leqq t \leqq B/C$, then since $|x_\mu| \leqq 1 + x^2$, the right hand side of (6″) is

$$\leqq K(1+x^2) + 4BKm(1+(m+1)x^2) + 4BKm + 16B^2Km^2x^2,$$

and thus is $< A + Cx^2$ if we set

$$A = K + 8BKm + 1, \qquad C = K + 4BKm(m+1) + 16B^2Km^2.$$

Hence the inequality (6) is valid, and so is (7) when we choose the free constant $B > K$. Thus by Theorem II, two solutions are identical for $0 \leqq t \leqq B/C$. Similarly if $T > B/C$ we prove their identity for $B/C \leqq t \leqq 2B/C, \ldots$.

If we have the equation

$$u_t = g(t, x, u) + \sum_{\mu=1}^{m} h_\mu(t, x) u_{x_\mu} + \sum_{\lambda,\mu} k_{\lambda\mu}(t, x) u_{x_\lambda x_\mu} \tag{10}$$

and if g has a partial derivative with respect to z, then the conditions (8) become

$$g_z(t, x, z) \leqq K(1+x^2), \; |h_\mu| \leqq K(1+|x_1| + \cdots + |x_m|), \; |k_{\lambda\mu}| \leqq K. \tag{11}$$

Hence, for example, if we consider only nonnegative solutions, the function

$$g(t, x, z) = \gamma z^m \qquad (\gamma \text{ constant}, m > 0)$$

"is allowed" for $m \leqq 1$ and arbitrary γ and for arbitrary $m > 0$ and negative γ (equations of this type arise in diffusion processes which are accompanied by chemical reactions; see 33 IX).

We summarize in the

III. Theorem. *The first boundary value problem for Eq.* (1) *has at most one solution belonging to the class Z and satisfying the growth condition* ($\bar{\eta}(x)$, *K fixed*)

$$|u(t, x) - \bar{\eta}(x)| \leqq e^{Kx^2} \quad \textit{for large } x \tag{12}$$

if the function g satisfies condition (5) *and if* (8) *holds. This is true in particular for Eq.* (10) *under the condition* (11).

This theorem was proved for linear differential equations by Krzyżański (1945).

IV. Super- and Subfunctions, the Maximum Principle. The considerations up to now were made from the special point of view of uniqueness. However the proofs covered more, namely, explicit bounds and hence assertions about the continuous dependence of the solution on the boundary values and (by the addition of a term $+\delta$ in (6)) on the right hand side of the differential equation. We can similarly sharpen the lemma of Nagumo 24 IV considerably for Eq. (1) with almost the same method.

Suppose that $f = g(t, x, z, p) + \sum_{\lambda,\mu} k_{\lambda\mu}(t, x) r_{\lambda\mu}$ *and that G is unbounded. For the functions* $v, w \in Z_0(f)$ *suppose we have*

(α) $v \leqq w$ *on* R_p^+,

(β) $Pv \leqq Pw$ *in* G_p.

Here let g and $k_{\lambda\mu}$ *satisfy the inequalities* (5), (8), *v and w the inequalities*

$$v \leqq e^{Kx^2} \quad \text{and} \quad w \geqq -e^{Kx^2} \quad \text{for large } x\,.$$

Then

$$v(t, x) \leqq w(t, x) \quad \text{in} \quad G_p\,.$$

The proof is obtained by applying Lemma 24 IV to the functions v and $\bar{w} = w + \varepsilon\varphi(t, x)$ ($\varepsilon > 0$), where φ is the function constructed in the proof of III. Obviously 24 IV (α) holds. Since

$$g(t, x, w + \varepsilon\varphi, w_x + \varepsilon\varphi_x) - g(t, x, w, w_x) \leqq \varepsilon a\varphi + \varepsilon \sum_{\mu} b_\mu |\varphi_{x_\mu}|$$

and by (6) we have $P\bar{w} > Pw$, hence 24 IV (β), and thus $v < \bar{w}$. The passage to the limit $\varepsilon \to 0$ leads directly to the assertion.

This contains, for example, the following sharpening of the maximum principle; it was proved in a special case by Krzyżański (1957) (see also (1959)).

Suppose the function $u \in Z_0$ *is a solution of* (1) ($Pu \leqq 0$ *is sufficient*) *and suppose that for two constants K and M*

$$u \leqq M \quad \text{on} \quad R_p^+ \quad \text{and} \quad u \leqq e^{Kx^2} \quad \text{for large } x\,.$$

Suppose the hypotheses (5) *and* (8) *hold for g and* $k_{\lambda\mu}$ *as well as*

$$g(t, x, M, 0) \leqq 0\,.$$

Then

$$u(t, x) \leqq M \quad \text{in} \quad G_p\,.$$

We can similarly sharpen the minimum principle for Eq. (1).

Moreover, Theorem III follows immediately from the above theorem (with a small modification in the case $\bar{\eta} \neq 0$). We should note, however, that the proof given for III can be generalized to parabolic systems while the above theorem holds only for systems with a right hand side which is quasimonotone increasing in z; see Section 32.

V. Discontinuous Boundary Values. Boundary value problems with discontinuous boundary values arise in many applications. Here we restrict ourselves to differential equations of the form

$$u_t = \Delta u + g(t, x, u) \tag{13}$$

and to points of discontinuity which lie in the plane $t=0$ [the case where they lie in finitely many planes $t=$ const. can be reduced to the above; see VIII (β) below]. Suppose again that G is a (not necessarily bounded) domain lying between the planes $t=0$ and $t=T$ and that N is a bounded, closed set in E^m of (m-dimensional Lebesgue) measure 0. We allow discontinuities on the set $\{0\} \times N$ but we require that the solution remains bounded as we approach $\{0\} \times N$. We shall see that in this case we always have uniqueness. For the case in which N consists of finitely many points and $g=0$ this fact has long been known; the general theorem just mentioned was first proved by Krzyżański (1950) for a linear equation in *one* space dimension. Existence theorems for boundary value problems with discontinuous boundary values were given for one space variable by John (1952) and Rektorys (1962), for several space variables by Weigel (1969), who uses essentially differential inequality methods, and by Schleinkofer (1970).

VI. Uniqueness Theorem. Suppose we have a first boundary value problem for a domain G and Eq. (13) with discontinuous boundary values in the following sense. Suppose a continuous function $\eta(t,x)$ is defined on $R^* = R_p - \{0\} \times N$ where $N \subset E^m$ is a bounded, closed null set. We seek a solution $u \in Z_0(f)$ of Eq. (13) which is continuous in $G_p + R^*$ and takes on the values $\eta(t,x)$ on R^*.

Suppose that the function $g(t,x,z)$ satisfies a condition

$$g(t,x,z) - g(t,x,\bar{z}) \leqq a(t,x)(z-\bar{z}) \quad \text{for} \quad z > \bar{z}, \tag{14}$$

that there exists a function $\psi(t,x) \in Z_0$ continuous in $G_p + R^*$ for which

$$\psi_t > a(t,x)\psi + \Delta\psi \quad \text{in} \quad G_p \tag{15}$$

and

$$\psi > 0 \text{ in } G_p + R^*, \quad \psi(t,x) \to \infty \text{ for } (t,x) \to (0,x_0),\ x_0 \in N,\ (t,x) \in G_p. \tag{16}$$

Then for bounded G there exists at most one solution u of the first boundary value problem for which $u(t,x)$ remains bounded as $(t,x) \to (0,x_0)$, $(t,x) \in G_p$, $x_0 \in N$. If G is unbounded and we also impose the condition (4) on the solutions, then again there is at most one solution if ψ also satisfies the condition

$$\psi(t,x)/\Phi(x) \to \infty \quad \text{for} \quad \|x\|_e \to \infty \quad \text{uniformly in } t. \tag{17}$$

This theorem can also be reduced to 25 IX as in II; here we have to set $\varrho = \varepsilon\psi$.

The construction of suitable functions ψ is somewhat more difficult here than in II. For this we need the following

VII. Lemma. Suppose N is a closed set of E^m of measure 0 which lies entirely in the sphere K_0: $\|x\|_e < r_0$ and let

$$\delta(x, N) = \inf(\|x-y\|_e \mid y \in N)$$

denote the distance of the point x from the set N. Then there is a continuous, positive and monotone decreasing function $\bar{\gamma}(s)$ for $s>0$ such that with the notation $\gamma(x) = \bar{\gamma}(\delta(x, N))$ we have[1]

$$\lim_{s \to +0} \bar{\gamma}(s) = \infty \quad \text{and} \quad \int_{K_0} \gamma(x)\, dx < \infty .$$

Then of course $\gamma(x) \to \infty$ as $x \to x_0 \in N$.

For the purposes of the proof we let N_ε $(\varepsilon > 0)$ denote the set of points which are at distance $<\varepsilon$ from N and let $e(\varepsilon)$ denote the measure of N_ε. According to a well-known theorem in measure theory then

$$e(\varepsilon) \to 0 \quad \text{as} \quad \varepsilon \to +0,$$

since for every null sequence ε_n $(\varepsilon_n > 0)$ the set N is equal to the intersection of the N_{ε_n} [see Saks (1937, p. 8); the fact that N is closed is used only at this point]. Hence there exists a strictly decreasing sequence ε_n with $e(\varepsilon_n) \leqq 1/n^3$ $(n = 1, 2, 3, \ldots)$ which converges to 0. We can construct a continuous, positive, monotone decreasing function $\bar{\gamma}(s)$ defined for positive arguments which takes on the values

$$\bar{\gamma}(\varepsilon_n) = n \quad \text{for} \quad n = 1, 2, 3, \ldots .$$

For the associated function $\gamma(x)$ we have (writing N_i instead of N_{ε_i})

$$\int_{K_0} \gamma(x)\, dx = \int_{K_0 - N_1} + \int_{N_1 - N_2} + \int_{N_2 - N_3} + \cdots \leqq \int_{K_0} 1 \cdot dx + e(\varepsilon_1) \cdot 2 + e(\varepsilon_2) \cdot 3 + \cdots$$

$$\leqq \|K_0\| + \frac{2}{1^3} + \frac{3}{2^3} + \cdots < \infty .$$

Thus the lemma is proved.

With the help of the function $\gamma(x)$ just constructed we now set

$$\psi(t, x) = \alpha(t) \int_{K_0} \gamma(\xi) e^{\beta(t)(x-\xi)^2}\, d\xi , \tag{18}$$

where we have assumed that N lies entirely in the sphere K_0: $\|x\|_e < r_0$. By differentiating under the integral sign we obtain

$$\frac{\alpha'}{\alpha} + \beta'(x-\xi)^2 > a + 2m\beta + 4\beta^2(x-\xi)^2 \tag{15'}$$

[1] The function $\gamma(x)$ is not defined on N; since N is a null set, this is of no consequence.

as a sufficient condition for the validity of (15) [see (6′)]. With $\alpha(t) = e^{At}/t^{m/2}$ and $\beta(t) = B - 1/4t$ this becomes

$$A - \frac{m}{2t} + \frac{(x-\xi)^2}{4t^2} > a + 2mB - \frac{m}{2t} + 4\left(B - \frac{1}{4t}\right)^2 (x-\xi)^2$$

or

$$A + \frac{2B}{t}(x-\xi)^2 > a + 2mB + 4B^2(x-\xi)^2 . \tag{15''}$$

Now if the function $a(t, x)$ satisfies the inequality (8), then since $x^2 \leqq 2(x-\xi)^2 + 2r_0^2$ the right hand side of (15″) is

$$\begin{aligned} &\leqq K(1+x^2) + 2mB + 4B^2(x-\xi)^2 \\ &\leqq K(1 + 2(x-\xi)^2 + 2r_0^2) + 2mB + 4B^2(x-\xi)^2 < A + \frac{2B}{t}(x-\xi)^2 , \end{aligned}$$

if we set

$$A = K(1 + 2r_0^2) + 2mB + 1, B = 1$$

and restrict t to the interval

$$0 \leqq t \leqq t_0 = 1/(K+2) .$$

We still have to verify that ψ also has the properties (16). To this end we replace ξ in the integral (18) by a new integration variable

$$\eta = (\xi - x)\sqrt{-\beta(t)} .$$

Then if we set $\gamma(x) = 0$ outside of K_0, we get

$$\left.\begin{aligned} \psi(t, x) &= \frac{e^{At}}{t^{m/2}} \int_{E^m} \gamma(x + (-\beta)^{-\frac{1}{2}}\eta)(-\beta)^{-m/2} e^{-\eta^2} d\eta \\ &= \frac{2^m e^{At}}{(1-4Bt)^{m/2}} \int_{E^m} \gamma\left(x + \frac{2\sqrt{t}}{\sqrt{1-4Bt}}\eta\right) e^{-\eta^2} d\eta . \end{aligned}\right\} \tag{19}$$

Now if $x_0 \in N$, $\|x - x_0\|_e < \varepsilon$ $(\varepsilon > 0)$, $2\sqrt{t}/\sqrt{1-4Bt} < \varepsilon$ and if we restrict consideration in the last integral to the domain $\|\eta\|_e < 1$, then the argument of γ in this integral is at most 2ε away from N and hence

$$\psi(t, x) \geqq \bar{\gamma}(2\varepsilon) \int_{\|\eta\|_e < 1} e^{-\eta^2} d\eta = C\bar{\gamma}(2\varepsilon) ,$$

from which, by VII, the second of the relations (16) follows immediately.

If on the other hand $x_0 \notin N$, then as $(t, x) \to (0, x_0)$ $\psi(t, x)$ approaches the value

$$2^m C_1 \gamma(x_0) \quad \text{with} \quad C_1 = \int_{E^m} e^{-\eta^2} d\eta$$

as we conclude from the integral representation (19) in the familiar fashion.

Now if G is a bounded domain and if the sphere K_0 is chosen so large that it also contains the projection of G on the x-space, then the first of conditions (16) is also verified since $\gamma(x)>0$ in K_0-N. Thus by VI we have uniqueness provided $T\leqq t_0=1/(K+2)$. If $T>t_0$, then we have proved uniqueness only for $0\leqq t\leqq t_0$. But the part of G lying between t_0 and T does not cause any difficulties; we could for example apply 25 V.

If G is unbounded, then we have only to add the function $\varphi(t,x)$ given in II (indeed that which corresponds to the case $b_\mu=0$, $k_{\lambda\mu}=\delta_{\lambda\mu}$) to the integral (18). This new function, which we again call ψ, then satisfies all the requirements (15), (16), and (17) with $\Phi(x)=e^{Kx^2}$.

Thus we have proved:

VIII. Theorem. *Suppose the function $g(t,x,z)$ satisfies a one-sided Lipschitz condition*

$$g(t,x,z)-g(t,x,\bar{z})\leqq K_1(1+x^2)(z-\bar{z}) \quad \textit{for} \quad z\geqq\bar{z}. \tag{20}$$

Then the first boundary value problem for Eq. (13) with discontinuous boundary values, stated in detail in VI, has at most one solution u which remains bounded with approach to N and, if G is unbounded, satisfies a bound ($\bar{\eta}(x)$, $K>0$ fixed)

$$|u(t,x)-\bar{\eta}(x)|\leqq e^{Kx^2} \quad \textit{for large } x.$$

IX. Remarks. (α) If the set N consists of only finitely many points, then VIII can be sharpened essentially. We then have uniqueness if we only require that in the neighborhood of a point $(0,x_0)$ with $x_0\in N$

$$|u(t,x)|\leqq K(\|x-x_0\|_e+\sqrt{t})^{-\alpha} \quad \text{with} \quad K>0, 0<\alpha<m$$

(with the same conditions as in VIII otherwise).

It suffices to verify this for the case of a single point; we assume $x_0=0$. For $\gamma(s)$ in (18) we can choose the function $s^{-\bar{\alpha}}$ with $\alpha<\bar{\alpha}<m$. For $t\leqq 1/8$ we have

$$\left\|x+\frac{2\sqrt{t}}{\sqrt{1-4t}}\eta\right\|_e\leqq\|x\|_e+4\sqrt{t}\|\eta\|_e.$$

It then follows from the integral representation (19) for ψ, if we again integrate only over $\|\eta\|_e<1$, that

$$\psi(t,x)>2^m C_1[\|x\|_e+4\sqrt{t}]^{-\bar{\alpha}}\geqq C_2(\|x\|_e+\sqrt{t})^{-\bar{\alpha}}$$

for (t,x) near $(0,0)$. Thus we can apply Theorem 25 IX in the same way as above to the functions $\varrho=\varepsilon\psi$ since the hypothesis (α) there is satisfied. This remark (for $m=1$ and for linear differential equations) is also found in Krzyżański (1950).

(β) Theorem VIII can be generalized to the case in which the set of boundary points at which discontinuities are allowed is the union of finitely many sets $\{t_k\} \times N_k$ where N_k denotes a bounded closed null set.

If the t_k are ordered by size, then we prove uniqueness separately for the part of G lying between 0 and t_1, between t_1 and $t_2, \ldots$ by the method given in VI.

Schleinkofer (1969) has proved a uniqueness theorem similar to VIII for Eq. (1). In the proof the auxiliary function ψ is more complicated; instead of the exponential term in (18) the fundamental solution of the linear part of (1) is used. See also Il'in, Kalashnikov and Oleinik (1962), Lu Li-Jiang (1965).

X. Problems Without Initial Values. By this we mean problems for a domain which does not lie in a half-space $t >$ const., hence for, say, an infinite cylinder $(-\infty, \infty) \times D, D \subset E^m$. A classical problem of this type, already investigated by Fourier, is: In the domain $-\infty < t < \infty$, $x > 0$ $(m = 1)$ we seek a solution of the heat equation

$$u_t = u_{xx},$$

which takes on given boundary values $u(t, 0) = \eta(t)$ for $x = 0$ $(-\infty < t < \infty)$.

Problems of this type arise in certain geophysical problems like questions about the cooling process of the earth and the temperature of the earth's interior, among others; they have been given much attention for this reason. A brief survey of the circle of problems and the literature can be found in Carslaw-Jaeger (1959), particularly in §2.14 and §9.14. The book of Tychonoff-Samarski (1959) also goes into such questions and contains a uniqueness theorem for the problem formulated above (hypothesis: bounded solutions). Now we want to derive a more general result.

XI. Uniqueness Theorem. *Let G be a domain which may extend from $t = -\infty$ to $t = \infty$ but which lies in a region $x_\mu \geq c_\mu$ $(\mu = 1, \ldots, m)$. Suppose we are given continuous boundary values $\eta(t, x)$ on R_p.*

If the function $g(t, x, z)$ is monotone decreasing in z, there is at most one solution u of Eq. (13) from the class Z which takes on the boundary values η and is bounded in the following sense: There exists a $K > 0$ such that

$$|u(t, x)| < K \quad \text{for all } (t, x) \in \bar{G} \text{ with } t < -K \quad \text{or} \quad \|x\|_e > K. \tag{21}$$

The proof is again based on applying Theorem 25 IX to the difference $v - u$ of two solutions and a suitable auxiliary function ψ. For this we use

$$\psi(t, x) = \sum_{\mu=1}^{m} \operatorname{erf} \frac{x_\mu}{2\sqrt{t}} \quad \text{with} \quad \operatorname{erf} z = \frac{2}{\sqrt{\pi}} \int_0^z e^{-s^2} ds.$$

This function is a solution of the heat equation

$$\psi_t = \Delta\psi \quad \textit{for} \quad t>0, x_\mu>0 \quad (\mu=1,\ldots,m) \tag{22}$$

with the initial values

$$\psi = m \quad \textit{for} \quad t=0. \tag{23}$$

For a fixed $\bar{t}$ we consider the domain $G(\bar{t})$ of all points of G with $t>\bar{t}$. The associated sets $R_p(\bar{t})$, $\partial_0 G(\bar{t})$, ... are defined according to 23 I. Now 25 IX is applied to the domain $G(\bar{t})$, two solutions u, v and the bound $\varrho(t,x) = (2K+1)\psi(t-\bar{t},x)$; here we assume that $\bar{t}<-\mathrm{K}$ and that G lies entirely in the region $x_\mu>1$. For $t=\bar{t}$ we have

$$|v-u| < 2K < (2K+1)m = \varrho;$$

at the remaining points of $R_p(\bar{t})$ we have $u=v$ and, since $x_\mu>1$,

$$\varrho \geqq m(2K+1)\,\mathrm{erf}\frac{1}{2\sqrt{t-\bar{t}}} > 0.$$

Likewise $|v-u|<\varrho$ on R_∞. For if $(t_k, x_{(k)})$ is a sequence like the one which occurs in 23 IV, then

$$\liminf \varrho(t_k, x_{(k)}) \geqq 2K+1,$$

since for at least one ν, the νth components of $x_{(k)}$ approach ∞.

Thus ϱ satisfies the hypothesis 25 IX (α) and by (22) also satisfies the hypothesis 25 IX (γ) with $\delta=a=b_\mu=0$, $k_{\mu\nu}=\delta_{\mu\nu}$. By this theorem we then have

$$|v-u| \leqq \varrho = (2K+1)\sum_\mu \mathrm{erf}\frac{x_\mu}{2\sqrt{t-\bar{t}}} \quad \text{in} \quad G(\bar{t}).$$

Hence we obtain the assertion for $\bar{t}\to-\infty$ with the point (t,x) held fixed since the right hand side of this inequality approaches 0.

A theorem of Hirschmann (1952) lies in a similar direction. Existence and uniqueness is treated for linear equations by Kulikov (1960, 1966); for more general equations by Fife (1964), Kono and Kusano (1966).

XII. Stability Problems. The problems well-known by this name for ordinary differential equations can be extended to parabolic differential equations in a natural fashion. Let $G=(0,\infty)\times D$, $D\subset E^m$, be a hypercylinder extending to infinity; by 23 I we have $G=G_p$, $R_p=\partial_0 G+\partial_1 G$ with $\partial_0 G=\{0\}\times D$, $\partial_1 G=(0,\infty)\times\partial D$. Suppose we have a differential equation (26.1) and $u\in Z(f)$ is a solution of this differential equation. Somewhat vaguely, we are here concerned with studying how "neighboring solutions," that is, solutions with (in a way to be made more precise) boundary values modified a little behave relative to u (say in G or as $t \to \infty$). By converting from f to a new right hand side $f^*(t, x, z, p, r) = f(t, x, u+z, u_x+p, u_{xx}+r) - f(t, x, u, u_x, u_{xx})$ we can – as for ordinary differential equations – always arrange that the solution to be examined for stability is the function $u=0$.

Two typical questions are ($u, v \in Z(f)$ are solutions): under what hypotheses does $u=v$ on $\partial_1 G$ imply the relationship $|v-u| \to 0$ as $t \to \infty$; when does there exist, for every $\varepsilon > 0$, a $\delta > 0$ such that $u=v$ on $\partial_1 G$ and $|v-u| < \delta$ on $\partial_0 G$ implies $|v-u| < \varepsilon$ in G_p? We shall not go into these problems in more detail here but only show in an example how we can handle them with our theorems.

Let $D \subset E^m$ be a bounded domain. In $G = (0, \infty) \times D$ suppose we are given the differential equation

$$u_t = g(t, x, u) + \sum_{\lambda, \mu = 1}^{m} k_{\lambda\mu}(x) u_{x_\lambda x_\mu} \tag{24}$$

where $g(t, x, 0) = 0$ and

$$g(t, x, z) - g(t, x, \bar{z}) \leqq L(z - \bar{z}) \quad \text{for} \quad z \geqq \bar{z}.$$

Suppose further that there exist two constants $\delta > 0$ and α as well as a function $\varphi(x) \in Z$ such that the inequalities

$$\sum_{\lambda, \mu} k_{\lambda\mu}(x) \varphi_{x_\lambda x_\mu} \leqq -\alpha \varphi(x) \quad \text{and} \quad \varphi(x) \geqq \delta > 0 \quad \text{in} \quad D \tag{25}$$

hold. Then for every solution $v \in Z$ of Eq. (24) *which vanishes on $\partial_1 G$ we have the bound*

$$|v(t, x)| \leqq \frac{1}{\delta} e^{(L-\alpha)t} \varphi(x) \max_{\bar{D}} |v(0, x)| \quad \text{in} \quad G. \tag{26}$$

If in particular $\alpha > L$, then v converges exponentially to 0 *as $t \to \infty$.*

For the proof we apply Theorem 25 IX to the functions $u=0$, v, $\varrho = A\varphi(x) e^{(L-\alpha)t}$, $a(t, x) = L$, $b_\mu = 0$. By (25) we have

$$\varrho_t = (L - \alpha)\varrho \geqq L\varrho + \sum_{\lambda, \mu} k_{\lambda\mu} \varrho_{x_\lambda x_\mu}.$$

Thus if $A = \max_{\bar{D}} |v(0, x)| / \delta$, all hypotheses of 25 IX are satisfied.

The special case $k_{\lambda\mu} = \delta_{\lambda\mu}$ leads to a result of Prodi (1951): *If L is smaller than the smallest positive eigenvalue λ_1 of the eigenvalue problem*

$$\Delta\varphi + \lambda\varphi = 0 \quad \text{in} \quad D, \qquad \varphi = 0 \quad \text{on} \quad \partial D,$$

then v converges exponentially to 0 *as $t \to \infty$.*

Since the eigenvalues depend continuously on the domain D, we can go from D to a domain $D^* \supset \bar{D}$ such that the smallest eigenvalue relative to

D^* is still $> L$. If we call this eigenvalue α and the associated first eigenfunction $\varphi(x)$, then (25) holds and hence so does the inequality (26).

Further contributions to this theme are contained in the works of Bellman (1948), Narasimhan (1954), Halilov (1956), Krzyżański (1956, 1957a, 1960, 1964a), Mlak (1957a), Friedman (1959, 1959a, 1967), Fulks and Maybee (1960), Protter (1961), Edmunds (1963), Kaplan (1963), Łojczyk-Królikiewicz (1963), Kanel' (1962, 1963), Gor'kov (1964), Lakshmikantham (1964), Besala and Fife (1966), Vaghi (1966) (this is only a sampling of a rapidly growing literature).

XIII. Nonlinear Equations. The results on unbounded domains proved in II – IV can be transferred to the nonlinear equation

$$u_t = f(t, x, u, u_x, u_{xx}) \tag{27}$$

provided the function f satisfies a Lipschitz condition in z, p and r. Indeed we have the following corollary of the estimation theorem 25 IX.

Let $f \in \mathscr{P}$, $u, v \in Z_0(f)$ and $\varrho \in Z_0$ satisfy:

(α) $|v-u| < \varrho$ *on* R_p^+ *and on* R_∞;

(β) $Pu = 0$ *and* $|Pv| \leqq \delta(t, x)$ *in* G_p;

(γ) $\varrho_t > \delta + a\varrho + \sum_\mu b_\mu |\varrho_{x_\mu}| + \sum_{\lambda,\mu} c_{\lambda\mu} |\varrho_{x_\lambda x_\mu}|$ *in* G_p;

(δ) $$f(t, x, z, p, r) - f(t, x, \bar{z}, \bar{p}, \bar{r}) \leqq a(z-\bar{z}) + \sum_\mu b_\mu |p_\mu - \bar{p}_\mu| + \sum_{\lambda,\mu} c_{\lambda\mu} |r_{\lambda\mu} - \bar{r}_{\lambda\mu}|$$

for $z \geqq \bar{z}$, where a, b_μ, $c_{\lambda\mu}$ are functions of (t, x).

Then

$$|v-u| < \varrho \quad \textit{in} \quad G_p\,.$$

The two inequalities $v - \varrho < u < v + \varrho$ are easily proved. They hold on R_p^+ and in R_∞ by (α). Using (β) – (δ), one shows that $P(v-\varrho) < Pu = 0 < P(v+\varrho)$. Hence the assertion follows from Nagumo's Lemma 24 IV.

We note a special case. *Let*

$$a = K(1+x^2), \quad b_\mu = K(1+\|x\|), \quad c_{\lambda\mu} = K\,, \tag{28}$$

where $\|x\| = |x_1| + \cdots + |x_m|$. Then the function

$$\varrho = \varrho(t, x; B) = e^{At + (B+Ct)x^2}\,, \qquad A = K(1+8Bm)+2, \quad C = K\{16B^2m^2 + 4Bm(m+1)+1\} \tag{29}$$

satisfies (γ) in the strip $0 \leqq t \leqq B/C$ for each $B \geqq K$, provided that $\delta \leqq \varrho$.

This follows from the calculation given in II.

As a result, the uniqueness theorem III remains true for Eq. (27) if f satisfies

$$f(t,x,z,p,r)-f(t,x,\bar{z},\bar{p},\bar{r}) \leqq K\{(1+x^2)(z-\bar{z})+(1+\|x\|)\|p-\bar{p}\|+\|r-\bar{r}\|\} \tag{30}$$

for $z \geqq \bar{z}$, where $\|p\| = \sum_{\mu} |p_\mu|$, $\|r\| = \sum_{\lambda,\mu} |r_{\lambda\mu}|$.

We shall now use the above estimation theorem to prove an

XIV. Existence Theorem for Unbounded Domains. *Let $f \in \mathscr{P}$ be defined in $G_p \times E^r$, $r = 1 + m + m^2$, and $\eta(t,x)$ be continuous in R_p. Suppose there exists a $K > 0$ such that* (30) *and*

$$|f(t,x,0,0,0)| \leqq K e^{Kx^2}, \quad |\eta(t,x)| \leqq K e^{Kx^2}$$

holds in G_p resp. R_p.

Let (D_i) be an increasing sequence of bounded domains in E^m such that D_i contains the ball $x^2 < i^2$, and

$$G_p^i = G_p \cdot ([0,T] \times D_i),$$

with R_p^i the parabolic boundary of G_p^i. Suppose that for $i = 1, 2, \ldots$ the first boundary value problem for Eq. (27) *in G_p^i with continuous boundary values prescribed on R_p^i has a solution of class Z.*

Then there exists exactly one function $u \in Z$ satisfying (27) *and $|u| \leqq L e^{Lx^2}$ (for some $L > 0$) in G_p and the boundary condition $u = \eta$ on R_p, provided that $T \leqq B/C$, where $B = K + 1$ and C is given by* (29).

The main assumption of this theorem is the one concerning the solvability of the first boundary value problem for certain bounded domains. Naturally the theorem applies also to the Cauchy problem, where $G = (0,T) \times E^m$.

For the proof we assume that $T \leqq B/C$. The function η possesses a continuous extension to the strip $[0,T] \times E^m$ (which we denote again by η) satisfying $|\eta| \leqq K e^{Kx^2}$ everywhere. According to our assumptions there exists a solution u_i of (27) in G_p^i with the boundary values

$$u_i = \eta \quad \text{on} \quad R_p^i.$$

Applying the above estimate XIII with u_i and 0 in place of u and v, we get

$$|u_i(t,x)| \leqq K\varrho(t,x;K) \quad \text{in} \quad G_p^i \tag{$*$}$$

(note that $|Pv| = |f(t,x,0,0,0)| \leqq K e^{Kx^2} \leqq K\varrho(t,x;K)$).

If $i < j$, then $u_i = u_j$ for those points of R_p^i which belong to R_p, while according to ($*$) $|u_i - u_j| \leqq 2K\varrho(t,x;K)$ for the points of R_p^i which do not belong to R_p. These latter points lie outside the ball $x^2 < i^2$. This remark, together with $\varrho(t,x;K) \leqq \varrho(t,x;K+1)e^{-x^2}$, yields (always for $i<j$)

$$|u_i - u_j| \leqq \varepsilon_i \varrho(t,x;K+1), \qquad \varepsilon_i = 2Ke^{-i^2} \tag{**}$$

on R_p^i. Now we again apply XIII, this time to u_i and u_j and the domain G_p^i. As a result we arrive at the inequality (**) in G_p^i.

The rest of the proof follows easily from (**). First

$$u = \lim_{i\to\infty} u_i$$

exists in $\bar{G}$, the convergence being uniform in compact subsets of $\bar{G}$. Thus u is continuous in $\bar{G}$, $u=\eta$ on R_p and $|u| \leqq K\varrho(t,x;K)$ in $\bar{G}$ due to (*). Now let w be a solution of (27) in G_p^k (k fixed) with the boundary values u on R_p^k. Letting $j\to\infty$ in (**) we get

$$|u_i - w| \leqq \varepsilon_i \varrho(t,x;K+1) \quad \text{for} \quad i \geqq k$$

on R_p^k, hence in G_p^k (by again applying XIII). This inequality shows that $u=w$ in $\bar{G}^k$, i.e. that u is a solution of (27). Since the uniqueness assertion is already contained in XIII, the theorem is completely proved.

This simple and ingenious method of proving existence theorems for unbounded domains, when a corresponding theorem for bounded domains is available, goes back to Krzyżański (1941). The method was further developed in subsequent papers by Krzyżański (1945, 1948, 1957, 1959), Besala (1961, 1963), Besala and Krzyżański (1962). See also the survey article by Krzyżański (1964), which contains an ample bibliography. Our presentation follows Weigel (1969).

In closing we state a corollary of Nagumo's lemma.

XV. Lemma. *Suppose $f \in \mathscr{P}$ satisfies the Lipschitz condition* (30), *and the functions $v, w \in Z_0(f)$ satisfy the inequalities*

$$v \leqq e^{Kx^2} \quad \textit{and} \quad w \geqq -e^{Kx^2} \quad \textit{for large } x.$$

Then

(α) $v \leqq w$ on R_p^+;

(β) $Pv \leqq Pw$ in G_p

implies

$$v \leqq w \quad \textit{in} \quad G_p.$$

The main point is that in (α) we do not assume $v \leqq w$ on R_∞. For the proof one considers the function $\bar{w} = w + \varepsilon\varphi(t,x)$, where $\varepsilon > 0$ and φ is the function constructed in III. It is easily seen that $v < \bar{w}$ on R_p^+ and R_∞ and that $Pw < P\bar{w}$. From the Lemma of Nagumo-Westphal 24 IV we get $v < \bar{w}$, whence the assertion follows for $\varepsilon \to 0$.

29. Heat Conduction as an Example

I. The Equation of Heat Conduction. The temperature distribution in a thin rod is described by the equation ($m=1$)

$$c\varrho u_t = \frac{\partial}{\partial x}(k u_x) + q . \tag{1}$$

Here c is specific heat capacity, ϱ is the density, k is the thermal conductivity, q is the quantity of heat arising in unit time per unit volume (density of heat sources) and $u=u(t,x)$ is the temperature at time t at the point x. If $q=0$ and the (positive) physical values c, ϱ, k are constant, then we have the simplified equation

$$u_t = a u_{xx}, \quad a = k/c\varrho > 0 . \tag{2}$$

By altering the unit of length, i.e., by introducing $\bar{x} = x/\sqrt{a}$, we can arrange that the constant a in (2) becomes 1.

If we now consider the case of an m-dimensional body, then (1) again holds if we use u_x as in 23 II for the vector grad u and $\partial/\partial x$ for the divergence operator. If the body is isotropic, then k is a scalar quantity; in the anisotropic case, which is realized for example for many crystals, $k=(k_{\lambda\mu})$ is a symmetric tensor (Frank-Mises II (1935, p. 529); Carslaw-Jaeger (1959, Chapter I)). Then in detail the heat equation is

$$\left.\begin{aligned} c\varrho u_t &= \sum_{\lambda=1}^{m} \frac{\partial}{\partial x_\lambda}\left(\sum_{\mu=1}^{m} k_{\lambda\mu} u_{x_\mu}\right) + q \\ &= \sum_{\lambda,\mu=1}^{m} \left(k_{\lambda\mu} u_{x_\lambda x_\mu} + \frac{\partial k_{\lambda\mu}}{\partial x_\lambda} u_{x_\mu} + \frac{\partial k_{\lambda\mu}}{\partial z} u_{x_\lambda} u_{x_\mu}\right) + q, \end{aligned}\right\} \tag{3}$$

where we are considering the most general case in which c, ϱ, q and $k_{\lambda\mu}$ are functions which depend on t, $x=(x_1,\ldots,x_m)$ and z (z corresponds to u); the existence of the partial derivatives of $k_{\lambda\mu}$ which are involved is assumed. The $m \times m$ matrix $k=(k_{\lambda\mu})$ is symmetric[1] and positive semi-definite. Under the assumption that the directions of the eigenvectors of the matrix $k=k(t,x,z)$ are independent of t, x and z, we can carry out a principal axis transformation in the x-space and thus make the mixed derivatives of second order in (3) vanish. Then (3) becomes

$$\left.\begin{aligned} c\varrho u_t &= \sum_{\mu=1}^{m} \frac{\partial}{\partial x_\mu}(k_\mu u_{x_\mu}) + q \\ &= \sum_{\mu=1}^{m} \left(k_\mu u_{x_\mu x_\mu} + \frac{\partial k_\mu}{\partial x_\mu} u_{x_\mu} + \frac{\partial k_\mu}{\partial z} u_{x_\mu}^2\right) + q . \end{aligned}\right\} \tag{4}$$

[1] Thus we assume that k is symmetric without getting involved with the difficult question of whether such a hypothesis is proper for all bodies. See Carslaw-Jaeger § 1.17, Frank-Mises p. 529.

The functions $c>0$, $\varrho>0$, $k_\mu>0$ $(\mu=1,\dots,m)$ and q depend on t, x and z; the isotropic body is characterized by the fact that all k_μ are equal. If (in the anisotropic case) the k_μ are constant, then we can reduce them to the value 1 by the transformation $\bar{x}_\mu=x_\mu/\sqrt{k_\mu}$ $(\mu=1,\dots,m)$ and we obtain the equation

$$c\varrho u_t=\Delta u+q \tag{5}$$

from (4).

Via a well-known transformation the more general isotropic case, where the thermal conductivity is not constant but depends on the temperature, can be reduced to an equation of the form (5). Suppose then that in (4) $k_\mu(t,x,z)=k(z)$ for $\mu=1,\dots,m$ [with scalar $k(z)$]. Here let $k(z)$ be continuous and let $K(z)$ be an indefinite integral of $k(z)$, i.e., $K'(z)=k(z)$. Since $k>0$ the function $\bar{z}=K(z)$ is continuous and strictly monotone increasing and therefore has a continuous inverse function $z=L(\bar{z})$. If we now consider the function $\bar{u}=K(u)$ in place of u, then from (4) we get the following equation for $\bar{u}$

$$\frac{c\varrho}{k}\bar{u}_t=\Delta\bar{u}+q\,. \tag{6}$$

Here to ensure that we really have a differential equation in $\bar{u}$, we substitute the expression $L(\bar{u})$ in c, ϱ, k and q in place of $z=u$ (indeed $u=L(\bar{u})$).

Moreover it is easy to see that a transformation to an equation of the form (6) is then also possible if $k_\mu(t,x,z)=\alpha_\mu k(z)$, α_μ is constant. A special case, where even a transformation to a linear differential equation is possible, is treated by Storm (1951).

The functions $f(t,x,z,p,r)$ corresponding to the equations (1) to (6) belong to the class $\mathscr{P}$.

It should be noted that in the case – almost always the one in concrete problems – of the equation (4) there are no mixed derivatives and accordingly the remarks 24 III and 24 V hold. For solutions (and also for super- and subfunctions) we have only to assume that the pure derivatives of second order in x_μ exist; this is true in particular for the uniqueness theorems.

II. Maximum-Minimum Principle. By 26 II this holds for the most general heat equation if $q=0$: *If there are neither heat sources nor sinks in the body, then the temperature takes on its maximum and its minimum at the boundary* (in the sense defined more precisely in 26 I). If $q\geqq 0$ or $q\leqq 0$ (heat is generated resp. used up in the body), then also by 26 II the minimum principle resp. the maximum principle holds.

According to 26 IV, V, we need additional hypotheses for the strong maximum-minimum principle. For simplicity let us assume that the functions c, ϱ, q, k are defined for $(t, x) \in G_p$ and $z > 0$[1]. Then a sufficient condition for the strong maximum principle to hold for $q \leqq 0$ and the strong minimum principle for $q \geqq 0$ (relative to the equation (3)) is:

There exist constants $\gamma_1, \ldots, \gamma_4$ such that

$$0 < \gamma_1 \leqq c\varrho \leqq \gamma_2, \quad |k_{\lambda\mu}|, \left|\frac{\partial k_{\lambda\mu}}{\partial x_\lambda}\right|, \left|\frac{\partial k_{\lambda\mu}}{\partial z}\right| \leqq \gamma_3,$$

$$\sum_{\lambda,\mu} k_{\lambda\mu} c_\lambda c_\mu \geqq \gamma_4 \sum_\mu c_\mu^2 \quad \text{for arbitrary } c_\mu \ (\gamma_4 > 0).$$

Indeed it suffices if for arbitrary $(\bar{t}, \bar{x}) \in G_p$ and arbitrary $M > 0$ we can find a lower half-neighborhood $U_-(\bar{t}, \bar{x})$ and constants $\gamma_1, \ldots, \gamma_4$ such that these inequalities hold if $(t, x) \in U_-$ and $0 < z < M$.

The assertions of 26 IV *hold relative to the equation* (3) *under these hypotheses.*

This follows in a simple way from 26 IV. If we consider a function $v \in Z_0$ for which (3) holds, for example, with the $\leqq$ sign and a point $(\bar{t}, \bar{x}) \in G_p$ with an associated lower half-neighborhood U_-^* in G_p held fixed, then there is a constant $M > 0$ such that $0 < v < M$ and $|v_{x_\mu}| < M$ in U_-^* (due to the continuity of the first derivatives). Now suppose that f is defined as in 26 IV but with

$$g(t, x, z, p) = \frac{1}{c(t, x, z)\,\varrho(t, x, z)} \left[q(t, x, z) + \sum_{\lambda,\mu} \left(\frac{\partial k_{\lambda\mu}(t, x, z)}{\partial x_\lambda} p_\mu + \frac{\partial k_{\lambda\mu}(t, x, z)}{\partial z} p_\lambda p_\mu \right) \right]$$

and with $k_{\lambda\mu}(t, x, v)/c(t, x, v)\,\varrho(t, x, v)$ in place of $k_{\lambda\mu}(t, x)$. Then we have $Pv \leqq 0$ with this f. Given M and $(\bar{t}, \bar{x})$, by hypothesis there is a neighborhood $U_-(\bar{t}, \bar{x})$, which we may assume to be a subset of U_-^*, and four constants γ_i such that the above inequalities hold. But they immediately imply, since we can assume $0 < z < M$, $|p_\mu| < M$, the three hypotheses 26 IV (α) to (γ).

The following fact should be noted. If in (1) we have $q(t, x, z) = a(t, x)\, z$ with $a > 0$ (i.e., that the heat generated is proportional to the temperature), then, perhaps surprisingly, the minimum principle does not hold. A counterexample is $u_t = u + u_{xx}$ with the solution $u = -e^t - \sin x$. The seeming contradiction is easily explained. In our interpretation as a temperature distribution we have only positive solutions in mind, while the differential equation also allows negative solutions. As our example shows, the latter may have minima in the interior of the domain, while with the restriction to positive solutions according to the above considerations the minimum principle holds.

[1] The restriction $z > 0$ is made for physical rather than mathematical reasons. These material "constants" are defined only for positive z since u denotes absolute temperature.

III. The First Boundary Value Problem. Physically this corresponds to the following situation. An m-dimensional body which occupies a region $\bar{D}=D+\partial D\subset E^m$ has at time $t=0$ a known initial temperature. Moreover its surface ∂D is held at a fixed, prescribed temperature during the time $0<t\leqq T$. We then have the case of the hypercylinder $G=(0,T)\times D$ described at the beginning of Section 23, on whose parabolic boundary R_p boundary values are prescribed.

Now we assume that G is bounded (but not necessarily a hypercylinder) and that the boundary values are continuous. From the uniqueness theorem 25 V we immediately obtain:

If c, ϱ and $k_{\lambda\mu}$ are independent of z, and thus are functions of t and x alone ($c\varrho>0, k_{\lambda\mu}$ positive semidefinite in G_p), and further if q satisfies a one-sided condition

$$q(t,x,z)-q(t,x,\bar{z})\leqq c(t,x)\,\varrho(t,x)\,\omega(t,z-\bar{z}) \quad \textit{for} \quad z\geqq\bar{z}$$

with $\omega\in\mathscr{E}_4$, then the first boundary value problem for the equation (3) *has at most one solution and it depends continuously on the boundary values and on the right hand side of the differential equation.*

However, the problem of uniqueness becomes far more difficult when the coefficients depend on z. Then in general a condition (25.2), as needed in 25 V, no longer exists. Nevertheless we can achieve special results in the way indicated in 25 VI. For the following we agree on the convention: A function $\varphi(t,x,z)$ defined for $(t,x)\in G_p, z>0$ "*satisfies a Lipschitz condition*" or "*satisfies a one-sided Lipschitz condition*" if for every $M>0$ there exists a constant L_M such that

$$|\varphi(t,x,z)-\varphi(t,x,\bar{z})|\leqq L_M|z-\bar{z}| \quad \text{for} \quad (t,x)\in G_p, 0<z,\bar{z}<M$$

or

$$\varphi(t,x,z)-\varphi(t,x,\bar{z})\leqq L_M(z-\bar{z}) \quad \text{for} \quad (t,x)\in G_p, 0<\bar{z}<z<M\,.$$

IV. Uniqueness Theorem. *Suppose that the first boundary value problem for the equation* (3) *for a bounded domain G and continuous boundary values η has a solution $\bar{u}\in Z$ for which there exists a function $h(t)$ that is continuous in J_0, integrable over J, and such that*

$$|\bar{u}_{x_\lambda x_\mu}|\leqq h(t) \quad \textit{in} \quad G_p \qquad (\lambda,\mu=1,\ldots,m)\,. \tag{7}$$

Suppose further that the functions $k_{\lambda\mu}/c\varrho$ satisfy a Lipschitz condition and the function $q/c\varrho$ satisfies a one-sided Lipschitz condition. Then for the case in which the $k_{\lambda\mu}$ depend only on t there exists no other solution of the first boundary value problem in the class Z.[1]

[1] Here we do not assume that a further solution which might exist satisfies the inequality (7).

If k also depends on x and z, if h(t) can be determined so that besides (7) *we have*

$$|\bar{u}_{x_\lambda}\bar{u}_{x_\mu}| \leqq h(t) \quad in \quad G_p \qquad (\lambda, \mu = 1, \ldots, m) \tag{8}$$

and if, moreover, the functions

$$\frac{1}{c\varrho}\,\frac{\partial}{\partial x_\lambda} k_{\lambda\mu} \quad and \quad \frac{1}{c\varrho}\,\frac{\partial}{\partial z} k_{\lambda\mu}$$

satisfy a Lipschitz condition, then again there is no other solution $u \in Z$.

In particular there exists (in the general case) at most one solution with bounded first and second derivatives in x_μ.

Indeed, if $u \in Z$ is another solution, then there exists a constant $M > 0$ such that $u, \bar{u} < M$. By 25 VI we have $u = \bar{u}$ if for the function f corresponding to the equation (3) we can find a function $h^*(t)$ which is continuous in J_0, integrable over J, and such that

$$f(t, x, z, \bar{u}_x, \bar{u}_{xx}) - f(t, x, \bar{z}, \bar{u}_x, \bar{u}_{xx}) \leqq h^*(t)\,(z - \bar{z}) \quad \text{for} \quad 0 \leqq \bar{z} \leqq z \leqq M$$

(indeed $D(f)$ can be restricted to $0 < z \leqq M$). But this is easily possible under our hypotheses. If all functions $k_{\lambda\mu}/c\varrho$ satisfy a Lipschitz condition with one and the same Lipschitz constant L_M, then for the first term of f in (3) we have

$$\left| \sum_{\lambda,\mu} \frac{k_{\lambda\mu}(t, x, z)}{c(t, x, z)\,\varrho(t, x, z)} \bar{u}_{x_\lambda x_\mu} - \sum_{\lambda,\mu} \frac{k_{\lambda\mu}(t, x, \bar{z})}{c(t, x, \bar{z})\,\varrho(t, x, \bar{z})} \bar{u}_{x_\lambda x_\mu} \right| \leqq m^2 L_M h(t)\,|z - \bar{z}|\,,$$

and we proceed similarly with the remaining terms of f.

In the above derivation we must keep in mind that the inequalities (7) and (8) are crucial and that hence, in particular, further regularity conditions going beyond continuity are imposed on the initial temperature $\eta(0, x)$. Below in VI (β) we illustrate this in an example. It should be noted in advance that in the most important case of an isotropic body with a thermal conductivity depending only on the temperature we can rid ourselves of the condition (8). For then a transformation to the equation (6) is possible in which no partial derivatives of first order in x occur. In the transformation which leads to (6), the boundary values η must also be transformed, i.e., replaced by $\bar{\eta} = K(\eta)$. We see immediately that the uniqueness of the original problem is identical with that of the transformed problem.

In order to be able to express ourselves more conveniently, we write the equation (6) in the new form

$$\bar{u}_t = \bar{\varkappa}(t, x, \bar{u})\,\Delta\bar{u} + \bar{q}(t, x, \bar{u})\,, \tag{9}$$

where $\varkappa = k/c\varrho$ is the so-called "temperature conduction number" and the functions $\bar{\varkappa}, \bar{q}$ are given by

$$\bar{\varkappa}(t, x, \bar{z}) = \varkappa(t, x, L(z)), \qquad \bar{q}(t, x, z) = \bar{\varkappa}(t, x, \bar{z})\, q(t, x, L(\bar{z}))\ ;$$

the function L was introduced in connection with (6) as the inverse function of K. The following uniqueness theorem shows that in special cases the estimate (7) can be replaced by a one-sided estimate.

V. Uniqueness Theorem. *Suppose the function $\bar{q}(t, x, \bar{z})$ satisfies a Lipschitz condition in $\bar{z}$. Suppose the first boundary value problem for the equation* (9) *(for a bounded domain G and continuous boundary values $\bar{\eta}$) has a solution $\bar{u}$ for which $\Delta\bar{u}$ can be bounded as follows (LC = Lipschitz condition):*

(α_1) *if $\bar{\varkappa}$ satisfies a LC in $\bar{z}$: $|\Delta\bar{u}| \leqq h(t)$ in G_p;*
(α_2) *if $\bar{\varkappa}$ is monotone increasing in $\bar{z}$: $\Delta\bar{u} \leqq 0$ in G_p;*
(α_3) *if $\bar{\varkappa}$ is monotone increasing in $\bar{z}$ and satisfies a LC: $\Delta\bar{u} \leqq h(t)$ in G_p;*
(α_4) *if $\bar{\varkappa}$ is monotone decreasing in $\bar{z}$: $\Delta\bar{u} \geqq 0$ in G_p;*
(α_5) *if $\bar{\varkappa}$ is monotone decreasing in $\bar{z}$ and satisfies a LC: $\Delta\bar{u} \geqq -h(t)$ in G_p.*

Here $h(t)$ is a function which is continuous and positive in J_0 and integrable over J.

In each of the five cases there is no other solution from the class Z; furthermore the solution depends continuously on the initial values and on the right hand side of the differential equation.

This assertion is proved just like the previous one relative to the equation (3). We have only to show that a one-sided Lipschitz condition

$$[\bar{\varkappa}(t, x, z) - \bar{\varkappa}(t, x, \bar{z})]\ \Delta\bar{u} \leqq h^*(t)\,(z - \bar{z}) \quad \text{for} \quad 0 < z < \bar{z} < M$$

holds and this is easy to do in all five cases.

VI. Additions and Remarks. (α) The hypothesis $\bar{\varkappa} > 0$ is unnecessarily sharp for V to hold; $\bar{\varkappa} \geqq 0$ is sufficient. Moreover it is permissible for both functions $\bar{\varkappa}$ and $\bar{q}$ to be defined in an arbitrary set D^* of the $(t, x, \bar{z})$ space. Then of course we consider only solutions $\bar{u}$ from the class $Z(f)$ with $f = \bar{\varkappa}(r_{11} + \cdots + r_{mm}) + \bar{q}$, i.e., those for which $(t, x, \bar{u}) \in D^*$ for $(t, x) \in G_p$. The formulation of the Lipschitz condition is then to be modified correspondingly.

(β) The meaning to be given to the hypothesis (7) and the hypotheses on $\Delta\bar{u}$ in Theorem V is indicated by a simple example. Suppose that for $m = 1$ we have the equation

$$\bar{u}_t = \bar{\varkappa}(t, x, u)\, \bar{u}_{xx} \tag{10}$$

in G_p: $0<t\leqq T$, $0<x<a$; as boundary values we then have

$$\bar{u}(0, x)=\eta_0(x), \quad \bar{u}(t, 0)=\eta_1(t), \quad \bar{u}(t, a)=\eta_2(t) \tag{11}$$

with continuous functions η_0, η_1, η_2 which satisfy the additional condition $\eta_0(0)=\eta_1(0)$, $\eta_0(a)=\eta_2(0)$. In the special case $a=\pi$, $\bar{\varkappa}$ constant, $\eta_1\equiv\eta_2\equiv 0$ the solution is

$$\bar{u}=\sum_{n=1}^{\infty} a_n \sin nx\, e^{-\bar{\varkappa}n^2 t}, \qquad a_n=\frac{2}{\pi}\int_0^{\pi} \eta_0(\xi)\sin n\xi\, d\xi$$

(see, e.g., Carslaw-Jaeger (1959), pp. 94–96)[1]. Now we make the further assumption that $\sum a_n$ converges absolutely; this is the case, for example, if $\eta_0(x)$ is piecewise continuously differentiable[2]. Then we have

$$|\bar{u}_{xx}|=|\sum a_n n^2 \sin nx\, e^{-\bar{\varkappa}n^2 t}|\leqq h(t),$$

where

$$h(t)=\sum_{n=1}^{\infty} n^2|a_n|\, e^{-\bar{\varkappa}n^2 t}$$

is a function which is continuous in J_0 and integrable over J.

Hence we can say as a result of this consideration that an inequality $|\bar{u}_{xx}|\leqq h(t)$ presents no serious restriction in the linear case and for the above special boundary values. Of course this is true to an even greater degree for one-sided inequalities. It is left to the reader to decide to what extent the assumption that similar relationships also hold if $\bar{\varkappa}$ depends on z in a "reasonable" fashion seems plausible. Nevertheless we shall learn two ways in which to arrive at guaranteed results.

(γ) Occasionally we will have a nonlinear problem (9) for which we know an explicit solution. Then we can verify directly whether the inequality required in V holds for this solution and if the answer is affirmative, we can conclude the uniqueness of this solution. It should again be noted that the bound relative to $\Delta\bar{u}$ in V is required not for *all* solutions but only for *one* solution.

Here is an example: Suppose we have the problem (10) (11) with $\eta_0=\alpha+\beta x$, $\eta_1=\alpha$, $\eta_2=\alpha+\beta a$, and the question is: Under what conditions is the stationary solution $u=\alpha+\beta x$ the only solution of this

[1] It is easy to see that this series and also the series produced by termwise differentiation converge uniformly in every domain $0\leqq x\leqq\pi$, $0<\delta\leqq t\leqq T$ if the a_n form a bounded sequence. Hence we obtain the validity of the differential equation as well as the assumption of the boundary values 0 on the lateral boundaries. On the other hand, if $\eta_0(x)$ is merely continuous, the proof that the initial values are assumed causes some difficulties; see Carslaw-Jaeger, loc. cit. p. 94, first footnote.

[2] Considerably weaker conditions suffice, for example, Hölder continuity of $\eta_0(x)$ (with arbitrary exponent) in connection with bounded variation; see Zygmund (1959), Chapter VI, particularly (3.13).

boundary value problem? Here each of the five cases of Theorem V is applicable: if $\varkappa$ has one of the properties named in V $(\alpha_1)-(\alpha_5)$, then the stationary solution is uniquely determined by its boundary values. However a far more general result follows from 25 VI (β):

If $m=1, f(t,x,z,p,0)=0$ *and* $u(t,x)=\alpha+\beta x$ *is a solution of a first boundary value problem for the equation* $u_t=f(t,x,u,u_x,u_{xx})$ *for a bounded domain G (hence* $\eta(t,x)=\alpha+\beta x$*), then this is the only solution of the boundary value problem.*

In particular if we are dealing with the equation (10), no hypotheses on $\bar{\varkappa}$ (except for $\bar{\varkappa}\geqq 0$) are necessary (by the way, we see this immediately from the inequality in the proof of Theorem V).

(δ) There are cases in which, on the basis of estimates, we can say a priori that for all solutions of a given first boundary value problem an inequality required in V holds. To derive such inequalities the assertions of Section 27 suggest themselves, among others. We illustrate in an example how to proceed.

We consider the first boundary value problem (10) (11) and use the notation of 27 I for the rectangle G. Since the condition (27.3) is satisfied, 27 VIII and IX hold. If $\eta_0(x)$ is convex from above, $\eta_1(t)$ and $\eta_2(t)$ are monotone decreasing, then $\bar{u}_{xx}\leqq 0$ in G_p by 27 IX. Together with V this leads to the following

VII. Uniqueness Theorem. *If* $\bar{\varkappa}(t,x,z)$ *is monotone increasing in* z, $\eta_0(x)$ *is convex from above and the functions* $\eta_1(t)$ *and* $\eta_2(t)$ *are monotone decreasing, then the first boundary value problem* (10) (11) *has at most one solution belonging to the class Z. The same holds if* $\bar{\varkappa}$ *is monotone decreasing,* η_0 *is convex from below,* η_1, η_2 *are monotone increasing.*

Here no further assumptions on $\bar{\varkappa}$ are needed.

That an assertion of this generality with respect to $\bar{\varkappa}$ is not true for arbitrary initial values $\eta_0(x)$ (not even if η_1 and η_2 are constant) is shown by the following

VIII. Counterexample. The equation

$$u_t=\frac{x(1-x)}{2tu}\,[x(1-x)-u]\,u_{xx} \quad \text{for} \quad 0<x<1, 0<t\leqq 1$$

with the boundary values $u(t,0)=u(t,1)=0$, $u(0,x)=x(1-x)$ has infinitely many solutions

$$u=x(1-x)(1-\beta t) \quad \text{for} \quad 0<\beta<1 .$$

As the set of definition D^* [see VI (α)] we have to take $0<u\leqq x(1-x)$ here [we could also extend $\bar{\varkappa}$ continuously for $u\geqq x(1-x)$].

In this section we have considered only heat conduction problems for bounded domains and continuous boundary values. The additional conditions that are needed at infinity for unbounded domains result from 25 V and 25 III. Moreover the results of the preceding section are at our disposal for discontinuous boundary values and/or unbounded domains. Here $k/c\varrho$ is to be assumed a function independent of z. Uniqueness assertions of this kind in the case $k=k(z)$, $m=1$ are given by Seyferth (1959, 1962). He uses other methods; the derivation of his results with our methods is not obvious.

30. Application to Boundary Layer Theory

This section was written in close collaboration with Professor K. Nickel. Most of the results given here have been taken from his work (1958).

The motion of a compressible viscous fluid is described by the well-known Navier-Stokes differential equations. By dropping terms, which is justified for flow near a fixed wall (in the so-called boundary layer), L. Prandtl (1904) derived

I. The Prandtl Boundary Layer Equations and thus founded a new field in fluid dynamics called boundary layer theory. Without going into the nature and the physical details of this derivation, we give these equations for the two-dimensional stationary case. They are (Schlichting (1955), p. 97)

$$u u_x + v u_y = U(x)\, U'(x) + \nu u_{yy}\,, \tag{1}$$

$$u_x + v_y = 0\,. \tag{2}$$

Here in the simplest case of a flow along a plane wall[1] the individual terms have the following physical meaning (Fig. 17): x and y are position coordinates in directions parallel and perpendicular to the wall respec-

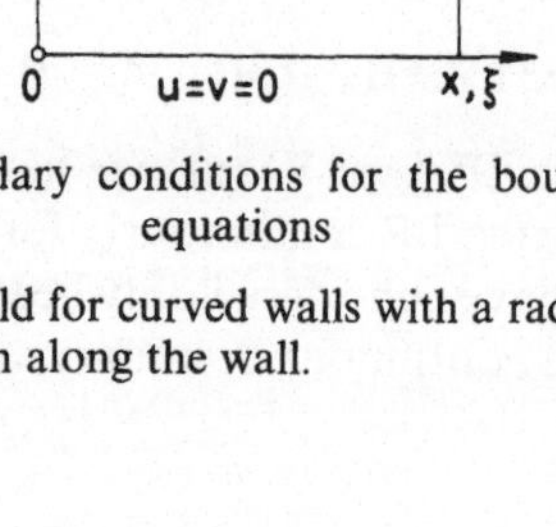

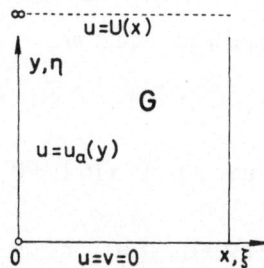

Fig. 17. Domain and boundary conditions for the boundary layer differential equations

[1] The equations also hold for curved walls with a radius of curvature not too small; then x is the arc length along the wall.

tively, u and v are the velocity components of the flow parallel and perpendicular to the wall respectively. Here the wall is taken to coincide with the x-axis, $U(x)>0$ is the velocity at infinity (velocity of the potential flow), $\nu>0$ the kinematic viscosity. The domain $\bar{G}$ is a half-strip $0\leqq x\leqq X$, $0\leqq y<\infty$ and as boundary conditions we have

$$u(x,0)=v(x,0)=0, \qquad u(0,y)=u_a(y), \qquad u(x,\infty)=U(x). \tag{3}$$

Here $u_a(y)$ is a given function, namely, the "initial velocity" on the y-axis. The last of the three boundary conditions is to be interpreted as

$$\lim_{y\to\infty} u(x,y)=U(x) \quad \text{uniformly for} \quad 0\leqq x\leqq X. \tag{4}$$

Accordingly we must have the compatibility condition between u_a and U, $u_a(y)\to U(0)$ as $y\to\infty$. Moreover, it can be shown that under certain hypotheses this third boundary condition is unnecessary (because it is automatically satisfied); see Nickel (1958a) and Walter (1970a).

The mathematical problem consists of finding two functions u, v for given outer and initial velocities $U(x)$ and $u_a(y)$ which satisfy the differential equations (1) (2) and the boundary conditions (3). Alhough this is considerably simpler than the corresponding problem for the Navier-Stokes differential equations – this is precisely the significance of boundary layer theory – it still presents considerable mathematical difficulties. Further simplifications are obtained via the following

II. Transformation of v. Mises (1927) to the new independent variables ξ, η

$$\xi=x, \qquad \eta=\eta(x,y)=\int_0^y u(x,s)\,ds, \qquad u(x,y)=\bar{u}(\xi,\eta). \tag{5}$$

With the help of the relations

$$\left.\begin{aligned} &\eta_x=\int_0^y u_x(x,s)\,ds=-\int_0^y v_y(x,s)\,ds=-v(x,y), \qquad \eta_y=u,\\ &u_x=\bar{u}_\xi-\bar{u}_\eta v, \qquad u_y=\bar{u}\bar{u}_\eta, \qquad u_{yy}=\bar{u}^2\bar{u}_{\eta\eta}+\bar{u}\bar{u}_\eta^2 \end{aligned}\right\} \tag{6}$$

(1) becomes the differential equation

$$\bar{u}_\xi=\frac{U(\xi)\,U'(\xi)}{\bar{u}}+\nu(\bar{u}\bar{u}_{\eta\eta}+\bar{u}_\eta^2). \tag{7}$$

The new boundary conditions are

$$\bar{u}(\xi,0)=0, \qquad \bar{u}(0,\eta)=\bar{u}_a(\eta), \qquad \bar{u}(\xi,\infty)=U(\xi). \tag{8}$$

Here $\bar{u}_a(\eta)$ is determined in accordance with (5) by $\bar{u}_a\left(\int_0^y u_a(s)\,ds\right)=u_a(y)$. The two equations (1) (2) are uncoupled by the v. Mises transformation.

Now we have only *one* differential equation (7). We obtain $u(x, y)$ by inverse transformation (see III) from a solution $\bar{u}(\xi, \eta)$ and from it we obtain $v(x, y)$ by (2), i.e., as an integral

$$v(x, y) = -\int_0^y u_x(x, s)\, ds\,. \tag{9}$$

We note that the transformation (5) represents only "half" of the v. Mises transformation; namely, v. Mises (1927) also introduces, besides the new independent variables ξ, η, a new function

$$g(\xi, \eta) = U^2(\xi) - \bar{u}^2(\xi, \eta)\,.$$

For the function g, the boundary value problem is

$$g_\xi = \nu g_{\eta\eta} \sqrt{U^2(\xi) - g}\,, \tag{6'}$$

$$g(\xi, 0) = U^2(\xi), \quad g(0, \eta) = U^2(0) - \bar{u}_a^2(\eta), \quad g(\xi, \infty) = 0\,. \tag{7'}$$

Which of the two forms one prefers depends on the problem. We prefer to work with $\bar{u}$, i.e., with (6) and (7).

We now go into more detail on the hypotheses needed for the preceding and subsequent considerations.

III. Notation and Hypotheses (Z_g, $u(x, \infty)$, $\bar{u}(x, \infty)$). The equation (7) is a parabolic differential equation but with the modified notation: instead of t and x we now have ξ and η. We have (in accordance with 23 I) $\bar{G} = [0, X] \times [0, \infty)$ and $G_p = (0, X] \times (0, \infty)$; this notation will be used in the (ξ, η)-plane as well as in the (x, y)-plane. Suppose the outer velocity $U(x)$ is a continuously differentiable and positive function on $[0, X]$ and the initial velocity $u_a(y)$ is continuously differentiable in $[0, \infty)$ and positive in $(0, \infty)$. Further let $u_a(y) \to U(0)$ as $y \to \infty$. The function $u(x, y)$ belongs to the class Z_g if it is continuous in $\bar{G}$ together with its derivatives u_x, u_y, if u_{yy} exists in G_p, $u > 0$ in $[0, X] \times (0, \infty)$ and if, as $y \to \infty$, $u(x, y)$ converges uniformly in $x \in [0, X]$ to a continuous and positive function, which we denote by $u(x, \infty)$[1]. Thus for a function $u \in Z_g$ the equation $u(x, \infty) = U(x)$ is identical with (4); furthermore u is bounded and for every $\delta > 0$ there exists a $\gamma > 0$ such that $u \geqq \gamma$ for $y \geqq \delta$.

For a function $u \in Z_g$ a unique invertible mapping of $\bar{G}$ *onto* itself is established by the transformation (5); the function $\eta(x, y)$ is indeed

[1] These hypotheses are unnecessarily sharp for the mathematical theory. However, they make a briefer formulation possible in various places and thus let the essential points of view be brought out more meaningfully. Moreover, they are such that no case of practical interest is excluded. The existence and continuity of the partial derivatives of first order in the *closed* domain $\bar{G}$ are mathematically justified in the case of u_x and physically in the case of u_y: For the former we need conditions for the existence of the integral (9), for the latter u_y for $y = 0$ represents a physically important quantity, the shearing stress at the wall.

strictly monotone increasing in y and, since u lies between two positive bounds for $y \geqq \delta > 0$, approaches ∞ like a linear function ($y \to \infty$). Hence, according to (5), it follows that the function $\bar{u}(\xi, \eta)$ associated with the function $u(x, y)$ also converges uniformly in ξ to $u(\xi, \infty)$ if $\eta \to \infty$. We set $u(\xi, \infty) = \bar{u}(\xi, \infty)$ and thus we have explained the third of the boundary conditions (8), again in the sense of uniform convergence.

If the function $\bar{u}(\xi, \eta)$ is formed from a function $u(x, y) \in Z_g$ via the transformation (5), then we also write $\bar{u} \in Z_g$. It is easy to see that such a function $\bar{u}$ is continuous and bounded in $\bar{G}$, positive in G_p and has continuous derivatives u_ξ, u_η in G_p as well as a derivative $u_{\eta\eta}$. Moreover we note that the mapping $u(x, y) \to \bar{u}(\xi, \eta)$ established by (5) has a *unique inverse* in the class Z_g. Indeed the function $\eta(x, y)$ satisfies the ordinary differential equation

$$\eta_y(x, y) = \bar{u}(x, \eta(x, y)) \quad \text{with} \quad \eta(x, 0) = 0\,. \tag{10}$$

(depending on a parameter x). Thus if we replace $y, \eta, \bar{u}$ by t, z, h, we are dealing with the initial value problem

$$z'(t) = h(z(t)) \quad \text{for} \quad 0 \leqq t < \infty, z(0) = 0\,. \tag{11}$$

For $z \geqq 0$ the function $h(z)$ is continuous and bounded, and positive for $z > 0$. We know that a solution $z(t)$ which is positive for $t > 0$ exists ($\bar{u}$ was obtained via the transformation (5)) and the question is whether this function $z(t)$ can again be obtained uniquely from (11). For $h(0) > 0$ this is trivial. For $h(0) = 0$, according to well-known theorems from the theory of ordinary differential equations (see Kamke (1945, p. 19f.)), there is exactly one solution $z(t)$ which is positive for positive t, namely, the maximal solution. Thus if $\bar{u} \in Z_g$ is given, then (10) determines a unique function $\eta(x, y)$, which is itself positive for positive y, and hence also determines the function $u(x, y) = \bar{u}(x, \eta(x, y))$.

We speak of the solution u of the problem (1) (2) (3) even though two unknown functions u, v appear there. We mean thereby that $u \in Z_g$ and satisfies the boundary conditions (3) [for given functions $U(x)$, $u_a(y)$] and that u together with the function v *defined* by (9) satisfies the differential equation (1) in G_p. Then (2) is satisfied automatically. If $u \in Z_g$ is a solution, then $\bar{u}$ is a solution of the transposed problem (6) (7).

The first uniqueness theorem for our problem was given by Nickel (1958). It goes:

IV. Uniqueness Theorem. *The boundary value problem* (1) (2) (3) *has at most one solution* $u \in Z_g$ *for which* $u_y(x, 0) > 0$ $(0 \leqq x \leqq X)$ *and* u_{yy} *is bounded from above in* G_p.

Because the correspondence $u \to \bar{u}$ is one-one (see III), it suffices to verify uniqueness of the problem (6) (7). To this end we use the uniqueness theorem 25 V in the sharpened version 25 VI (α). Accordingly it suffices to show that the f-difference satisfies a one-sided Lipschitz condition

$$\frac{UU'}{\bar{u}+z}+v((\bar{u}+z)\,\bar{u}_{\eta\eta}+\bar{u}_\eta^2)-\frac{UU'}{\bar{u}}-v(\bar{u}\bar{u}_{\eta\eta}+\bar{u}_\eta^2)$$
$$=-\frac{zUU'}{\bar{u}(\bar{u}+z)}+vz\bar{u}_{\eta\eta}\leqq Lz \quad \text{for small positive } z$$

with constant L. This is surely true if the inequality

$$v\bar{u}_{\eta\eta}+\frac{|U(\xi)\,U'(\xi)|}{\bar{u}^2}\leqq L \tag{12}$$

or, rewritten in terms of u, x and y,

$$\frac{1}{u^3}\,[vuu_{yy}-vu_y^2+u|U(x)\,U'(x)|]\leqq L \tag{12'}$$

holds. The quantities $|UU'|$ and vu_{yy} are bounded from above and, as $y\to+0$, the function u converges uniformly to 0 and the function u_y converges uniformly to $u_y(x,0)\geqq\delta>0$. Hence the expression in square brackets in (12′) is negative for small y, say $0<y<\gamma$ ($\gamma>0$). But for all $y\geqq\gamma$ the left hand side of (12′) is bounded from above since u lies between two positive bounds and, as we have already observed, vu_{yy} and $|UU'|$ are bounded from above. Thus there indeed exists a constant L satisfying the inequality (12′), whereby the theorem is completely proved.

V. Remarks. (α) The uniqueness theorem can be generalized in an obvious way to the case in which $U(x)$ is only piecewise continuously differentiable. If $0<x_0<X$ and if $U(x)$ is continuously differentiable in each of the two intervals $[0,x_0]$ and $[x_0,X]$ (at the end points we have to take the one-sided derivative directed into the interior of the appropriate interval), then we apply the uniqueness theorem first to the part of $\bar{G}$ lying between 0 and x_0 and then to the part between x_0 and X. Here the differential equation need not hold on the line $x=x_0$ (we first consider the domain $0\leqq x\leqq\bar{x}<x_0$ and then let $\bar{x}\to x_0$). The corresponding statement is true when U' has several points of discontinuity.

(β) The hypotheses on u_y and u_{yy} in IV can be weakened in some cases. Only the bound preceding (12) is crucial for the proof. Here we can work with a generalized Lipschitz condition rather than the simple one used above, i.e., we replace L by a function $l(x)$ which is continuous in $(0,X]$ and integrable over this interval. Thus a bound

$$\frac{1}{u^3}\left[vuu_{yy}-vu_y^2-\frac{u^2}{u+z}\,UU'\right]\leqq l(x) \quad \text{(for small positive } z) \tag{12''}$$

is sufficient for uniqueness. For example we can allow u_{yy} to become infinite as we approach the wall if the product uu_{yy} approaches 0.

(γ) The further hypothesis $u>0$ means physically that the uniqueness is proven only as long as no "back flow" is involved. We should compare this with IX; there we show that under certain hypotheses absolutely no back flow occurs.

The situation is similar with the hypothesis $u_y(x,0)>0$. The uniqueness is shown only up to "points of separation"; again compare IX.

Next we shall show that Nagumo's Lemma can also be sharpened, as in 24 VI, for the boundary value problem of boundary layer theory.

VI. The Sharpened Lemma of Nagumo. *Suppose $U(x)$ is a continuously differentiable function in $[0,X]$ and $u, w \in Z_g$. Suppose further that $w_y(x,0)>0$ and that w_{yy} is bounded above [more generally: a bound (12″) holds for w]. Then*

(α) $\bar{u}(\xi,0) \leqq \bar{w}(\xi,0)$, $\quad \bar{u}(\xi,\infty) \leqq \bar{w}(\xi,\infty)$ *in* $[0,X]$,
$\bar{u}(0,\eta) \leqq \bar{w}(0,\eta)$ *in* $(0,\infty)$;

(β) $P\bar{u} \leqq P\bar{w}$ *in* G_p *(where $P\bar{u}=\bar{u}_\xi - v(\bar{u}\bar{u}_{\eta\eta}+\bar{u}_\eta^2) - UU'/\bar{u}$) implies the inequality*

$$\bar{u} \leqq \bar{w} \quad \text{in} \quad \bar{G}.$$

If moreover one of the two functions $\bar{u}$, $\bar{w}$ is monotone increasing in η for every fixed $\xi \in [0,X]$, then we also have

$$u \leqq w \quad \text{in} \quad \bar{G}.$$

The proof is similar to that of 24 VI. If $\bar{w}$ satisfies the inequality (12) (with $\bar{w}$ in place of $\bar{u}$ of course), then the Lemma of Nagumo 24 IV can be applied to the two functions $\bar{u}$ and $\bar{w}+\varrho$, $\varrho=\varepsilon e^{2L\xi}$ ($\varepsilon>0$). The hypothesis 24 VI (α) follows from (α) because $\varrho \geqq \varepsilon > 0$, the hypothesis 24 VI ($\beta$) from ($\beta$) because of the inequalities

$$\begin{aligned} P(\bar{w}+\varrho) &= \bar{w}_\xi + 2L\varrho - \frac{UU'}{\bar{w}+\varrho} - v\bar{w}_\eta^2 - v(\bar{w}+\varrho)\,\bar{w}_{\eta\eta} \\ &\geqq P\bar{w} + \varrho\left(2L - v\bar{w}_{\eta\eta} + \frac{UU'}{\bar{w}(\bar{w}+\varrho)}\right) \geqq P\bar{w} + L\varrho > P\bar{w}. \end{aligned}$$

Here the inequality (12) was used. Accordingly $\bar{u} < \bar{w} + \varepsilon e^{2L\xi}$ whence we obtain the assertion for $\varepsilon \to 0$.

Moreover VI contains the uniqueness theorem IV. Nevertheless we do not have a new proof since the essential idea is the same as in the proof of IV.

The inequality $u \leqq w$ still lacking is obtained from the following general remark.

If one of the two functions $u, w \in Z_g$ is monotone increasing in y and $u \leqq w$, then $\bar{u} \leqq \bar{w}$; if one of the functions $\bar{u}, \bar{w} \in Z_g$ is monotone increasing in η and $\bar{u} \leqq \bar{w}$, then $u \leqq w$.

The first part is very easy to see while the second needs a little thought. If $\eta(x, y)$ satisfies the differential equation (10) and $\eta^*(x, y)$ satisfies the corresponding differential equation

$$\eta_y^* = \bar{w}(x, \eta^*(x, y)) \quad \text{with} \quad \eta^*(x, 0) = 0,$$

then by 8 X, $\bar{u} \leqq \bar{w}$ implies the inequality $\eta \leqq \eta^*$ (according to the procedure in III we must take the maximal solution for η^*). Thus, for example, if $\bar{u}$ is monotone increasing in η, then we have

$$u(x, y) = \bar{u}(x, \eta) \leqq \bar{u}(x, \eta^*) \leqq \bar{w}(x, \eta^*) = w(x, y).$$

We proceed similarly if w is monotone increasing.

Lemma VI is of great practical significance. It provides the theoretical foundation for constructing super- and subfunctions. In particular, since the equality signs are not excluded in (β), we can use special solutions of the boundary layer equations as super- and subfunctions. We show in two examples how to proceed here and what general conclusions can be obtained using such simple functions.

VII. The Velocity of the Potential Flow as Solution. The function $U(\xi)$ represents the simplest solution of the equation (7). At the same time this function is a superfunction for all problems with $\bar{u}_a(\eta) \leqq U(0)$. If $\bar{u}$ is a solution of such a problem and $\bar{w} = U(\xi)$, then the hypotheses of VI hold. Hence we have proved the following theorem.

(α) *If we have a problem* (1) (2) (3) *with* $u_a(y) \leqq U(0)$ $(0 \leqq y < \infty)$, *then*

$$u(x, y) \leqq U(x) \quad \text{in} \quad \bar{G}$$

for every[1] *solution* $u \in Z_g$.

At first glance this result is highly paradoxical. Suppose we take an outer velocity $U(x)$ which first proceeds "smoothly" but then suddenly falls off sharply! The above inequality shows that the solution "conforms" to this braking without "lagging behind" (we can object to this terminology that here a time property is imposed on the variables ξ while we consider stationary solutions). As the mathematical basis for this behavior we can consider the fact that the boundary condition also appears in the differential equation.

(β) We see immediately that for arbitrary positive c the function $\bar{w} = \sqrt{U^2(\xi) + c}$ also represents a solution of the equation (7). Exactly as above, this implies:

If we have a problem (1) (2) (3) *with* $u_a(y) \leqq \sqrt{U^2(0) + c}$, *then*

$$u(x, y) \leqq \sqrt{U^2(x) + c} \quad \text{in} \quad \bar{G}$$

for every solution $u \in Z_g$.

[1] The additional hypotheses of the uniqueness theorem are not used; it then cannot be asserted that only one solution exists.

This contains an interesting bound for the

(γ) *Velocity excess.* We say that there is a velocity excess of size $e(x)$ at the point x if

$$e(x) = \sup_{0 \leqq y < \infty} u(x, y) - U(x) > 0 .$$

Since $u_a(y) \leqq U(0) + e(0)$, we can rewrite the inequality in (β) as

$$u(x, y) \leqq \sqrt{U^2(x) + 2e(0)\, U(0) + e^2(0)} .$$

We note two special cases: *We have*

$$e(x) \leqq e(0) \qquad \text{if} \quad U(0) \leqq U(x),$$
$$u(x, y) \leqq e(0) + U(0) \quad \text{if} \quad U(0) \geqq U(x)$$

(in both cases nothing is assumed about the behavior of $U(x)$ between the points 0 and x).

VIII. The Similarity Solution of the Boundary Layer along a Plate. H. Blasius (1908) has shown that when $U(x) \equiv U_\infty$ is constant we obtain a special solution $u = \varphi$ of equation (1) from the following boundary value problem for $h = h(s)$:

$$h''' + h h'' = 0 \text{ for } 0 \leqq s < \infty,\ h(0) = h'(0) = 0,\ h'(s) \to 1 \text{ for } s \to \infty, \quad (13)$$

where we set

$$u(x, y) = \varphi(x, y) = U_\infty h'\left(y \sqrt{\frac{U_\infty}{2 \nu x}}\right). \quad (14)$$

The solution $h(s)$[1] has been computed very precisely by several authors. A table can be found in Schlichting (1955, p. 107). Fig. 18 shows the behavior of h' and therefore that of φ. For later purposes we need the formulas

$$\left.\begin{aligned} \varphi_y(x, y) &= U_\infty \sqrt{\frac{U_\infty}{2 \nu x}}\, h''\left(y \sqrt{\frac{U_\infty}{2 \nu x}}\right), \\ \varphi_y(x, 0) &= \frac{C}{\sqrt{x}}, \quad C = U_\infty \sqrt{\frac{U_\infty}{2 \nu}}\, h''(0) > 0 ; \end{aligned}\right\} \quad (15)$$

here $h''(0) \approx 0.4696$.

Physically the solution φ represents the flow along a thin flat plate.

[1] Existence and uniqueness were proved by Weyl (1942); see 4 VIII. Elementary proofs using differential inequalities are given by Coppel (1960) and Hartman (1964; Chapter XIV, Part II).

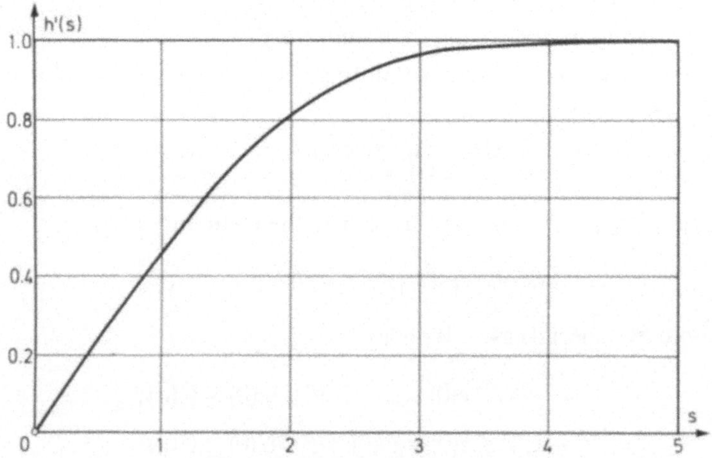

Fig. 18. The function $h'(s)$

This solution can be used as an upper bound for problems with $U'(x) \leqq 0$ and as lower bound for problems with $U'(x) \geqq 0$. We shall consider the latter case in more detail.

IX. Back Flow and Separation. *If* $U'(x) \geqq 0$, $u_a'(0) > 0$, $u_a(y) > 0$ *for* $y > 0$, *if* $u(x, y)$ *is a solution of the problem* (1) (2) (3) *which is not necessarily* > 0 *in* G_p *but otherwise has all the properties of* Z_g *and if also the derivative* u_{yy} *is bounded, then*

$$u(x, y) > 0 \quad \text{in} \quad G_p \quad \text{and} \quad u_y(x, 0) > 0 \quad \text{for} \quad 0 \leqq x \leqq X. \tag{16}$$

In other words there is no back flow and no separation.

The proof is based on using the function φ as a subfunction. For constant positive c, $\varphi(x+c, y)$ is also a solution of the equation (1) with $U' = 0$. Thus (in the notation of VI) $P\bar{\varphi}(\xi + c, \eta) \leqq 0$ (since $U' \geqq 0$) for the transformed function $\bar{\varphi}$. Now if we choose $U_\infty = 1$ and c so large that $\varphi_y(c, 0) < u'(0)$, then we can arrange (by making U_∞ smaller if necessary) that $\varphi(c, y) \leqq u_a(y)$ for all $y \geqq 0$ and $\varphi(x+c, \infty) \equiv U_\infty \leqq U(x)$ for $0 \leqq x \leqq X$.

If the assertion is not true, then there exists a first point $x_0 > 0$ such that both inequalities (16) hold for $0 \leqq x < x_0$ but not for $x = x_0$. In the case of u_y this follows from the fact that $u_y(x, 0)$ is continuous and $u_y(0, 0) > 0$, and in the case of the function u from the fact that because of the boundary conditions at infinity the zeros of u form a bounded set. Now we can apply VI (with $\bar{\varphi}(\xi + c, \eta)$, $\bar{u}$ instead of $\bar{u}$, $\bar{w}$) for every $X_1 < x_0$ and we obtain

$$u(x, y) \geqq \varphi(x+c, y) \quad \text{for} \quad 0 \leqq x \leqq X_1,$$

and thus also for $0 \leqq x \leqq x_0$. Hence we have obtained a contradiction since $\varphi(x_0 + c, y) > 0$ for $y > 0$ and $\varphi_y(x_0 + c, 0) > 0$.

X. An Estimation Theorem. Suppose $u \in Z_g$ is a solution and $w \in Z_g$ is an approximate solution of a problem (1) (2) (3) with $U'(x) \geqq 0$. Suppose the approximate solution satisfies the inequalities

$$w w_{yy} \leqq w_y^2, \quad \text{i.e.,} \quad \overline{w}_{\eta\eta} \geqq 0 \quad \text{in} \quad G_p$$

$$-\overline{\delta}(\xi) \leqq P\overline{w} \equiv \overline{w}_\xi - \frac{U U'}{\overline{w}} - v(\overline{w}\,\overline{w}_{\eta\eta} + w_\eta^2) \leqq \delta(\xi) \quad \text{in} \quad G_p,$$

where $\delta(\xi)$, $\overline{\delta}(\xi)$ are two continuous functions in $[0, X]$. We denote their integrals by $\Delta(\xi)$, $\overline{\Delta}(\xi)$:

$$\Delta(\xi) = \int_0^\xi \delta(s)\,ds, \qquad \overline{\Delta}(\xi) = \int_0^\xi \overline{\delta}(s)\,ds.$$

Then

$$-\overline{\Delta}(x) - \overline{c} \leqq w(x, y) - u(x, y) \leqq c + \Delta(x) \quad \text{in} \quad \overline{G}, \tag{17}$$

provided the constants c, $\overline{c}$ are chosen so that the left hand side of (17) is $\leqq 0$, the right hand side of (17) is $\geqq 0$ and so that the inequalities (17) hold at the wall (i.e., for $0 \leqq x \leqq X$, $y = 0$), on the initial profile (for $x = 0$, $y > 0$), and at infinity (for $0 \leqq x \leqq X$, $y = \infty$).

This is the following special case of 25 III: $\omega = \overline{\omega} = 0$, $\varrho(\xi) = c + \Delta(\xi)$, $\overline{\varrho}(\xi) = \overline{c} + \overline{\Delta}(\xi)$. We can enforce the $>$ symbol in 25 III (α) and (γ) in the usual way, for example, by adding another term $\varepsilon(1 + \xi)$ to ϱ and $\overline{\varrho}$.

The examples given here are simple throughout. They are meant above all to demonstrate the fruitfulness of this method as applied to boundary layer theory. A basic work in this regard is that of Nickel (1958) in which the Nagumo Lemma is applied to boundary layer theory for the first time. It already contains the results proved in IV to IX and a series of further results for whose derivation, however, we have to use transformations other than (5). These ideas have been further developed by Nickel (1959, 1960, 1961, 1962a, 1963) and Velte (1960). Moreover, Görtler (1950, 1951) referred to the possibility of handling boundary layer equations with the help of the Nagumo Lemma even earlier. A detailed theory of existence, uniqueness and asymptotic behavior for the boundary layer equations was given by Oleinik (1962, 1963, 1963a, 1969). An existence proof based on the line method and asymptotic results (as $y \to \infty$) were found by Walter (1970, 1970a).

31. The Third Boundary Value Problem

A simple but typical problem of the type we now have in mind is:

We seek a function $u(t, x)$ depending on two real variables t and x which represents a solution of the heat equation $u_t = u_{xx}$ in the rectangle $0 < t \leqq T$, $0 < x < 1$ and satisfies the following boundary conditions:

$u(0, x)=0$ for $0 \leqq x \leqq 1$, $u(t, 0)=1$ for $0<t \leqq T$, $\partial u(t, 1)/\partial n_a + u(t, 1)=0$ for $0<t \leqq T$.

Problems of this type differ from the ones handled earlier in that a new kind of boundary condition is prescribed on a part of the boundary, namely, a relation between the outer normal derivative $\partial u/\partial n_a$ and the function u. We are often led to such boundary conditions, which usually allow a simple physical interpretation.

Let us consider the heat equation (29.3). The density of the outward heat flow through the surface of the body (i.e., the quantity of heat leaving in unit time per unit area) is given by $-k \dfrac{\partial u}{\partial n_a}$. If we set this heat flow density equal to $h(u-U)$ according to "Newton's law of cooling" (Frank-Mises (1935, p. 528)) – here U is the "outer temperature", $h>0$ is the "outer thermal conductivity" – then the boundary condition is

$$\frac{\partial u}{\partial n_a} + \lambda(u-U)=0, \qquad \lambda = h/k > 0. \tag{1}$$

Thus the problem mentioned at the beginning characterizes a thin rod of length 1 which is at temperature $u=0$ at time $t=0$ and is then held at the constant temperature $u=1$ at the left end while there is a heat loss proportional to the temperature at the right end (the remaining part of the surface in $0<x<1$ is thermally insulated).

Nonlinear boundary conditions also arise in practice. Thus the boundary condition

$$k \frac{\partial u}{\partial n_a} + \sigma(u^4 - U^4)=0 \tag{2}$$

corresponds to a heat radiation in a neighborhood of the temperature U according to the Stefan-Boltzmann law; see Tychonoff-Samarski (1959, p. 181). Collatz (1956) quotes a boundary condition $\partial u/\partial n_a + u^2 = 0$ in connection with a diffusion problem.

In order to be able to formulate the problem as generally as possible, we begin with the notation of 23 I. The new type of boundary condition which contains the normal derivative is given on a certain part of the boundary R_p which we call R_n while, as before, function values $\eta(t, x)$ are given on the remaining part $R_p - R_n$. Here we always have $R_n \subset \partial_1 G$; in the case of a hypercylinder G this means that the boundary conditions with normal derivatives are given on the lateral surface or a part of it but not on the bottom surface.

In order to be able to speak of a normal derivative, we must require that a normal exist at the points of R_n. However this is often not the case for the simplest domains. If G is the unit cube $0<t<1$, $0<x_1<1$, $0<x_2<1$ in the three-dimensional (t, x)-space $(m=2)$, then there is no normal on the edge $x_1 = x_2 = 0$, $0 < t \leqq 1$. However we can overcome the difficulty in this case by simply requiring, if $(\bar{t}, \bar{x})$ is a point in R_n, that there be a straight line segment in the hyperplane $t = \bar{t}$ which has the point $(\bar{t}, \bar{x})$ as an end point and otherwise lies entirely in G_p. This means that there should be a unit vector $p = (p_1, \ldots, p_m)$ so that the points $(\bar{t}, \bar{x} + \alpha p)$ lie in G_p for $0 < \alpha < \delta$ $(\delta > 0)$. This line segment or the associated half-line is also called the *interior normal*. In this sense, as we see immediately, every point on the side of the unit cube described above has an interior normal (in fact infinitely many of them). But if instead of the unit square we take a crescent-shaped domain for the basic surface in the (x_1, x_2)-space, say the open point set D described by the three inequalities

$$x_1^2 + x_2^2 < 1, \qquad x_1^2 + 4x_2^2 > 1, \qquad x_2 > 0$$

and again set up the cylinder $G = (0, 1) \times D$, then the points (t, x) with $x_1 = 1$, $x_2 = 0$, $0 < t \leqq 1$ also have no interior normal in this broader sense.

In order to overcome such difficulties we give the following very general

I. Definition *(normal, $\partial\varphi/\partial n_a$)*. We say that an interior normal exists at the point $(\bar{t}, \bar{x}) \in \partial_1 G$ if there is a sequence of points $(\bar{t}, x_{(k)}) \in G_p$ $(k = 1, 2, \ldots)$ with $x_{(k)} \to \bar{x}$ $(k \to \infty)$. The *outer* normal derivative relative to this normal is then defined by[1]

$$\frac{\partial \varphi(\bar{t}, \bar{x})}{\partial n_a} = -\lim_{k \to \infty} \frac{\varphi(\bar{t}, x_{(k)}) - \varphi(\bar{t}, \bar{x})}{\|x_{(k)} - \bar{x}\|_e}. \tag{3}$$

In this sense, for example, every hypercylinder has an interior normal at every point of its lateral surface. We note that the normal also lies in the hyperplane $t = \bar{t}$ if G is not a hypercylinder.

II. Definition $(R_n, Z(f, R_n), Z_0(f, R_n))$. Suppose we are given a domain G (lying between $t = 0$ and $t = T$), a set $R_n \subset \partial_1 G$ and a function $f \in \mathscr{P}$. Suppose that at every point of R_n there exists an interior normal (thus, a sequence of points as it occurs in I); we choose one. The class $Z(f, R_n)$ includes all functions $\varphi(t, x) \in Z(f)$ which have an outer normal derivative at the points of R_n (relative to these fixed normals), and the class $Z_0(f, R_n)$ includes all functions $\varphi(t, x) \in Z_0(f)$ which are defined and con-

[1] At first glance, it seems strange that we first define an *interior* normal and then an *outer* normal derivative artificially by writing in a minus sign. Of course, we could just as well use the inner normal derivative, then there would be minus signs in many places, including the Gaussian integral theorem. This is evidently considered inelegant by most authors and the outer normal derivative (obtained by means of interior points) considered as the lesser evil. A survey of a dozen German textbooks on analysis revealed a proportion of 10 : 1 in favor of the outer normal, while one author uses both concepts; the outer normal is also preferred almost everywhere in the physics literature.

tinuous in $G_p + R_n$ and have an outer normal derivative at the points of R_n. These classes still depend on the special choice of normals, a fact not expressed in the notation. If the relation to an f is clear from context or unimportant, then we also write $Z(R_n)$ and $Z_0(R_n)$.

III. The Second and Third Boundary Value Problems. We are given $G, R_n \subset \partial_1 G$, $f \in \mathscr{P}$, a function $\eta(t, x)$ defined on $R_p - R_n$ and a function $\vartheta(t, x, z)$ defined for $(t, x) \in R_n$ and arbitrary z. We seek a function $u \in Z(f, R_n)$ which satisfies the differential equation

$$u_t = f(t, x, u, u_x, u_{xx}) \quad \text{in} \quad G_p \tag{4}$$

and the boundary conditions

$$u = \eta \quad \text{on} \quad R_p - R_n, \quad \frac{\partial u}{\partial n_a} + \vartheta(t, x, u) = 0 \quad \text{on} \quad R_n. \tag{5}$$

Every such function u is called a *solution* of the third boundary value problem.

Obviously this problem is meaningful only if η is continuous on $R_p - R_n$. In the following discussions we shall for the most part restrict consideration to this case. For a problem with discontinuous η we call u a solution if $u \in Z_0(f, R_n)$ satisfies the differential equation (4), the second boundary condition (5) and the inequalities[1]

$$\eta_* \leqq u_* \leqq u^* \leqq \eta^* \quad \text{on} \quad R_p - R_n. \tag{6}$$

The problem noted at the beginning of this section is of discontinuous type. The second boundary value problem is a special case of the third boundary value problem; here ϑ is a function which depends only on t and x.

It would be possible to base the following theorems on a boundary condition which is defined more generally than in (5) by a relationship "$\lambda(t, x, u, \partial u/\partial n_a) = 0$ on R_p" where the function λ is monotone non-decreasing in the last variable and is otherwise completely arbitrary. Hence we could combine the two equations (5) into *one* equation. However the more specialized form (5) suffices for all cases of practical importance. Writing the new boundary condition in the form $\partial u/\partial n_a + \vartheta = 0$ (instead of $\partial u/\partial n_a = \vartheta$) is completely optional. It has the advantage that in most problems of mathematical physics $\vartheta(t, x, z)$ is increasing in z. Moreover, some authors prefer to write the boundary condition in the form $u = \zeta(t, x, \partial u/\partial n_a)$. This representation is less general than in (5) (we then need a monotonicity assumption relative to ζ, while ϑ in (5) is completely arbitrary) and it also has other drawbacks.

Now we carry over the most important of the previous theorems to the new problem. We begin with the lemma of Nagumo-Westphal.

[1] In the definition 25 I of η^*, η_*, R_p is to be replaced by $R_p - R_n$.

IV. Lemma. *Suppose $f \in \mathscr{P}$, v, $w \in Z_0(f, R_n)$, $\vartheta(t, x, z)$ is defined in $R_n \times E^1$. Suppose further that*

(α) $v < w$ on[1] $(R_p - R_n)^+$ and on R_∞,
$\partial v/\partial n_a + \vartheta(t, x, v) < \partial w/\partial n_a + \vartheta(t, x, w)$ on R_n;

(β) $Pv < Pw$ in G_p.

Then

$$v < w \quad in \quad G_p + R_n.$$

For $R_n = \emptyset$, IV coincides exactly with Lemma 24 IV. The proof given there needs just a small addition. We want to show that 24 I (β) is inconsistent with the present hypothesis (α). This is immediately clear for $(\bar{t}, \bar{x}) \in R_p - R_n$, exactly as in 24 IV. But if we had $(\bar{t}, \bar{x}) \in R_n$, we would have $w - v = 0$ at this point and $w - v \geqq 0$ on the hyperplane $t = \bar{t}$, i.e., $\partial(w - v)/\partial n_a \leqq 0$ at this point. On the other hand $\vartheta(\bar{t}, \bar{x}, w) = \vartheta(\bar{t}, \bar{x}, v)$ and thus, by (α), $\partial v/\partial n_a < \partial w/\partial n_a$ at the point $(\bar{t}, \bar{x})$. Thus in every case we obtain a contradiction from 24 I (β).

Since we will still occasionally need this simple argument, we summarize it in a somewhat different form:

(γ) *For two functions $\varphi, \psi \in Z_0(R_n)$ and arbitrary $\vartheta(t, x, z)$, the statement*

$$\varphi < \psi \quad on\ (R_p - R_n)^+ \ and\ on\ R_\infty,$$

$$\partial \varphi/\partial n_a + \vartheta(t, x, \varphi) < \partial \psi/\partial n_a + \vartheta(t, x, \varphi) \quad on\ R_n$$

is inconsistent with 24 I (β).

Remark 24 V holds without change. Corollary 24 VI corresponds to the following

V. Corollary. Suppose $v, w \in Z_0(f, R_n)$, $\vartheta(t, x, z)$ is strictly monotone increasing in z, and the function f satisfies the hypothesis of 24 VI. Then

(α) $v \leqq w$ on $(R_p - R_n)^+$ and on R_∞,
$\partial v/\partial n_a + \vartheta(t, x, v) \leqq \partial w/\partial n_a + \vartheta(t, x, w)$ on R_n;

(β) $Pv \leqq Pw$ in G_p

imply the inequality

$$v \leqq w \quad \text{in} \quad G_p + R_n.$$

The details of the extension of the proof of 24 VI made necessary by the new type of boundary condition are left to the reader. For this the inequality

$$\partial v/\partial n_a + \vartheta(t, x, v) \leqq \partial w/\partial n_a + \vartheta(t, x, w) < \partial(w + \varrho)/\partial n_a + \vartheta(t, x, w + \varrho)$$

is needed together with IV (γ).

The restriction that $\vartheta(t, x, z)$ be *strictly* monotone increasing in z is essential for the proof to go through. This does not include, among

[1] Of course this means that now in 23 IV we consider only sequences with $(t_k, x_{(k)}) \to (\bar{t}, \bar{x}) \in R_p - R_n$.

others, the important case of a *boundary condition of second kind* in which the values of the normal derivative are given on a part of the boundary. With regard to such boundary conditions we prove a further

VI. Corollary. For two functions $v, w \in Z_0(f, R_n)$ and a function $\vartheta(t, x, z)$ which is monotone increasing in the variable z, suppose that

(α) $v \leqq w$ on $(R_p - R_n)^+$ and on R_∞,
$\partial v/\partial n_a + \vartheta(t, x, v) \leqq \partial w/\partial n_a + \vartheta(t, x, w)$ on R_n;

(β) $Pv \leqq Pw$ in G_p.

Then again we have

$$v \leqq w \quad \text{in} \quad G_p + R_n,$$

provided the following three conditions can be satisfied:

(γ_1) We have (as in 24 VI)

$$f(t, x, w+z, w_x, w_{xx}) - f(t, x, w, w_x, w_{xx}) \leqq \omega(t, z) \quad \text{for} \quad z > 0,$$

but with a function $\omega \in \mathscr{E}_4$.

(γ_2) There is a bounded function $h(x) \in Z_0(R_n)$ with bounded first and second derivatives and independent of t with the property

$$\partial h/\partial n_a > 0 \quad \text{on} \quad R_n.$$

(γ_3) For every $\delta > 0$ there exists a $\beta > 0$ such that

$$f(t, x, w+z+\lambda h, w_x + \lambda h_x, w_{xx} + \lambda h_{xx}) - f(t, x, w+z, w_x, w_{xx}) < \delta$$

for $(t, x) \in G_p$, $z > 0$ and $0 < \lambda < \beta$ [as long as the arguments lie in $D(f)$].

For the proof we refer back to 24 I and let $\varphi = v$, $\psi = w + \varrho + \lambda h$. First we choose $\varepsilon > 0$ arbitrarily, then $\delta > 0$ and ϱ according to 10 I, $h(x)$ according to (γ_2) and then $\beta > 0$ according to (γ_3); moreover let β be so small that $0 < \delta/2 \leqq \varrho + \lambda h \leqq 2\varepsilon$ for $0 < \lambda < \beta$. Thus $\varphi < \psi$ on $(R_p - R_n)^+$ and on R_∞. Further, by (α), (γ_2) and the monotonicity of ϑ,

$$\begin{aligned} \partial\varphi/\partial n_a + \vartheta(t, x, \varphi) &\leqq \partial w/\partial n_a + \vartheta(t, x, w) < \partial(w + \lambda h)/\partial n_a + \vartheta(t, x, \psi) \\ &= \partial\psi/\partial n_a + \vartheta(t, x, \psi). \end{aligned}$$

Then, according to IV (γ), the case 24 I (β) does not hold.

We still have to verify the hypothesis of 24 I. If $\varphi = \psi$, $\varphi_x = \psi_x$, $\varphi_{xx} \leqq \psi_{xx}$, and thus $v = w + \varrho + \lambda h$, $v_x = w_x + \lambda h_x$, $v_{xx} \leqq w_{xx} + \lambda h_{xx}$ and $0 < \lambda < \beta$, then by (β) (γ_1) (γ_3) we have

$$\begin{aligned} \varphi_t &= Pv + f(t, x, v, v_x, v_{xx}) \leqq Pw + f(t, x, v, v_x, v_{xx}) \\ &\leqq w_t + f(t, x, w + \varrho + \lambda h, w_x + \lambda h_x, w_{xx} + \lambda h_{xx}) - f(t, x, w, w_x, w_{xx}) \\ &\leqq w_t + f(t, x, w + \varrho, w_x, w_{xx}) - f(t, x, w, w_x, w_{xx}) + \delta \\ &\leqq w_t + \delta + \omega(t, \varrho) < w_t + \varrho' = \psi_t, \end{aligned}$$

q.e.d. Thus we have the case 24 I (α), i.e., we have

$$v < w + \varrho + \lambda h \leqq w + 2\varepsilon ,$$

whence the assertion is obtained immediately.

VII. Remarks. In order to extend the Nagumo Lemma to *weakly monotone increasing* ϑ and hence to the second boundary value problem in VI, we needed three new hypotheses (γ_1) to (γ_3) about whose nature we make several remarks.

For (γ_1). This sharpening of the corresponding hypothesis of Theorem V is insignificant.

For (γ_2). In many cases it is very simple to find a suitable function $h(x)$. Let us examine the case of the hypercylinder $G_p = (0, T] \times D$ with $D \subset E^m$ somewhat more closely. If $m = 1$ and $D = (a, b)$, then we can take $h(x) = (x - x')^2$, $a < x' < b$; if $D = (a, \infty)$, then we can take

$$h(x) = -\arctan(x - a) .$$

Both hold for arbitrary R_n and arbitrarily defined interior normals.

The function $h(x) = (x - x')^2 = (x_1 - x_1')^2 + \cdots + (x_m - x_m')^2$ can also be put to good use for $m > 1$. Let us consider the "normal case" where the normal points associated with the point $(\bar{t}, \bar{x}) \in R_n$ lie on a straight line and are thus of the form $(\bar{t}, x_{(k)}) = (\bar{t}, \bar{x} + \alpha_k p)$ $(p \in E^m, \alpha_k \to +0)$. If we succeed in finding a point x' such that for all points $(\bar{t}, \bar{x}) \in R_n$ the two vectors $x' - \bar{x}$ and p form an angle smaller than $\pi/2$ (the scalar product $(x' - \bar{x})\, p > 0$), then $h = (x - x')^2$ is a suitable function. Among the situations in which this approach is possible are the cases of convex domains D where for every point $(\bar{t}, \bar{x}) \in R_n$ we choose a p such that the hyperplane perpendicular to p through the point $\bar{x}$ is a supporting plane for D (i.e., lies completely to one side of the domain D). In this case we can even choose $x' \in D$ arbitrarily.

For (γ_3). If f is linear,

$$f(t, x, z, p, r) = g + hz + \sum_{\mu=1}^{m} h_\mu p_\mu + \sum_{\lambda, \mu=1}^{m} k_{\lambda\mu} r_{\lambda\mu} , \tag{7}$$

and if the coefficients are bounded functions of t and x, then (γ_3) holds.

To demonstrate the difficulties in the nonlinear case, we consider the example

$$f(t, x, z, p, r) = \varkappa(t, x, z)\,(r_{11} + \cdots + r_{mm}) . \tag{8}$$

Then the inequality under consideration is (without the arguments t, x)

$$\varkappa(w + z + \lambda h)\,(\Delta w + \lambda \Delta h) - \varkappa(w + z)\,\Delta w < \delta .$$

If $\varkappa$ is bounded and continuous, then the term $\lambda\varkappa(w+z+\lambda h)\,\Delta h$ can be made arbitrarily small. Now it is a matter of bounding the difference

$$[\varkappa(w+z+\lambda h)-\varkappa(w+z)]\,\Delta w$$

for positive z. The hypothesis (γ_1) is quite similar in our case, namely,

$$[\varkappa(w+z)-\varkappa(w)]\,\Delta w \leqq \omega(t,z) \quad \text{for} \quad z>0\,.$$

Then for (γ_1) and (γ_3) we have almost the same difficulties already discussed thoroughly in the uniqueness theorem 29 V. Thus each of the three hypotheses (α_2) to (α_4) of 29 V is sufficient for the validity not only of (γ_1) but also of (γ_3); the same is true for 29 V (α_1) and (α_5) if the function $h(t)$ occurring there is constant.

We describe briefly another approach which also leads to a result useful for the second boundary value problem. We use an idea due to Szarski (1955) which consists of requiring validity of the differential inequality on R_n [Nickel (1959, Example 5) uses it for the Crocco boundary layer differential equation].

VIII. Corollary. Let $\bar{G}=[0,T]\times\bar{D}, D\subset E^m$, be a hypercylinder and $R_n=[0,T]\times E$, $E\subset\partial D$. Suppose the domain D has a tangent plane at least at the boundary points belonging to E and has an inwardly directed normal in the classical sense[1]. The hypotheses of V are modified as follows. We let the functions v, w have partial derivatives of first order in t and x_μ and derivatives of second order in x_μ on R_n. Let 23 III (β) and V (β) hold on R_n (where in 23 III (β), h is restricted by $x+h\in D+E$). Let f also be defined on R_n and satisfy there the condition of 24 VI. On the other hand let the function $\vartheta(t,x,z)$ be only weakly monotone increasing in z. Then the assertion of V holds.

To be convinced that this assertion is true, we return to the main idea of the proof of V. As before we need to show only that the inequality $\partial v/\partial n_a<\partial w/\partial n_a=\partial(w+\varrho)/\partial n_a$ holds at a boundary point $(\bar{t},\bar{x})\in R_n$ at which $v=w+\varrho$. In V this inequality was a consequence of the strict monotonicity of $\vartheta(t,x,z)$; here, on the contrary, ϑ is weakly monotone increasing and we obtain only $\partial v/\partial n_a\leqq\partial(w+\varrho)/\partial n_a$. With the help of the new hypotheses we now show that we cannot have the equality sign here. Since the function $w+\varrho-v$ has a minimum (with respect to x) at the point $(\bar{t},\bar{x})$, all the derivatives are zero in the tangential direction. If the outer normal derivative were also 0, then we would have $(w+\varrho-v)_x=0$ and hence (see 23 III (β)) $(w+\varrho-v)_{xx}\geqq 0$ and moreover $(w+\varrho-v)_t\leqq 0$. However with the help of V (β) this yields the inequality $(w+\varrho-v)_t>0$, and thus a contradiction, exactly as in the proof of 24 VI.

Thus the essential difference between VI and VIII is that on the one hand we have two new hypotheses VI (γ_2) (γ_3) (of which the first is not serious), and on the other hand we require in addition the existence of the tangent plane and differentiability conditions at the points of R_n. Here we should take note of the fact, already significant in the discussion under VII, that in many cases, particularly for quasi-linear differential equations, the condition VI (γ_3) requires no new restrictions (or only insignificant ones) because the principal difficulty is already contained in the uniqueness condition VI (γ_1) which indeed must hold in both cases.

[1] We need only: For every $\bar{x}\in E$ there is an inner normal vector $p=(p_1,\ldots,p_m)\neq 0$ with the property that for every vector $q\in E^m$ which encloses an angle $<\pi/2$ with p (scalar product $pq>0$), there exists a sequence of points $\bar{x}+\delta_k q\in D$, $\delta_k\to+0$ $(k\to\infty)$.

Conversion of the result of Section 25 to the third boundary value problem encounters no essential new difficulties. It can be done in different ways corresponding to the various methods of proof in IV, V, VI and VIII.

We give only a sample of the possible theorems.

IX. Theorem. Theorem 25 III also holds for functions $u, v \in Z_0(f, R_n)$ if the hypotheses (β), (γ), (δ) there are retained and the hypothesis (α) is replaced by

(α) $-\bar{\varrho} < v - u < \varrho$ *on* $(R_p - R_n)^+$ *and on* R_∞,
$\partial u/\partial n_a + \vartheta(t, x, u) = 0$ *on* R_n,
$\partial v/\partial n_a + \vartheta(t, x, v + \bar{\varrho}) > 0$, $\partial v/\partial n_a + \vartheta(t, x, v - \varrho) < 0$ *on* R_n.

Here the function $\vartheta(t, x, z)$ is arbitrary.

The proof given there needs just an addition. By (α) for $\varphi = v - \varrho$, $\psi = u$ we have

$$\partial\varphi/\partial n_a + \vartheta(t, x, \varphi) < 0 = \partial\psi/\partial n_a + \vartheta(t, x, \psi),$$

and thus again IV (γ).

X. Uniqueness Theorem. *The uniqueness theorem* 25 V, *including the assertions about unbounded domains and the continuous dependence of the solution on the boundary values and on the right hand side of the differential equation, also remains valid for the third boundary value problem if* $\vartheta(t, x, z)$ *is strictly monotone increasing in z and if solutions or approximate solutions from the class* $Z(f, R_n)$ *are considered (this means that* η *must be uniformly continuous on bounded subsets of* $R_p - R_n$*). However it is assumed here that the approximate solution v also satisfies the boundary condition on* R_n *exactly*[1] (the definition of continuous dependence given in 25 V is to be modified as follows: $|v - u| < \delta$ on $(R_p - R_n)$ and on R_∞, $\partial v/\partial n_a + \vartheta(t, x, v) = 0$ on R_n, $|Pv| = 0$ or $< \delta$ imply $|v - u| < \varepsilon$).

Of course the sharpened versions 25 VI and Remark 25 VII also hold here.

Thus for strictly monotone increasing $\vartheta(t, x, z)$ the uniqueness theorem 25 V can be carried over to the third boundary value problem in full generality. For in Theorem 25 III only the condition (α) was replaced by the new condition IX (α) and the latter is true if u and v are solutions of the boundary value problem and ϱ, $\bar{\varrho}$ are positive functions.

Moreover, Theorem X is already contained in essence in Corollary V.

If we want to extend the uniqueness assertions to weakly monotone increasing ϑ and hence to the second boundary value problem, we first have to change the estimation theorem IX appropriately, namely, in such

[1] Under special assumptions on ϑ, say $\vartheta(t, x, z) = \zeta(t, x,)z$ with $\zeta(t, x) \geqq \alpha > 0$, we also have continuous dependence when $|\partial v/\partial n_a + \vartheta(t, x, v)| < \delta$ is allowed.

a way that equality signs are allowed in the last two conditions of IX (α):

$$\partial v/\partial n_a + \vartheta(t, x, v+\bar{\varrho}) \geqq 0, \qquad \partial v/\partial n_a + \vartheta(t, x, v-\varrho) \leqq 0 .$$

Then sharper requirements are needed in other places. We can accomplish this aim either with the method of proof used in VI or that used in VIII. We shall be satisfied with this remark and just state the following theorem (already contained in VI).

XI. Uniqueness Theorem. *The third boundary value problem has at most one solution in the class $Z(f, R_n)$ under the following hypotheses: The function $\vartheta(t, x, z)$ is (weakly) monotone increasing in z;* VI (γ_2) *holds; for every solution $w \in Z(f, R_n)$,* VI (γ_1) *and* (γ_3) *hold; for an unbounded domain conditions are prescribed at infinity such that $u = w$ on R_∞ for two solutions u, w.*

Theorem 25 VIII can also be easily adapted to the new situation. We restrict ourselves to demonstrating this in the special case 25 IX.

XII. Theorem. *Suppose the function $\vartheta(t, x, z)$ is strongly* [*or weakly*] *monotone increasing in z. Suppose the hypotheses of* 25 IX *hold for the functions $f \in \mathscr{P}$, $u, v \in Z_0(f, R_n)$ and $\varrho \in Z_0(R_n)$, where only* (α) *is changed:*

(α) $|v-u| \leqq \varrho$ *on* $(R_p - R_n)^+$ *and on* R_∞;
$\partial u/\partial n_a + \vartheta(t, x, u) = 0$, $\partial v/\partial n_a + \vartheta(t, x, v) = 0$ *on* R_n,
$\partial \varrho/\partial n_a \geqq 0$ [*or* > 0] *on* R_n.

Then we have

$$|v-u| \leqq \varrho \quad in \quad G_p + R_n .$$

We obtain this immediately from the reasoning in 25 IX together with the modified hypothesis (α) and Remark IV (γ).

Finally we turn to the results of Sections 28 and 29. On the basis of the preceding theorems it is possible to carry them over to the third boundary value problem in great generality.

Let us consider the uniqueness theorems 28 II, III, VI, VIII. Their proofs were all based on an estimate for the difference of two solutions $|v-u| < \varrho(t, x)$ by Theorem 25 IX, where the difficulty was in constructing a suitable function ϱ. Because of Theorem XII we can immediately carry over the proofs to the third boundary value problem, but because of XII (α') there is a new condition

$$\partial \varrho/\partial n_a > 0 \quad \text{on} \quad R_n \tag{9}$$

to be imposed on ϱ [provided $\vartheta(t, x, z)$ is weakly monotone increasing]. Thus we will examine the functions ϱ constructed in Section 28 to see for which domains and boundary sets R_n they satisfy the inequality (9). Of course this procedure can be improved in many ways such as by modifying ϱ suitably, say, by additional additive terms; we shall however not go into this.

XIII. Unbounded Domain. *Theorem* 28 II *remains valid for the third boundary value problem (and of course for solutions belonging to the class* $Z(R_n)$*) if* $\varphi(t, x)$ *also satisfies the inequality* $\partial\varphi/\partial n_a > 0$ *on* R_n *and if:*

(α) *The function* $\vartheta(t, x, z)$ *is (weakly) monotone increasing in* z; *at every point* $(\bar{t}, \bar{x}) \in R_n$ *the interior normal is defined by a sequence of points* $x_{(k)} = \bar{x} + \alpha_k p$ $(p \in E^m, \alpha_k > 0, \alpha_k \to 0)$ *lying on a straight line (see* I*).*

We can give $\partial\varphi/\partial n_a$ explicitly for Theorem 28 III:

$$\partial\varphi/\partial n_a = -2xp\beta(t)\,\varphi(t, x) \qquad (xp = x_1 p_1 + \cdots + x_m p_m).$$

Thus Theorem 28 III *also holds for a third boundary value problem defined according to* (α) *if*

$$p\bar{x} < 0 \quad on \quad R_n \tag{10}$$

holds for the interior normal direction p belonging to the point $(\bar{t}, \bar{x}) \in R_n$.

If for example $m = 1$ and $G = (0, T) \times (a, \infty)$, then $\bar{x} = a$ for all $(\bar{t}, \bar{x}) \in R_n$ and further $p > 0$. Thus we have to take care (via parallel shift) that $a < 0$. Hence the uniqueness assertions 28 III also hold (in the case $m = 1$) for the second and third boundary value problem provided $G = (0, T) \times (a, \infty)$ or, more generally, provided G is defined by the inequalities $0 < t < T$, $a(t) < x < \infty$, where the function $a(t)$ is continuous in $[0, T]$.

The condition (10) is also easy to verify in multidimensional cases and can often be fulfilled by a translation. Geometrically it means that the vector directed from the point $\bar{x}$ to the origin forms an angle $< \pi/2$ with the interior normal vector. This is surely true, for example, if $G = (0, T) \times D$ with $0 \in D \subset E^m$ and if the hyperplane perpendicular to p through the point $\bar{x}$ has no point in common with D (say D convex).

XIV. Discontinuous Boundary Values. Here by differentiating the function $\psi(t, x)$ of (28.18) we obtain

$$\partial\psi(t, x)/\partial n_a = 2\alpha(t) \int_{K_0} \gamma(\xi)\, e^{\beta(t)(x-\xi)^2}\, \beta(t)\,(x - \xi)\, p\, d\xi\,,$$

and thus the following sufficient condition for positivity of the normal derivative

$$(\xi - \bar{x})\, p > 0 \quad \text{for} \quad (\bar{t}, \bar{x}) \in R_n, \quad \xi \in K_0\,. \tag{11}$$

For an unbounded domain the condition (10) noted in XIII is added; however, it is contained in (11) if K_0 contains the origin.

Theorem 28 VIII *remains valid for a third boundary value problem with discontinuous boundary values if* (11) *and* XIII (α) *hold.*

Here the definition of a solution given in 28 VI obviously must be changed. We now have $R^* = R_p - R_n - \{0\} \times N$ and $u \in Z_0(f, R_n)$ continuous in $G_p + R^*$.

If $G=(0, T)\times D$ is a hypercylinder and K_0 is entirely contained in D, then the criterion given at the end of XIII is also sufficient for the validity of (11). *In particular, if* $m=1$ *and* $D=(a, b)$ *or* $D=(a, \infty)$, *then the assertions* 28 VIII *are also valid for the second and third boundary value problem for arbitrary* R_n.

XV. The Heat Equation was thoroughly investigated in Section 29. Most of those results remain valid for the third boundary value problem.

The uniqueness theorems 29 IV for the general heat equation and 29 V for the equation (29.9) also hold for the third boundary value problem if the function $\vartheta(t, x, z)$ is strictly monotone increasing in z. In physical terms: If a heat exchange law is prescribed on part of the surface of a body such that at every point a higher temperature corresponds to a greater outflow of heat (or a smaller inflow of heat) while the temperature is prescribed on the rest of the surface, then the previous uniqueness assertions hold.

These assertions were special cases of the uniqueness theorem 25 V, VI which, however, remains valid in full generality for the third boundary value problem, according to X.

On the other hand if we are dealing with the second boundary value problem (or more generally the third boundary value problem with weakly monotone increasing ϑ), then according to Theorem XI we need the additional hypothesis VI (γ_2), and in 29 IV and 29 V the inequalities in which a function $h(t)$ occurs must hold even for a constant function $h(t)$. The latter is needed for the proof of 29 VI (γ_3).

Estimation and uniqueness theorems similar to our theorems IX to XI were proved by Szarski (1955, 1959); further results were proved by Mlak (1958) for boundary value problems of third type. Both authors consider systems of differential equations but linear boundary conditions. Collatz (1956) extended the Lemma of Nagumo to nonlinear boundary conditions (see IV) and gave corresponding estimation theorems. Krzyżański (1960) proved uniqueness theorems for linear differential equations and unbounded domains. There one finds a condition which bears a similarity to VI (γ_2) but is considerably sharper; see also Kaplan (1963). Boundary conditions in which u_t also occurs were considered by Ficken (1952).

XVI. Other Boundary Value Problems. We outline briefly two boundary value problems which arise from physical situations.

(a) *Heat conduction in multicomponent bodies.* The heat flow in a body consisting of several components with different physical properties gives rise to boundary value problems with "inner boundaries" or "interfaces". These inner boundaries correspond to the surfaces separating the different components. The conditions at inner boundary points are obtained from the law of conservation of energy. A simple example in the case of two components is an equation

$$k_1 \frac{\partial u}{\partial n_1} + k_2 \frac{\partial u}{\partial n_2} = 0,$$

where the k_i are constants and the derivatives are interior normal derivatives with respect to the first resp. second component.

Monotonicity and uniqueness theorems for problems of this kind were proved by Walter (1968 b).

(b) *Heat conduction of a body in a fluid.* A physical interpretation of a first boundary value problem with constant boundary values U on $\partial_1 G$ is a body immersed in a fluid of constant temperature U. Here it is assumed that the heat flow does not change the temperature of the fluid, i.e., that the fluid has infinite thermal capacity. If this assumption is dropped (that is, if the bathtub is finite), the effect of the body on the fluid can no longer be neglected. There are then two unknown functions, the temperature $u(t, x)$ of the body and the temperature $U(t)$ of the fluid.

In the corresponding mathematical problem one has a parabolic differential equation for u, together with initial conditions $u(0, x) = \eta(x)$ and boundary conditions on $(0, T] \times \partial D$ such as $u(t, x) = U(t)$ in the case of the first boundary value problem. There are two additional equations for U, an initial condition $U(0) = U_0$ and an integro-differential equation

$$\int_{\partial D} \frac{\partial u(t, x)}{\partial n} do + \gamma U'(t) = Q(t),$$

which relates the flow of heat through the surface of the body D with the change of temperature $U'(t)$.

Linear problems of this kind were first formulated by Freud (1955) and investigated by him and Adler (1956, 1959). Becker (1967) proved monotonicity and uniqueness theorems for nonlinear problems; see also Walter (1968 b).

32. Systems of Parabolic Differential Equations

I. Notation ($f \in \mathscr{P}^n$, *defect* $\mathbf{P}, Z(f), Z_0(f), Z(f, R_n), Z_0(f, R_n), \varphi < \psi$ *on* $R_p^+, \ldots$). Let $\boldsymbol{u} = (u_1, \ldots, u_n)$ be a vector in E^n $(n \geqq 1)$ whose components u_ν are functions of t and $x = (x_1, \ldots, x_m) \in E^m$ $(m \geqq 1)$. Let $u_{\nu,t}$ denote the partial derivative $\partial u_\nu(t, x)/\partial t$, $u_{\nu,x}$ the gradient vector of the component u_ν,

$$u_{\nu,x}(t, x) = (\partial u_\nu(t, x)/\partial x_\mu) \quad (\mu = 1 \ldots, m),$$

and $u_{\nu,xx}$ the $m \times m$-matrix of the partial derivatives of second order

$$u_{\nu,xx} = (\partial^2 u_\nu(t, x)/\partial x_\lambda \partial x_\mu) \quad (\lambda, \mu = 1, \ldots, m).$$

Here we exclusively consider systems of parabolic differential equations of the form

$$\left.\begin{array}{c} u_{\nu,t}=f_\nu(t,x,u_1,\ldots,u_n,u_{\nu,x},u_{\nu,xx})=f_\nu(t,x,\boldsymbol{u},u_{\nu,x},u_{\nu,xx}) \\ (\nu=1,\ldots,n)\,, \end{array}\right\} \tag{1}$$

which we abbreviate as

$$\boldsymbol{u}_t=\boldsymbol{f}(t,x,\boldsymbol{u},\boldsymbol{u}_x,\boldsymbol{u}_{xx})\,. \tag{1'}$$

They are of special form in that, although all functions $u_1,\ldots,u_n$ may appear in each equation, the derivatives of the νth component u_ν may appear only in the νth equation. The "defect" $\boldsymbol{P u}$ with respect to the equation (1) is now a vector function whose components $(\boldsymbol{P u})_\nu$ are given by the equation

$$(\boldsymbol{P u})_\nu=u_{\nu,t}-f_\nu(t,x,\boldsymbol{u},u_{\nu,x},u_{\nu,xx})\,. \tag{2}$$

Suppose that every component $f_\nu=f_\nu(t,x,z,p,r)$ – here z is an n-dimensional vector, in contrast to 23 II – is monotone increasing in r in the sense of 23 II and has a domain of definition $D(f_\nu)=M_\nu\times M_{\nu r}$ where, as in 23 II, M_ν is a set in (t,x,z,p)-space and $M_{\nu r}$ is a set in r-space. We abbreviate this as $\boldsymbol{f}\in\mathscr{P}^n$. The notation $\boldsymbol{\varphi}\in Z$ or Z_0, $Z(R_n)$, $Z_0(R_n)$ means that every component $\varphi_\nu(t,x)$ is in the class defined according to 23 III or 31 II; if, moreover, $D(f_\nu)$ is such that we can "substitute" $\boldsymbol{\varphi}$ in f_ν, then $\boldsymbol{\varphi}\in Z(\boldsymbol{f})$ or $Z_0(\boldsymbol{f})$, $Z(\boldsymbol{f},R_n)$, $Z_0(\boldsymbol{f},R_n)$.

The inequalities defined in 23 IV are correspondingly defined for vector functions. Thus for example "$\boldsymbol{\varphi}\leqq\boldsymbol{\psi}$ on R_∞" means that for every sequence of points $(t_k,x_{(k)})\in G_p$ with $t_k\to\bar{t}+0$, $\|x_{(k)}\|_e\to\infty$ $(k\to\infty)$ and for every ν the inequality $\limsup_{k\to\infty}[\psi_\nu(t_k,x_{(k)})-\varphi_\nu(t_k,x_{(k)})]\geqq 0$ holds. Finally, we recall the laws for manipulating with vectors, especially 6 I.

In the problem before us, the experience of Chapter II, where (starting with Section 11) we faced the same situation for ordinary differential equations, will be of great use in extending the present results to systems of parabolic differential equations. From the numerous possible theorems, we will make a useful selection below. The fundamental results for parabolic systems were discovered by Szarski (1955, 1959) and Mlak (1957). Further extensions were made by Besala (1961a, 1963, 1963/64, 1965), Szarski (1964), Stys' (1965), Kusano (1965) and Łojczyk-Królikiewicz (1966); Besala and Kusano also gave applications to existence theory. Literature on systems arising in the mathematical theory of diffusion is quoted in 33 IX, X.

We begin with the n-dimensional setting of 24 I.

II. Lemma. *For two functions $\boldsymbol{\varphi},\boldsymbol{\psi}\in Z_0$ suppose that: If $\boldsymbol{\varphi}\leqq\boldsymbol{\psi}$, $\varphi_\nu=\psi_\nu$, $\varphi_{\nu,x}=\psi_{\nu,x}$, $\varphi_{\nu,xx}\leqq\psi_{\nu,xx}$ for a fixed ν and a fixed point in G_p, then we have $\varphi_{\nu,t}<\psi_{\nu,t}$ at this point.*

Then we have exactly one of the following two cases:

(α) $\varphi < \psi$ *in* G_p;

(β) *there exists a maximal* $\bar{t}$, $0 \leqq \bar{t} < T$, *such that* $\varphi < \psi$ *for all points in* G_p *with* $0 < t \leqq \bar{t}$;

thus there is an index ν *and a sequence of points* $(t_k, x_{(k)}) \in G_p$ *where the* t_k *are strictly decreasing to* $\bar{t}$ *and*

$$\text{either} \quad (t_k, x_{(k)}) \to (\bar{t}, \bar{x}) \in R_p \quad \text{or} \quad \|x_{(k)}\|_e \to \infty$$

such that we have

$$\varphi_\nu(t_k, x_{(k)}) \geqq \psi_\nu(t_k, x_{(k)}) \quad \text{for} \quad k = 1, 2, 3, \ldots.$$

The changes in the proof of 24 I are insignificant and are left to the reader.

We can also convert 24 II, III correspondingly. We only make the remark that in Lemma II the following hypothesis suffices in place of $\varphi, \psi \in Z_0$: φ, ψ are continuous in G_p and their νth components satisfy the differentiability conditions 23 III (α) (β) at all points at which $\varphi \leqq \psi$ and $\varphi_\nu = \psi_\nu$.

III. The Third Boundary Value Problem. We give the following theorems in a form applicable to the third boundary value problem with boundary conditions

$$u_\nu(t, x) = \eta_\nu(t, x) \quad \text{on } R_p - R_n, \qquad \partial u_\nu / \partial n_a + \vartheta_\nu(t, x, \boldsymbol{u}) = 0 \quad \text{on } R_n \tag{3}$$

($\nu = 1, \ldots, n$) or, briefly,

$$\boldsymbol{u} = \boldsymbol{\eta} \quad \text{on } R_p - R_n, \qquad \partial \boldsymbol{u} / \partial n_a + \boldsymbol{\vartheta}(t, x, \boldsymbol{u}) = 0 \quad \text{on } R_n. \tag{3'}$$

This contains the first boundary value problem for $R_n = \emptyset$. Here we make the assumption that the set R_n is the same for all n components. This is not necessary for the validity of our theorems and was assumed only for the sake of convenient notation.

In applications ϑ_ν usually depends only on u_ν. Since the more general case, in which $\vartheta_\nu(t, x, z)$ depends on all $z_\varkappa$ ($\varkappa = 1, \ldots, n$), presents no difficulties, we shall treat this situation. However we always assume that ϑ_ν is monotone decreasing in the variables $z_\varkappa$ with $\varkappa \neq \nu$, i.e., that $-\boldsymbol{\vartheta}(t, x, z)$ is quasi-monotone increasing in z.

The following three theorems correspond to the results of Section 12.

IV. Theorem. *If* $\boldsymbol{f} \in \mathscr{P}^n$, $\boldsymbol{v}, \boldsymbol{w} \in Z_0(\boldsymbol{f}, R_n)$ *and the function* $-\boldsymbol{\vartheta}(t, x, z)$ *is quasi-monotone increasing in* $\boldsymbol{z}$, *then*

(α) $\boldsymbol{v} < \boldsymbol{w}$ *on* $(R_p - R_n)^+$ *and on* R_∞,
$\partial \boldsymbol{v} / \partial n_a + \boldsymbol{\vartheta}(t, x, \boldsymbol{v}) < \partial \boldsymbol{w} / \partial n_a + \boldsymbol{\vartheta}(t, x, \boldsymbol{w})$ *on* R_n;

(β) $v_{\nu,t} - f_\nu(t, x, \boldsymbol{z}, v_{\nu,x}, v_{\nu,xx}) < w_{\nu,t} - f_\nu(t, x, \boldsymbol{w}, w_{\nu,x}, w_{\nu,xx})$,

provided that at the point $(t, x) \in G_p$ $\mathbf{v} \leqq \mathbf{z} \leqq \mathbf{w}$, $z_\nu = v_\nu$ *and* $(t, x, \mathbf{z}, v_{\nu,x}, v_{\nu,xx}) \in D(f_\nu)$ $(\nu = 1, \ldots, n)$;
imply the inequality

$$\mathbf{v} < \mathbf{w} \quad \text{in} \quad G_p + R_n.$$

The theorem remains valid with the hypothesis (β') instead of (β):

$$(\beta') \quad v_{\nu,t} - f_\nu(t, x, \mathbf{v}, v_{\nu,x}, v_{\nu,xx}) < w_{\nu,t} - f_\nu(t, x, \mathbf{z}, w_{\nu,x}, w_{\nu,xx}),$$

provided that at the point $(t, x) \in G_p$ $\mathbf{v} \leqq \mathbf{z} \leqq \mathbf{w}$, $z_\nu = w_\nu$ *and* $(t, x, \mathbf{z}, w_{\nu,x}, w_{\nu,xx}) \in D(f_\nu)$ $(\nu = 1, \ldots, n)$.

The t-derivatives can also be replaced by Dini derivatives under the same circumstances as for ordinary differential equations.

The application of Lemma II with $\boldsymbol{\varphi} = \mathbf{v}$, $\boldsymbol{\psi} = \mathbf{w}$ to the proof is allowed since, by (β) and the monotonicity of f_ν in the last variable, $v_{\nu,t} < w_{\nu,t}$ follows from $\mathbf{v} = \mathbf{w}$, $v_\nu = w_\nu$, $v_{\nu,x} = w_{\nu,x}$, $v_{\nu,xx} \leqq w_{\nu,xx}$. Thus the theorem is proved if we can show the following:

(γ) *For two functions* $\boldsymbol{\varphi}, \boldsymbol{\psi} \in Z_0(R_n)$ *and a function* $\boldsymbol{\vartheta}(t, x, \mathbf{z})$ *for which* $-\boldsymbol{\vartheta}(t, x, \mathbf{z})$ *is quasi-monotone increasing in* $\mathbf{z}$, *the assertion*
$\boldsymbol{\varphi} < \boldsymbol{\psi}$ *on* $(R_p - R_n)^+$ *and on* R_∞,
$\partial \boldsymbol{\varphi}/\partial n_a + \boldsymbol{\vartheta}(t, x, \boldsymbol{\varphi}) < \partial \boldsymbol{\psi}/\partial n_a + \boldsymbol{\vartheta}(t, x, \boldsymbol{\psi})$ *on* R_n
is incompatible with II (β).

This is obtained from reasoning analogous to 31 IV (γ). If $\boldsymbol{\varphi} \leqq \boldsymbol{\psi}$ and $\varphi_\nu = \psi_\nu$ at a boundary point $(\bar{t}, \bar{x}) \in R_n$, then the second inequality in (γ) implies

$$\partial \varphi_\nu/\partial n_a < \partial \psi_\nu/\partial n_a + \vartheta_\nu(\bar{t}, \bar{x}, \boldsymbol{\psi}) - \vartheta_\nu(\bar{t}, \bar{x}, \boldsymbol{\varphi}) \leqq \partial \psi_\nu/\partial n_a,$$

the latter due to the quasi-monotonicity of $-\boldsymbol{\vartheta}$. However this is inconsistent with the fact that the function $\psi_\nu - \varphi_\nu$ vanishes at the point $(\bar{t}, \bar{x}) \in R_n$ and has a minimum relative to x for fixed $t = \bar{t}$.

What we said for 12 II is true for the applicability of this theorem. The reader can easily draw a conclusion for the boundary value problem analogous to 12 III. Of much more importance is the following theorem which extends the theorem of M. Müller 12 IV to parabolic systems.

V. Super- and Subfunctions. Suppose that we are given a third boundary value problem (with continuous or discontinuous boundary values), that $\mathbf{v}, \mathbf{w} \in Z_0(\mathbf{f}, R_n)$ and that $-\boldsymbol{\vartheta}(t, x, \mathbf{z})$ is quasi-monotone increasing in $\mathbf{z}$. Then a criterion for $\mathbf{v}$ to be a subfunction and $\mathbf{w}$ a superfunction (it should be noted that we must have both simultaneously) is:

(α) $\mathbf{v} < \mathbf{u} < \mathbf{w}$ *on* $(R_p - R_n)^+$ *and on* R_∞ *for every solution* u *of the boundary value problem,*
$\partial \mathbf{v}/\partial n_a + \boldsymbol{\vartheta}(t, x, \mathbf{v}) < 0 < \partial \mathbf{w}/\partial n_a + \boldsymbol{\vartheta}(t, x, \mathbf{w})$ *on* R_n;

(β) $v_{\nu,t} < f_\nu(t, x, \mathbf{z}, v_{\nu,x}, v_{\nu,xx})$ *if* $\mathbf{v} \leqq \mathbf{z} \leqq \mathbf{w}$, $z_\nu = v_\nu$,
$w_{\nu,t} > f_\nu(t, x, \mathbf{z}, w_{\nu,x}, w_{\nu,xx})$ *if* $\mathbf{v} \leqq \mathbf{z} \leqq \mathbf{w}$, $z_\nu = w_\nu$
[and the arguments lie in $D(f_\nu)$ $(\nu = 1, \ldots, n)$].

Then every solution $\boldsymbol{u} \in Z_0(\boldsymbol{f}, R_n)$ lies between $\boldsymbol{v}$ and $\boldsymbol{w}$,

$$\boldsymbol{v} < \boldsymbol{u} < \boldsymbol{w} \quad \text{in} \quad G_p + R_n .$$

For the proof let $\boldsymbol{u}$ be a fixed solution and T_1 the largest number such that $\boldsymbol{v} \leqq \boldsymbol{u} \leqq \boldsymbol{w}$ at the points of $G_p + R_n$ with $0 \leqq t \leqq T_1$. According to the first line of (α), we have the inequality $\boldsymbol{v}(t_k, x_{(k)}) < \boldsymbol{u}(t_k, x_{(k)}) < \boldsymbol{w}(t_k, x_{(k)})$ for every sequence $(t_k, x_{(k)}) \in G_p$ with $t_k \to +0$ for sufficiently large k. Thus $T_1 > 0$. By applying IV twice to the subdomain of G_p for whose points $t \leqq T_1$, we also obtain $\boldsymbol{v} < \boldsymbol{u} < \boldsymbol{w}$ in this subdomain (in particular for $t = T_1$). But now it follows that $T_1 = T$ and thus we have the assertion. Otherwise there would exist a sequence $(t_k, x_{(k)}) \in G_p$ with $t_k \to T_1 + 0$ for which, e.g., $v_\nu(t_k, x_{(k)}) \geqq u_\nu(t_k, x_{(k)})$ for one and the same ν. However such a sequence or a subsequence thereof can neither converge to a point $(T_1, \bar{x}) \in G_p$ (because $v_\nu < u_\nu$ at this point) nor to a point in R_p nor to infinity (by Remark IV (γ)).

Exactly as for the ordinary differential equations, the systems with a right hand side $\boldsymbol{f}$ quasi-monotone increasing in the variable $\boldsymbol{z}$ play an outstanding role here. Many of the theorems at present set up for *one* parabolic differential equation hold for them without change. A first example is the lemma of Nagumo-Westphal 24 IV or 31 IV.

VI. Lemma. *If the functions $\boldsymbol{f} \in \mathscr{P}^n$ and $-\boldsymbol{\vartheta}(t, x, \boldsymbol{z})$ are quasi-monotone increasing in $\boldsymbol{z}$ and if $\boldsymbol{v}, \boldsymbol{w} \in Z_0(\boldsymbol{f}, R_n)$, then*

(α) *$\boldsymbol{v} < \boldsymbol{w}$ on $(R_p - R_n)^+$ and on R_∞,*
$\partial \boldsymbol{v}/\partial n_a + \boldsymbol{\vartheta}(t, x, \boldsymbol{v}) < \partial \boldsymbol{w}/\partial n_a + \boldsymbol{\vartheta}(t, x, \boldsymbol{w})$ on R_n;

(β) *$\boldsymbol{P}\boldsymbol{v} < \boldsymbol{P}\boldsymbol{w}$ in G_p*

imply the inequality

$$\boldsymbol{v} < \boldsymbol{w} \quad \text{in} \quad G_p + R_n .$$

In this lemma – which goes back to Mlak (1957) and follows immediately from IV – the special structure of systems with a quasi-monotone right hand side becomes especially significant. We have a super- or a sub-function for a boundary value problem for such a system if there is a $>$ or a $<$ sign everywhere in the differential equation and in the boundary conditions, i.e., VI is a monotonicity theorem.

In Section 31, starting with 31 IV, we arrived at similar theorems, but allowing equality signs, by various means. These were especially important for the uniqueness problem. It should be noted that there are no obstacles to extending such results to parabolic systems. Such theorems yield uniqueness assertions but only for systems with quasi-monotone increasing right hand sides. For this reason we do not go further into this and instead prove a theorem which generalizes Theorem 25 VIII.

VII. Theorem. *Suppose that* $f \in \mathscr{P}^n$, $\boldsymbol{u}, \boldsymbol{v} \in Z_0(\boldsymbol{f}, R_n)$, $\boldsymbol{\delta}(t, x)$ *is defined in* G_p *and* $\boldsymbol{\varepsilon}(t, x)$ *is defined on* R_n. *Suppose that* $\boldsymbol{\varrho}(t, x) \in Z_0(\boldsymbol{\omega}, R_n)$, *where* $\boldsymbol{\omega}(t, x, \boldsymbol{z}, \boldsymbol{p}, \boldsymbol{r}) \in \mathscr{P}^n$ *and* ω_ν *is defined and quasi-monotone increasing in* $\boldsymbol{z}$ *at least for the points* $(t, x, \boldsymbol{z}, \varrho_{\nu,x}, \varrho_{\nu,xx})$ *with* $\boldsymbol{0} \leqq \boldsymbol{z} \leqq \boldsymbol{\varrho}(t, x)$, *and that* $-\boldsymbol{\vartheta}(t, x, \boldsymbol{z})$ *is quasi-monotone increasing in* $\boldsymbol{z}$.

Suppose further that

(α) $|\boldsymbol{v} - \boldsymbol{u}| < \boldsymbol{\varrho}$ *on* $(R_p - R_n)^+$ *and on* R_∞,
$\partial \boldsymbol{u}/\partial n_a + \boldsymbol{\vartheta}(t, x, \boldsymbol{u}) = 0$, $|\partial \boldsymbol{v}/\partial n_a + \boldsymbol{\vartheta}(t, x, \boldsymbol{v})| \leqq \boldsymbol{\varepsilon}(t, x)$ *on* R_n,
$\partial \boldsymbol{\varrho}/\partial n_a > \boldsymbol{\varepsilon}(t, x) + \boldsymbol{\vartheta}(t, x, \boldsymbol{v} - \boldsymbol{\varrho}) - \boldsymbol{\vartheta}(t, x, \boldsymbol{v})$ *and*
$\partial \boldsymbol{\varrho}/\partial n_a > \boldsymbol{\varepsilon}(t, x) + \boldsymbol{\vartheta}(t, x, \boldsymbol{v}) - \boldsymbol{\vartheta}(t, x, \boldsymbol{v} + \boldsymbol{\varrho})$ *on* R_n;

(β) $\boldsymbol{P u} = \boldsymbol{0}$ *and* $|\boldsymbol{P v}| \leqq \boldsymbol{\delta}(t, x)$ *in* G_p;

(γ) $\boldsymbol{\varrho}_t > \boldsymbol{\delta}(t, x) + \boldsymbol{\omega}(t, x, \boldsymbol{\varrho}, \boldsymbol{\varrho}_x, \boldsymbol{\varrho}_{xx})$ *in* G_p[1];

(δ) $f_\nu(t, x, \boldsymbol{v}, v_{\nu,x}, v_{\nu,xx}) - f_\nu(t, x, \boldsymbol{u}, v_{\nu,x} - \varrho_{\nu,x}, v_{\nu,xx} - \varrho_{\nu,xx})$
$\leqq \omega_\nu(t, x, |\boldsymbol{v} - \boldsymbol{u}|, \varrho_{\nu,x}, \varrho_{\nu,xx})$ *if* $|\boldsymbol{v} - \boldsymbol{u}| \leqq \boldsymbol{\varrho}$, $v_\nu - u_\nu = \varrho_\nu$,
$f_\nu(t, x, \boldsymbol{u}, v_{\nu,x} + \varrho_{\nu,x}, v_{\nu,xx} + \varrho_{\nu,xx}) - f_\nu(t, x, \boldsymbol{v}, v_{\nu,x}, v_{\nu,xx})$
$\leqq \omega_\nu(t, x, |\boldsymbol{v} - \boldsymbol{u}|, \varrho_{\nu,x}, \varrho_{\nu,xx})$ *if* $|\boldsymbol{v} - \boldsymbol{u}| \leqq \boldsymbol{\varrho}$, $u_\nu - v_\nu = \varrho_\nu$
$(\nu = 1, \ldots, n)$.

Then we have

$$|\boldsymbol{v} - \boldsymbol{u}| < \boldsymbol{\varrho} \quad \text{in} \quad G_p + R_n.$$

For the proof let T_1 be the largest number such that $|\boldsymbol{v} - \boldsymbol{u}| \leqq \boldsymbol{\varrho}$ in the intersection of G_p with $(0, T_1] \times E^m$. As in V we obtain $T_1 > 0$. Now we apply Lemma II to the functions $\boldsymbol{\varphi} = \boldsymbol{v} - \boldsymbol{u}$, $\boldsymbol{\psi} = \boldsymbol{\varrho}$ and the subdomain of G_p with $t \leqq T_1$. First we check the hypotheses of II! From[2] $|\boldsymbol{v} - \boldsymbol{u}| \leqq \boldsymbol{\varrho}$, $v_\nu - u_\nu = \varrho_\nu$, $(v_\nu - u_\nu)_x = \varrho_{\nu,x}$, $(v_\nu - u_\nu)_{xx} \leqq \varrho_{\nu,xx}$, with the help of (β), the quasi-monotonicity of $\boldsymbol{\omega}$ in $\boldsymbol{z}$, and (γ) (we suppress the variables t, x), we obtain

$$\begin{aligned}(v_\nu - u_\nu)_t &\leqq \delta_\nu + f_\nu(\boldsymbol{v}, v_{\nu,x}, v_{\nu,xx}) - f_\nu(\boldsymbol{u}, u_{\nu,x}, u_{\nu,xx})\\ &\leqq \delta_\nu + f_\nu(\boldsymbol{v}, v_{\nu,x}, v_{\nu,xx}) - f_\nu(\boldsymbol{u}, (v_\nu - \varrho_\nu)_x, (v_\nu - \varrho_\nu)_{xx})\\ &\leqq \delta_\nu + \omega_\nu(|\boldsymbol{v} - \boldsymbol{u}|, \varrho_{\nu,x}, \varrho_{\nu,xx}) \leqq \delta_\nu + \omega_\nu(\boldsymbol{\varrho}, \varrho_{\nu,x}, \varrho_{\nu,xx}) < \varrho_{\nu,t},\end{aligned}$$

which is what we wanted to show.

Further we must show that we do not have the case II (β). By the first line of (α) it suffices to prove that: if $\boldsymbol{\varphi} = \boldsymbol{v} - \boldsymbol{u} \leqq \boldsymbol{\psi} = \boldsymbol{\varrho}$, $\varphi_\nu = \psi_\nu$ at a point in R_n, then $\partial \varphi_\nu/\partial n_a < \partial \psi_\nu/\partial n_a$ at this boundary point. This follows from the remaining hypotheses of (α). Indeed by the quasi-monotonicity of $-\boldsymbol{\vartheta}$ we have

$$\partial(v_\nu - u_\nu)/\partial n_a \leqq \vartheta_\nu(\boldsymbol{u}) - \vartheta_\nu(\boldsymbol{v}) + \varepsilon_\nu \leqq \vartheta_\nu(\boldsymbol{v} - \boldsymbol{\varrho}) - \vartheta_\nu(\boldsymbol{v}) + \varepsilon_\nu < \partial \varrho_\nu/\partial n_a$$

(the variables t, x were again suppressed). Hence we have the case II (α), that is, $\boldsymbol{v} - \boldsymbol{u} < \boldsymbol{\varrho}$ in the part of $G_p + R_n$ lying between 0 and T_1. We show

[1] Here we use the notation of (1′). Thus, on the right hand side of the νth equation we have $\omega_\nu(t, x, \boldsymbol{\varrho}, \varrho_{\nu,x}, \varrho_{\nu,xx})$.

[2] We introduced the number T_1 only so that we could start with the sharper inequality $|\boldsymbol{v} - \boldsymbol{u}| \leqq \boldsymbol{\varrho}$ rather than with $\boldsymbol{v} - \boldsymbol{u} \leqq \boldsymbol{\varrho}$.

in exactly the same way that also $\boldsymbol{u}-\boldsymbol{v}<\boldsymbol{\varrho}$ and thus $|\boldsymbol{v}-\boldsymbol{u}|<\boldsymbol{\varrho}$ in this part of G_p+R_n. If $T_1=T$, then the theorem is now proven. But the assumption $T_1<T$ leads easily – as at the end of the proof of V – to a contradiction.

This theorem is of prime importance for future considerations. First we go into the special case of a bound depending only on t.

VIII. Special Case. Let $f\in\mathscr{P}^n$, $\boldsymbol{u},\boldsymbol{v}\in Z_0(f,R_n)$, let $\boldsymbol{\delta}(t)$ and $\boldsymbol{\varepsilon}(t)$ be defined in J_0, let $-\boldsymbol{\vartheta}(t,x,\boldsymbol{z})$ be quasi-monotone increasing in $\boldsymbol{z}$, and let $\boldsymbol{\varrho}(t)\in Z_0(\boldsymbol{\omega})$. We assume that $\boldsymbol{\omega}=\boldsymbol{\omega}(t,\boldsymbol{z})$ is defined at least for $0<t\leqq T$, $\boldsymbol{0}\leqq\boldsymbol{z}\leqq\boldsymbol{\varrho}(t)$, and is quasi-monotone increasing in $\boldsymbol{z}$. Suppose further that

(α) $|\boldsymbol{v}-\boldsymbol{u}|<\boldsymbol{\varrho}$ on $(R_p-R_n)^+$ and on R_∞,
$\partial\boldsymbol{u}/\partial n_a+\boldsymbol{\vartheta}(t,x,\boldsymbol{u})=0$, $\quad|\partial\boldsymbol{v}/\partial n_a+\boldsymbol{\vartheta}(t,x,\boldsymbol{v})|\leqq\boldsymbol{\varepsilon}(t)$ on R_n,
$\boldsymbol{\vartheta}(t,x,\boldsymbol{v})-\boldsymbol{\vartheta}(t,x,\boldsymbol{v}-\boldsymbol{\varrho})>\boldsymbol{\varepsilon}(t)$ and
$\boldsymbol{\vartheta}(t,x,\boldsymbol{v}+\boldsymbol{\varrho})-\boldsymbol{\vartheta}(t,x,\boldsymbol{v})>\boldsymbol{\varepsilon}(t)$ on R_n;

(β) $\boldsymbol{P}\boldsymbol{u}=\boldsymbol{0}$ and $|\boldsymbol{P}\boldsymbol{v}|\leqq\boldsymbol{\delta}(t)$ in G_p;

(γ) $\boldsymbol{\varrho}'>\boldsymbol{\delta}(t)+\boldsymbol{\omega}(t,\boldsymbol{\varrho})$ in J_0;

(δ) $f_\nu(t,x,z,v_{\nu,x},v_{\nu,xx})-f_\nu(t,x,\bar{z},v_{\nu,x},v_{\nu,xx})\leqq\omega_\nu(t,|z-\bar{z}|)$,
if $z_\nu\geqq\bar{z}_\nu$ and $|z-\bar{z}|\leqq\varrho(t)$ $(\nu=1,\dots,n)$.

Then we have
$$|\boldsymbol{v}-\boldsymbol{u}|<\boldsymbol{\varrho}\quad\text{in}\quad G_p+R_n.$$

IX. Remarks. (α) If, as often happens in applications, ϑ_ν depends only on z_ν and is linear in z_ν

$$\vartheta_\nu(t,x,z)=\varkappa_\nu(t,x)\,z_\nu\,,$$

then the hypothesis VII (α), as far as it relates to R_n and ϱ, reads

$$\partial\varrho_\nu/\partial n_a+\varkappa_\nu(t,x)\,\varrho_\nu>\varepsilon_\nu(t,x)\quad\text{on}\quad R_n\qquad(\nu=1,\dots,n)$$

and the corresponding part of VIII (α) is simply

$$\varepsilon_\nu(t)<\varkappa_\nu(t,x)\,\varrho_\nu(t)\quad\text{on}\quad R_n\qquad(\nu=1,\dots,n)\,.$$

(β) There is no difficulty involved in modifying VIII so that the bound is determined not from a differential inequality but from the differential equation
$$\boldsymbol{\sigma}'=\boldsymbol{\delta}(t)+\boldsymbol{\omega}(t,\boldsymbol{\sigma})$$
and, in fact, as the maximal solution of this system. Here the procedure corresponds exactly to the conversion from 13 I to 13 II. Then equality signs can be allowed everywhere in the inequalities of VIII (α) (with $\boldsymbol{\sigma}$ in place of $\boldsymbol{\varrho}$). However, if we have a boundary value problem of third kind, $\vartheta_\nu(t,x,\boldsymbol{z})$ must be strictly monotone increasing in z_ν $(\nu=1,\dots,n)$. Szarski (1955) first proved VIII in a similar form.

(γ) We have referred several times, at similar opportunities, to two important points in connection with the condition VII (δ) or VIII (δ).

First – this is important for nonlinear problems – we have not arbitrary p and r but rather $v_{\nu,x}$ and $v_{\nu,xx}$ in the last two places of $f_\nu(t, x, z, p, r)$. Second, VII (δ) is a *one-sided* condition in the same sense as for ordinary systems. We repeated it in VIII (δ) in a cruder form which makes this state of affairs particularly clear[1].

We became acquainted with another kind of bound for systems of ordinary differential equations, in Section 11, in which we go over to a single equation by using a K-norm. The conversion of this principle to parabolic systems causes difficulties when we permit arbitrary K-norms, on whose nature we do not dwell. On the other hand, the proof becomes very simple if we restrict consideration to one of the K-norms $\|z\| = \max|z_\nu|$, $\|z\| = \max z_\nu$ used in 11 III. Thus with the first of these norms we obtain the following

X. Theorem. *Suppose* $f \in \mathscr{P}^n$, $-\boldsymbol{\vartheta}(t, x, z)$ *is quasi-monotone increasing in* z *and* $\boldsymbol{u}, \boldsymbol{v} \in Z_0(f, R_n)$. *Suppose the scalar functions* $\delta(t, x)$, $\varepsilon(t, x)$ *are defined in* G_p *and* R_n *respectively and the functions* $\omega(t, x, z, p, r) \in \mathscr{P}$ *and* $\varrho(t, x) \in Z_0(\omega, R_n)$ *are also scalar. In order to obtain a bound*

$$|v_\nu - u_\nu| < \varrho \quad in \quad G_p + R_n \qquad (\nu = 1, \ldots, n),$$

we need the following four hypotheses (the index ν *runs from* 1 *to* n *everywhere;* $\boldsymbol{e}$ *is the vector* $\boldsymbol{e} = (1, 1, \ldots, 1) \in E^n$*):*

(α) $|v_\nu - u_\nu| < \varrho$ *on* $(R_p - R_n)^+$ *and on* R_∞,
$\partial \boldsymbol{u}/\partial n_a + \boldsymbol{\vartheta}(t, x, \boldsymbol{u}) = \boldsymbol{0}$, $\quad |\partial v_\nu/\partial n_a + \vartheta_\nu(t, x, \boldsymbol{v})| \leqq \varepsilon(t, x)$ *on* R_n,
$\partial \varrho/\partial n_a > \varepsilon(t, x) + \vartheta_\nu(t, x, \boldsymbol{v} - \boldsymbol{e}\varrho) - \vartheta_\nu(t, x, \boldsymbol{v})$ *and*
$> \varepsilon(t, x) + \vartheta_\nu(t, x, \boldsymbol{v}) - \vartheta_\nu(t, x, \boldsymbol{v} + \boldsymbol{e}\varrho)$ *on* R_n;

(β) $\boldsymbol{P}\boldsymbol{u} = \boldsymbol{0}$ *and* $|(\boldsymbol{P}\boldsymbol{v})_\nu| = |v_{\nu,t} - f_\nu(t, x, \boldsymbol{v}, v_{\nu,x}, v_{\nu,xx})| \leqq \delta(t, x)$ *in* G_p;

(γ) $\varrho_t > \delta(t, x) + \omega(t, x, \varrho, \varrho_x, \varrho_{xx})$ *in* G_p;

(δ) $f_\nu(t, x, \boldsymbol{v}, v_{\nu,x}, v_{\nu,xx}) - f_\nu(t, x, \boldsymbol{u}, v_{\nu,x} - \varrho_x, v_{\nu,xx} - \varrho_{xx})$
$\leqq \omega(t, x, \max_{\varkappa=1,\ldots,n} |v_\varkappa - u_\varkappa|, \varrho_x, \varrho_{xx})$ *if* $|v_\varkappa - u_\varkappa| \leqq v_\nu - u_\nu = \varrho$ $(\varkappa \neq \nu)$,

$f_\nu(t, x, \boldsymbol{u}, v_{\nu,x} + \varrho_x, v_{\nu,xx} - \varrho_{xx}) - f_\nu(t, x, \boldsymbol{v}, v_{\nu,x}, v_{\nu,xx})$
$\leqq \omega(t, x, \max_{\varkappa=1,\ldots,n} |v_\varkappa - u_\varkappa|, \varrho_x, \varrho_{xx})$ *if* $|u_\varkappa - v_\varkappa| \leqq u_\nu - v_\nu = \varrho$ $(\varkappa \neq \nu)$.

The proof of VII needs only a trivial modification which is left to the reader[2].

---

[1] These points are not taken into account in the theorems of Szarski (1955).

[2] The idea of reducing X completely to VII by introducing vectors $\boldsymbol{\varepsilon}, \boldsymbol{\delta}, \boldsymbol{\varrho}$ whose components are all equal and indeed equal to the corresponding scalars of X and by defining a vector $\boldsymbol{\omega}$ with $\omega_\nu = \omega(t, x, \max|z_\lambda|, p, r)$ $(\nu = 1, \ldots, n)$ fails because the function $\boldsymbol{\omega}$ thus defined is not quasi-monotone increasing in $\boldsymbol{z}$. However, if we restrict the domain of definition $\boldsymbol{\omega}$ in Hypothesis VII as much as necessary, i.e., if for $D(\omega_\nu)$ we do not require $\boldsymbol{0} \leqq \boldsymbol{z} \leqq \boldsymbol{\varrho}$ as in VII but only $0 \leqq z_\varkappa \leqq \varrho_\varkappa$ $(\varkappa \neq \nu)$, $z_\nu = \varrho_\nu$, then the reduction to VII is in fact possible. The function $\boldsymbol{\omega}$ just defined is indeed quasi-monotone increasing on this set even if $\omega(t, x, z, p, r)$ is not monotone increasing in z.

It is possible, as in VIII, to specialize to bounds $\varrho=\varrho(t)$ which satisfy an ordinary differential inequality $\varrho'>\delta(t)+\omega(t,\varrho)$ (resp. are maximal solutions of the corresponding differential equation). As the new condition (δ) we obtain

$$(\delta')\ f_\nu(t,x,z,v_{\nu,x},v_{\nu,xx})-f_\nu(t,x,\bar z,v_{\nu,x},v_{\nu,xx})\leqq\omega(t,\max_{\varkappa=1,\dots,n}|z_\varkappa-\bar z_\varkappa|)$$

for $z_\nu\geqq\bar z_\nu$ and $|z_\varkappa-\bar z_\varkappa|\leqq\varrho(t)$ $(\varkappa,\nu=1,\dots,n)$.

33. Uniqueness Problems for Parabolic Systems

With the estimation theorems of the preceding sections we have furnished the resources for extending the previously obtained uniqueness assertions to parabolic systems. As in Section 32 we consider the third boundary value problem

$$\boldsymbol{u}_t=\boldsymbol{f}(t,x,\boldsymbol{u},\boldsymbol{u}_x,\boldsymbol{u}_{xx})\quad\text{in}\quad G_p, \tag{1}$$

$$\boldsymbol{u}=\boldsymbol{\eta}(t,x)\quad\text{on}\quad R_p-R_n,\quad \partial\boldsymbol{u}/\partial n_a+\boldsymbol{\vartheta}(t,x,\boldsymbol{u})=\boldsymbol{0}\quad\text{on}\quad R_n, \tag{2}$$

where conditions at infinity are added for an unbounded domain.

Generalizing 25 V and 31 X, we obtain the

I. Uniqueness Theorem. *Suppose that the function $\vartheta_\nu=\vartheta_\nu(t,x,z_\nu)$, independent of $z_\varkappa$ $(\varkappa\neq\nu)$, is strictly monotone increasing in z_ν. For $\boldsymbol{f}\in\mathscr{P}^n$ suppose:*

(α) *we have*

$$f_\nu(t,x,z,p,r)-f_\nu(t,x,\bar z,p,r)\leqq\omega_\nu(t,|z-\bar z|)\quad\textit{if}\quad z_\nu\geqq\bar z_\nu$$

with a function $\boldsymbol{\omega}(t,z)\in\mathscr{E}_5^n$ $[\mathscr{E}_4^n]$, and (t,x,z,p,r), $(t,x,\bar z,p,r)\in D(f_\nu)$ $(\nu=1,\dots,n)$.

Then the boundary value problem (1) (2) *with bounded domain G has at most one solution* $\boldsymbol{u}$ *in the class $Z(\boldsymbol{f},R_n)$*[1]. *The same is also true for an unbounded domain G if conditions are given at infinity such that "$\boldsymbol{u}-\bar{\boldsymbol{u}}=\boldsymbol{0}$ on R_∞" for any two solutions $\boldsymbol{u},\bar{\boldsymbol{u}}$ of the boundary value problem.*

The solution $\boldsymbol{u}$ *depends continuously on the boundary values [on the boundary values and on the right hand side of the differential equation],* i.e., for every $\varepsilon>0$ there is a $\delta>0$ such that $|\boldsymbol{v}-\boldsymbol{u}|<\varepsilon$ in $\bar G$ if $\boldsymbol{v}\in Z(\boldsymbol{f},R_n)$ satisfies the inequalities

$$|v_\nu-u_\nu|<\delta\quad\text{on } R_p-R_n \text{ and on } R_\infty,\quad \partial\boldsymbol{v}/\partial n_a+\boldsymbol{\vartheta}(t,x,\boldsymbol{v})=0 \text{ on } R_n,$$

$$\boldsymbol{P}\boldsymbol{v}=\boldsymbol{0}\quad[|(\boldsymbol{P}\boldsymbol{v})_\nu|<\delta\ (\nu=1,\dots,n)]\quad\text{in}\quad G_p.$$

[1] Hence the problem is only meaningful if $\eta(t,x)$ is uniformly continuous on every bounded subset of R_p-R_n.

If $\varrho(t)$ is a function as described in the definition of $\mathscr{E}_5^n$ or $\mathscr{E}_4^n$ and if $\boldsymbol{u}$ is a solution, $\boldsymbol{v}$ an approximate solution, then from 32 VIII we obtain the inequality $|\boldsymbol{v}-\boldsymbol{u}|<\varrho$. Here we should note that for $\varepsilon(t)=0$ the last two conditions in 32 VIII (α) follow from our monotonicity assumptions on $\boldsymbol{\vartheta}$.

The hypothesis (α) can be weakened as in 25 VI. For example, it suffices if (α') holds:

(α') *For one fixed solution* $\boldsymbol{u}$ *there is an* $\alpha>0$ *and a function* $\omega(t,z)\in\mathscr{E}_5^n\,[\mathscr{E}_4^n]$ *such that*

$$\left.\begin{aligned} f_\nu(t,x,\boldsymbol{u}+z,u_{\nu,x},u_{\nu,xx})-f_\nu(t,x,\boldsymbol{u},u_{\nu,x},u_{\nu,xx})\\ f_\nu(t,x,\boldsymbol{u},u_{\nu,x},u_{\nu,xx})-f_\nu(t,x,\boldsymbol{u}-z,u_{\nu,x},u_{\nu,xx})\end{aligned}\right\}\leqq\omega_\nu(t,|z|)$$

for $0\leqq z_\nu\leqq\alpha$ *and* $|z_\varkappa|\leqq\alpha\ (\varkappa\neq\nu)$ (as long as the arguments lie in $D(f_\nu)$; $\nu=1,\ldots,n$).

When the remarks about unbounded domains in 25 VII are given a vector interpretation they also turn out to be true.

Quite analogously from Theorem 32 X we obtain the

II. Uniqueness Theorem. *Theorem* I *remains valid if* I (α) *is replaced by* (α) *we have*

$$f_\nu(t,x,z,p,r)-f_\nu(t,x,\bar z,p,r)\leqq\omega(t,\max_\varkappa|z_\varkappa-\bar z_\varkappa|)\quad\textit{if}\quad z_\nu\geqq\bar z_\nu$$

with a function $\omega(t,z)\in\mathscr{E}_5\,[\mathscr{E}_4]$ *and* (t,x,z,p,r), $(t,x,\bar z,p,r)\in D(f_\nu)$ ($\nu=1,\ldots,n$).

Of course this hypothesis also can be put in a form (α') analogous to I (α').

Both theorems go back to Szarski (1955) [however, here they are sharpened in several respects – allowable domains and boundary conditions as well as one-sided conditions in (α)]; see also Montaldo (1958).

It should be noted that in Theorem II we can also allow the more general boundary condition in which ϑ_ν depends on all $z_\varkappa$. Then we need the hypothesis

$$\vartheta_\nu(t,x,z+\alpha\boldsymbol{e})-\vartheta_\nu(t,x,z)>0\quad\text{for}\quad\alpha>0\,,$$

$\boldsymbol{e}=(1,1,\ldots,1)\in E^n$. If for example $\vartheta_\nu=a_{\nu1}z_1+\cdots+a_{\nu n}z_n$, then $a_{\nu\nu}>0$ must be so large that $a_{\nu1}+\cdots+a_{\nu n}>0$. We see immediately that this inequality suffices for the proof of the last two conditions of 32 X (α) (with $\partial\varrho/\partial n_a=\varepsilon(t,x)=0$). This remark does not hold for Theorem I.

Now we proceed to the theorems of Section 28. These concerned a differential equation linear in the second derivatives. Analogously, we set our sights on functions $\boldsymbol{f}\in\mathscr{P}^n$ of the form

$$f_\nu(t,x,z,p,r)=g_\nu(t,x,z,p)+\sum_{\lambda,\mu=1}^{m}k_{\lambda\mu}^\nu(t,x)\,r_{\lambda\mu}\qquad(\nu=1,\ldots,n)\,.\tag{3}$$

Here $(k^{\nu}_{\lambda\mu})$ is a positive semi-definite $m \times m$ matrix for every ν and at every point in G_p.

First we specialize Theorem 32 X to such differential equations and to the first boundary value problem.

III. Theorem. *Let f be given by* (3). *Suppose that $\boldsymbol{u}, \boldsymbol{v} \in Z_0(f)$, a function $\varrho(t, x) \in Z_0$ bounded below in G_p by a positive number, and the functions $a(t, x)$, $b_\mu(t, x)$, $\delta(t, x)$ defined in G_p (ν always runs from 1 to n) satisfy*

(α) $|v_\nu - u_\nu| \leqq \varrho$ *on* R_p^+ *and on* R_∞;

(β) $\boldsymbol{P u} = \boldsymbol{0}$ *and* $|(\boldsymbol{P v})_\nu| \leqq \delta$ *in* G_p;

(γ) $\varrho_t \geqq \delta + a\varrho + \sum_\mu b_\mu |\varrho_{x_\mu}| + \sum_{\lambda,\mu} k^{\nu}_{\lambda\mu} \varrho_{x_\lambda x_\mu}$ *in* G_p;

(δ) $g_\nu(t, x, z, p) - g_\nu(t, x, \bar{z}, p) \leqq a \max_\varkappa |z_\varkappa - \bar{z}_\varkappa| + \sum_{\mu=1}^{m} b_\mu |p_\mu - \bar{p}_\mu|$ *for* $z_\nu \geqq \bar{z}_\nu$.

Then we have

$$|v_\nu - u_\nu| \leqq \varrho \quad \text{in} \quad G_p \qquad (\nu = 1, \ldots, n).$$

For $n = 1$ this theorem turns into 25 IX. Except for the equality signs in (α) and (γ), it represents a special case of 32 X. To justify these equality signs we adopt the reasoning used in 25 IX.

Since all the results of Section 28 were obtained via the estimation theorem 25 IX by constructing suitable functions ϱ and since the conditions to be imposed on ϱ, (α) and (γ) in III and in 25 IX, are identical, *all these assertions also hold for parabolic systems.* We state the most important one as briefly as possible.

IV. Unbounded Domains. Suppose that we fix a vector function $\bar{\boldsymbol{\eta}}(x)$ and a $K > 0$ and that the inequalities

$$\left.\begin{aligned} &g_\nu(t, x, z, p) - g_\nu(t, x, \bar{z}, p) \leqq K(1 + x^2) \max_\varkappa |z_\varkappa - \bar{z}_\varkappa| \\ &\quad + K(1 + |x_1| + \cdots + |x_m|) \sum_{\mu=1}^{m} |p_\mu - \bar{p}_\mu| \quad \text{for} \quad z_\nu \geqq \bar{z}_\nu \quad (\nu = 1, \ldots, n), \end{aligned}\right\} \tag{4}$$

$$|k^{\nu}_{\lambda\mu}| \leqq K \qquad (\lambda, \mu = 1, \ldots, m;\ \nu = 1, \ldots, n) \tag{5}$$

hold. Then the first boundary value problem for the system

$$u_{\nu,t} = g_\nu(t, x, \boldsymbol{u}, u_{\nu,x}) + \sum_{\lambda,\mu=1}^{m} k^{\nu}_{\lambda\mu}(t, x)\, u_{\nu, x_\lambda x_\mu} \qquad (\nu = 1, \ldots, n) \tag{6}$$

has at most one solution $\boldsymbol{u} \in Z$ which satisfies the growth condition

$$|u_\nu(t, x) - \bar{\eta}_\nu(x)| \leqq e^{K x^2} \quad \text{for large } x \qquad (\nu = 1, \ldots, n). \tag{7}$$

V. Discontinuous Boundary Values. We consider the first boundary value problem for the system

$$u_{\nu,t} = g_\nu(t, x, \boldsymbol{u}) + \Delta u_\nu \qquad (\nu = 1, \dots, n) \tag{8}$$

in the sense made more precise in 28 V, VI. Thus we consider solutions $\boldsymbol{u} \in Z_0$ of the equation (8) which (with the notation of 28 VI) are continuous in $R_p + R^*$, have the values of the given continuous function $\boldsymbol{\eta}(t, x)$ on R^*, and remain bounded with approach to the set $\{0\} \times N$[1].

The boundary value problem has at most one solution in this sense if the condition

$$g_\nu(t, x, z) - g_\nu(t, x, \bar{z}) \leqq K(1 + x^2) \max_{\varkappa} |z_\varkappa - \bar{z}_\varkappa| \quad \text{for} \quad z_\nu \geqq \bar{z}_\nu \tag{9}$$

holds and the domain is bounded. For an unbounded domain, if we also require the condition (7), then we again have uniqueness.

This follows from 28 III or 28 VIII together with III. The remark 28 IX (α), which relates to the case in which the set N of exceptional points consists of only finitely many points, can also be carried over easily.

VI. Remark. To a great extent the theorems IV, V can also be preserved for boundary conditions of the third kind. To include this case, we have to change Theorem III in the first hypothesis only. If we assume that $\boldsymbol{u}$ and $\boldsymbol{v}$ satisfy the boundary conditions on R_n, thus $\partial \boldsymbol{u}/\partial n_a + \boldsymbol{\vartheta}(t, x, \boldsymbol{u}) = \partial \boldsymbol{v}/\partial n_a + \boldsymbol{\vartheta}(t, x, \boldsymbol{v}) = \mathbf{0}$, and that $\boldsymbol{\vartheta}(t, x, z)$ has the property

$$\vartheta_\nu(t, x, z + \alpha \boldsymbol{e}) - \vartheta_\nu(t, x, z) \geqq 0 \quad \text{for} \quad \alpha > 0, \boldsymbol{e} = (1, 1, \dots, 1) \in E^n, \tag{10}$$

then, as we learn from a glance at 32 X, the hypothesis III (α) is to be changed to

$$|v_\nu - u_\nu| \leqq \varrho \quad \textit{on } (R_p - R_n)^+ \textit{ and on } R_\infty; \quad \partial\varrho/\partial n_a > 0 \quad \textit{on } R_n.$$

Under these hypotheses on the boundary conditions on R_n, IV *and* V *also hold for the third boundary value problem under the circumstances described in detail in* 31 XIII *and* 31 XIV[2].

From these remarks it is clear that we do not always have to proceed as in 31 VI or 31 VIII in order to obtain assertions which also hold for the second boundary value problem. This becomes necessary only when we work with a bound $\varrho = \varrho(t)$ which is independent of x, because then $\partial\varrho/\partial n_a = 0$. For parabolic systems we have so far avoided operating with an auxiliary function $h(x)$ as in 31 VI. As a consequence, the uniqueness theorems I and II do not hold for the second boundary value problem. This gap will now be closed.

---

[1] This condition is consistent with the boundary value problem only if $\boldsymbol{\eta}$ is continuous on R^* and, moreover, bounded in every bounded subset of R^*.

[2] In place of the weak monotonicity of $\vartheta(t, x, z)$ in z, which is required in 31 XIII (α), we now have condition (10).

VII. Uniqueness Theorem. *All the assertions of the uniqueness theorem I referring to $\mathscr{E}_4^n$ also hold for the second boundary value problem and more generally for those third boundary value problems in which ϑ_ν depends only on z_ν and is (weakly) monotone increasing in z_ν, if the following three assumptions hold:*

(α) *like* II (α) (*or* II (α'));

(β) *there is a function* $h(x)$ *with the properties of* 31 VI (γ_2);

(γ) *for a (fixed) solution* $\boldsymbol{u}$ *we have: for every* $\delta > 0$ *there is a* $\beta > 0$ *such that*

$$|f_\nu(t, x, z, u_{\nu,x}, u_{\nu,xx}) - f_\nu(t, x, \bar{z}, u_{\nu,x} + p, u_{\nu,xx} + r)| < \delta$$

for $|z_\nu - \bar{z}_\nu| < \beta$, $|p_\mu| < \beta$, $|r_{\lambda\mu}| < \beta$ $(\nu = 1, \ldots, n;\ \lambda, \mu = 1, \ldots, m)$ [*as long as both arguments lie in* $D(f_\nu)$].

The remarks adjoined to (γ_3) in 31 VII also hold for (γ). For the proof we first take $\varepsilon > 0$, then determine a $\varrho(t)$ and a $\delta > 0$ such that $\varrho' > 3\delta + \omega(t, \varrho)$ and $3\delta < \varrho < \varepsilon$, and then a $\beta > 0$ according to hypothesis (γ). Suppose $\boldsymbol{u}$ is a solution of the boundary value problem and $|(\boldsymbol{P}\boldsymbol{v})_\nu| < \delta$ in G_p, $|v_\nu - u_\nu| < \delta$ on $R_p - R_n$ and on R_∞, $\partial \boldsymbol{u}/\partial n_a + \boldsymbol{\vartheta}(t, x, \boldsymbol{u}) = \partial \boldsymbol{v}/\partial n_a + \boldsymbol{\vartheta}(t, x, \boldsymbol{v}) = \boldsymbol{0}$ on R_n. Suppose $\lambda_0 > 0$ is so small that $\lambda_0 h(x)$ and its derivatives of first and second order are in magnitude $< \beta$ and $< \delta$, and that $\boldsymbol{e} = (1, 1, \ldots, 1) \in E^n$. It obviously suffices to prove the inequality $|v_\nu - u_\nu - \lambda h| < \varrho$ $(\nu = 1, \ldots, n)$ or, equivalently, $|\boldsymbol{v} - \boldsymbol{u} - \lambda \boldsymbol{e} h| < \boldsymbol{e}\varrho$ for $0 < \lambda < \lambda_0$. It is true on $R_p - R_n$ and on R_∞ because of $\varrho \geqq 3\delta$, $|\lambda h| < \delta$. If T_1 is the greatest number such that $|\boldsymbol{v} - \boldsymbol{u} - \lambda \boldsymbol{e} h| \leqq \boldsymbol{e}\varrho$ for $t \leqq T_1$, then (see the proof in 32 V) $T_1 > 0$. Now we apply Lemma 32 II to the functions $\boldsymbol{\varphi} = \boldsymbol{v} - \boldsymbol{u} - \lambda \boldsymbol{e} h$, $\boldsymbol{\psi} = \boldsymbol{e}\varrho$ and the part of G lying below the hyperplane $t = T_1$. By (α) and (γ) $|\boldsymbol{v} - \boldsymbol{u} - \lambda \boldsymbol{e} h| \leqq \boldsymbol{e}\varrho$, $v_\nu - u_\nu - \lambda h = \varrho$, $(v_\nu - u_\nu - \lambda h)_x = 0$, $(v_\nu - u_\nu - \lambda h)_{xx} \leqq 0$ (for a fixed ν and a fixed $(t, x) \in G_p$, which is omitted in the following inequalities) imply

$$\begin{aligned}
(v_\nu - u_\nu)_t &\leqq \delta + f_\nu(\boldsymbol{v}, v_{\nu,x}, v_{\nu,xx}) - f_\nu(\boldsymbol{u}, u_{\nu,x}, u_{\nu,xx}) \\
&\leqq \delta + f_\nu(\boldsymbol{v}, u_{\nu,x} + \lambda h_x, u_{\nu,xx} + \lambda h_{xx}) - f_\nu(\boldsymbol{v} - \lambda \boldsymbol{e} h, u_{\nu,x}, u_{\nu,xx}) \\
&\qquad + f_\nu(\boldsymbol{v} - \lambda \boldsymbol{e} h, u_{\nu,x}, u_{\nu,xx}) - f_\nu(\boldsymbol{u}, u_{\nu,x}, u_{\nu,xx}) \\
&\leqq \delta + \delta + \omega(t, \varrho) < \varrho',
\end{aligned}$$

and thus $\varphi_{\nu,t} < \psi_{\nu,t}$.

We still have to prove that the case 32 II (β) is excluded. Since $\boldsymbol{\varphi} < \boldsymbol{\psi}$ on $R_p - R_n$ and on R_∞, it suffices to show: if $\boldsymbol{\varphi} \leqq \boldsymbol{\psi}$ and $\varphi_\nu = \psi_\nu$ at a point of R_n, then $\partial\varphi_\nu/\partial n_a < \partial\psi_\nu/\partial n_a$ at this point. The inequalities (note that $v_\nu > u_\nu$)

$$\partial(v_\nu - u_\nu - \lambda h)/\partial n_a \leqq \vartheta_\nu(u_\nu) - \vartheta_\nu(v_\nu) - \lambda \partial h/\partial n_a \leqq -\lambda \partial h/\partial n_a < 0 = \partial\varrho/\partial n_a$$

show that this is the case. Hence by 32 II (α) we have $\boldsymbol{v}-\boldsymbol{u}-\lambda \boldsymbol{e}h<\boldsymbol{e}\varrho$ for $t \leqq T_1$. In the same way we also obtain the inequality $-\boldsymbol{e}\varrho<\boldsymbol{v}-\boldsymbol{u}-\lambda \boldsymbol{e}h$ for $t \leqq T_1$. From this $T_1=T$ and hence the assertion of the theorem follows easily (see the end of the proof of 32 V).

VIII. The Maximum-Minimum Principle. We use the definitions supplied by an obvious conversion of 26 I. Thus, for example, $M(\boldsymbol{\varphi})$ is the vector with components $M(\varphi_\nu)$. The maximum principle holds if $\boldsymbol{u} \leqq M(\boldsymbol{u})$ for every solution $\boldsymbol{u} \in Z_0(\boldsymbol{f})$ of the equation (1). Analogously to 26 II, we have the following situation:

Suppose $D(f_\nu)=M_\nu \times M_{\nu r}$ and the set $M_{\nu r}$ contains the null matrix ($\nu=1, \ldots, n$). Then

the maximum principle holds if $f_\nu(t, x, z, 0, 0) \leqq 0$,
the minimum principle holds if $f_\nu(t, x, z, 0, 0) \geqq 0$,
the maximum-minimum principle holds if $f_\nu(t, x, z, 0, 0)=0$

for $(t, x, z, 0, 0) \in D(f_\nu)$ ($\nu=1, \ldots, n$).

The sharpened version 26 III can also be carried over.

The proof is trivial. If $\bar{\boldsymbol{u}}$ is a given solution, we can take its νth component as a solution u of the scalar equation

$$u_t=f^*(t, x, u, u_x, u_{xx})$$

with

$$f^*(t, x, z, p, r)=f_\nu(t, x, \bar{u}_1, \ldots, \bar{u}_{\nu-1}, z, \bar{u}_{\nu+1}, \ldots, \bar{u}_n, p, r)$$

and apply the results of Section 26.

It can certainly happen that in the inequalities of VIII we have the $\leqq$ sign for some components f_ν and $\geqq$ for others, i.e., that for some solution components the maximum principle holds, for others the minimum principle holds.

IX. Example. If the system (1) has the form

$$u_{\nu,t}=k_\nu \Delta u_\nu+\psi_\nu(t, x, u_1, \ldots, u_n) \quad (\nu=1, \ldots, n;\, k_\nu \geqq 0)$$

and the ψ_ν satisfy a Lipschitz condition

$$|\psi_\nu(t, x, \boldsymbol{z})-\psi_\nu(t, x, \bar{\boldsymbol{z}})| \leqq L\|\boldsymbol{z}-\bar{\boldsymbol{z}}\|_e,$$

then for the third boundary value problem (and thus in particular for the first boundary value problem) we have uniqueness and continuous dependence on the boundary values in the context described in I. A corresponding statement is also true for the second boundary value problem if we can determine a function $h(x)$ according to 31 VI (γ_2).

Problems of this kind – in many cases the boundary conditions involve normal derivatives – arise in electrochemistry. The equation above describes an n-component mixture; u_ν is the concentration of the νth component. The change in concentration in a unit time ($u_{\nu,t}$) is accomplished first by diffusion ($k_\nu \Delta u_\nu$), second by chemical reaction of the individual components with each other (ψ_ν). Since the rate of chemical reaction is often proportional to the concentration of each component involved, ψ_ν is usually nonlinear (for $n=2$, for example, $\psi_\nu = l_\nu u_1 u_2$). See Dračka (1961) and the bibliography therein.

X. Systems Involving Ordinary Differential Equations. According to our general concept of parabolicity, an equation

$$u_t = f(t, x, u)$$

is also parabolic. But since this is an ordinary differential equation depending on a parameter x, it is natural to give boundary values only on $\partial_0 G$ and not on $\partial_1 G$.

Now suppose we have a system with two kinds of equations

$$\begin{aligned} u_{\alpha,t} &= f_\alpha(t, x, \boldsymbol{u}, u_{\alpha,x}, u_{\alpha,xx}) \quad \text{for } \alpha = 1, \ldots, n' \\ u_{\beta,t} &= f_\beta(t, x, \boldsymbol{u}) \quad \text{for } \beta = n'+1, \ldots, n. \end{aligned} \tag{11}$$

Here again it is natural to have boundary conditions on R_p for the components u_α and boundary conditions on $\partial_0 G$ for the components u_β. There are problems in diffusion theory and also in reactor physics which give rise to systems of the form (11). See, e.g., Wilson (1948), Crank (1952) (1956; p. 132f.). It is not difficult to formulate the basic theorems of this section in such a way that these problems are covered. Instead of going into detail, which requires lengthy definitions, we shall give a monotonicity theorem similar to Nagumo's lemma. For the sake of simplicity we restrict ourselves to cylindrical domains.

XI. Lemma. Let $G = (0, T) \times D$, where D is an open subset of E^m. Let the function $f \in \mathscr{P}^n$ corresponding to the system (11) be quasimonotone increasing in $\boldsymbol{z}$. Suppose that $v_\nu, w_\alpha \in Z_0$, and v_β, w_β are continuous in $\bar{G}$, $v_{\beta,t}$, $w_{\beta,t}$ exist in $(0, T] \times \bar{D}$. Then

(α) $v_\alpha < w_\alpha$ on R_p^+ and R_∞,
$v_\beta < w_\beta$ on $\partial_0 G$ and R_∞;

(β) $(\boldsymbol{P v})_\alpha < (\boldsymbol{P w})_\alpha$ in $R_p = (0, T] \times D$,
$(\boldsymbol{P v})_\beta < (\boldsymbol{P w})_\beta$ in $(0, T] \times \bar{D}$

implies

$$\boldsymbol{v} < \boldsymbol{w} \quad \text{in} \quad G_p.$$

It is assumed that $D(f_\nu)$ contains all arguments occurring in (β).

The proof is not difficult and follows the usual pattern. In conclusion we remark that the parabolic boundary can be defined in such a way the above Lemma XI is contained in Lemma VI. Roughly speaking, the parabolic boundary is different for the different components. A point $(\bar{t}, \bar{x}) \in \partial_1 G$ is taken to the parabolic interior for the νth equation if $f_\nu(\bar{t}, \bar{x}, z, p, r)$ is independent of p and r, and to the parabolic boundary otherwise. With this definition the basic theorems still hold. See also Nickel (1960). Estimation, uniqueness and existence for diffusion systems are treated by McNabb (1961) and Maslennikova (1963, 1963a, 1963b, 1963c).

34. Generalizations and Supplements. The Nonstationary Boundary Layer Equations

Now we consider an implicit parabolic differential equation

$$F(t, x, u, u_t, u_x, u_{xx}) = 0. \tag{1}$$

Let $F(t, x, z, q, p, r)$ (below we always use the letter q for u_t) be (weakly) monotone increasing in q and monotone decreasing in r (in the sense of 23 II). For $F = q - f(t, x, z, p, r)$ we obtain the type of equation encountered up to now. The principal lemma 24 IV can be extended to the equation (1); it has already been given in this form by Nagumo-Simoda (1951). This development is not only of a formal nature since only *weak* monotonicity in q is required.

We go a step further and also allow $t = (t_1, \ldots, t_l) \in E^l$ to be a vector ($l \geqq 1$). Then in (1) we have $u_t = (u_{t_1}, \ldots, u_{t_l})$ and $q = (q_1, \ldots, q_l)$. The function $F(t, x, z, q, p, r)$ belongs to the class $\mathcal{R}$ if it is monotone decreasing in r (in the sense of 23 II) and (weakly) monotone increasing in q ($q \leqq \bar{q}$ thus means $q_\lambda \leqq \bar{q}_\lambda$, $\lambda = 1, \ldots, l$). Furthermore the definitions 23 I are taken over in the obvious way. If $\|t\|_e$, $\|x\|_e$ are the Euclidean norms in E^l and E^m respectively, then we define a neighborhood U of the point $(\bar{t}, \bar{x})$ by the inequalities $\|t - \bar{t}\|_e^2 + \|x - \bar{x}\|_e^2 < \delta^2$, and the lower half-neighborhood U_- by the additional inequality $t < \bar{t}$ (and thus $t_\lambda < \bar{t}_\lambda$ for $\lambda = 1, \ldots, l$). Furthermore we take over the concepts G_p, R_p, $\partial_0 G, \ldots, Z, Z_0$ as well as $Z(F)$, $Z_0(F)$ in the obvious translation of 23 III. Thus G lies between 0 and T ($0 < T$; $0, T \in E^l$). For example, if $l = 2$ and $G = (0, T_1) \times (0, T_2) \times D$ with an open set $D \subset E^m$, then

$$\left.\begin{aligned} \partial_0 G &= [0, T_1] \times \{0\} \times \bar{D} + \{0\} \times [0, T_2] \times \bar{D} \\ \partial_1 G &= (0, T_1] \times (0, T_2] \times \partial D \\ \partial_2 G &= \{T_1\} \times (0, T_2] \times D + (0, T_1] \times \{T_2\} \times D \\ G_p &= (0, T_1] \times (0, T_2] \times D. \end{aligned}\right\} \tag{2}$$

For the concepts "$\varphi < \psi$ on R_p^+" ... of 24 IV, instead of the sequences $(t_k, x_{(k)}) \in G_p$ for which the t_k are monotone decreasing, we now have sequences $(t_{(k)}, x_{(k)}) \in G_p$ for which $s(t_{(k)})$ is (weakly) monotone decreasing; here we use the notation

$$s(t) = t_1 + \cdots + t_l . \tag{3}$$

The t_λ are also called "*time*" variables, the x_μ "*space*" variables.

The l-dimensional analog of 24 I is the following

I. Lemma. *For two functions $\varphi, \psi \in Z_0$ suppose: If $\varphi = \psi$, $\varphi_x = \psi_x$, $\varphi_{xx} \leqq \psi_{xx}$ at a point in G_p, then $\varphi_{t_\lambda} < \psi_{t_\lambda}$ at this point for at least one λ. Then exactly one of the two statements is true:*

(α) $\varphi < \psi$ in G_p;

(β) there is a maximal c, $0 \leqq c < s(T)$, such that

$$\varphi < \psi \text{ in } G_p \cdot \{(t, x) | s(t) \leqq c\} \text{ and}$$

$$\varphi(t_{(k)}, x_{(k)}) \geqq \psi(t_{(k)}, x_{(k)}) \text{ for } k = 1, 2, 3, \ldots$$

where $s(t_{(k)})$ decreases strictly monotonely to c and $(t_{(k)}, x_{(k)}) \to (\bar{t}, \bar{x}) \in R_p$ or $\|x_{(k)}\|_e \to \infty$.

For the proof, as in 24 I we choose c to be as large as possible with $\varphi < \psi$ for $s(t) < c$ and then show that this inequality still holds for the points of G_p lying on the hyperplane $s(t) = c$. Indeed, if we had $\varphi = \psi$ at a point of G_p with $s(t) = c$, then we would have (since we have a minimum of $\psi - \varphi$ relative to x) $\varphi = \psi$, $\varphi_x = \psi_x$, $\varphi_{xx} \leqq \psi_{xx}$ and further $\varphi_{t_\lambda}^- \geqq \psi_{t_\lambda}^-$ at this point for all left-sided t_λ-derivatives. However this is impossible by hypothesis. Hence if $c = s(T)$ we obtain the case (α), otherwise the case (β).

From this we easily obtain the following central lemma. It was given by Nagumo-Simoda (1951) for $l = 1$.

II. Lemma. *If $F \in \mathcal{R}$ and $v, w \in Z_0(F)$ and we have*

(α) $v < w$ on R_p^+ and on R_∞;

(β) $F(t, x, v, v_t, v_x, v_{xx}) < F(t, x, w, w_t, w_x, w_{xx})$ in G_p,

then

$$v < w \quad \text{in} \quad G_p .$$

Indeed if $v = w$, $v_x = w_x$, $v_{xx} \leqq w_{xx}$ at a point, then (β) (we omit t, x) implies

$$F(v, v_t, v_x, v_{xx}) < F(w, w_t, w_x, w_{xx}) \leqq F(v, w_t, v_x, v_{xx}) .$$

But this is inconsistent with the inequality $v_t \geqq w_t$ because of the monotonicity of F in q. Thus the hypothesis of I holds and hence, since I (β) is eliminated due to II (α), we have the assertion I (α).

We shall not pursue the theory further systematically, but just refer to several consequences.

III. Remarks. (α) No trouble arises in extending Lemma II to systems of implicit parabolic differential equations

$$F_\nu(t, x, u_1, \ldots, u_n, u_{\nu,t}, u_{\nu,x}, u_{\nu,xx}) = 0 \qquad (\nu = 1, \ldots, n) . \tag{4}$$

If $F=(F_1, \ldots, F_n)$ and $F_\nu = F_\nu(t, x, z, q, p, r)$ is quasi-monotone increasing in z, then the "vector" Lemma II holds. It represents a generalization of 32 VI[1].

(β) There seems to be the following objection to introducing several "time" variables. If $l>1$, we can leave t_1 alone as the only time variable and consider the remaining t_λ as "space" variables, $t_2 = x_{m+1}, \ldots, t_l = x_{m+l-1}$. Since F only has to be *weakly* monotone in r, after such a renaming F is allowable in Lemma II. However with this – and here is the essential point – the sets G_p and R_p are changed and hence so are the boundary conditions.

As an example of the application of Lemma II we consider

IV. The Nonstationary Boundary Layer Equations, namely, the two-dimensional case treated in Section 30 but with dependence on time. They are [Schlichting (1955), p. 97]

$$u_t + u u_x + v u_y = U_t + U U_x + \nu u_{yy}, \tag{5}$$

$$u_x + v_y = 0. \tag{6}$$

The physical meaning of the quantities x, y, U, u, v, ν is the same as in Section 30; the time t appears as a new independent variable, and we have $u = u(t, x, y)$, $v = v(t, x, y)$, $U = U(t, x)$ (we note that t, x, y are scalars and have a meaning different from that at the beginning of the section). The domain is obtained as the topological product of a time interval $0 \leqq t \leqq T$ with the domain of Section 30, i.e.,

$$\bar{G} = [0, T] \times [0, X] \times [0, \infty).$$

The boundary conditions are

$$\left.\begin{aligned} u(t,x,0) = v(t,x,0) = 0,\ u(t,0,y) = u_a(t,y),\ u(t,x,\infty) = U(t,x)^2 \\ u(0,x,y) = u_b(x,y), \quad v(0,x,y) = v_b(x,y) \end{aligned}\right\} \tag{7}$$

$(0 \leqq t \leqq T, 0 \leqq x \leqq X, 0 \leqq y \leqq \infty)$ with given functions U, u_a, u_b, v_b. We assume consistency conditions among these four functions which guarantee that there are indeed functions u continuous in $\bar{G}$ with the boundary values (7), for example, $u_a(t, y) \to U(t, 0)$ for $y \to \infty$ uniformly in t.

The handling of the system (5) (6) is made more difficult by the fact that this system is not of the form (4), since a derivative of u appears in the continuity equation (6). While we avoided these difficulties in Section 30 by means of the von Mises transformation, we remove them here by a relatively primitive device: we simply ignore the continuity equation. Thus the equation (5), where we have to consider v as a given function, is

[1] For reasons of brevity we do not go into boundary conditions of the third kind; II (α) can be replaced by 32 VI (α).

[2] In the sense of uniform convergence as in Section 30.

handled by itself. Here we consider t and x as two time variables and y as a space variable. Then boundary values are given by (7) just on R_p and at infinity; see (2). If we use the abbreviation (for given v)

$$Hu \equiv H(u, u_t, u_x, u_y, u_{yy}) \equiv u_t + uu_x + vu_y - U_t - UU_x - \nu u_{yy}, \tag{8}$$

then $H \in \mathscr{R}$ as long as $u \geqq 0$, and (5) is identical with $Hu = 0$. For H we have (always with t, x as time and y as space variables) Lemma II which, however, can be sharpened as in 30 VI to the

V. Lemma. *If (for given v) for two nonnegative functions $u, w \in Z(H)$ we have*

(α) $u \leqq w$ *on* R_p *and on* R_∞;

(β) $Hu \leqq Hw$ *in* G_p,

then

$$u \leqq w \quad in \quad \bar{G}$$

provided w_x *is bounded below.*

This follows by applying II to $F = H$ and the functions u and $\bar{w} = w + \varrho$, $\varrho = \varepsilon e^{Kt}$ ($\varepsilon > 0$). Then, as was required in II (α), we have $v < \bar{w}$ on R_p and on R_∞, and further

$$H\bar{w} - Hw = \bar{w}_t + \bar{w}\bar{w}_x - w_t - ww_x = \varrho' + \varrho w_x = \varepsilon e^{Kt}(K + w_x) > 0$$

for a sufficiently large K. From this and from (β) we have $Hu \leqq Hw < H\bar{w}$ and thus by II, $u < \bar{w}$, from which we obtain the assertion for $\varepsilon \to 0$.

As an application we consider as in 30 VII

VI. The Velocity of the Potential Flow as Solution. The velocity of the potential flow $U(t, x)$ is itself a solution of the equation (5) *for arbitrary* v. From this and from V we obtain:

(α) *For a solution* $u \geqq 0$ *of the problem* (5) (6) (7) *we have*

$$u(t, x, y) \leqq U(t, x) \quad in \quad \bar{G},$$

if $U \geqq 0$, U_x *is bounded below and*

$$u_a(t, y) \leqq U(t, 0), \qquad u_b(x, y) \leqq U(0, x)$$

[Nickel (1960); Velte (1960) for $U = U(x)$].

The assertion 30 VII (β) is not so easy to carry over since $\sqrt{U^2 + c}$ is not a solution in general.

(β) Suppose $U > 0$, U_x is bounded below and $c(t)$ is a differentiable function in $[0, T]$. We seek a bound w of the form

$$w(t, x) = \sqrt{U^2(t, x) + c(t)}\,.$$

It is easy to verify that the inequality $Hw \geqq 0$ is equivalent to

$$c' \geqq 2U_t\,[\sqrt{U^2 + c} - U]\,. \tag{9}$$

For a solution $u \geqq 0$ *of the problem* (5) *to* (7) *we have*

$$u(t, x, y) \leqq w = \sqrt{U^2(t, x) + c(t)}\,, \tag{10}$$

if the inequality (9) *holds and if*

$$u_a(t, y) \leqq w(t, 0) \quad \textit{and} \quad u_b(x, y) \leqq w(0, x)\,.$$

For example, if $U_t \geqq 0$, then we can set

$$c(t) = c_0 e^{\int_0^t h(s)\,ds} \quad \text{with} \quad h(t) = \max_{0 \leqq x \leqq X} \frac{U_t(t, x)}{U(t, x)}\,.$$

In particular we may take a constant $c(t) = c_0$ for $U = U(x)$; we find the bound for this case in Velte (1960).

For $U_t \leqq 0$, $c(t) = \text{const.}$ is also allowed by (10). However we can often give better functions, i.e., functions $c(t)$ monotone decreasing in t. Finally we note that we obtain an ordinary differential equation for c

$$c' = 2U'(\sqrt{U^2 + c} - U)$$

in the case of a velocity U depending only on t.

VII. Comment. We were able to eliminate the continuity equation for the stationary boundary layer differential equations in Section 30 by the von Mises transformation. We can indeed achieve the same thing in the nonstationary case with a "nonstationary von Mises transformation," but complications do arise here. We should emphasize the procedure used in IV to VI of simply suppressing a "disturbing" equation [here the continuity equation (6)] and treating the remaining equations with the help of Lemma II (or the corresponding lemma for systems), even though it seems very crude at first glance, because it allows us to handle other fundamental equations of hydrodynamics. Here we are thinking particularly of the three-dimensional stationary and nonstationary boundary layer differential equations and the (incompressible) Navier-Stokes differential equations. Here the continuity equation is always disturbing, the remaining equations are of the form (4). These thoughts have already been expressed in the work of Velte (1960) which represents to some extent a first step in this direction. However it seems improbable that we can also prove uniqueness theorems in this way[1]. On the other hand, Nickel (1963) succeeded in using this method to show in a mathematically satisfactory way that the boundary layer equations can be considered as a limiting case of the Navier-Stokes equations (for a stationary, incom-

[1] The trouble is already evident in the system (5) (6). If we have two solutions u_1, u_2, then we associate two different functions v_1, v_2 with them, i.e., if we consider (5) alone, we have two different differential equations!

pressible, two-dimensional boundary layer). He also gave a uniqueness theorem for nonstationary boundary layers (1964) using the Crocco transformation.

VIII. Partial Differential Equations of First Order. Differential equations of the form

$$u_t = f(t, x, u, u_x) \quad \text{for} \quad u = u(t, x) \tag{11}$$

and systems of such differential equations can be considered as special cases of parabolic differential equations [it was assumed only that $f(t, x, z, p, r)$ is *weakly* monotone increasing in r]. However since the boundary conditions involved in the equation (11) are of a different type than those in the parabolic differential equations, our theorems are not directly applicable. But with minor modifications in statement and proof they can be put into a form suitable to the initial value problem for the equation (11). The most important older results in this direction are contained in the work of Haar (1928), Ważewski (1933, 1933a), Nagumo (1938). Newer results have been obtained by many authors, including Szarski (1948, 1949, 1954), Pagni (1951), Scorza Dragoni and Volpato (1951), Volpato (1951), and Mlak (1956a). These works contain extensive bibliographies.

35. The Line Method for Parabolic Equations

The one-dimensional heat equation

$$u_t = u_{xx}$$

in the rectangle $G_p = (0, T] \times (0, a)$ (see 27 I) will serve as an introductory example. By using a discretization in the spatial variable only, i.e., by putting

$$x_\nu = \nu h \quad \text{for} \quad \nu = 0, 1, \ldots, n,$$

where $h = a/n$ (n fixed), and replacing $u(t, x_\nu)$ by $v_\nu(t)$,

$$u_{xx}(t, x_\nu) \quad \text{by} \quad \delta^2 v_\nu = \frac{1}{h^2}(v_{\nu+1} + v_{\nu-1} - 2v_\nu),$$

the heat equation is replaced by

$$v_\nu'(t) = \delta^2 v_\nu(t) = \frac{v_{\nu+1} + v_{\nu-1} - 2v_\nu}{h^2} \qquad (\nu = 1, \ldots, n-1).$$

If a first boundary value problem

$$u(0, x) = \eta(x) \text{ in } [0, a], \quad u(t, 0) = \eta_0(t), \quad u(t, a) = \eta_1(t) \text{ in } (0, T]$$

is given, the corresponding conditions for the new problem are

$$v_\nu(0)=\eta(x_\nu) \quad \text{for} \quad \nu=1,\dots,n-1,$$
$$v_0(t)=\eta_0(t) \quad \text{and} \quad v_n(t)=\eta_1(t).$$

The new problem is an initial value problem for a system of $n-1$ ordinary differential equations. Note that the initial values $v_\nu(0)$ are the initial values η of the original problem, while the boundary values η_0 and η_1 appear in the system of differential equations (in the first and last equations).

The above approximation scheme is a special differential-difference method which is called the line method or straight-line method or method of lines. This method should not be confused with another line method consisting of the discretization of the variable t only. It transforms the original problem into a sequence of boundary value problems for ordinary differential equations of second order and is also called the "method of Rothe". Liskovets (1965) distinguishes between the "transversal" line method (Rothe's method) and the "longitudinal" line method (the method discussed here).

We will now turn to the general case.

I. The Problems A and A_n. We consider the parabolic differential equation

$$u_t=f(t,x,u,u_x,u_{xx}) \quad (m=1) \tag{1}$$

in the rectangle $G_p=(0,T]\times(0,a)$ $(a>0)$ with given boundary values

$$u(0,x)=\eta(x) \text{ in } [0,a],\ u(t,0)=\eta_0(t),\ u(t,a)=\eta_1(t) \text{ in } (0,T]. \tag{2}$$

The problem (1) (2) is called *problem A* for short.

Let n be fixed, $h=a/n$ and

$$x_\nu=\nu h \quad \text{for} \quad \nu=0,1,\dots,n.$$

For a given vector function $\boldsymbol{v}(t)=(v_1(t),\dots,v_{\nu-1}(t))$ we define the difference expressions corresponding to the first and second derivatives with respect to x

$$\delta v_\nu=\frac{v_{\nu+1}-v_{\nu-1}}{2h}, \quad \delta^2 v_\nu=\frac{v_{\nu+1}+v_{\nu-1}-2v_\nu}{h^2} \quad (\nu=1,\dots,n-1). \tag{3}$$

In this definition it is understood that $v_0(t)=\eta_0(t)$ and $v_n(t)=\eta_1(t)$, i.e.,

$$\begin{aligned} \delta v_1&=\frac{v_2-\eta_0}{2h}, & \delta v_{n-1}&=\frac{\eta_1-v_{n-2}}{2h},\\ \delta^2 v_1&=\frac{v_2+\eta_0-2v_1}{h^2}, & \delta^2 v_{n-1}&=\frac{\eta_1+v_{n-2}-2v_{n-1}}{h^2}. \end{aligned} \tag{4}$$

With this notation *problem* A_n is the following:

$$v_\nu'(t) = f(t, x_\nu, v_\nu, \delta v_\nu, \delta^2 v_\nu) \quad (\nu = 1, \ldots, n-1), \tag{5}$$

$$v_\nu(0) = \eta(x_\nu) \quad (\nu = 1, \ldots, n-1). \tag{6}$$

It is again an initial value problem for a system of $n-1$ ordinary differential equations.

II. Quasimonotonicity of the Problem A_n. In the above example of the heat equation it is immediately seen that the corresponding system of ordinary differential equations has the quasimonotonicity property defined in 6 II. Indeed the right hand side of the ν-th equation is monotone increasing in $v_{\nu-1}$ and $v_{\nu+1}$. The same turns out to be true for the general case. This fact is fundamental for our treatment of the line method.

We use the notation $\boldsymbol{z} = (z_1, \ldots, z_{n-1})$ and abbreviate the system (5) as

$$\boldsymbol{v}' = \boldsymbol{g}(t, \boldsymbol{v}) \tag{5'}$$

$$g_\nu(t, \boldsymbol{z}) = f\left(t, x_\nu, z_\nu, \frac{z_{\nu+1} - z_{\nu-1}}{2h}, \frac{z_{\nu+1} + z_{\nu-1} - 2z_\nu}{h^2}\right)$$

($\nu = 1, \ldots, n-1$), where again $z_0 = \eta_0(t)$ and $z_n = \eta_1(t)$.

Let f be defined in $D(f) = M \times M_r$, and $\boldsymbol{g}$ in a corresponding set $D(\boldsymbol{g})$ according to (5').

(α) *If $f(t, x, z, p, r)$ is (weakly) increasing in r and independent of p, then $\boldsymbol{g}(t, \boldsymbol{z})$ is quasimonotone increasing in $\boldsymbol{z}$.*

(β) *If f satisfies in $D(f)$*

$$f(t, x, z, p, r) - f(t, x, z, p, \bar{r}) \geqq \alpha(r - \bar{r}) \quad \textit{for } r \geqq \bar{r}, \tag{7}$$

$$|f(t, x, z, p, r) - f(t, x, z, \bar{p}, r)| \leqq L|p - \bar{p}|, \tag{8}$$

where α and L are positive constants, then $\boldsymbol{g}$ is quasimonotone increasing in $\boldsymbol{z}$ provided $h \leqq 2\alpha/L$.

The statement (α) is evident; the proof of (β) will be indicated. Let $\boldsymbol{z}$ and ν be fixed and $\bar{p}$, $\bar{r}$ be such that $g_\nu(t, \boldsymbol{z}) = f(t, x_\nu, z_\nu, \bar{p}, \bar{r})$, i.e., $\bar{p} = \delta z_\nu$, $\bar{r} = \delta^2 z_\nu$. Then

$$g_\nu(t, z_{\nu-1}, z_\nu, z_{\nu+1} + \varepsilon) = f\left(t, x_\nu, z_\nu, \bar{p} + \frac{\varepsilon}{2h}, \bar{r} + \frac{\varepsilon}{h^2}\right).$$

This equation, together with (7) (8), shows that g_ν is increasing in the variable $z_{\nu+1}$. In the same way it is shown that g_ν increases in $z_{\nu-1}$.

We note that this result remains essentially true if, instead of the central difference δv_ν, one uses the forward or backward difference,

$$\delta v_\nu = \frac{v_{\nu+1} - v_\nu}{h} \quad \text{or} \quad \delta v_\nu = \frac{v_\nu - v_{\nu-1}}{h}. \tag{9}$$

Indeed the following two statements are easily proved.

(γ) *If* (7) (8) *holds and* δv_ν *is the forward or backward difference, then* $\boldsymbol{g}(t, \boldsymbol{z})$ *is quasimonotone increasing in* $\boldsymbol{z}$ *for* $h \leqq \alpha/L$.

(δ) *If* f *is increasing in* r *and increasing [resp. decreasing] in* p *and if* δv_ν *is the forward [resp. backward] difference, then* $\boldsymbol{g}$ *is quasimonotone increasing in* $\boldsymbol{z}$.

These facts make it possible to give an error estimate for the line method using the theory of differential inequalities developed in Section 12. The first theorem is very general. We make no regularity assumptions on f and therefore *assume* that the problems A and A_n have solutions. Let us recall that $\boldsymbol{v} \in Z(\boldsymbol{g})$ means that $\boldsymbol{v}$ is continuous in J and differentiable in J_0 and that $(t, x_\nu, v_\nu, \delta v_\nu, \delta^2 v_\nu) \in D(f)$ for $\nu = 1, \ldots, n-1$ and $t \in J_0$.

III. Error Estimation for the Line Method. Theorem. *Suppose* $\boldsymbol{v} = (v_1, \ldots, v_{n-1}) \in Z(\boldsymbol{g})$ *is a solution of problem* A_n, $u \in Z(f)$ *is a solution of problem* A *and* $(u_1, \ldots, u_{n-1}) \in Z(\boldsymbol{g})$, *where* $u_\nu(t) = u(t, x_\nu)$. *It is assumed that* f, h *and the difference* δv_ν *are defined in such a way that one of the cases* II (α)–(δ) *applies. If*

(α) $|u_x(t, x_\nu) - \delta u_\nu| \leqq \alpha(t)$, $|u_{xx}(t, x_\nu) - \delta^2 u_\nu| \leqq \beta(t)$ *in* J_0 *for* $\nu = 1, \ldots, n-1$;

(β) $f(t, x, z, p, r) - f(t, x, \bar{z}, \bar{p}, \bar{r}) \leqq \omega(t, z - \bar{z}, |p - \bar{p}|, r - \bar{r})$ *for* $z \geqq \bar{z}, r \geqq \bar{r}, (t, x, z, p, r), (t, x, \bar{z}, \bar{p}, \bar{r}) \in D(f)$;

(γ) $\varrho' > \omega(t, \varrho, \alpha(t), \beta(t))$ in J_0, $\varrho(0) > 0$,

where $\alpha(t)$, $\beta(t)$ *are defined in* J_0, $\omega(t, z, p, r)$ *is defined in*

$$J_0 \times \{(z, p, r) : z \geqq 0, p \geqq 0, r \geqq 0\}$$

and increasing in p *and* r, *then*

$$|u(t, x_\nu) - v_\nu(t)| < \varrho(t) \text{ in } J \text{ for } \nu = 0, \ldots, n.$$

The proof of the inequality $v_\nu < u_\nu + \varrho$ will be given; the inequality $u_\nu - \varrho < v_\nu$ is proved similarly. We apply the monotonicity theorem 12 V to the functions v_ν and $w_\nu = u_\nu + \varrho$ and to the function $\boldsymbol{g}(t, \boldsymbol{z})$ defined by (5′) which is quasimonotone increasing in $\boldsymbol{z}$. Since $u_\nu(0) = v_\nu(0)$ and $\varrho(0) > 0$, we have $v_\nu(0) < w_\nu(0)$, and it remains to prove that

$$w_\nu' = u_\nu' + \varrho' > f(t, x_\nu, w_\nu, \delta w_\nu, \delta^2 w_\nu) \quad \text{for} \quad \nu = 1, \ldots, n-1.$$

According to (α) (β) (γ) we have

$$\begin{aligned} u_\nu' + \varrho' &> f(t, x_\nu, u_\nu, u_x(t, x_\nu), u_{xx}(t, x_\nu)) + \omega(t, \varrho, \alpha(t), \beta(t)) \\ &\geqq f(t, x_\nu, u_\nu + \varrho, \delta u_\nu, \delta^2 u_\nu). \end{aligned}$$

Hence the theorem is proved if the inequality

$$(*) \qquad f(t, x_\nu, w_\nu, \delta u_\nu, \delta^2 u_\nu) \geqq f(t, x_\nu, w_\nu, \delta w_\nu, \delta^2 w_\nu)$$

holds. This is trivial for $\nu=2, \ldots, n-2$, since $\delta u_\nu=\delta w_\nu$ and $\delta^2 u_\nu=\delta^2 w_\nu$ in these cases. In the remaining cases $\nu=1$ and $\nu=n-1$

$$\delta^2 u_\nu=\delta^2 w_\nu+\varrho/h^2 ,$$
$$\delta u_\nu=\delta w_\nu \quad \text{or} \quad \delta u_\nu=\delta w_\nu \pm \varrho/h$$

depending on the definition of δ; see (4) and (9). It is easy to see that in each of the cases II $(\alpha)-(\delta)$ the inequality $(*)$ also follows for $\nu=1, n-1$.

The following special case is important for applications.

IV. Error Estimate in the Case of a Lipschitz Condition. *If, under the general hypotheses of* III,

(α) $|u_x(t, x_\nu)-\delta u_\nu| \leqq \alpha(t)$, $|u_{xx}(t, x_\nu)-\delta^2 u_\nu| \leqq \beta(t)$ *in* J_0;

(β) $f(t, x, z, p, r)-f(t, x, \bar{z}, \bar{p}, \bar{r}) \leqq A(t)(z-\bar{z})+B(t)|p-\bar{p}|+C(t)(r-\bar{r})$ *for* $z \leqq \bar{z}$, $r \leqq \bar{r}$ *and* $(t, x, z, p, r), (t, x, \bar{z}, \bar{p}, \bar{r}) \in D(f)$;

where the functions $A(t)$ *and* $B(t)\,\alpha(t)+C(t)\,\beta(t)$ *are continuous in* J_0 *and integrable over* J, *then*

$$|u(t, x_\nu)-v_\nu(t)| \leqq \int_0^t \{B(s)\,\alpha(s)+C(s)\,\beta(s)\}\, e^{\int_s^t A(r)\,dr}\, ds \quad \text{in} \quad J. \tag{10}$$

If the functions α, β, A, B, C *are constant, this inequality reduces to*

$$|u(t, x_\nu)-v_\nu(t)| \leqq \frac{\alpha B+\beta C}{A}(e^{At}-1). \tag{10'}$$

This result is easily reduced to III. The condition III (γ) now reads

$$\varrho' > A(t)\,\varrho+B(t)\,\alpha(t)+C(t)\,\beta(t), \qquad \varrho(0)>0.$$

Hence a bound ϱ is obtained as the solution of the initial value problem

$$\varrho'=A(t)\,\varrho+B(t)\,\alpha(t)+C(t)\,\beta(t)+\varepsilon, \qquad \varrho(0)=\varepsilon \qquad (\varepsilon>0).$$

From the known integral representation of ϱ, the inequality (10) follows for $\varepsilon \to 0$.

We remark that the Lipschitz condition (β) is a uniqueness condition for both problems A and A_n.

Now we investigate the question of under what conditions the solution of the problem A_n tends to the solution of the original problem A when the step size h tends to zero. This *convergence problem* is of great importance from the theoretical and practical points of view. Our previous estimates yield general results in this direction.

Let us recall that $\bar{G}=J \times [0, a]$, $h=a/n$ and $x_\nu=\nu h$. If u is a solution of problem A, we let $\boldsymbol{u}^n$ denote the vector function

$$\boldsymbol{u}^n(t)=\big(u(t, x_1), \ldots, u(t, x_{n-1})\big),$$

while a solution of problem A_n will be denoted by $\boldsymbol{v}^n = (v_1^n, \ldots, v_{n-1}^n)$ in order to emphasize the dependence on n. For vector functions $\boldsymbol{w}(t) = (w_1, \ldots, w_{n-1})$ we introduce the maximum norm

$$\|\boldsymbol{w}\|_n = \max\{|w_\nu(t)| : t \in J, \nu = 1, \ldots, n-1\}.$$

Using this notation, we investigate the behavior of $\|\boldsymbol{u}^n - \boldsymbol{v}^n\|_n$ for $n \to \infty$, i.e., uniform convergence.

V. Convergence Theorem. *Let $u(t, x)$ be a solution of problem A with u, u_x, u_{xx} continuous in $\bar{G}$ and $D < u(t, x) < D'$ in $\bar{G}$. Suppose $f(t, x, z, p, r)$ is continuous in $G \times [D, D'] \times E^2$, and one of the cases* II (α)–(δ) *applies to f and the first order difference δ. If*

(α) $f(t, x, z, p, r) - f(t, x, \bar{z}, \bar{p}, \bar{r}) \leqq A(t)(z - \bar{z}) + B(t)|p - \bar{p}| + C(t)(r - \bar{r})$ *for* $D \leqq \bar{z} < z < D'$, $-\infty < p, \bar{p} < \infty$, $-\infty < \bar{r} < r < \infty$,

where the functions A, B, C are continuous in J_0 and integrable over J, then the problem A_n has, for large n, exactly one solution $\boldsymbol{v}^n$ existing in J, and

$$\|\boldsymbol{u}^n - \boldsymbol{v}^n\|_n \to 0 \quad \textit{for} \quad n \to \infty.$$

Furthermore, if δ is the central difference and one of the cases II (α), (β) *applies,*

$$\|\boldsymbol{u}^n - \boldsymbol{v}^n\|_n = \begin{cases} o(h) \text{ if } u_{xxx} \text{ is continuous in } \bar{G} \\ o(h^2) \text{ if } u_{xxxx} \text{ is bounded in } \bar{G}. \end{cases}$$

The proof follows easily from IV. If, for fixed n, α_n and β_n are the smallest constants such that IV (α) holds with $\alpha(t) = \alpha_n$, $\beta(t) = \beta_n$, then $\alpha_n \to 0$, $\beta_n \to 0$ for $n \to \infty$, since u_x and u_{xx} are uniformly continuous. Let γ_n be the maximum in J of the right hand side of (10) for $\alpha(s) = \alpha_n$, $\beta(s) = \beta_n$. Again $\gamma_n \to 0$ for $n \to \infty$.

According to (α) the function $\boldsymbol{g}(t, z)$, defined by (5′), satisfies a generalized Lipschitz condition

$$g_\nu(t, z) - g_\nu(t, \bar{z}) \leqq \left(|A(t)| + \frac{B(t)}{h} + \frac{2C(t)}{h^2}\right) \|z - \bar{z}\| \quad \text{for} \quad z_\nu \geqq \bar{z}_\nu,$$

where $\|z\|$ is the maximum norm. From 11 II, III or 14 II it follows that A_n has a unique solution. There exists an $\varepsilon > 0$ such that $D + \varepsilon \leqq u \leqq D' - \varepsilon$ in $\bar{G}$. If $\gamma_n < \varepsilon$ and if the solution $\boldsymbol{v}^n$ of problem A_n has the property that $D \leqq v_\nu^n(t) \leqq D'$ ($\nu = 1, \ldots, n-1$) for $0 \leqq t \leqq t_0$, then from (10) we obtain the inequality

$$|u(t, x_\nu) - v_\nu^n(t)| \leqq \gamma_n \quad \text{for} \quad 0 \leqq t \leqq t_0$$

and hence $D + (\varepsilon - \gamma_n) \leqq v_\nu^n(t) \leqq D' - (\varepsilon - \gamma_n)$ for $0 \leqq t \leqq t_0$. From this fact we conclude in the usual way that $\boldsymbol{v}^n$ exists in J and that $D < v_\nu^n(t) < D'$ in J. Then we can apply the estimate (10) in J and get

$$\|\boldsymbol{u}^n - \boldsymbol{v}^n\| \leqq \gamma_n \to 0.$$

If, furthermore, u_{xxx} is continuous in $\bar{G}$ or u_{xxxx} is bounded in G, we have α_n, $\beta_n = o(1/n)$ or $o(1/n^2)$, and the corresponding assertions again follow from (10).

In Theorem V we required that the solution have continuous derivatives up to the boundary. This implies that the initial and boundary values $\eta(x)$, $\eta_0(t)$, $\eta_1(t)$ have corresponding differentiability properties and that the following compatibility conditions

$$\eta_0(0) = \eta(0), \qquad \eta_1(0) = \eta(a), \tag{11}$$

$$\eta_0'(0) = f(0, 0, \eta(0), \eta'(0), \eta''(0)), \qquad \eta_1'(0) = f(0, a, \eta(a), \eta'(a), \eta''(a)) \tag{12}$$

hold (if f is defined and continuous up to the boundary of G). The next convergence theorem deals with the case where the boundary values are assumed only to be continuous.

VI. Convergence Theorem. *Suppose $f(t, x, z, p, r)$ is continuous in $\bar{G} \times E^3$ and satisfies the condition* V (α), *but for $-\infty < \bar{z} < z < \infty$. Suppose further that problem A has a solution u with continuous derivatives u_x, u_{xx} in $\bar{G}$, if $\eta(x)$, $\eta_0(t)$, $\eta_1(t)$ are of class C^∞ and satisfy* (11) (12).

Then, if $u \in Z$ is a solution of problem A, where η, η_0, η_1 are continuous and satisfy (11), *problem A_n has exactly one solution $\mathbf{v}^n$ existing in J, and*

$$\|\mathbf{u}^n - \mathbf{v}^n\|_n \to 0 \quad \textit{for} \quad n \to \infty .$$

The proof uses Theorem V and the following approximation technique. For given $\varepsilon > 0$ there are C^∞-functions $\bar{\eta}$, $\bar{\eta}_0$, $\bar{\eta}_1$ satisfying (11) (12) and

$$|\eta(x) - \bar{\eta}(x)| \leqq \varepsilon \text{ in } [0, a], \ |\eta_0(t) - \bar{\eta}_0(t)| \leqq \varepsilon, \ |\eta_1(t) - \bar{\eta}_1(t)| \leqq \varepsilon \text{ in } J .$$

If $\bar{u}$ is the solution of problem A with the boundary values $\bar{\eta}$, $\bar{\eta}_0$, $\bar{\eta}_1$, then according to 25 XI

$$(*) \qquad |u - \bar{u}| \leqq \varepsilon_1 = \varepsilon e^{\int_0^T A(s)\,ds} \quad \text{in} \quad G .$$

Similarly, if $\mathbf{v}^n$ and $\bar{\mathbf{v}}^n$ are the solutions of problem A_n with respect to the boundary values $\eta, \ldots$ resp. $\bar{\eta}, \ldots$, the inequality

$$(**) \qquad \|\mathbf{v}^n - \bar{\mathbf{v}}^n\|_n \leqq \varepsilon_1$$

holds. For the proof of this inequality we use the auxiliary function

$$\varrho(t) = \varepsilon e^{\int_0^t A(s)\,ds} + \alpha(1 + t) \quad \text{with} \quad \alpha > 0$$

and show that (we suppress the upper index n)

$$v_\nu' + \varrho' > f(t, x_\nu, v_\nu + \varrho, \delta(v_\nu + \varrho), \delta^2(v_\nu + \varrho)) ,$$

$$v_\nu' - \varrho' < f(t, x_\nu, v_\nu - \varrho, \delta(v_\nu - \varrho), \delta^2(v_\nu - \varrho)) .$$

Here, as in the preceding proof, the cases $\nu=1$ and $\nu=n-1$ require special care. It then follows from 12 V that

$$\bar{v}_\nu - \varrho < v_\nu < \bar{v}_\nu + \varrho \quad \text{in } J \quad \text{for} \quad \nu = 1, \ldots, n-1;$$

hence the inequality (**) holds.

Since on the right hand side of the inequality

$$\|\boldsymbol{u}^n - \boldsymbol{v}^n\|_n \leqq \|\boldsymbol{u}^n - \bar{\boldsymbol{u}}^n\|_n + \|\bar{\boldsymbol{u}}^n - \bar{\boldsymbol{v}}^n\|_n + \|\bar{\boldsymbol{v}}^n - \boldsymbol{v}^n\|_n$$

the first and third terms are small by virtue of (*) and (**), while the second term tends to zero for $n \to \infty$ according to Theorem V, the assertion of the theorem follows.

VII. Remarks. We illustrate the previous results by way of examples.

(α) In the linear case

$$u_t = a(t, x)\, u_{xx} + b(t, x)\, u_x + c(t, x)\, u + d(t, x)$$

we have convergence according to Theorem V if the coefficients a, b, c, d are continuous in G and $a(t, x) \geqq \alpha > 0$ (this corresponds to the case II (β); according to II (α) the assumptions $a(t, x) \geqq 0$, $b(t, x) = 0$ are also admissible). If furthermore the coefficients are uniformly Hölder-continuous in G, then Theorem VI is applicable, i.e., convergence holds in the case of continuous boundary values. This follows from a well-known existence theorem; cf. Friedman (1964; Chapter 3, Theorem 7).

(β) We have the same situation for the equation

$$u_t = a(t, x)\, u_{xx} + b(t, x)\, u_x + c(t, x, u)$$

if c satisfies a Lipschitz condition

$$c(t, x, z) - c(t, x, \bar{z}) \leqq C(z - \bar{z}) \quad \text{for} \quad z \geqq \bar{z}$$

(this condition can be weakened). If the functions a, b, c are continuous [uniformly Hölder continuous] in $\bar{G}$, the conclusions of Theorem V [Theorem VI] hold.

(γ) The general quasilinear case

$$u_t = a(t, x, u, u_x)\, u_{xx} + b(t, x, u, u_x)$$

is more involved. Here we get convergence only in a possibly smaller interval $0 \leqq t \leqq t_0$, as is seen from the following considerations.

We consider equation (1) and assume that f is continuous in $\bar{G} \times E^3$ and satisfies a local Lipschitz condition: To each $M > 0$ there corresponds an $L = L(M)$ such that

$$f(t, x, z, p, r) - f(t, x, \bar{z}, \bar{p}, \bar{r}) \leqq L\{(z - \bar{z}) + |p - \bar{p}| + (r - \bar{r})\}$$

for $|z|, |\bar{z}|, |p|, |\bar{p}|, |r|, |\bar{r}| \leqq M$ and $z \geqq \bar{z}, r \geqq \bar{r}$. We assume further that to each M there corresponds an $\alpha = \alpha(M)$ such that

$$f(t, x, z, p, r) - f(t, x, z, p, \bar{r}) \geqq \alpha(r - \bar{r}) \quad \text{for} \quad r \geqq \bar{r}$$

and $|z|, |p|, |r|, |\bar{r}| \leqq M$.

Now let u be a solution of problem A whose derivative u_{xxxx} is bounded in G, and let M be a constant such that $|u|, |u_x|, |u_{xx}|, |u_{xxx}|, |u_{xxxx}| \leqq M$ in $\bar{G}$. We choose $N > M$ and set

$$f_N(t, x, z, p, r) = \alpha r + f(t, x, \varphi_N(z), \varphi_N(p), \varphi_N(r)) - \alpha \varphi_N(r),$$

where $\varphi_N(s)$ is a monotone increasing function of class C^∞ such that

$$\varphi_N(s) = s \quad \text{for} \quad |s| \leqq N \quad \text{and} \quad \varphi_N(s) = \pm(N+1) \quad \text{for} \quad \pm s \geqq N+2,$$

and $\alpha = \alpha(N+1)$. The function f_N satisfies II (β) with $L = L(N+1)$, $\alpha = \alpha(N+1)$ and IV (β) with $A(t) = B(t) = C(t) = L = L(N+1)$ globally, i.e., in $\bar{G} \times E^3$.

The solution u is also a solution of the differential equation with right hand side f_N. The same is true for a solution $\boldsymbol{v}$ of problem A_n, but only as long as $|v_\nu|, |\delta v_\nu|, |\delta^2 v_\nu| \leqq N$, i.e., for small t. For the functions $\alpha(t)$, $\beta(t)$ in IV (α) we can choose $\alpha(t) = Mh^2/24$, $\beta(t) = Mh^2/6$. Then (10′) yields the inequality

$$|u(t, x_\nu) - v_\nu(t)| \leqq \frac{5}{24} Mh^2(e^{Lt} - 1)$$

for $h \leqq 2\alpha/L$ and small t (we assume that δ is the central difference). As long as this inequality is valid, we have the estimate

$$|\delta v_\nu - u_x(t, x_\nu)| \leqq |\delta v_\nu - \delta u_\nu| + |\delta u_\nu - u_x| \leqq \frac{5}{12} Mh(e^{Lt} - 1) + \frac{1}{6} Mh^2$$

$$|\delta^2 v_\nu - u_{xx}(t, x_\nu)| \leqq |\delta^2 v_\nu - \delta^2 u_\nu| + |\delta^2 u_\nu - u_{xx}| \leqq \frac{5}{6} M(e^{Lt} - 1) + \frac{1}{24} Mh^2.$$

Now let t_0 be a positive number such that the right hand side of these two inequalities is $\leqq N - M$ for $0 \leqq t \leqq t_0$ and small h. Then reasoning similar to that used in the proof of V shows that v_ν, δv_ν and $\delta^2 v_\nu$ are bounded in absolute value by N for $0 \leqq t \leqq t_0$. Summarizing, we have:

If f satisfies the conditions stated above (local Lipschitz condition and local strong parabolicity) and if u is a solution of problem A with a bounded derivative u_{xxxx}, then there exists a positive t_0 such that

$$u(t, x_\nu) - v_\nu^n(t) = 0(h^2) \quad \textit{uniformly in} \quad 0 \leqq t \leqq t_0.$$

This result applies in particular to the general quasilinear equation.

(δ) In the general nonlinear case one needs, as the proof in (γ) exhibits, an a priori estimate for δv_ν and $\delta^2 v_\nu$. It was obtained from an estimate of order $o(h^2)$ for the difference $|u_\nu - v_\nu|$. This is a very coarse procedure which yields a convergence theorem only "in the small." A proper way of obtaining such a priori estimates is to derive and investigate differential equations for δv_ν and $\delta^2 v_\nu$. We will not discuss this procedure here since we come back to it in the next section when we deal with the existence problem. There stronger results are derived, namely, convergence of v_ν to u, of δv_ν to u_x and of $\delta^2 v_\nu$ to u_{xx}; see Theorem 36 XII.

There is an ample literature on the line method. A survey article by Liskovets (1965) contains a large bibliography. The line method is also treated in the books by Tychonov and Samarski (1967), where a convergence proof for the heat equation is given, and by Berezin and Zhidkov (1965; p. 598–607). Linear equations are investigated in the works of Wadsworth and Wragg (1964, 1965) and Price, Varga and Warren (1966), among others. Midly nonlinear and quasilinear equations are treated by Kamynin (1961), Budak (1961, 1962), and Morozov (1964). Our approach differs from that of all the other authors in so far as it makes systematic use of the theory of ordinary differential inequalities; it was first published in Walter (1968).

VIII. The Cauchy Problem. The nonlinear Cauchy problem

$$u_t = f(t, x, u, u_x, u_{xx}) \quad \text{in} \quad G_p = J_0 \times E^1, \tag{1'}$$

$$u(0, x) = \eta(x) \quad \text{for} \quad x \in E, \tag{13}$$

where $\eta(x)$ is a given function, is called problem A^*. Applying the line method, we arrive at problem A_h^*:

$$v_\nu' = f(t, x_\nu, v_\nu, \delta v_\nu, \delta^2 v_\nu), \tag{14}$$

$$(\nu = 0, \pm 1, \pm 2, \ldots)$$

$$v_\nu(0) = \eta(x_\nu) \tag{15}$$

Here h is a positive number, $x_\nu = \nu h$, and the differences δv_ν, $\delta^2 v_\nu$ are defined as in (3) or (9).

Problem A_h^* is an initial value problem for an infinite system of ordinary differential equations. It differs essentially in two ways from problem A_n. First, there are no more boundary values $\eta_0(t)$, $\eta_1(t)$ which makes the problem easier. Second, we are dealing with an infinite system which makes the problem more difficult. We now need the theory of infinite systems of differential inequalities developed in Section 12.

We restrict consideration to the case of bounded initial values η and consider equation (14) as an ordinary differential equation in the Banach space B, where B consists of all elements $z = (z_\nu)_{\nu=-\infty}^{\infty}$ with finite maximum norm

$$\|z\| = \sup\{|z_\nu| : \nu = 0, \pm 1, \pm 2, \ldots\}.$$

IX. Convergence Theorem. *Suppose f is continuous in $\bar{G} \times E^3$, $\bar{G} = J \times E^1$, and uniformly continuous in t in the set $\bar{G} \times [-M, M] \times E^2$ for each $M > 0$. Suppose further that for each $M > 0$ there exist two positive constants L and α such that*

(α) $f(t, x, z, p, r) - f(t, x, \bar{z}, \bar{p}, \bar{r}) \leqq L\{(z - \bar{z}) + |p - \bar{p}| + (r - \bar{r})\}$ *for* $z \geqq \bar{z}, r \geqq \bar{r}$

(β) $f(t, x, z, p, r) - f(t, x, z, p, \bar{r}) \geqq \alpha(r - \bar{r})$ *for* $r \geqq \bar{r}$,

as long as the arguments belong to $\bar{G} \times [-M, M] \times E^2$.

Let u be a solution of problem A^* which is, together with its derivatives u_x and u_{xx}, bounded and uniformly continuous in $\bar{G}$. Then the problem A_h^*, considered as initial value problem for ordinary differential equations in the Banach space B, has exactly one solution for each positive h. This solution $\mathfrak{v}^h$ exists in J for small h, and

$$\sup\{|u(t, x_\nu) - v_\nu^h(t)| : t \in J, \nu = 0, \pm 1, \pm 2, \ldots\} \to 0 \quad \textit{for} \quad h \to 0 .$$

Furthermore, if δ is the central difference and u_{xxx} is uniformly continuous in $\bar{G}$ [u_{xxxx} is bounded in $\bar{G}$], the convergence is of order $o(h)$ [$o(h^2)$].

The proof runs along the same lines as in the finite case. It is even simpler since we have no boundary values and hence no exceptional cases. Let M be a constant such that $|u| \leqq M - 1$, and L, α the positive constants corresponding to $M + 1$. The function

$$f^*(t, x, z, p, r) = f(t, x, \varphi_M(z), p, r),$$

where $\varphi_M(s)$ is defined as in VII (γ), satisfies (α) (β) in $\bar{G} \times E^3$ (with these values of L and α). The function $\mathfrak{g}(t, \mathfrak{z})$ from $J \times B$ to B, where

$$g_\nu(t, \mathfrak{z}) = f^*(t, x_\nu, z_\nu, \delta z_\nu, \delta^2 z_\nu),$$

is quasimonotone increasing in $\mathfrak{z}$ for $h \leqq \alpha/L$ and satisfies a uniform Lipschitz condition

$$\|\mathfrak{g}(t, \mathfrak{z}) - \mathfrak{g}(t, \bar{\mathfrak{z}})\| \leqq L\left(1 + \frac{1}{h} + \frac{2}{h^2}\right) \|\mathfrak{z} - \bar{\mathfrak{z}}\| \quad \text{in} \quad J \times B$$

(with respect to the maximum norm in B). The problem (14) (15), with f^* in place of f, has exactly one solution $\mathfrak{v}^h(t)$ according to 7 XII.

Now let α_h, β_h be two constants such that IV (α) holds for all $\nu = 0, \pm 1, \pm 2, \ldots$, and

$$\varrho(t) = (\alpha_h + \beta_h)(e^{Lt} - 1).$$

We want to prove that (h is fixed)

$$u(t, x_\nu) - \varrho(t) \leqq v_\nu(t) \leqq u(t, x_\nu) + \varrho(t).$$

This is done by reduction to the monotonicity theorem 12 XIV, where now $w_\nu(t) = u(t, x_\nu) + \varrho(t)$ (we consider only the second inequality, since the first one can be proved similarly). The function ϱ satisfies the equation $\varrho' = L(\varrho + \alpha_h + \beta_h)$, $\varrho(0) = 0$. Hence we obtain, using assumption (α) and the abbreviation $u_\nu = u(t, x_\nu)$,

$$w_\nu' = u_\nu' + \varrho' = f_M(t, x_\nu, u_\nu, u_x(t, x_\nu), u_{xx}(t, x_\nu)) + L(\varrho + \alpha_h + \beta_h)$$

$$\geqq f_M(t, x_\nu, u_\nu + \varrho, \delta u_\nu, \delta^2 u_\nu) = f_M(t, x, w_\nu, \delta w_\nu, \delta^2 w_\nu)\,.$$

This inequality shows that 12 XIV can be applied and that $u_\nu \leqq w_\nu$. Now for small h the numbers α_h and β_h are small and hence $\varrho < 1$ in J. Therefore we have

$$|u(t, x_\nu) - v_\nu^h(t)| \leqq (\alpha_h + \beta_h)(e^{Lt} - 1) < 1 \quad \text{in} \quad J\,.$$

But then v_ν^h is bounded in absolute value by M, i.e. $\boldsymbol{v}^h$ is also a solution of the original problem (14) (15). From here the conclusions of the theorem follow in the same way as in V.

X. Remarks. (α) These results can be generalized in several directions. First the remarks in VII (γ) carry over to the Cauchy problem. In particular convergence also holds for the general quasilinear equation, but only for small t. Better results in this direction can be achieved by a more thorough investigation of the quantities δv_ν and $\delta^2 v_\nu$. This is done by establishing differential equations for these terms; see the existence proof in the next section.

(β) It is easy to prove a convergence theorem similar to Theorem VI when the initial values are only continuous.

(γ) The case in which the initial values are not bounded but satisfy a certain growth condition can be treated with the same method. Then instead of the maximum norm a weighted maximum norm as in 12 XI is used. Convergence theorems under the assumption that η satisfies an inequality $|\eta(x)| \leqq M e^{\alpha|x|}$ are given by Walter (1969 c). Similar results for the heat equation $u_t = u_{xx}$ were first proved by Kamynin (1952, 1953). The question of whether convergence holds when $\eta(x)$ is allowed to grow like $e^{\alpha x^2}$ is still open. – Our presentation is based on Walter (1969 a).

36. Existence Theorems Based on the Line Method

This section contains existence theorems for the first boundary value problem and for the Cauchy problem with respect to the nonlinear parabolic differential equation

$$u_t = f(t, x, u, u_x, u_{xx})\,. \tag{1}$$

As in the previous section they are called problem A and problem A^*. The proofs are self-contained, elementary and constructive in so far as the solution is obtained as the limit of line method approximations. In order to give a readable presentation we have to compromise between generality and an easy method of proof. Yet our assumptions are still such that the quasilinear case is covered under fairly general conditions.

I. Definitions and Assumptions. We will treat both problems simultaneously. Thus $\bar{G}$ is the set $J \times [0, a]$ in problem A and the set $J \times E^1$ in problem A^*. The boundary conditions are defined as in (35.2) and (35.13). We introduce the following assumptions.

(α) f is defined and twice continuously differentiable in all variables in the set $\bar{G} \times E^3$. In the case of problem A^*, f and its first and second order derivatives are bounded and uniformly continuous in each set $\bar{G} \times M$, where $M \subset E^3$ is bounded. Hence there exists in both cases a function $h(s)$, monotone increasing for $s \geqq 0$, such that

$$|Df(t, x, z, p, r)| \leqq h(|z| + |p| + |r|),$$

where Df stands for f or any first or second order derivative of f.

(β) There exists a constant C and a function $\mu(s)$, positive and continuous for $s \geqq 0$ and with the property that

$$\mu(s) \to \infty \quad \text{for} \quad s \to \infty \quad \text{and} \quad \int^{\infty} \frac{ds}{s\mu(s)} = \infty,$$

such that

$$\left.\begin{array}{r} f(t, x, z, 0, 0) \\ -f(t, x, -z, 0, 0) \end{array}\right\} \leqq C + z\mu(z) \quad \text{for} \quad z \geqq 0.$$

(γ) There exists a positive monotone decreasing function $\delta(s)$ such that

$$f_r(t, x, z, p, r) \geqq \delta(|z|) \qquad \text{(problem } A)$$

$$f_r(t, x, z, p, r) \geqq \delta(|z| + |p|) \quad \text{(problem } A^*).$$

(δ) To each $M > 0$ there corresponds a constant C_0 and a function $\mu_0(s)$ with the properties of μ such that

$$\left.\begin{array}{r} f_x(t, x, z, p, 0) + pf_z(t, x, z, p, 0) \\ -f_x(t, x, z, -p, 0) + pf_z(t, x, z, -p, 0) \end{array}\right\} \leqq p\mu_0(p) + C_0 \quad \text{for} \quad |z| \leqq M, p \geqq 0$$

$$|f_x + pf_z|, |f_t + ff_z| \leqq C_0(1 + r^2) \quad \text{for} \quad |z|, |p| \leqq M,$$

$$|f(t, x, z, p, 0)| \leqq C_0(1 + |p|) \text{ for } |z| \leqq M \text{ (for problem } A \text{ only)}.$$

(ε) $|\eta|, |\eta'|, |\eta''|, |\eta_0|, |\eta_0'|, |\eta_1|, |\eta_1'| \leqq C.$

Since we are dealing with a nonlinear equation we first need an

II. A priori Estimate. *Let* I (α)–(ε) *be satisfied and let u be a solution of problem A or A^* which is, together with its derivatives u_t, u_x, u_{xx}, bounded and uniformly Lipschitz continuous in $\bar{G}$. Then*

$$|u|, |u_x|, |u_{xx}|, |u_t| \leqq M_1 \quad in \quad \bar{G},$$

where the constant M_1 depends only on a, T, C, $\mu(s)$, $\delta(s)$, $h(s)$ and $C_0 = C_0(M)$, $\mu_0(s) = \mu_0(s; M)$.

A function $\varphi(t, x)$ is said to be uniformly Lipschitz continuous in $\bar{G}$ (or to satisfy a uniform Lipschitz condition in $\bar{G}$) if

$$|\varphi(t, x) - \varphi(\bar{t}, \bar{x})| \leqq L(|x - \bar{x}| + |t - \bar{t}|) \quad \text{for} \quad (t, x), (\bar{t}, \bar{x}) \in \bar{G}.$$

We begin with some obvious remarks. For M, P, $Q > 0$, using the function $\varphi_N(s)$ defined in 35 VII (γ), we set

$$f_{M,P,Q}(t, x, z, p, r) = f(t, x, \varphi_M(z), \varphi_P(p), \varphi_Q(r)) + \delta(r - \varphi_Q(r)),$$

where

$$\delta = \delta(M+1) \text{ (problem } A), \qquad \delta = \delta(M+P+2) \text{ (problem } A^*).$$

All first order derivatives of $f_{M,P,Q}$ are bounded in absolute value by the constant $h(M+P+Q+3)$, according to I (α), and $f_{M,P,Q}$ has all the properties of f stated in I (α)–(δ). Furthermore

$$\frac{\partial}{\partial r} f_{M,P,Q}(t, x, z, p, r) \geqq \delta \quad \text{in} \quad \bar{G} \times E^3.$$

We assumed that $|u|$, $|u_x|$, $|u_{xx}|$ are bounded, say, by the constant N. Hence we are allowed to replace f by $f_{N,N,N}$. In the subsequent considerations we shall replace f, along with improved estimates for u, u_x and u_{xx}, by other functions $f_{M,P,Q}$. Thus we may assume from the beginning that f satisfies a uniform Lipschitz condition with respect to z, p and r. Accordingly, for both problems the Lemma of Nagumo-Westphal takes on the simple form

$$v \leqq w \text{ on } R_p \text{ and } Pv \leqq Pw \text{ in } G_p \quad \text{implies } v \leqq w \text{ in } \bar{G};$$

see 24 VI and 28 XV. Naturally we have to take care that the constant N is not involved in the following estimates.

Another remark proves helpful. The function $\bar{u} = -u$ is a solution of

$$\bar{u}_t = \bar{f}(t, x, \bar{u}, \bar{u}_x, \bar{u}_{xx}), \quad \text{where} \quad \bar{f}(t, x, z, p, r) = -f(t, x, -z, -p, -r).$$

It is easily checked that all assumptions on f also hold for $\bar{f}$. Hence if we want an estimate for the absolute value of u or a derivative of u, it suffices to construct an upper bound since we then automatically have an upper bound for $\bar{u} = -u$.

The first part of the proof, the estimate of $|u|$, is very simple. If $w(t, x) = w(t)$ is the solution of

$$w' = C + w\mu(w) \quad \text{in} \quad J, \qquad w(0) = C,$$

then $u \leqq w$ on R_p due to I (ε) and

$$f(t, x, w, w_x, w_{xx}) = f(t, x, w, 0, 0) \leqq C + w\mu(w) = w_t$$

due to I (β). Hence $u \leqq w$. From the assumption on μ and the well-known integral representation of w it follows that

$$|u| \leqq M_2 = w(T) \quad \text{in} \quad \bar{G},$$

where M_2 depends on C, $\mu(s)$ and T.

The next step, the

III. Estimate of u_x, is more difficult. Consider the function

$$u^h(t, x) = \frac{1}{h}(u(t, x+h) - u(t, x)) \qquad (h > 0).$$

From (1), applying the mean value theorem, we obtain a differential equation

$$u_t^h = f_x + f_z u^h + f_p u_x^h + f_r u_{xx}^h. \tag{2}$$

The argument in $f_x, f_z, \ldots$ is of the form $(t, x^*, u^*, u_x^* \, u_{xx}^*)$, where x^*, $u^*, \ldots$ is a value between x and $x+h$, $u(t, x)$ and $u(t, x+h), \ldots$.

We treat the Cauchy problem A^* first and assume that in (2), f stands for the function $f_{M_2, N, N}$. In (2) we consider f_p and f_r as given functions of (t, x), while in f_x and f_z we replace the argument by

$$(t, x, u, u^h, u_x^h).$$

Since the derivatives of f and the derivatives u_x and u_{xx} are uniformly continuous and bounded, this replacement changes the right hand side of (2) by a number not exceeding 1 in absolute value, if h is sufficiently small, say $0 < h \leqq h_0$. Here h_0 depends on N, but this will not affect the validity of our estimate. In other words, if we set

$$g(t, x, z, p, r) = f_x(t, x, u, z, p) + z f_z(t, x, u, z, p) + p f_p(t, x) + r f_r(t, x),$$

then $g \in \mathscr{P}$ satisfies a uniform Lipschitz condition with respect to z, p and r, and

$$u_t^h < g(t, x, u^h, u_x^h, u_{xx}^h) + 1, \qquad |u^h(0, x)| \leqq C.$$

Let $w(t, x) = w(t)$ be the solution of

$$w' = C_0 + w\mu_0(w) + 1, \qquad w(0) = C,$$

where C_0 and μ_0 correspond, according to I (δ), to the constant $M = M_2 + 1$.

Then I (δ) implies

$$\begin{aligned}1+g(t,x,w,w_x,w_{xx}) &= g(t,x,w,0,0)+1\\ &= f_x(t,x,u,w,0)+wf_z(t,x,u,w,0)+1\\ &\leqq 1+C_0+w\mu_0(w)=w'=w_t\,.\end{aligned}$$

Hence we have $u^h \leqq w$ for small h. Passing to the limit we obtain

$$|u_x| \leqq M_3 \quad \text{in} \quad \bar{G}\,,$$

where $M_3 = w(T)$ depends only on M_2 and the data in I (δ) (ε).

In the case of problem A the function u^h is defined in $J \times [0, a-h]$. Here the difficulty arises that the value of u^h is not known at the lateral boundary of this rectangle. In order to get an estimate for $u^h(t,0)$, we show that the function

$$w(t,x) = \eta_0(t) + \cosh Ba - \cosh B(x-a)$$

is, for sufficiently large B, a superfunction for problem A. Indeed $w(t,0) = u(t,0)$ and $w \geqq u$ on R_p for $B \geqq B_0$, where B_0 depends on a and C. The assumptions I (γ) (δ) yield

$$\begin{aligned}Pw = w_t - f = \eta_0' - f\,(t,x,w,\, -B\sinh B(x-a),\, -B^2\cosh B(x-a))\\ \geqq -C+\delta B^2\cosh B(x-a) - f\,(t,x,w,\, -B\sinh B(x-a),0)\\ \geqq -C+\delta B^2\cosh B(x-a) - C_0(1+B|\sinh B(x-a)|)\,,\end{aligned}$$

where $f = f_{M_2,N,N}$ and $\delta = \delta(M_2+1)$, and C_0 corresponds to $M = M_2+1$. Therefore $Pw \geqq 0$ for a fixed $B \geqq B_0$, which again depends only on the data in I. Since the function w is a superfunction, we have

$$u^h(t,0) \leqq \frac{1}{h}(w(t,h)-w(t,0)) \leqq w_x(t,0)\,.$$

Applying this method also to the function $\bar{u} = -u$, we arrive at a bound $|u^h| \leqq C_1$ at the left side of the rectangle and in the same way also on the right side. Now we can repeat the first part of the proof, where we have only to change the initial condition for the bound w in $w(0) = C_1$. The inequality $|u_x| \leqq M_3$ is thus established for both cases.

IV. Estimate of u_t. Both problems A and A^* are treated simultaneously. In the following $f = f_{M_2,M_3,N}$, $\delta = \delta(M_2+M_3+2)$, C_0 is the constant corresponding to $M = \max(M_2+1, M_3+1)$ in I (δ), and C_2 is a constant such that $|u_t| \leqq C_2$ on R_p. Since $u_t(0,x)$ can be calculated from the differential equation, C_2 depends only on C and $h(s)$.

Let

$$U^k(t,x) = \frac{1}{k}(u(t+k,x)-u(t,x)) \qquad (k>0)$$

and

$$v(t,x) = U^k(t,x) + \gamma(u^h(t,x)), \qquad \gamma(s) = \beta e^{\alpha s}$$

($\alpha, \beta > 0$ to be determined). Applying the mean value theorem, we get a differential equation for v

$$\begin{aligned} v_t = U_t^k + \gamma' u_t^h = f_t' + U^k f_z' + U_x^k f_p' + U_{xx}^k f_r' \\ + \gamma'(f_x'' + u^h f_z'' + u_x^h f_p'' + u_{xx}^h f_r'') . \end{aligned} \tag{3}$$

Here and below $f_z = f_z(t, x, u, u_x, u_{xx})$, while f_z' and f_z'' denote the value of the derivatives at a neighbouring argument, and similarly for the other derivatives. Since all the derivatives of f are bounded and uniformly continuous and all other terms are bounded uniformly in h and k (here we use the assumption that the derivatives of u are Lipschitz-continuous), we may change the arguments and also replace U^k by u_t and u_x^h by u_{xx} without committing a large error if h and k are small, say $0 < h, k < h_1$. Here again h_1 depends on N and the constants α, β which are not yet determined, but this is irrelevant for our conclusions. In this way, for h and k sufficiently small, we obtain

$$\begin{aligned} |f_t' + U^k f_z'| &\leqq 1 + |f_t + f f_z| \leqq 1 + C_0(1 + u_{xx}^2) , \\ |f_x'' + u^h f_z''| &\leqq 1 + |f_x + u_x f_z| \leqq 1 + C_0(1 + u_{xx}^2) , \\ U_x^k f_p' + \gamma' u_x^h f_p'' &= v_x f_p + \pi , \qquad |\pi| < 1 , \\ U_{xx}^k f_r' + \gamma' u_{xx}^h f_r'' &= (v_{xx} - \gamma'' u_{xx}^2) f_r + \pi_1 , \qquad |\pi_1| < 1 , \end{aligned}$$

and hence from (3),

$$v_t < C_0(1 + u_{xx}^2)(1 + \gamma') + 3 + \gamma' + f_p v_x + f_r v_{xx} - \gamma'' u_{xx}^2 \delta .$$

If α is determined from $\alpha\delta = C_0 + 1$ and then β from $\gamma'(-M_3) = C_0$, the inequalities $C_0(1 + \gamma') \leqq \delta\gamma''$ and

$$v_t < C_3 + f_p v_x + f_r v_{xx}, \qquad C_3 = C_0(1 + \gamma'(M_3)) + 3 + \gamma'(M_3)$$

follow. In the last inequality the right hand side is linear if we consider f_p and f_r as functions of (t, x). The function v is defined on the set $[0, T-k] \times [0, a-h]$ (problem A) resp. $[0, T-k] \times E^1$ (problem A^*). For small h and k we have $|v| \leqq C_4 = C_2 + 1 + \gamma(M_3)$ on the parabolic boundary of this set, since $|u_t| \leqq C_2$ on R_p and u_t is uniformly continuous in $\bar{G}$. Therefore the function $w = C_4 + C_3 t$ is an upper bound for v by Nagumo's Lemma, since

$$w_t = C_3 + f_p w_x + f_r w_{xx} .$$

This estimate $v \leqq w$ gives an upper bound for U^k and also for u_t. Together with the corresponding lower bound we have

$$|u_t| \leqq M_4 \quad \text{in} \quad \bar{G} ,$$

where M_4 depends only on the data in assumption I.

The a priori estimate of u_{xx} follows from the estimates already obtained and the differential equation, since

$$u_t = f = f(t, x, u, u_x, 0) + f_r u_{xx},$$

where $f_r \geqq \delta > 0$. Theorem II is now completely proved.

V. Existence Theorem for the First Boundary Value Problem. *Let $\eta_0(t)$, $\eta_1(t)$ be three times continuously differentiable in J, $\eta(x)$ three times differentiable in $[0, a]$ and η''' uniformly Lipschitz-continuous in $[0, a]$. Suppose furthermore that*

$$\begin{aligned} \eta_0'(0) &= f(0, 0, \eta(0), \eta'(0), \eta''(0)), \\ \eta_1'(a) &= f(0, a, \eta(a), \eta'(a), \eta''(a)). \end{aligned} \tag{4}$$

Then under the assumptions I (α)–(δ) *there is exactly one solution of problem A such that u, u_t, u_x, u_{xx} are uniformly Lipschitz-continuous in $\bar{G} = J \times [0, a]$.*

VI. Existence Theorem for the Cauchy Problem. *Let η, η', η'', η''' be uniformly Lipschitz-continuous and bounded in E^1. Then under the assumptions* I (α)–(δ) *there is exactly one solution of problem A^* such that u, u_t, u_x, u_{xx} are bounded and uniformly Lipschitz-continuous in $\bar{G} = J \times E^1$.*

The proof of these theorems is based essentially on the following

VII. A Priori Estimates for Systems of Ordinary Differential Equations. We consider the system

$$y_\nu' = a_\nu(t)\,\delta^2 y_\nu + b_\nu(t)\,\delta y_\nu + c_\nu(t)\,y_\nu + d_\nu(t) + e_\nu(t)\,y_\nu\,\delta y_\nu \quad \text{in } J, \tag{5}$$

where h is a given positive number, $\delta^2 y_\nu$ is the second order central difference defined in (35.3), while δy_ν is always understood to be the forward difference

$$\delta y_\nu = \frac{y_{\nu+1} - y_\nu}{h}. \tag{6}$$

Two cases are distinguished. In the "finite case" ν runs from 1 to $n-1$ and $y_0(t)$, $y_n(t)$ are given functions, in the "infinite case" $\nu = 0, \pm 1, \pm 2, \ldots$. A solution is a function $\mathbf{y}(t) = (y_\nu(t))$ continuously differentiable in J which satisfies (5) in J, where in the infinite case (5) is considered as a differential equation in the Banach space B defined in 35 VIII.

(α) *Infinite case. If $a_\nu(t) \geqq \delta > 0$, $c_\nu(t) \leqq D$, $|d_\nu(t)| \leqq D$, $|y_\nu(0)| \leqq D$, $|b_\nu(t)| \leqq D$, $|e_\nu(t)| \leqq D_1$, then*

$$|y_\nu(t)| \leqq N = N(D, T) \qquad \text{in } J \text{ for } 0 < h \leqq h_0 = h_0(\delta, D, D_1). \tag{7}$$

(β) *Finite case. If in addition to the assumptions in* (α) *$|y_0(t)| \leqq D$, $|y_n(t)| \leqq D$, then* (7) *holds.*

(γ) *Finite case. If in addition to the assumptions in* (β) $e_\nu(t)=0$, $|y_0'(t)|\leqq D$, $|y_n'(t)|\leqq D$, $|\delta y_\nu(0)|\leqq D$, *then*

$$|\delta y_0(t)|,\ |\delta y_{n-1}(t)|\leqq N=N(D,T) \quad \text{in } J \text{ for } 0<h\leqq h_0(\delta,D,D_1). \qquad (8)$$

For the proof we remark first that the function $-\boldsymbol{y}(t)$ satisfies the same differential equation as $\boldsymbol{y}(t)$, but with d_ν, e_ν replaced by $-d_\nu$, $-e_\nu$. Therefore it is sufficient to find upper bounds. The proof is based on the theorems on differential inequalities 12 V (finite case) and 12 XIV (infinite case).

(α) We abbreviate (5) as $\boldsymbol{y}'=\boldsymbol{g}(t,\boldsymbol{y})$ and assume for the moment that $\boldsymbol{g}(t,\boldsymbol{z})$ is quasimonotone increasing in $\boldsymbol{z}$. Then the function $\boldsymbol{w}(t)$ with the components $w_\nu=\varrho$ for all ν, where

$$\varrho'=D\varrho+D+1, \qquad \varrho(0)=D+1,$$

is a superfunction. Indeed we have $\boldsymbol{w}(0)>\boldsymbol{y}(0)$ and $\boldsymbol{w}'>\boldsymbol{g}(t,\boldsymbol{w})$ because of $\delta w_\nu=\delta^2 w_\nu=0$. Hence we obtain (under the quasimonotonicity assumption)

$$|y_\nu(t)|\leqq N=\varrho(T).$$

Now we replace the term $y_\nu\delta y_\nu$ in (5) by $\varphi_N(y_\nu)\,\delta y_\nu$, where φ_N is the function defined in 35 VII (γ). The new right hand side is quasimonotone increasing if $h<\delta/((N+1)\,D_1+D)$, since $|\varphi_N|\leqq N+1$. Then by the usual reasoning $|y_\nu(t)|\leqq\varrho(t)$ as long as $|y_\nu(t)|\leqq N$, i.e., in all of J.

(β) The proof is essentially the same as in (α). Yet we have to check that the differential inequality for w_ν is also satisfied in the cases $\nu=1$ and $\nu=n-1$. Now

$$w_1'=\varrho'>a_1\,\frac{y_0-\varrho}{h^2}+c_1\varrho+d_1$$

$$w_{n-1}'=\varrho'>a_{n-1}\,\frac{y_n-\varrho}{h^2}+b_{n-1}\,\frac{y_n-\varrho}{h}+c_{n-1}\varrho+d_{n-1}$$
$$+\,e_{n-1}\varrho\,\frac{y_n-\varrho}{h}$$

for $h<\delta/(D+D_1N)$, because $h^{-1}\,a_{n-1}+b_{n-1}+e_{n-1}\varrho$ is positive for these h. Hence (β) is also proved.

(γ) Since $e_\nu=0$, the right hand side of (5) satisfies a Lipschitz condition and we may use Theorem 12 V in the form 12 V (a) with equality signs permitted (see 12 X). Here we take the auxiliary function

$$\varrho(t,x)=y_0(t)+\cosh Ba-\cosh B(x-a)$$

already used in III and set

$$w_\nu(t)=\varrho(t,x_\nu) \quad \text{with} \quad x_\nu=\nu h, \quad a=nh.$$

At the point (t, x_ν) we have $(\xi = B(x_\nu - a))$

$$a_\nu \varrho_{xx} + b_\nu \varrho_x + c_\nu \varrho + d_\nu \leqq -\delta B^2 \cosh\xi + DB|\sinh\xi| + D\cosh\xi + D$$
$$\leqq \cosh\xi(D + DB - \delta B^2) + D\,.$$

For sufficiently large B the last expression is $\leqq -D-1 \leqq \varrho_t - 1$. Also, for sufficiently large B, $w_\nu(0) \geqq y_\nu(0)$ and $w_0(t) = y_0(t)$, $w_n(t) \geqq y_n(t)$. If for fixed B, we replace ϱ_x and ϱ_{xx} by δw_ν and $\delta^2 w_\nu$ in the above inequality, then there is an $h_1 = h_1(B)$ such that for $0 < h < h_1$ this replacement changes the expression $a_\nu \varrho_{xx} + b_\nu \varrho_x$ by at most 1 in absolute value. Thus we have

$$w_\nu' = y_0' > a_\nu \delta^2 w_\nu + b_\nu \delta w_\nu + c_\nu w_\nu + d_\nu\,,$$

i.e., $\boldsymbol{w} = (w_\nu)$ is a superfunction. Hence the inequality

$$\delta y_0 = \frac{y_1 - y_0}{h} < \frac{w_1 - w_0}{h} \leqq \varrho_x(t, 0)$$

follows. The corresponding lower bound and the bounds for δy_n are derived similarly.

We start the proof of the existence theorems V and VI with some preliminary remarks. Let M_1 be the constant of the a priori estimate II depending only on the data in I (including I (ε)). In what follows

$$f = f_{M_1, M_1, M_1}, \quad \delta = \delta(2M_1 + 2), \quad D = h(3M_1 + 3)\,. \tag{9}$$

This means that all first and second order derivatives of f are uniformly continuous in $\bar{G} \times E^3$ and bounded in absolute value by the constant D, and that $f_r \geqq \delta > 0$.

The uniqueness of the solution follows, e.g., from 25 V (problem A) and 28 XV (problem A^*). The existence proof is based on the line method. Its main part consists of five estimates of the line method approximations corresponding to the functions u, u_x, u_t, u_{xt}, u_{tt}. Our notation is somewhat different from that of Section 35. We let

$$u_\nu \quad p_\nu \quad v_\nu \quad q_\nu \quad w_\nu$$

correspond to

$$u \quad u_x \quad u_t \quad u_{xt} \quad u_{tt}\,.$$

In problem A^*, h is an arbitrary positive number, while in problem A it is of the form $h = a/n$, n a positive integer. As was already stated in VII, $\delta^2 u$ is the central second order difference, δu is the forward first order difference. Thus in problem A, as in 35 I, δu_ν is defined for $\nu = 0, \ldots, n-1$, $\delta^2 u_\nu$ is defined for $\nu = 1, \ldots, n-1$, where $u_0(t) = \eta_0(t)$ and $u_n(t) = \eta_1(t)$.

In the case of problem A^*, the line method approximations are solutions of infinite systems of ordinary differential equations. They are considered (as in VII) as differential equations in the Banach space B defined in 35 VIII.

VIII. Estimate of u_ν, p_ν and v_ν. The initial value problem

$$u'_\nu = f(t, x_\nu, u_\nu, \delta u_\nu, \delta^2 u_\nu), \qquad u_\nu(0) = \eta(x_\nu)\,, \tag{A_h}$$

has exactly one solution in both cases (see 35 IX for the infinite case) denoted by $\boldsymbol{u}^h = (u_\nu^h)$ or simply (u_ν). Using the mean value theorem we get

$$u'_\nu = f(t, x_\nu, 0, 0, 0) + f_z u_\nu + f_p \delta u_\nu + f_r \delta^2 u_\nu\,, \tag{A_h^*}$$

where the values of f_z, f_p and f_r are taken at an intermediate point $(t, x_\nu, z_\nu, p_\nu, r_\nu)$ which is of no interest for further considerations.

Now we define functions $p_\nu(t)$, $v_\nu(t)$ by

$$p_\nu = \delta u_\nu (\approx u_x), \qquad v_\nu = u'_\nu (\approx u_t)\,.$$

They satisfy the following equations

$$p'_\nu = f_x + f_z p_\nu + f_p \delta p_\nu + f_r \delta^2 p_\nu, \qquad p_\nu(0) = \delta\eta(x_\nu) \tag{B_h}$$

$$v'_\nu = f_t^\nu + f_z^\nu v_\nu + f_p^\nu \delta v_\nu + f_r^\nu \delta^2 v_\nu,\;\; v_\nu(0) = f(0, x_\nu, \eta(x_\nu), \eta'(x_\nu), \eta''(x_\nu))\,. \tag{C_h}$$

In equation (B_h), which is derived from (A_h) with the help of the mean value theorem, the argument of the derivatives of f is irrelevant. In equation (C_h), obtained from (A_h) by differentiation with respect to t, the upper index ν indicates that the argument is

$$(t, x_\nu, u_\nu, \delta u_\nu, \delta^2 u_\nu)\,.$$

The three differential equations (A_h) (B_h) (C_h) are of the type

$$y'_\nu = a_\nu(t)\, \delta^2 y_\nu + b_\nu(t)\, \delta y_\nu + c_\nu(t)\, y_\nu + d_\nu(t)\,,$$

where

$$a_\nu(t) \geqq \delta \quad \text{and} \quad |b_\nu(t)|, |c_\nu(t)|, |d_\nu(t)| \leqq D$$

and $|y_\nu(0)| \leqq C$ in (A_h) (B_h), $|y_\nu(0)| \leqq D$ in (C_h).

Hence for the Cauchy problem, from VII (α) we obtain the first three of the following estimates

$$|u_\nu(t)|, |p_\nu(t)|, |v_\nu(t)|, |\delta^2 u_\nu| \leqq N_1 \quad \text{for} \quad 0 < h \leqq h_0(\delta, D)\,, \tag{10}$$

where N_1 depends on C, D, and T. The estimate for $\delta^2 u_\nu$ then follows from (A_h^*).

In the finite case we have $y_0 = \eta_0$, $y_n = \eta_1$ in (A_h), $y_0 = \eta'_0$, $y_n = \eta'_1$ in (C_h) and therefore by VII (β) we have bounds for u_ν and v_ν, by VII (γ) bounds for $\delta u_0 = p_0$, $\delta u_{n-1} = p_{n-1}$ and δv_0, δv_{n-1}. In problem (B_h), ν runs only from 1 to $n-2$. Using the bounds for p_0 and p_{n-1} just obtained, we deduce the estimate for p_ν from VII (β). Summarizing, we have proved that (10) also holds in the finite case and that

$$|\delta v_0|, |\delta v_{n-1}| \leqq N_1 \quad \text{for} \quad 0 < h \leqq h_0(\delta, D)\,. \tag{11}$$

IX. Estimate of q_ν. The estimate of

$$q_\nu = \delta v_\nu = p_\nu' \ (\approx u_{tx})$$

presents some difficulties. Using the identities $\delta^2 v_\nu = \delta q_{\nu-1}, \delta(\delta^2 u_\nu) = \delta^2 p_\nu$ and the "product rule" $\delta(a_\nu b_\nu) = a_{\nu+1}\delta b_\nu + b_\nu \delta a_\nu$, from (C_h) we derive the equations

$$\begin{aligned} q_\nu' = \delta v_\nu' &= \delta(f_t^\nu + f_z^\nu v_\nu + f_p^\nu q_\nu + f_r^\nu \delta q_{\nu-1}) \\ &= \delta f_t^\nu + f_z^{\nu+1} q_\nu + v_\nu \delta f_z^\nu + f_p^{\nu+1}\delta q_\nu + q_\nu \delta f_p^\nu + f_r^\nu \delta^2 q_\nu + \delta q_\nu \delta f_r^\nu . \end{aligned}$$

If $g(t, x, z, p, r)$ is a first order derivative of f and $g^\nu = g(t, x_\nu, u_\nu, \delta u_\nu, \delta^2 u_\nu)$, then

$$\delta g^\nu = g_x + g_z p_\nu + g_p \delta p_\nu + g_r \delta^2 p_\nu .$$

In the subsequent calculations $B, B_0, B_1, \ldots$ denote functions of t and ν which are uniformly bounded in t and ν, where the bound is independent of h (for small h). In this terminology

$$\delta g^\nu = B + B_0 q_\nu ,$$

where $\delta^2 p_\nu$ was replaced by q_ν by means of (B_h). Thus the above equation for q_ν reduces to

$$q_\nu' = B_1 + B_2 q_\nu + B_3 q_\nu^2 + B_4 \delta q_\nu + B_5 q_\nu \delta q_\nu + f_r^\nu \delta^2 q_\nu . \tag{D_h}$$

Now we want to derive a differential equation for

$$\psi_\nu = q_\nu + \gamma(v_\nu) \quad \text{with} \quad \gamma(s) = \beta e^{\alpha s};$$

the constants α and β will be determined later. We write $\gamma_\nu = \gamma(v_\nu)$, $\gamma_+ = \gamma(s)$ with s between v_ν and $v_{\nu+1}$, $\gamma_- = \gamma(s)$ with s between v_ν and $v_{\nu-1}$ (similarly for the derivatives of γ). From (C_h) we have

$$\begin{aligned} \psi_\nu' = q_\nu' + \gamma_\nu' v_\nu' &= B_1 + B_2 q_\nu + B_3 q_\nu^2 + B_4 \delta q_\nu + B_5 q_\nu \delta q_\nu \\ &\quad + f_r^\nu(\delta^2 q_\nu + \gamma_\nu' \delta q_{\nu-1}) + \gamma_\nu'(B_6 + B_7 q_\nu) . \end{aligned}$$

Since

$$\delta\psi_\nu = \delta q_\nu + \gamma_+' q_\nu, \qquad \delta^2\psi_\nu = \delta^2 q_\nu + \gamma_\nu' \delta q_{\nu-1} + \tfrac{1}{2}\gamma_+'' q_\nu^2 + \tfrac{1}{2}\gamma_-'' q_{\nu-1}^2 ,$$

this yields

$$\begin{aligned} \psi_\nu' &= B_1 + \gamma_\nu' B_6 + (\psi_\nu - \gamma_\nu)(B_2 + \gamma_\nu' B_7 - \gamma_+' B_4) + q_\nu^2 (B_3 - \gamma_+' B_5 - \tfrac{1}{2}\gamma_+'' f_r^\nu) \\ &\quad + (\delta\psi_\nu - \gamma_\nu' \psi_\nu + \gamma_+' \gamma_\nu)(B_4 + B_5 \psi_\nu - \gamma_\nu B_5) + f_r^\nu(\delta^2 \psi_\nu - \tfrac{1}{2}\gamma_-'' q_{\nu-1}^2) . \end{aligned}$$

Now we choose positive numbers α, β in such a way that

$$B_3 - \gamma_\nu' B_5 - \tfrac{1}{2}\gamma_\nu'' f_r^\nu \leqq -1 \tag{12}$$

and assume for the moment that

$$B_3 - \gamma_+' B_5 - \tfrac{1}{2}\gamma_+'' f_r^\nu \leqq 0 . \tag{13}$$

Under this assumption

$$\psi_\nu' \leqq B_8 + B_9\psi_\nu + B_{10}\delta\psi_\nu + B_{11}\psi_\nu\delta\psi_\nu + f_r^\nu\delta^2\psi_\nu .$$

From the last inequality we obtain, exactly as in VII (α) and VII (β), an upper bound for ψ_ν independent of h (for small h), since the initial values $\psi_\nu(0) = \delta v_\nu(0) + \gamma(v_\nu(0))$ [and in the finite case by (11) also the boundary values $\psi_0(t)$, $\psi_{n-1}(t)$] are bounded uniformly in h. We then have also an upper bound for q_ν and, according to the remark in II, a lower bound. Yet this estimate, say $|q_\nu| < N_2$ for small h, was derived under the assumption (13).

We get rid of this assumption in the following way. We may assume that $|q_\nu(0)| \leqq N_2$. Then for each fixed h there is a largest $t_h \in (0, T]$ such that $|q_\nu(t)| \leqq N_2 + 1$ for $0 \leqq t \leqq t_h$. This means that

$$|v_\nu - v_{\nu-1}|, \quad |v_\nu - v_{\nu+1}| \leqq (N_2 + 1)\, h \quad \text{for} \quad 0 \leqq t \leqq t_h .$$

There is an $\varepsilon > 0$ such that (13) follows from (12) if $|v_\nu - v_{\nu+1}|, |v_\nu - v_{\nu-1}| \leqq \varepsilon$. Hence there is an $h' > 0$ such that (13) is valid for $0 < h \leqq h'$ and $0 < t < t_h$. But as long as (13) holds, the above considerations are justified, yielding the bound $|q_\nu| \leqq N_2$. Hence one derives, in the usual way, that $t_h = T$, i.e., that there are two positive constants N_2, h_2 such that

$$|q_\nu| \leqq N_2 \quad \text{for} \quad 0 < h \leqq h_2 .$$

X. Estimate of w_ν. For

$$w_\nu(t) = v_\nu' \quad (\approx u_{tt})$$

(C_h) yields (D_t indicates total differentiation with respect to t)

$$\begin{aligned} w_\nu' &= D_t(f_t^\nu + f_z^\nu v_\nu + f_p^\nu \delta v_\nu + f_r^\nu \delta^2 v_\nu) \\ &= D_t f_t^\nu + v_\nu D_t f_z^\nu + \delta v_\nu D_t f_p^\nu + \delta^2 v_\nu D_t f_r^\nu + f_z^\nu w_\nu + f_p^\nu \delta w_\nu + f_r^\nu \delta^2 w_\nu . \end{aligned}$$

If $g(t, x, z, p, r)$ is a first order derivative of f, then

$$D_t g^\nu = g_t^\nu + g_z^\nu v_\nu + g_p^\nu \delta v_\nu + g_r^\nu \delta^2 v_\nu = B + B_0 \delta^2 v_\nu$$

($B, B_0, \ldots$ again denote functions which are uniformly bounded in t, ν and h). Hence, with the notation

$$s_\nu = \delta q_\nu = \delta^2 v_{\nu+1} ,$$

$$w_\nu' = B_1 + B_2 s_{\nu-1} + B_3 s_{\nu-1}^2 + B_4 \delta w_\nu + f_r^\nu \delta^2 w_\nu ; \qquad (E_h)$$

here we used the fact that, by (C_h), w_ν can be expressed in terms of $s_{\nu-1}$. The procedure here is somewhat different from IX. In the infinite case let

$$\psi_\nu = w_\nu + \beta q_\nu^2 ,$$

where the constant $\beta > 0$ will be chosen later. Then

$$\begin{aligned}\psi_\nu' &= w_\nu' + 2\beta q_\nu q_\nu' \\ &= B_1 + B_2 s_{\nu-1} + B_3 s_{\nu-1}^2 + B_4 \delta w_\nu + f_r^\nu \delta^2 w_\nu + 2\beta q_\nu (B_5 + B_6 s_\nu + f_r^\nu \delta^2 q_\nu)\end{aligned}$$

and

$$\delta\psi_\nu = \delta w_\nu + \beta(q_\nu + q_{\nu+1}) s_\nu, \quad \delta^2\psi_\nu = \delta^2 w_\nu + \beta(s_{\nu-1}^2 + s_\nu^2 + 2q_\nu \delta^2 q_\nu).$$

Therefore

$$\begin{aligned}\psi_\nu' = f_r^\nu(\delta^2\psi_\nu - \beta s_{\nu-1}^2 - \beta s_\nu^2) + B_1 + B_2 s_{\nu-1} + B_3 s_{\nu-1}^2 + 2\beta q_\nu (B_5 + B_6 s_\nu) \\ + B_4(\delta\psi_\nu - \beta s_\nu(q_\nu + q_{\nu+1})) \\ \leqq f_r \delta^2 \psi_\nu + B_4 \delta\psi_\nu + B_7\end{aligned}$$

for sufficiently large $\beta > 0$, e.g. $\beta = \frac{1}{\delta}(\sup|B_3| + 1)$. Furthermore

$$w_\nu(0) = f_t + f_z v_\nu + f_p q_\nu + f_r \delta^2 v_\nu|_{t=0}$$

and $v_\nu(0) = f(0, x_\nu, \eta(x_\nu), \eta'(x_\nu), \eta''(x_\nu))$, which yields $|w_\nu(0)| \leqq B_8$. We conclude in the same way as in IX that there exist two positive constants N_3 and h_3 such that

$$|w_\nu| \leqq N_3 \quad \text{for} \quad 0 < h \leqq h_3.$$

In the finite case a difficulty arises since w_ν is defined for $\nu = 0, \ldots, n$, while q_ν is defined only for $\nu = 0, \ldots, n-1$. In the following we assume (without loss of generality) that $a = 1$, $h = 1/n$. The difficulty is overcome by introducing the auxiliary function

$$\psi_\nu = w_\nu + \beta(\alpha_\nu q_{\nu-1}^2 + (1 - \alpha_\nu) q_\nu^2) \quad \text{with} \quad \alpha_\nu = \nu/n.$$

This function is indeed defined for $\nu = 0, \ldots, n$. The identities (c_ν arbitrary)

$$\begin{aligned}\delta(\alpha_\nu c_{\nu-1} + (1-\alpha_\nu) c_\nu) &= \alpha_\nu \delta c_{\nu-1} + (1 - \alpha_{\nu+1}) \delta c_\nu \\ \delta^2(\alpha_\nu c_{\nu-1} + (1-\alpha_\nu) c_\nu) &= \alpha_{\nu-1} \delta^2 c_{\nu-1} + (1 - \alpha_{\nu+1}) \delta^2 c_\nu\end{aligned}$$

yield

$$\begin{aligned}\delta\psi_\nu &= \delta w_\nu + \beta\{\alpha_\nu(q_{\nu-1} + q_\nu) s_{\nu-1} + (1 - \alpha_{\nu+1})(q_\nu + q_{\nu+1}) s_\nu\} \\ \delta^2\psi_\nu &= \delta^2 w_\nu + \beta\{\alpha_{\nu-1}(s_{\nu-2}^2 + s_{\nu-1}^2 + 2q_{\nu-1}\delta^2 q_{\nu-1}) \\ &\quad + (1 - \alpha_{\nu+1})(s_{\nu-1}^2 + s_\nu^2 + 2q_\nu \delta^2 q_\nu)\}.\end{aligned}$$

With the relations

$$q_\nu' = \delta w_\nu \quad \text{and} \quad \delta^2 q_\nu = B + B_1 s_\nu + B_2 \delta w_\nu \quad (\text{by } (D_h))$$

we obtain

$$\begin{aligned}\psi'_\nu &= w'_\nu + 2\beta(\alpha_\nu q_{\nu-1} q'_{\nu-1} + (1-\alpha_\nu)\, q_\nu q'_\nu) \\ &= B_3 + B_4 s_{\nu-1} + B_5 s^2_{\nu-1} + B_6 \delta w_\nu + f^\nu_r \delta^2 w_\nu + B_7 \delta w_{\nu-1} \\ &\leqq f^\nu_r(\delta^2 \psi_\nu - \beta(1-2h)\, s^2_{\nu-1}) + B_8 + B_9 s_{\nu-1} + B_{10} s_\nu \\ &\quad + B_5 s^2_{\nu-1} + B_{11} \delta w_{\nu-1} + B_{12} \delta w_\nu\end{aligned}$$

and finally, with the relations

$s_{\nu-1} = B_{13} + B_{14}\psi_\nu$ (by (C_h)) and $\delta w_\nu = \delta\psi_\nu + B_{15}\psi_\nu + B_{16}\psi_{\nu+1} + B_{17}$,

$$\psi'_\nu \leqq f_r \delta^2 \psi_\nu + B_{11}\delta\psi_{\nu-1} + B_{12}\delta\psi_\nu + B_{18}\psi_{\nu-1} + B_{19}\psi_\nu + B_{20}\psi_{\nu+1} + B_{21}\,.$$

Together with this inequality for $\nu = 1, \ldots, n-1$ we have bounded initial values $\psi_\nu(0)$ and bounded functions $\psi_0 = \eta''_0 + \beta q^2_0$, $\psi_n = \eta''_1 + \beta q^2_{n-1}$. Since the right hand side of the above differential inequality is quasi-monotone increasing for small h, we can employ the method of VII and construct a superfunction $\boldsymbol{\varphi} = (\varphi_\nu)$ where $\varphi_\nu = \varrho$ and $\varrho(t)$ is a solution of the linear differential equation $\varrho' = A\varrho + A_1$, $A > \sup\{|B_{18}| + |B_{19}| + |B_{20}|\}$, $A_1 > \sup|B_{21}|$, with an initial value $\varrho(0) > \psi_\nu(0)$ and such that $\varrho(t) > \psi_0(t)$, $\varrho(t) > \psi_n(t)$. By the theorem on differential inequalities 12 V we have $\varrho(t) > \psi_\nu(t)$, i.e., we have an upper bound for ψ_ν and therefore also for w_ν independent of h. Together with the corresponding lower bound we obtain the previous estimate $|w_\nu| \leqq N_3$ for $0 < h < h_3$ also in the finite case.

We summarize the results of VIII–X in the following statement:

In the first boundary value problem A and in the Cauchy problem A there exist two positive constants N_4 and h_4 such that*

$$|u_\nu|,\ |v_\nu|,\ |w_\nu|,\ |p_\nu|,\ |q_\nu|,\ |\delta^2 u_\nu|,\ |\delta^2 p_\nu|,\ |\delta^2 v_\nu| \leqq N_4 \quad \textit{for} \quad 0 < h < h_4\,. \tag{14}$$

The equations (A^*_h) (B_h) (C_h) were used for the estimate of the second order differences.

XI. Construction of the Solution. From the function $(u_\nu) = (u^h_\nu)$ we construct, by linear interpolation in the x-direction, a function $U^h(t, x)$ defined in $\bar{G}$. More explicitly, we set

$$U^h(t, \alpha x_\nu + (1-\alpha)\, x_{\nu+1}) = \alpha u^h_\nu(t) + (1-\alpha)\, u^h_{\nu+1}(t) \quad \text{for} \quad 0 \leqq \alpha \leqq 1\,. \tag{15}$$

The functions $V^h(t, x)$, $P^h(t, x)$, $R^h(t, x)$ are constructed in a similar fashion, starting from (v^h_ν), (p^h_ν), $(\delta^2 u^h_\nu)$. In the infinite case these functions are again defined on $\bar{G}$, while in the finite case P^h is defined only for $0 \leqq x \leqq a-h$, R^h is defined only for $h \leqq x \leqq a-h$. Therefore we define R^h in the strip $0 \leqq x \leqq h$ as a linear function in x with the same slope as in the neighboring strip $h \leqq x \leqq 2h$, i.e., in the corresponding equation (15) for R^h and $\nu = 1$ we let α vary between 0 and 2. A similar procedure is applied to P^h and R^h in the strip $a-h \leqq x \leqq a$ (α varies between -1 and 1 in the last equation).

The four functions U^h, V^h, P^h, R^h have continuous derivatives with respect to t and piecewise continuous derivatives with respect to x, and these derivatives are, according to (14), bounded in $\bar{G}$, uniformly in h for $0 < h < h_4$. Therefore, according to the Ascoli-Arzelà theorem, there exists a sequence (h_i) such that $h_i \to 0$ for $i \to \infty$ and [with the notation $U^{h_i}(t, x) = U^i(t, x), \ldots$]

$$U^i(t, x) \to U(t, x), \quad V^i(t, x) \to V(t, x), \quad P^i(t, x) \to P(t, x),$$
$$R^i(t, x) \to R(t, x)$$

for $i \to \infty$, uniformly in compact subsets of $\bar{G}$. The four limiting functions are bounded in absolute value by the constant N_4 and uniformly Lipschitz-continuous in x and t with Lipschitz constant N_4. Furthermore, $U_t^i = V^i$ and hence

$$U^i(t, x) = U^i(0, x) + \int_0^t V^i(\tau, x)\, d\tau,$$

which yields

$$U(t, x) = \eta(x) + \int_0^t V(\tau, x)\, d\tau,$$

i.e., $U_t = V$ in $\bar{G}$. Similarly, for $x_\nu < x < x_{\nu+1}$, $U_x^i(t, x) = P^i(t, x_\nu) = P^i(t, x) + \varepsilon$, $|\varepsilon| < N_4 h_i$. Hence, for $\alpha < \beta$,

$$\int_\alpha^\beta P^i(t, \xi)\, d\xi = U^i(t, \beta) - U^i(t, \alpha) + \varepsilon_1, \qquad |\varepsilon_1| < N_4 h_i(\beta - \alpha),$$

whence

$$\int_\alpha^\beta P(t, \xi)\, d\xi = U(t, \beta) - U(t, \alpha).$$

This means that $U_x = P$ and, as can be proved similarly, $P_x = R$ in $\bar{G}$.

Letting M_i denote the set of points $x_\nu = \nu h_i$ ($\nu = 1, \ldots, n_i - 1$, $n_i = a/h_i$ in the finite case, $\nu = 0, \pm 1, \pm 2, \ldots$ in the infinite case), from (A_h) we obtain

$$U_t^i = f(t, x, U^i, P^i, R^i) \quad \text{for} \quad (t, x) \in J \times M_i.$$

Now, if (t, x) is an arbitrary point in $\bar{G}$, there is a sequence (x^i), $x^i \in M_i$, such that $x^i \to x$ for $i \to \infty$. Letting $i \to \infty$ in the above differential equation, we get

$$U_t = f(t, x, U, U_x, U_{xx}) \quad \text{in} \quad \bar{G}.$$

Since U also satisfies the initial and (in the finite case) boundary conditions, it is a solution of problem A or A^*, and it is, together with its derivatives U_t, U_x, U_{xx}, bounded and uniformly Lipschitz-continuous in

$\bar{G}$. But we have to remember that f is in fact the function f_{M_1,M_1,M_1} defined in (9). Since this right hand side has all the properties of the original f stated in I, the a priori estimate II holds for U. This means that $|U|$, $|U_x|$, $|U_{xx}| \leqq M_1$ and that U is also a solution of the original differential equation (1). This remark completes the proof of the existence theorems V and VI.

XII. Convergence Theorem. *Let the conditions of* I *be satisfied and let u be the unique solution of problem A or problem A^*. Then the corresponding problem (A_h), where $h = a/n$ in the case of the first boundary value problem and $h > 0$ in the case of the Cauchy problem, has a unique solution, and the differences*

$$|u(t, h\nu) - u_\nu^h(t)|, \quad |u_x(t, h\nu) - \delta u_\nu^h(t)|,$$

$$|u_{xx}(t, h\nu) - \delta^2 u_\nu^h(t)|, \quad \left|u_t(t, h\nu) - \left(\frac{d}{dt} u_\nu^h(t)\right)\right|$$

tend to zero for $h \to 0$. The convergence is uniform in problem A and uniform on bounded subsets of $\bar{G}$ in problem A^*.

The proof follows immediately from the preceding considerations. If (h_i) is an arbitrary sequence with $h_i \to 0$ for $i \to \infty$ (naturally, h_i is of the form a/n_i in problem A), then there exists, according to the proof in XI, a subsequence (h_i') such that the four differences in question tend to zero for $h = h_i' \to 0$, uniformly on bounded subsets of $\bar{G}$. Hence, by familiar reasoning, convergence also holds for $h \to 0$.

XIII. Remarks. First we discuss the previous results and possible simplifications of their proofs in special cases.

(α) *Linear and mildly nonlinear equations.* Let

$$u_t = a(t, x)\, u_{xx} + b(t, x)\, u_x + c(t, x)\, u + d(t, x)\,.$$

The assumptions I (β)–(δ) are implied by the inequality $a(t, x) \geqq \delta > 0$. Here the existence proof becomes very simple. The a priori estimate II is not needed, and the estimates of q_ν and w_ν are derived directly from VII, since the problems (D_h) and (E_h) are linear. Thus all five estimates for $u_\nu, p_\nu, v_\nu, q_\nu, w_\nu$ follow immediately from VII.

The same is true for the mildly nonlinear equation

$$u_t = a(t, x)\, u_{xx} + b(t, x)\, u_x + c(t, x, u)\,.$$

Here $c(t, x, z)$ must satisfy

$$\left.\begin{array}{r} c(t, x, z) \\ -c(t, x, -z) \end{array}\right\} \leqq C + z\mu(z) \quad \text{for} \quad z \geqq 0\,;$$

see I (β). Again the a priori estimate II can be skipped and the problems (D_h), (E_h) are linear.

(β) *Quasilinear equations.* For the equation

$$u_t = a(t, x, u)\, u_{xx} + b(t, x, u, u_x)$$

the proof is also considerably simplified. Here we have to assume that

$$\left.\begin{array}{r} b(t,x,z,0) \\ -b(t,x,-z,0) \end{array}\right\} \leqq C + z\mu(z) \quad \text{for} \quad z \geqq 0\,,$$

$$a(t,x,z) \geqq \delta(|z|) \qquad (\delta(s) > 0)\,,$$

$$|p|\, b_z(t,x,z,p) \leqq C_0 + |p|\, \mu_0(|p|) \quad \text{for} \quad |z| \leqq M\,,$$

$$|b(t,x,z,p)| \leqq C_0(1+|p|) \quad \text{for} \quad |z| \leqq M \qquad \text{(for problem } A \text{ only)}$$

(C_0 and μ_0 depend on M). The a priori estimate II is needed in this case, but the difficult parts IX and X of the proof are reduced to VII, since in (D_h) and (E_h) we have $B_3 = 0$, i.e., these equations are of the type (5).

More generally, we have $B_3 = 0$ in (D_h) whenever $f_{pr} = 0$ and $B_3 = 0$ in (E_h) whenever $f_{rr} = 0$.

(γ) *The general quasilinear equation*

$$u_t = a(t,x,u,u_x)\, u_{xx} + b(t,x,u,u_x)$$

is treated in the book by Ladyzenskaja, Solonnikov, and Ural'ceva (1968). The relevant existence theorem for problem A is Theorem 5.2 in Chapter VI of their book. The assumptions made there differ considerably from ours. Apart from the parabolicity condition $a(t,x,z,p) \geqq \delta(|z|)$ we have no growth conditions on $a(t,x,z,p)$ whatsoever, while in the cited Theorem 5.2 it is assumed that $0 < \nu \leqq a(t,x,z,p) \leqq \mu$ and that, for $|z| \leqq M$, $|a_z|$ is bounded, $|a_x|$ is growing at most like $|p|$ and $|a_p|$ is smaller than $C/|p|$ for large $|p|$ (this corresponds to the Remark 5.1, case $m = 2$). On the other hand, the conditions on $b(t,x,z,p)$ in Theorem 5.2 are less restrictive than ours. See also Kružkov (1966).

(δ) We note that the differentiability conditions on f can be relaxed. The second derivatives of f are involved only in parts IX and X of the proof. It is easily seen that in IX it suffices to assume that the first derivatives of f are uniformly Lipschitz-continuous. The same is true for X, if the proof is slightly changed, i.e., if w_ν is not defined as the derivative of v_ν, but as a finite difference in the t-direction of v_ν. Therefore the existence theorems V and VI and the convergence theorem XII hold under the assumption that the first derivatives of f are uniformly Lipschitz-continuous in all variables.

(ε) The author has used the line method as a tool for proving existence theorems in two papers, Walter (1968a) (problem A) and (1969b) (problem A^*). These investigations are continued in the present section. Nevertheless the line method is far from being fully developed. In principle our method can be used for other boundary value problems as well as for problems in several space variables, and we hope that our presentation stimulates further research.

While there is, to the author's knowledge, no other literature on the line method with respect to existence proofs (this remark does not apply to the "transversal" line method mentioned at the beginning of Section 35), problems of estimation and convergence have been treated by many authors. Reference is made to the literature quoted in 35 VII (δ). The convergence theorem XII imposes stronger regularity assumptions than the theorems of Section 35. On the other hand it goes far beyond these theorems (and beyond all other published results) with regard to the type of equations involved.

APPENDIX

Elliptic Differential Equations

Here we give a brief survey of the basic theorems on partial differential inequalities of elliptic type.

I. Definition ($G, R=\partial G, \mathscr{Q}, Z, Z_0, \varphi<\psi$ on R^+ and R_∞, monotonicity in r). Our notation and definitions are very similar to those introduced in Section 23 for the parabolic case. Let G be an open set in E^m with boundary $R=\partial G$ ($m \geqq 1$); the points of E^m are denoted by $x=(x_1, \ldots, x_m)$. We consider nonlinear elliptic differential equations of the form

$$f(x, u, u_x, u_{xx})=0 \quad \text{in} \quad G. \tag{1}$$

Here u_x is the gradient vector of u, u_{xx} is the matrix of the second order derivatives of u (23 II). The ellipticity of the equation (1) is defined by the requirement that $-f \in \mathscr{P}$ and that f be independent of t; in this case we write $f \in \mathscr{Q}$. More explicitly, $f=f(x, z, p, r) \in \mathscr{Q}$ means that f is defined in a set $D(f)=M \times M_r$, where M is a set in the $(2m+1)$-dimensional (x, z, p)-space, M_r is a set in the m^2-dimensional r-space, and that f is decreasing in r, i.e.,

$$f(x, z, p, r) \geqq f(x, z, p, \bar{r}) \quad \text{for} \quad r \leqq \bar{r}. \tag{2}$$

We recall that $r \geqq 0$ means that (the symmetric matrix) r is positive semidefinite and that $r \leqq \bar{r}$ is identical with $\bar{r}-r \geqq 0$ (23 II). In short, ellipticity is defined by the inequality (2).

The reader should notice that, according to our definition, f has to be *decreasing* in r rather than increasing. Thus in the case of the Laplace equation $\Delta u=0$ we are in fact dealing with the operator $-\Delta$, i.e., with the function $f(x, z, p, r)=-(r_{11}+r_{22}+\cdots+r_{mm})$. The reason for this will soon become obvious.

The class Z_0 consists of all functions $\varphi(x)$ twice continuously differentiable in G; if furthermore φ is continuous in $\bar{G}$, then $\varphi \in Z$. If $\varphi \in Z_0$ [Z] and $(x, \varphi, \varphi_x, \varphi_{xx}) \in D(f)$ for $x \in G$, then $\varphi \in Z_0(f)$ [$Z(f)$]. We remark here that the differentiability assumptions can be relaxed without affecting the validity of the following theorems; see 23 III (β) and (β').

Finally we define, in much the same way as in 23 IV, inequalities on the boundary of G for functions which are defined only in G. Let φ, ψ be defined in G. Then we write

$$\varphi < \psi \text{ on } R^+ \quad \text{if} \quad \limsup_{k\to\infty} \{\psi(x_{(k)}) - \varphi(x_{(k)})\} > 0$$

for every sequence of points $x_{(k)} \in G$ which converges to a boundary point, $x_{(k)} \to \bar{x} \in R$. Similarly, $\varphi < \psi$ on R_∞ if the above lim sup is positive for every sequence of points $x_{(k)} \in G$ such that $\|x_{(k)}\| \to \infty$. This definition is needed in order to formulate the following theorems in such a way that they apply both to bounded and unbounded G. We remark that for bounded G the statement "$\varphi < \psi$ on R_∞" is always true. The relations $\varphi \leqq \psi$, $\varphi = \psi$ on R^+ or on R_∞ are defined similarly; see 23 IV.

The following simple facts are given without proof.

(α) Let φ, ψ be continuous in $\bar{G}$. Then $\varphi < \psi$ on R^+ if and only if $\varphi(x) < \psi(x)$ for all $x \in R$, and $\varphi \leqq \psi$ on R^+ if and only if $\varphi(x) \leqq \psi(x)$ for all $x \in R$.

(β) Let φ, ψ be continuous in G. If $\varphi < \psi$ on R^+ and R_∞, then the set of points in G with $\varphi(x) \geqq \psi(x)$ is a compact subset of G (which may be empty).

(γ) Let φ be continuous in G and let $\varphi \geqq 0$ on R^+ and R_∞. If $\varphi(x) \leqq 0$ for some point $x \in G$, then there is a point $x_0 \in G$ where φ assumes its minimum, i.e., $\varphi(x_0) \leqq \varphi(x)$ for all $x \in G$.

II. Monotonicity Theorem. *Let $f \in \mathcal{Q}$ be monotone increasing in z and $v, w \in Z_0(f)$. Then*

(α) $v \leqq w$ on R^+ and R_∞,

(β) $f(x, v, v_x, v_{xx}) < f(x, w, w_x, w_{xx})$ in G

implies

$$v < w \quad \text{in} \quad G.$$

For the proof we consider the difference $d = w - v$. According to (α), $d \geqq 0$ on R^+ and R_∞. If, contrary to the conclusion of the theorem, $d(x) \leqq 0$ for some $x \in G$, then by I (γ) there is a point $x_0 \in G$ where d assumes its minimum. At the point x_0 we then have [see 23 III (γ)]

$$d \leqq 0, \quad d_x = 0, \quad d_{xx} \geqq 0, \quad \text{i.e.,} \quad v \geqq w, \quad v_x = w_x, \quad v_{xx} \leqq w_{xx}$$

and hence

$$f(x_0, v, v_x, v_{xx}) \geqq f(x_0, w, w_x, v_{xx}) \geqq f(x_0, w, w_x, w_{xx}).$$

Thus we have arrived at a contradiction to the assumption (β), whence the theorem follows.

This theorem is the elliptic analog of the Nagumo-Westphal Lemma 24 IV. The proof is also very similar. For some applications is is important to note that the conclusion of the theorem holds under much weaker assumptions.

(γ) *Let* $v, w \in Z_0(f)$. *Suppose that at each point* $x \in G$ *for which* $v \geqq w$, $v_x = w_x$, $v_{xx} \leqq w_{xx}$, *the inequalities*

$f(x, w+z, w_x, w_{xx}) \geqq f(x, w, w_x, w_{xx})$ *for* $z > 0$

$f(x, z, w_x, w_{xx} - r) \geqq f(x, z, w_x, w_{xx})$ *for* $r \geqq 0$, $z \geqq w$

or the inequalities

$f(x, v-z, v_x, v_{xx}) \leqq f(x, v, v_x, v_{xx})$ *for* $z \geqq 0$,

$f(x, z, v_x, v_{xx} + r) \leqq f(x, z, v_x, v_{xx})$ *for* $r \geqq 0$, $z \leqq v$

hold (as long as the arguments are in $D(f)$*).*

Then (α) (β) *implies* $v < w$ *in* G.

Here the monotonicity of f in z is replaced by the first inequality, while the ellipticity of f is replaced by the second inequality. For the proof we again consider a point x_0 in G where $v \geqq w$, $v_x = w_x$, $v_{xx} \leqq w_{xx}$ and obtain a contradiction to (β) by means of the inequalities

$$f(x_0, w, w_x, w_{xx}) \leqq f(x_0, v, v_x, w_{xx}) \leqq f(x_0, v, v_x, v_{xx})$$

(the reasoning is similar if the assumptions mentioned in the second place apply).

Remark. The assumption, introduced in I, that $D(f)$ is a product set, has been used in several places. It guarantees that, e.g., $(x, v, v_x, w_{xx}) \in D(f)$.

We now give several other versions of the monotonicity theorem, where the equality sign is permitted in the assumption (β).

III. Corollary. *Let* $v, w \in Z_0(f)$ *and*

(α) $v \leqq w$ *on* R^+ *and* R_∞,

(β) $f(x, v, v_x, v_{xx}) \leqq f(x, w, w_x, w_{xx})$ *in* G.

Then

$$v \leqq w \quad \text{in} \quad G,$$

if one of the following conditions on f *is satisfied:*

(γ_1) $f \in \mathcal{Q}$ *is strictly increasing in* z;

(γ_2) $f = f(x, z, r)$ *is independent of* p, *weakly increasing in* z *and strictly decreasing in* r, *i.e.*, $f(x, z, r) < f(x, z, \bar{r})$ *for* $r > \bar{r}$;

(γ_3) f *is weakly increasing in* z, *and there exist two constants* $L, \alpha > 0$ *such that*

$$|f(x, z, p, r) - f(x, z, \bar{p}, r)| \leqq L \|p - \bar{p}\|$$

$$f(x, z, p, \bar{r}) - f(x, z, p, r) \geqq \alpha \operatorname{tr}(r - \bar{r}) \quad \text{for} \quad r > \bar{r},$$

where $\operatorname{tr}(r) = r_{11} + r_{22} + \cdots + r_{mm}$ *and* $\|.\|$ *is the Euclidean norm in* E^m;

(γ_4) $f \in \mathcal{Q}$ *is weakly increasing in* z, *and there exists an index* μ *such that* f *is independent of* p_μ *and strictly decreasing in* $r_{\mu\mu}$;

(γ_5) *$f \in \mathscr{Q}$ is weakly increasing in z, and there exist two constants L, $\alpha > 0$ and an index μ such that*

$$|f(x, z, p + \lambda e^{\mu}, r) - f(x, z, p, r)| \leqq L|\lambda|$$

$$f(x, z, p, r) - f(x, z, p, r + \lambda r^{\mu}) \geqq \alpha\lambda \quad \textit{for} \quad \lambda > 0,$$

where e^{μ} is the μ-th unit vector ($e^{\mu}_{\lambda} = 1$ for $\mu = \lambda$ and $= 0$ otherwise), and $(r^{\mu})_{\mu\mu} = 1$, $(r^{\mu})_{\lambda\nu} = 0$ otherwise.

For some applications it is important to know that the conclusion of Corollary III holds in fact under much weaker assumptions. Only the following conditions on f are actually needed in the proof (notation as before, $\alpha > 0$):

(γ'_1) $f(x, w + z, w_x, w_{xx}) > f(x, w, w_x, w_{xx})$ for $z > 0$,
$f(x, v, v_x, v_{xx}) \geqq f(x, v, v_x, v_{xx} + r)$ for $r \geqq 0$;

(γ'_2) f is independent of p and
$f(x, w + z, w_{xx}) \geqq f(x, w, w_{xx})$ for $z \geqq 0$,
$f(x, v, v_{xx}) > f(x, v, v_{xx} + r)$ for $r > 0$;

(γ'_3) $f(x, w, w_x, w_{xx}) \leqq f(x, w + z, w_x + p, w_{xx}) + L\|p\|$ for $z \geqq 0$,
$f(x, v, v_x, v_{xx}) - f(x, v, v_x, v_{xx} + r) \geqq \alpha\,\mathrm{tr}(r)$ for $r \geqq 0$

(γ'_4) f is independent of p_{μ} and
$f(x, w + z, w_x, w_{xx}) \geqq f(x, w, w_x, w_{xx})$ for $z \geqq 0$,
$f(x, v, v_x, v_{xx}) > f(x, v, v_x, v_{xx} + r + \lambda r^{\mu})$ for $r \geqq 0$, $\lambda > 0$;

(γ'_5) $f(x, w, w_x, w_{xx}) \leqq f(x, w + z, w_x + \lambda e^{\mu}, w_{xx}) + L|\lambda|$ for $z \geqq 0$,
$f(x, v, v_x, v_{xx}) - f(x, v, v_x, v_{xx} + r + \lambda r^{\mu}) \geqq \alpha\lambda$ for $r \geqq 0$, $\lambda \geqq 0$.

We mention a further generalization. In (γ'_3) and (γ'_5) it is sufficient that for each compact subset H of G there exist two positive constants L and α such that the above inequalities hold.

For the proof we assume that the conclusion is false, i.e., that the difference $d = w - v$ has a negative infimum, say

$$\inf d = \inf(w - v) = -2\beta < 0.$$

Since $d > -\beta$ on R^+ and R_∞, the set H of all $x \in G$ such that $d(x) \leqq -\beta$ is, by I (β), a nonempty compact subset of G.

Now let $\eta(x)$ be a twice continuously differentiable function and $0 \leqq \eta(x) \leqq M$. Then the function $d + \varepsilon\eta$ has, for $0 < \varepsilon < \beta/M$, a minimum at a point x_0 in the interior of H, since on the boundary of H we have $d = -\beta$, $d + \varepsilon\eta \geqq -\beta$, while at a point in H where d assumes its minimum, we have $d = -2\beta$, $d + \varepsilon\eta \leqq -2\beta + M\varepsilon < -\beta$. At the point x_0 we have (see the proof of II) $d + \varepsilon\eta < 0$, $d_x + \varepsilon\eta_x = 0$, $d_{xx} + \varepsilon\eta_{xx} \geqq 0$, and hence

$$v > w, \qquad v_x = w_x + \varepsilon\eta_x, \qquad v_{xx} \leqq w_{xx} + \varepsilon\eta_{xx}. \tag{3}$$

The essential part of the proof consists in the construction of suitable functions η in such a way that a contradiction of the assumption (β) is obtained.

Case (γ_1). We take $\eta(x)=0$. From the assumptions on f and (3) we obtain

$$f(x_0, w, w_x, w_{xx}) < f(x_0, v, v_x, w_{xx}) \leqq f(x_0, v, v_x, v_{xx}),$$

which contradicts (β).

Case (γ_2). Here we take $\eta(x)=A-x^2$, where A is so large that η is positive in H. Then, at the point x_0, $w_{xx}=v_{xx}+r$, where $r \geqq -\varepsilon\eta_{xx}=2\varepsilon I>0$ (I the unit matrix), and hence

$$f(x_0, w, w_{xx}) \leqq f(x_0, v, w_{xx}) < f(x_0, v, v_{xx}).$$

This again contradicts (β).

Case (γ_3). Here the function

$$\eta(x)=A-\varphi(x), \qquad \varphi(x)=e^{B(x_1+\cdots+x_m)},$$

is used. The positive constants A, B will be determined later. Simple computations show that

$$\eta_x=Be\varphi, \qquad \eta_{xx}=-B^2E\varphi,$$

where $e=(1, 1, \ldots, 1)$ and $E=(e_{\lambda\mu})$ with $e_{\lambda\mu}=1$ for all $\lambda, \mu=1, \ldots, m$. We note that $E \geqq 0$, $\operatorname{tr}(E)=m$. Then by (3) the relations

$$w_{xx}-v_{xx} \geqq -\varepsilon\eta_{xx}=\varepsilon B^2E\varphi \geqq 0, \qquad \operatorname{tr}(w_{xx}-v_{xx}) \geqq \operatorname{tr}(-\varepsilon\eta_{xx})=\varepsilon B^2m\varphi$$

hold at the point x_0. The conditions imposed on f then yield

$$\begin{aligned} f(x_0, w, w_x, w_{xx}) &\leqq f(x_0, v, v_x, w_{xx})+L\varepsilon\|\eta_x\| \\ &\leqq f(x_0, v, v_x, v_{xx})+L\varepsilon\|\eta_x\|-\alpha\operatorname{tr}(w_{xx}-v_{xx}). \end{aligned}$$

Now we choose B in such a way that $L<\alpha B\sqrt{m}$, and then A so large that $\eta(x)\geqq 0$ in H. Then the inequalities derived for f contradict the assumption (β), since

$$L\varepsilon\|\eta_x\|-\alpha\operatorname{tr}(w_{xx}-v_{xx}) \leqq L\varepsilon\varphi B\sqrt{m}-\alpha\varepsilon B^2\varphi m<0.$$

Cases (γ_4) *and* (γ_5). The proof is left to the reader. One may use the auxiliary function η given by

$$\eta(x)=A-x_\mu^2 \quad \text{and} \quad \eta(x)=A-e^{Bx_\mu}$$

respectively.

IV. The First Boundary Value Problem (Dirichlet Problem). Sub- and Superfunctions. Let G be bounded, $f\in\mathcal{Q}$ and $\eta(x)$ a continuous function defined on R (discontinuous boundary values will be discussed later). The Dirichlet problem consists of finding a function $u\in Z(f)$ satisfying the two equations

$$f(x, u, u_x, u_{xx}) = 0 \quad \text{in} \quad G \quad \text{and} \quad u = \eta \quad \text{on} \quad R.$$

In accordance with the previous terminology (see introduction, IV) a function v is called a subfunction if $v \leqq u$ in G for all solutions u of the given Dirichlet problem. According to II, $v \in Z(f)$ is a subfunction if

$$f(x, v, v_x, v_{xx}) < 0 \quad \text{in} \quad G \quad \text{and} \quad v \leqq \eta \quad \text{on} \quad R.$$

In this case $v < u$ in G for every solution u of the boundary value problem. If the assumptions of III are valid, then equality signs are allowed, i.e., $v \leqq u$ in G for all solutions u if

$$f(x, v, v_x, v_{xx}) \leqq 0 \quad \text{in} \quad G \quad \text{and} \quad v \leqq \eta \quad \text{on} \quad R.$$

Similar definitions and statements hold for superfunctions.

Since in the case of Corollary III each solution is simultaneously a subfunction and a superfunction, we immediately obtain the following

V. Uniqueness Theorem. *Under the assumptions of Corollary* III *the Dirichlet problem has at most one solution.*

VI. Remarks and Examples. Suppose the differential equation is linear in the second derivatives,

$$-\sum_{\lambda,\mu=1}^{m} k_{\lambda\mu}(x)\, u_{x_\lambda x_\mu} + g(x, u, u_x) = 0, \tag{4}$$

where the symmetric matrix $k = (k_{\lambda\mu})$ is assumed to be positive semidefinite for each $x \in G$. Then the corresponding function f,

$$f(x, z, p, r) = -\sum_{\lambda,\mu=1}^{m} k_{\lambda\mu}(x)\, r_{\lambda\mu} + g(x, z, p), \tag{4'}$$

is decreasing in r, i.e., f belongs to $\mathcal{Q}$. The simple calculation leading to this conclusion was carried out in 23 II (α). We shall give two more special results.

(α) If the (positive semidefinite) matrix $k(x)$ is different from the zero matrix for each $x \in G$, then $f(x, z, p, r)$ is strictly decreasing in r in the sense of III (γ_2).

(β) If the equation (4) is uniformly elliptic, i.e., if there exists a constant $\alpha > 0$ such that

$$\sum_{\lambda,\mu} k_{\lambda\mu}(x)\, \xi_\lambda \xi_\mu \geqq \alpha \xi^2 \quad \text{for all} \quad \xi \in E^m, \quad x \in G, \tag{5}$$

then

$$f(x, z, p, \bar{r}) - f(x, z, p, r) \geqq \alpha \operatorname{tr}(r - \bar{r}) \quad \text{for} \quad r \geqq \bar{r}.$$

Both results follow readily from 23 II (α). Observe that, in the notation of 23 II (α), all elements $d_{\mu\mu}$ are nonnegative and at least one of them is positive if (α) applies, while $d_{\mu\mu} \geqq \alpha$ for all μ if (β) applies.

(γ) According to (α) (β), Corollary III is applicable to the equation (4) if $k(x)$ is positive semidefinite in G, $g(x, z, p)$ is weakly increasing in z and if one of the following conditions is satisfied:

(γ_1) $g(x, z, p)$ is strictly increasing in z;

(γ_2) $g = g(x, z)$ is independent of p, $k(x)$ is different from the zero matrix for each $x \in G$;

(γ_3) $g(x, z, p)$ satisfies a Lipschitz condition in p, $k(x)$ satifies (5);

(γ_4) there exists an index μ such that g is independent of p_μ and $k_{\mu\mu}(x) > 0$ in G;

(γ_5) there exists an index μ such that g satisfies a Lipschitz condition with respect to p_μ and $k_{\mu\mu}(x) \geqq \alpha > 0$ in G.

The above conditions are listed in such a way that (γ_i) corresponds to III (γ_i) for $i = 1, \ldots, 5$. *In particular, each one of these conditions guarantees uniqueness for the Dirichlet problem with respect to equation* (4). These uniqueness conditions are in fact very general, as can be seen by constructing counterexamples. We mention only one very simple counterexample. If $k(x) = 0$, $g(x, z, p) = 0$, then II does apply, but III does not. Indeed there is no uniqueness, since every function $u \in Z$ having the boundary values η is a solution of the Dirichlet problem.

(ε) As we have already pointed out, the auxiliary function $\eta(x)$ plays a basic role in the proof of III. The conditions (γ_i) in III were formulated in such a way that the proof goes through with relatively simple functions η. By more elaborate constructions one can arrive at somewhat more general conditions on f. For example, by making use of ideas used by Redheffer (1960) in connection with the maximum principle, one can replace the Lipschitz condition in (γ_3) and (γ_5) by a more general Osgood condition.

The method of proof presented here goes back essentially to Paraf (1892). It was subsequently used by many authors, mostly in connection with the maximum principle rather than with monotonicity theorems. Yet there is an intimate connection between these two subjects, which will become clear from our treatment of

VII. The Maximum Principle. The definitions are essentially the same as in the parabolic case in 26 I. For a function φ defined in G, $M(\varphi)$ is the infimum of all real numbers A such that $\varphi < A$ on R^+ and R_∞. The maximum principle holds [for the equation (1) in G] if $u \leqq M(u)$ for every solution $u \in Z_0(f)$ of (1). The strong maximum principle holds if for every solution $u \in Z_0(f)$ of (1) we have either $u < M(u)$ in G or $u = M(u)$ in G.

The number $m(\varphi)$ and the minimum principle are defined analogously.

Suppose $f \in \mathfrak{Q}$ *satisfies the assumptions of Corollary* III. *If* $u \in Z_0(f)$ *and*

$$f(x, u, u_x, u_{xx}) \leqq 0, \qquad f(x, M(u), 0, 0) \geqq 0,$$

then

$$u \leqq M(u) \quad in \quad G.$$

In particular, the maximum principle holds if $f(x, z, 0, 0) \geqq 0$ *for* $(x, z, 0, 0) \in D(f)$.

This result follows immediately when III is applied to the functions $v = u$ and $w = M(u)$. We note that a similar statement is valid for the minimum principle; see also 26 II.

The above considerations apply in particular to linear equations and to mildly nonlinear equations of the form (4), but in general they do not apply to quasilinear equations

$$-\sum_{\lambda, \mu = 1}^{m} k_{\lambda\mu}(x, u, u_x)\, u_{x_\lambda x_\mu} + b(x, u, u_x) = 0, \tag{6}$$

since the conditions (γ_i) of III are not satisfied except in very special cases. In order to get useful results for quasilinear equations, one has to consider the refined version of Corollary III with the hypotheses (γ_i') instead of (γ_i). For reasons which will be clear from the next theorem we shall deal with the equation

$$-\sum_{\lambda, \mu} k_{\lambda\mu}(x, u, u_x)\, u_{x_\lambda x_\mu} + \sum_{\mu} b_\mu(x, u, u_x)\, u_{x_\mu} + c(x, u, u_x) = 0, \tag{7}$$

though it is formally not more general than (6).

VIII. Theorem (Maximum Principle). *Let* f *be the function corresponding to the left side of* (7) *and let* $u \in Z_0(f)$. *Then*

(α) $f(x, u, u_x, u_{xx}) \leqq 0$ *in* G,

(β) $c(x, M(u) + z, 0) \geqq c(x, M(u), 0) \geqq 0$ *in* G *for* $z \geqq 0$

implies

$$u \leqq M(u) \quad in \quad G,$$

if the matrix $k(x, u, u_x) = (k_{\lambda\mu}(x, u, u_x))$ *is positive semidefinite in* G *and if one of the following assumptions holds:*

(γ_1) $c(x, M(u) + z, 0) > c(x, M(u), 0)$ *for* $z > 0$;

(γ_2) $b_\mu = 0$ *for* $\mu = 1, \ldots, m$, $c = c(x, z)$ *is independent of* p, $k(x, u, u_x)$ *is different from the zero matrix for* $x \in G$;

(γ_3) *to each compact subset* H *of* G *there correspond two constants* L, $\alpha > 0$ *such that* $|b_\mu(x, u, u_x)| \leqq L$ *for* $\mu = 1, \ldots, m$, $x \in H$,

$$|c(x, z, p) - c(x, z, \bar{p})| \leqq L \|p - \bar{p}\| \quad for \ z \geqq M(u),\ x \in H$$

$$\sum_{\lambda, \mu} k_{\lambda\mu}(x, u, u_x)\, \xi_\lambda \xi_\mu \geqq \alpha \xi^2 \quad for \ \xi \in E^m,\ x \in H;$$

(γ_4) *there is an index* μ *such that* $b_\mu = 0$, c *is independent of* p_μ, *and* $k_{\mu\mu}(x, u, u_x) > 0$ *in* G;

(γ_5) *to each compact subset* H *of* G *there correspond an index* μ *and two constants* $L, \alpha > 0$ *such that* $|b_\mu(x, u, u_x)| \leqq L$ *and* $k_{\mu\mu}(x, u, u_x) \geqq \alpha$ *for* $x \in H$.

For the proof we set

$$\bar{k}_{\lambda\mu}(x) = k_{\lambda\mu}(x, u, u_x), \quad \bar{b}_\mu(x) = b_\mu(x, u, u_x)$$

and

$$\bar{f}(x, z, p, r) = -\sum_{\lambda,\mu} \bar{k}_{\lambda\mu}(x)\, r_{\lambda\mu} + \sum_\mu \bar{b}_\mu(x)\, p_\mu + c(x, z, p).$$

Then the conclusion follows readily from Corollary III if we replace v by u, w by $M(u)$ and f by $\bar{f}$ there, since

$$f(x, u, u_x, u_{xx}) = \bar{f}(x, u, u_x, u_{xx}) \leqq 0 \leqq c(x, M(u), 0) = \bar{f}(x, M(u), 0, 0).$$

The assumption (γ_i) in our theorem corresponds to the assumption (γ_i') in III.

The following results on the maximum principle are found, perhaps in less general form, in several textbooks. Our presentation is somewhat different from the usual one. In order to give a better understanding of our method as compared with other methods, the proofs are written in such a way that they do not rely on previous results. We are dealing with the mildly nonlinear equation (4). The basic result is the following

IX. Lemma. *We consider the function* $f(x, z, p, r)$ *defined by* (4′) *in the open ball* B: $\|x - a\| < R$ $(a \in E^m, R > 0)$. *Suppose the function* $u \in Z(f)$ *satisfies*

(α) $u(x) \leqq M$ *on* ∂B, $u(a) < M$,

(β) $f(x, u, u_x, u_{xx}) = -\sum_{\lambda,\mu} k_{\lambda\mu}(x)\, u_{x_\lambda x_\mu} + g(x, u, u_x) \leqq 0$ *in* B,

where f *has the properties*

(γ_1) f *is uniformly elliptic in* B, *i.e.*, (5) *is satisfied with* $\alpha > 0$;

(γ_2) $|k_{\mu\mu}(x)| \leqq L$ *in* B *for* $\mu = 1, \ldots, m$;

(γ_3) $g(x, z, p) \geqq -L\|p\|$ *for* $z \geqq M$;

(γ_4) $g(x, z, p) \geqq -L\|p\| - L(M - z)$ *for* $z < M$.

Then

$$u < M \quad \text{in} \quad B.$$

Furthermore, if $u(\bar{x}) = M$ *for a boundary point* $\bar{x} \in \partial B$, *then*

$$\liminf_{t \to +0} \frac{u(\bar{x}) - u(\bar{x} - te)}{t} > 0 \quad \text{for} \quad \|e\| = 1, \quad e \cdot (\bar{x} - a) > 0; \tag{8}$$

that is to say, every outer directional derivative not tangent to B *is positive, if it exists.*

For the proof we determine ϱ, $0<\varrho<R$, in such a way that $u<M$ for $\|x-a\|\leqq\varrho$. Now we consider the auxiliary function

$$w(x)=M-\varepsilon(\varphi(x)-1),\qquad \varphi(x)=e^{\beta(R^2-r^2)},\qquad r=\|x-a\|\,.$$

The constant $\beta>0$, which depends only on the constants L and α in assumption (γ) and on ϱ, will be determined below. From (α) we have $u(x)\leqq M=w(x)$ for $\|x-a\|=R$, and we choose $\varepsilon>0$ so small that $u(x)\leqq w(x)$ for $\|x-a\|=\varrho$. We claim that $u(x)\leqq w(x)$ for $\varrho\leqq\|x-a\|\leqq R$.

For the proof of this inequality we proceed in the usual way. Suppose it does not hold. Then there is a point x_0, $\varrho<\|x_0-a\|<R$, at which

$$u>w,\qquad u_x=w_x,\qquad u_{xx}\leqq w_{xx}\,.$$

Since

$$w_{x_\mu}=2\beta\varepsilon(x_\mu-a)\,\varphi,\qquad w_{x_\lambda x_\mu}=2\beta\varepsilon\varphi\,\{\delta_{\lambda\mu}-2\beta(x_\lambda-a)\,(x_\mu-a)\}\,,$$

we obtain

$$f(x_0,u,u_x,u_{xx})\geqq f(x_0,u,w_x,w_{xx})=-\sum k_{\lambda\mu}(x_0)\,w_{x_\lambda x_\mu}+g(x_0,u,w_x)\,,$$

which is, by (5) and (γ_2),

$$\geqq 2\beta\varepsilon\varphi(2\alpha\beta r^2-Lm)+g(x_0,u,w_x)\,.$$

By (γ_3) and (γ_4) the quantity $g(x_0,u,w_x)$ has the lower bound $-L\|w_x\|$ if $u(x_0)\geqq M$, and the lower bound $-L\|w_x\|-L(M-u)$ if $u(x_0)<M$. Since $M-u\leqq M-w<\varepsilon\varphi$, in both cases we have

$$g(x_0,u,w_x)\geqq -L\|w_x\|-L\varepsilon\varphi=\varepsilon\varphi(2\beta rL-L)$$

and hence

$$f(x_0,u,u_x,u_{xx})\geqq\varepsilon\varphi(4\alpha\beta^2\varrho^2-2\beta mL-2\beta RL-L)>0$$

for sufficiently large β. This contradiction to the assumption (β) proves the inequality $u(x)\leqq w(x)$ for $\varrho\leqq r\leqq R$. Since $u<M$ for $r\leqq\varrho$ and $w<M$ for $r\leqq R$, we have $u<M$ in B. The second conclusion about the normal derivative follows readily from the fact that the scalar product $e\cdot w_x(\bar{x})$ $=e\cdot 2\beta\varepsilon(\bar{x}-a)\,\varphi$ is positive.

X. Theorem (Strong Maximum Principle). *Let M be the supremum in G of the function $u\in Z_0(f)$ and*

$$f(x,u,u_x,u_{xx})\leqq 0\quad\textit{in}\quad G\,,$$

where f is the function defined by (4′) and G is a connected open set. Then

$$u<M\quad\textit{in}\quad G\quad\textit{or}\quad u=M\quad\textit{in}\quad G$$

if for each open ball B with $\bar{B}\subset G$ there exist two constants L, $\alpha>0$ such that IX (γ_1)-(γ_4) holds in B.

Furthermore, if $\overline{x} \in R$, $u(\overline{x}) = M$, *and if there exists an open ball* $B \subset G$ *with center* a *such that* $\overline{x} \in \partial B$ *and that* IX (γ_1)-(γ_4) *holds in* B, *then* (8) *holds. In particular, the outer normal derivative at* $\overline{x}$ *is positive, if it exists.*

For the proof we suppose that there are two points $x_1, x_2 \in G$ such that $u(x_1) < M$ and $u(x_2) = M$. There is an arc in G connecting the points x_1, x_2 with distance $r > 0$ from the boundary of G. If the center a of a ball with radius $r/2$ is moved along that arc, it is possible to find a position for a such that $u(a) < M$ and $u(x) = M$ for some point x in the interior of the ball. Hence we arrive at a contradiction with IX.

The last statement about the directional derivative is a direct consequence of IX.

XI. Remarks. (α) The first part of Theorem X is due to Hopf (1927), while the statement about directional derivatives was found independently by Hopf (1952) and Oleinik (1952). There are numerous papers on the maximum principle for elliptic equations which sharpen and generalize these results. We mention here Simoda and Nagumo (1951), Simoda (1956), Pucci (1952, 1953, 1957, 1958, 1966), Collatz (1958), Redheffer (1958, 1960, 1962, 1963, 1964, 1967), Redheffer and Straus (1964), McNabb (1961), Kusano (1963), Vyborny (1964), Bochenek (1966). Further references and an excellent historical review are given in the book by Protter and Weinberger (1967; pp. 156–158).

(β) There are important examples of elliptic equations where the uniform ellipticity requirement (5) holds in compact subsets of G, but fails to hold near the boundary. In such cases the first part of Theorem X does apply, but the assumptions for the second part are not valid. Actually the second part, the statement about directional derivatives at the boundary, holds under much weaker assumptions on the matrix $k(x)$. This becomes clear from the following reasoning.

By passing to a smaller ball, if necessary, we can assume that the closed ball $\overline{B}$ has only the point $\overline{x}$ in common with the boundary R of G. Therefore, if $U(\overline{x})$ is an arbitrary neighborhood of $\overline{x}$ and $u < M$ in G, then $u \leqq M - \delta$ in $\overline{B} - U(\overline{x})$ for a positive δ. Therefore the auxiliary function w constructed in the proof of IX satisfies $u \leqq w$ in $\overline{B} - U(\overline{x})$ if $\varepsilon > 0$ is sufficiently small. This means that a point x_0 such that $u(x_0) > w(x_0)$ is necessarily in $U(\overline{x})$. Therefore the proof of Lemma IX goes through if to each neighborhood $U(\overline{x})$ there corresponds an $\alpha > 0$ such that

$$\sum_{\lambda,\mu} k_{\lambda\mu}(x)\,(x_\lambda - a_\lambda)\,(x_\mu - a_\mu) \geqq \alpha\,\|x - a\|^2 \quad \text{for} \quad x \in G \cdot U(\overline{x}). \tag{9}$$

For example, this is the case if the matrix $k(x)$ is defined and continuous at $\overline{x}$ and if

$$\sum_{\lambda,\mu} k_{\lambda\mu}(\overline{x})\,(\overline{x}_\lambda - a_\lambda)\,(\overline{x}_\mu - a_\mu) > 0. \tag{9'}$$

(γ) Let us consider, as an illustrative example, the two-dimensional equation (x, y real)

$$k(x, y)\, u_{xx} + u_{yy} = 0\,, \tag{10}$$

where $k(x, y)$ is continuous for $y \geqq 0$, $k(x, 0) = 0$ and $k(x, y) > 0$ for $y > 0$. If G is an open set lying entirely in the upper half plane $y > 0$, then both statements of Theorem X are valid. Thus, if a solution u assumes its maximum at a boundary point on the x-axis, then the outer normal derivative at this point is positive. This in an example of an equation which is not uniformly elliptic in G (if R contains points on the x-axis), but for which (9′) can always be satisfied. We remark that the Tricomi equation is a special case of (10).

(δ) We add the rather trivial remark that Theorem X also applies to quasilinear equations. For if f is the function corresponding to the quasilinear equation (7) and $f(x, u, u_x, u_{xx}) \leqq 0$, we construct, as we did in the proof of VIII, a mildly nonlinear function $\bar{f}$ such that $f(x, u, u_x, u_{xx}) = \bar{f}(x, u, u_x, u_{xx})$, and apply Theorem X to $\bar{f}$.

Some more special questions will now be briefly discussed.

XII. Discontinuous Boundary Values. If we want to apply a monotonicity theorem, such as Theorem II or III, to boundary value problems with discontinuous boundary values, we must allow an "exceptional set" $N \subset R$, where we have no knowledge (or only a very vague knowledge such as boundedness) of the functions involved. In other words, we have to replace the assumption that $v \leqq w$ on R^+ by the weaker assumption that $v \leqq w$ on $(R-N)^+$, where the latter relation is defined as in I, but with the provision that the sequence $x_{(k)}$ tends to a point $\bar{x} \in R - N$. The following theorem concerns the mildly nonlinear equation (4).

XIII. Theorem. *Let $v, w \in Z_0(f)$, where f is given by (4′). Suppose that N is a subset of R such that v is bounded above and w is bounded below in a neighborhood of N and that*

(α) *$v \leqq w$ on $(R-N)^+$ and R_∞,*

(β) *$f(x, v, v_x, v_{xx}) \leqq f(x, w, w_x, w_{xx})$ in G.*

Then

$$v \leqq w \quad in \quad G\,,$$

if the following conditions are satisfied:

(γ_1) *the matrix $k(x)$ is positive semidefinite in G;*

(γ_2) *$|g(x, z, p) - g(x, z, \bar{p})| \leqq L\,\|p - \bar{p}\|$, g is weakly increasing in z;*

(γ_3) *there exists a function $\varphi(x) \in Z_0$ such that $\varphi(x) > 0$ in G,*

$$\sum_{\lambda,\mu} k_{\lambda\mu}(x)\, \varphi_{x_\lambda x_\mu} + L\,\|\varphi_x\| < 0 \quad in \quad G\,,$$

$$\varphi(x) \to \infty \quad for \quad x \to \bar{x} \in N\ (x \in G)\,.$$

For the proof it is sufficient to show that $v \leqq \overline{w}$ in G, where $\overline{w} = w + \varepsilon\varphi$ ($\varepsilon > 0$ arbitrary). Let us assume that this inequality does not hold. Since $v \leqq \overline{w} = w + \varepsilon\varphi$ on R^+ and R_∞, there is a point $x_0 \in G$ at which $v > \overline{w}$, $v_x = \overline{w}_x$, $v_{xx} \leqq \overline{w}_{xx}$. The inequalities

$$\begin{aligned} f(x_0, v, v_x, v_{xx}) &\geqq f(x_0, w, \overline{w}_x, \overline{w}_{xx}) = -\sum k_{\lambda\mu}(x_0)\,\overline{w}_{x_\lambda x_\mu} + g(x_0, w, w_x + \varepsilon\varphi_x) \\ &\geqq -\sum k_{\lambda\mu}(x_0)\, w_{x_\lambda x_\mu} + g(x_0, w, w_x) - L\varepsilon\,\|\varphi_x\| - \varepsilon\sum k_{\lambda\mu}(x_0)\,\varphi_{x_\lambda x_\mu} \\ &> f(x_0, w, w_x, w_{xx}) \end{aligned}$$

contradict the assumption (β).

This theorem is not more than a starting point for the investigation of problems with discontinuous boundary values. The main assumption is (γ_3). Here we do not discuss the rather delicate problem of under what conditions and in which way functions φ with the properties of (γ_3) can be constructed. The papers of Krzyżański (1950) and Adler (1962) are devoted to this question. Roughly speaking, if the equation is uniformly elliptic and $k(x)$ is Hölder-continuous, then sets consisting of a finite number of points and hypersurfaces of dimension $\leqq m-3$ are admissible for N.

XIV. Problems with Mixed Boundary Conditions (*normal*, $\partial\varphi/\partial n_a$, $Z(R_n)$, $Z(f, R_n)$). We use the definitions of 31 I. The interior normal at a point $\overline{x} \in R$ is defined by a fixed sequence of points $x_{(k)} \in G$ such that $x_{(k)} \to \overline{x}$ for $k \to \infty$, the *outer* normal derivative with respect to this normal is given by

$$\frac{\partial\varphi(\overline{x})}{\partial n_a} = \lim_{k\to\infty} \frac{\varphi(\overline{x}) - \varphi(x_{(k)})}{\|\overline{x} - x_{(k)}\|}.$$

An interior normal in this sense exists for each boundary point of an arbitrary open set. The first of the following two theorems is valid with this general concept of a normal, while for the second one we have additional requirements (making the normal more normal).

For the sake of simplicity we restrict consideration to bounded sets G. We assume that a subset R_n of R is fixed and that to each $\overline{x} \in R_n$ a normal is assigned. A function $\varphi \in Z\,[Z(f)]$, which has an outer normal derivative at each point of R_n, is of class $Z(R_n)$ $[Z(f, R_n)]$. We write the boundary conditions in the form $\vartheta(x, u, \partial u/\partial n_a) = 0$, where $\vartheta(x, z, q)$ is a given function with some monotonicity properties stated below. It is assumed that $\vartheta(x, z, q)$ does not depend on q for $x \in R - R_n$. Mixed boundary conditions of this general form contain the special cases where u or $\partial u/\partial n_a$ or a linear combination of both is prescribed at the boundary (boundary conditions of the first, second and third kind).

Corresponding to Theorem II we have the following

XV. Monotonicity Theorem. Let G be bounded, $f \in \mathcal{Q}$ weakly increasing in z, $\vartheta(x, z, q)$ weakly increasing in z and q, and $v, w \in Z(f, R_n)$. Then

(α) $\vartheta(x, v, \partial v/\partial n_a) < \vartheta(x, w, \partial w/\partial n_a)$ on R,

(β) $f(x, v, v_x, v_{xx}) < f(x, w, w_x, w_{xx})$ in G

implies

$$v < w \quad \text{in} \quad \bar{G}.$$

For the proof we assume that the last inequality is false. Then there is a point $x_0 \in \bar{G}$ where $d = v - w$ attains a nonnegative maximum. The assumption $x_0 \in G$ leads to a contradiction as in the proof of II. But if $x_0 \in R$, then $d(x_0) \geqq 0$ [and $\partial d(x_0)/\partial n_a \geqq 0$, if $x_0 \in R_n$], or

$$v \geqq w \text{ [and } \partial v/\partial n_a \geqq \partial w/\partial n_a] \text{ at the point } x_0,$$

whence $\vartheta(x_0, v, \partial v/\partial n_a) \geqq \vartheta(x_0, w, \partial w/\partial n_a)$ follows. The last inequality is incompatible with (α).

In the following monotonicity theorem, where equality is permitted in the assumptions, we restrict consideration to the mildly nonlinear case (4).

XVI. Monotonicity Theorem. *Suppose G is bounded, f is the function defined by (4′), and $v, w \in Z(f, R_n)$. Then*

(α) $\vartheta(x, v, \partial v/\partial n_a) \leqq \vartheta(x, w, \partial w/\partial n_a)$ *on* R,

(β) $f(x, v, v_x, v_{xx}) \leqq f(x, w, w_x, w_{xx})$ *in* G

implies

$$v \leqq w \quad \text{in} \quad G$$

if the following assumptions hold:

(γ_1) $\vartheta(x, z, q) < \vartheta(x, \bar{z}, \bar{q})$ *for* $z < \bar{z}$, $q < \bar{q}$;

(γ_2) *for each* $\bar{x} \in R_n$ *there exists an open ball* $B = B(\bar{x}) \subset G$ *such that* $\bar{x} \in \partial B$; *the points defining the interior normal at* $\bar{x}$ *are on a straight line and in* B;

(γ_3) *for each ball* $B(\bar{x})$, $\bar{x} \in R_n$, *with the properties stated in* (γ_2), *and also for each ball* B *whose closure is in* G, *there are two constants* L, $\alpha > 0$ *such that the condition* (5) *is satisfied in* B, $|k_{\mu\mu}(x)| \leqq L$ *in* B *for* $\mu = 1, \ldots, m$, *and*

$$g(x, w+z, w_x+p) - g(x, w, w_x) \geqq \begin{cases} -L\,\|p\| & \text{for } z \geqq 0,\ x \in B \\ -L\,\|p\| - L|z| & \text{for } z < 0,\ x \in B. \end{cases}$$

We remark that, according to XI (β), condition (γ_3) can be weakened as far as boundary points are involved; it is sufficient that (9) or (9′) hold instead of (5). We remark further that the assertion of the theorem can be sharpened as follows: $v = w$ in G or $v < w$ in G.

The proof of this theorem relies on the strong maximum principle. The function $d = v - w$ is a solution of the inequality

$$\bar{f}(x, d, d_x, d_{xx}) = -\sum k_{\lambda\mu}(x)\, d_{x_\lambda x_\mu} + \bar{g}(x, d, d_x) \leqq 0 \quad \text{in} \quad G,$$

where

$$\bar{g}(x, z, p) = g(x, w + z, w_x + p) - g(x, w, w_x).$$

Let M be the supremum of d in G. We assume that, contrary to the conclusion of the theorem, $M > 0$. Now we apply Theorem X to the functions d and $\bar{f}$; it is easily seen that the hypotheses of this theorem are true (for $M = 0$ and, a fortiori, for $M > 0$). Since d is continuous in $\bar{G}$, there is, according to Theorem X, a point $\bar{x} \in R$ such that $d(\bar{x}) = M > 0$ and, if $\bar{x} \in R_n$, $\partial d/\partial n_a > 0$ (here the assumption (γ_2) on the direction of the interior normal was used). But this implies $v(\bar{x}) > w(\bar{x})$ and, if $\bar{x} \in R_n$, $\partial v(\bar{x})/\partial n_a > \partial w(\bar{x})/\partial n_a$. Hence from (γ_1) we obtain $\vartheta(x, v, \partial v/\partial n_a) > \vartheta(x, w, \partial w/\partial n_a)$. This contradiction of the assumption (α) shows that $M \leqq 0$. If $M = 0$, then X can again be applied, i.e., we have $v = w$ in G or $v < w$ in G.

XVII. Remarks. (α) It is immediately clear that the preceding monotonicity theorem contains a uniqueness theorem for the equation (4) with the boundary condition $\vartheta(x, u, \partial u/\partial n_a) = 0$. If the boundary condition is linear,

$$\vartheta(x, u, \partial u/\partial n_a) = \alpha(x)\, u + \beta(x)\, \partial u/\partial n_a + \gamma(x),$$

then the condition XV (γ_1) is satisfied if $\alpha(x), \beta(x) \geqq 0$ and $\alpha(x) + \beta(x) > 0$, i.e., in all classical cases.

(β) In the second part of the Appendix (IX–XV) we started with a maximum principle and then we derived a monotonicity theorem from that maximum principle, while in Chapter IV on parabolic equations and also in the first part of the Appendix we proceeded in the opposite direction. We could as easily have proved the monotonicity theorem XV first and then derived the corresponding maximum principle X as a special case. The reason for not doing so was to show that both ways are possible and that maximum principles and monotonicity theorems are two aspects of the same thing.

(γ) The theorems proved so far can be used for many other investigations on elliptic differential equations. Among others, problems of the Phragmèn-Lindelöf type, Harnack's inequality, and the investigation of singularities of solutions are topics where the methods derived here are used with great success. For these and other questions we refer to the book by Protter and Weinberger (1967), which contains numerous references, to the survey articles by Landis (1959, 1963), and to Besala (1964), Kusano (1965, 1965 a), Akô and Kusano (1964), Miheeva (1966), Nitsche (1966), Oddson (1967).

(δ) Most of the results presented here can be generalized to elliptic systems of the form

$$f_\nu(x, u_1, \ldots, u_n, u_{\nu,x}, u_{\nu,xx}) = 0 \quad \text{for} \quad \nu = 1, \ldots, n.$$

We refer to papers by Besala (1964) and Stys (1963/64). These systems, sometimes called weakly coupled systems, are of the same special kind as the parabolic systems treated in Sections 32 and 33.

List of Symbols

Intervals. As usual, open, closed, and half-open (one-dimensional) intervals are denoted by (a, b), $[a, b]$, $(a, b]$, $[a, b)$.

Sets and Set Operations. We abbreviate $\{x \mid x \in E^m, x > 0\}$, $\{z \mid z \in E^n, z \geqq 0\}$ as $\{x > 0\}$, $\{z \geqq 0\}$. The operations of sum, product, and difference of sets are denoted by the symbols $+, \cdot, -$. $\emptyset$ is the empty set.

Inequalities between vectors and vector functions should always be interpreted component-wise; see 6 I. The following are defined in particular:

$\varphi(a+) < \psi(a+)$	1 V, 17
$\varphi(a+) < \psi(a+)$	6 II, 45
$\varphi < \psi$ on R_v^+	17 I, 125
$\varphi < \psi$ on R_v^+	17 IV, 127
$\varphi < \psi$, $\varphi \leqq \psi$, $\varphi = \psi$ on R_p^+ and on R_∞	23 IV, 184; 34 I, 270
$\varphi < \psi$, $\varphi \leqq \psi$, $\varphi = \psi$ on $(R_p - R_n)^+$	31 IV, 247
$\varphi < \psi$, $\varphi \leqq \psi$, $\varphi = \psi$ on R_p^+ and on R_∞	32 I, 256
$\varphi < \psi$ on R^+ and R_∞	App. I, 305
$r \leqq \bar{r}$ (for matrices)	23 II, 181

Point Sets in Euclidean space (domains, boundary sets, neighborhoods):

J, J_0, E^n	1 I, 13
$G, \partial G, \bar{G}, R_v, G_v, G(x)$ (Chapter III)	17 I, 124
G_0	21 VIII, 165
$U, U_-, G, \bar{G}, \partial G, \partial_0 G, \partial_1 G, \partial_2 G, G_p, R_p$ (parabolic differential equations)	23 I, 180; 34 I, 270
$H(t), H(t, x), H^*(t, x)$	26 I, 198
R_n	31 II, 245
$D(f) = M \times M_r$	23 II, 182; App. I, 304
$D(f_v) = M_v \times M_{vr}$	32 I, 256
$G, R = \partial G$ (elliptic differential equations)	App. I, 304

Function Classes. The class of continuous (resp. Lebesgue-integrable) functions is denoted by C (resp. L); see 1 I. The following classes refer to solutions, approximate solutions, ... of operator equations, integral equations, differential equations. A subscript c indicates continuous functions and ac absolutely continuous functions. If X is a class of scalar functions, then $\varphi \in X$ means that every component of φ is in X.

$Z_c(k)$ (one-dimensional integral equations)   1 I, 13

$Z_c(K)$ (one-dimensional operator equations)   1 IV, 16

$Z, Z_{ac}, Z(f), Z_{ac}(f)$ (ordinary differential equations)   5 I, 39

$Z_c(K), Z_c(k)$ (one-dimensional systems)   6 II, 45

$Z(f), Z_{ac}(f)$ (systems of ordinary differential equations) 6 VI, 48

$Z(f)$ (ordinary differential equations of nth order)   15 I, 110

$Z_c(k), Z_c(K)$ (multidimensional equations)   17 II, 125

$Z_c(k), Z_c(K)$ (multidimensional systems)   17 IV, 127; 19 I, 141

$Z^*, Z^*(f)$ (hyperbolic differential equations)   20 I, 150; 21 II, 161

Z_1^*, Z_2^*, Z_3^* (hyperbolic differential equations)   22 I, 175

$Z, Z_0, Z(f), Z_0(f)$ (parabolic differential equations) 23 III, 183

Z_g (boundary-layer differential equations)   30 III, 236

$Z(f, R_n), Z_0(f, R_n)$ (boundary value problems of the   31 II, 245
    third kind)

$Z(f), Z_0(f), Z(f, R_n), Z_0(f, R_n)$ (parabolic systems)   32 I, 256

$Z(F), Z_0(F)$ (implicit parabolic differential equations) 34 I, 270

Z, Z_0 (elliptic differential equations)   App. I, 304

The following function classes refer to operators, kernels of integral equations, "right hand sides" of differential equations.

$\mathscr{E}, \mathscr{E}_1$ 4 II, 33

$\mathscr{E}_2, \mathscr{E}_3$ 5 VI, 41

$\mathscr{E}_4, \mathscr{E}_5, \mathscr{E}_6$ 10 I, 81

$\mathscr{E}_7$ 10 V, 84

$\mathscr{E}^n, \mathscr{E}_1^n$ 6 V, 47

$\mathscr{E}_2^n, \mathscr{E}_3^n$ 6 IX, 50

$\mathscr{E}_4^n, \mathscr{E}_5^n, \mathscr{E}_6^n, \mathscr{E}_7^n$ 14 I, 106

$\bar{\mathscr{E}}_4^n, \bar{\mathscr{E}}_5^n, \bar{\mathscr{E}}_6^n$ 15 X, 114

$\mathscr{E}[o(g(t))], \mathscr{E}[O(g(t))]$ 10 III, 82

$\mathscr{K}, \mathscr{K}_0$ 2 I, 23

$\mathscr{K}_1$ 2 VI, 27

$\mathscr{K}^n, \mathscr{K}_0^n$ 6 II, 45

$\mathscr{F}, \mathscr{F}_0$ 8 VII, 66

$\mathscr{F}^n, \mathscr{F}_0^n$ 12 VII, 94; 15 I, 110

$\mathscr{H}, \mathscr{H}_0$ 18 I, 130

$\mathscr{D}^n$ 17 VIII, 128

Further Notation

Bibliography

Adler, G.: Über Wärmeleitungs- und Diffusionsprobleme mit zusammengesetzten Randbedingungen II. Magy. Tud. Akad. Mat. Kut. Int. Közlemen. **1**, 167—183 (1956).

— Un type nouveau des problèmes aux limites de la conduction de la chaleur. Magy. Tud. Akad. Mat. Kut. Int. Közlemen. **4**, 109—127 (1959).

— Principes du maximum relatifs aux équations du type elliptique et parabolique dans le cas des conditions aux limites et des conditions initiales respectivement non-continues et non-limitées. Acta Math. Acad. Sci. Hung. **13**, 289—297 (1962).

— Une idée concernant la majoration numérique de la solution du problème de Neumann relatif à l'équation de la chaleur. Ann. Polon. Math. **17**, 119—128 (1965).

Agaev, G. N., Mamazov, G. K.: Solution of a mixed problem for a non-linear parabolic equation. Dokl. Akad. Nauk Azerb. SSR **14**, 501—510 (1958).

Agmon, S., Nirenberg, L., Protter, M. H.: A maximum principle for a class of hyperbolic equations and applications to equations of mixed elliptic-hyperbolic type. Commun. Pure Appl. Math. **6**, 455—470 (1953).

Akô, K., Kusano, T.: On bounded solutions of second order elliptic differential equations. J. Fac. Sci. Univ. Tokyo Sect. I **11**, 29—37 (1964).

Alexiewicz, A., Orlicz, W.: Some remarks on the existence and uniqueness of solutions of the hyperbolic equation $\frac{\partial^2 z}{\partial x \partial y} = f\left(x, y, z, \frac{\partial z}{\partial x}, \frac{\partial z}{\partial y}\right)$. Studia Math. **15**, 201—215 (1956).

Antosiewicz, H. A.: A survey of Lyapunov's second method. Contributions to the theory of nonlinear oscillations, Vol. 4, 141—166 (Annals of Mathematics Studies No. 41) (1958).

— Lyapunov-like functions and approximate solutions of ordinary differential equations. Symposium on the numerical treatment of ordinary differential equations, integral and integro-differential equations. Basel: Birkhäuser 1960.

— An inequality for approximate solutions of ordinary differential equations. Math. Z. **78**, 44—52 (1962).

Arnese, G.: Sull'approssimazione, col metodo di Tonelli, delle soluzioni del problema di Darboux per l'equazione $s = f(x, y, z, p, q)$. Ric. Mat. **11**, 61—75 (1962).

— Sul problema di Darboux in ipotesi di Carathéodory. Ric. Mat. **12**, 13—43 (1963).

Aronson, D. G.: Uniqueness of solutions of the initial value problem for parabolic systems of differential equations. J. Math. Mech. **11**, 403—420 (1962).

— Uniqueness of positive weak solutions of second order parabolic equations. Ann. Polon. Math. **16**, 285—303 (1965).

— Besala, P.: Uniqueness of solutions of the Cauchy problem for parabolic equations. J. Math. Anal. Appl. **13**, 516—526 (1966).

— — Uniqueness of positive solutions of parabolic equations with unbounded coefficients. Colloq. Math. **18**, 125—135 (1967).

— Serrin, J.: Local behavior of solutions of quasilinear parabolic equations. Arch. Rational Mech. Anal. **25**, 81—122 (1967).

Aronson, D. G., Serrin, J.: A maximum principle for nonlinear parabolic equations. Ann. Scuola Norm. Sup. Pisa (3) **21**, 291—305 (1967a).

Aziz, A. K., Maloney, J. P.: An application of Tychonoff's fixed point theorem to hyperbolic partial differential equations. Math. Ann. **162**, 77—82 (1965/66).

Babkin, B. N.: On a generalization of a theorem of academican S. A. Čaplygin on a differential inequality. Molotov. Gos. Univ. Uč. **8**, 3—6 (1953).

Baiada, E.: Confronto e dipendenza dai parametri degli integrali delle equazioni differenziali, I. Atti Accad. Naz. Lincei. Rend. Classe Sci. Fis. Mat. Nat. (8) **3**, 258—263 (1947).

— Lorefice, M.: I metodi di approssimazione nello studio dei sistemi differenziali ordinari. Att Accad. Sci. Lettere Arti Palermo (4) **16**, 193—223 (1957).

Banach, S.: Théorie des opérations linéaires. Monografje Matematyczne, t. 1. Varsovie 1932.

Bantas, Gh.: On bounding the solutions of a system of nonlinear Volterra integral equations. (Romanian. Russian and French summaries) An. Sti. Univ. "Al. I. Cuza" Iasi Sect. I a Mat. (N.S.) **11a**, 291—298 (1965).

Bebernes, J., Fulks, W., Meisters, G. H.: Differentiable paths and the continuation of solutions of differential equations. J. Diff. Equations **2**, 102—106 (1966).

— Meisters, G. H.: An elementary proof of some differential inequalities. J. Math. Anal. Appl. **17**, 92—98 (1967).

Beckenbach, E. F., Bellman, R.: Inequalities. Ergebnisse der Mathematik und ihrer Grenzgebiete, N.F., H. 30. Berlin-Göttingen-Heidelberg: Springer 1961.

Becker, H.: Anwendung der Theorie der Differentialungleichungen auf zwei neue Randwertaufgaben für parabolische Differentialgleichungen. Studia Sci. Math. Hung. **2**, 53—61 (1967).

Bellman, R.: The stability of solutions of linear differential equations. Duke Math. J. **10**, 643—647 (1943).

— On the existence and boundedness of solutions of nonlinear partial differential equations of parabolic type. Trans. Amer. Math. Soc. **64**, 21—44 (1948).

— Stability theory of differential equations. New York-Toronto-London: Mc Graw-Hill Book Co. 1953.

Berezin, I. S., Zhidkov, N. P.: Computing methods, Vol. II. New York: Pergamon Press 1965.

Besala, P.: Solutions of a non-linear equation of parabolic type in an unbounded region. (Polish. English and Russian summaries). Zeszyty Nauk. Uniw. Jagiel. Prace Mat. No. 6, 83—92 (1961).

— On solutions of non-linear parabolic equations defined in unbounded domains. Bull. Acad. Polon. Sci. Ser. Sci. Math. Astron. Phys. **9**, 531—535 (1961a).

— On solutions of Fourier's first problem for a system of non-linear parabolic equations in an unbounded domain. Ann. Polon. Math. **13**, 247—265 (1963).

— Concerning solutions of an exterior boundary-value problem for a system of non-linear parabolic equations. Ann. Polon. Math. **14**, 289—301 (1963/64).

— On solutions of non-linear second order elliptic equations defined in unbounded domains. Atti Seminar Mat. Fis. Univ. Modena **13**, 74—86 (1964).

— Limitations of solutions of non-linear parabolic equations in unbounded domains. Ann. Polon. Math. **17**, 25—47 (1965).

— Krzyżański, M.: Un théorème d'unicité de la solution du problème de Cauchy pour l'équation linéaire normale parabolique du second ordre. Atti Accad. Naz. Lincei Rend. Classe Sci. Fis. Mat. Nat. (8) **33**, 230—236 (1962).

— Fife, P.: The unbounded growth of solutions of linear parabolic differential equations. Ann. Scuola Norm. Sup. Pisa (3) **20**, 719—732 (1966).

Bielecki, A.: Une remarque sur l'application de la méthode de Banach-Caccioppoli-Tikhonov dans la théorie de l'équation $s = f(x, y, z, p, q)$. Bull. Acad. Polon. Sci. Classe III, **4**, 265—268 (1956).

Bihari, I.: A generalization of a lemma of Bellman and its application to uniqueness problems of differential equations. Acta Math. Acad. Sci. Hung. **7**, 81—94 (1956).

— Researches of the boundedness and stability of the solutions of non-linear differential equations. Acta Math. Acad. Sci. Hung. **8**, 261—278 (1957).

Birkhoff, G., Kotik, J.: Note on the heat equation. Proc. Amer. Math. Soc. **5**, 162—167 (1954).

Blasius, H.: Grenzschichten in Flüssigkeiten mit kleiner Reibung. Z. Math. Phys. **56**, 1 (1908).

Bochenek, J.: On a modification of a theorem of O. Olejnik and on its applications. Ann. Polon. Math. **18**, 121—126 (1966).

Bompiani, E.: Un teorema di confronto ed un teorema di unicità per l'equazione differenziale $y' = f(x, y)$. Atti Accad. Naz. Lincei. Rend. Classe Sci. Fis. Mat. Nat. (6) **1**, 298—302 (1925).

Brauer, F.: A note on uniqueness and convergence of successive approximations. Can. Math. Bull. **2**, 5—8 (1959).

— Some results on uniqueness and successive approximations. Can. J. Math. **11**, 527—533 (1959a).

— Global behaviour of solutions of ordinary differential equations. J. Math. Anal. Appl. **2**, 145—158 (1961).

— Sternberg, S.: Local uniqueness, existence in the large, and the convergence of successive approximations. Amer. J. Math. **80**, 421—430 (1958).

— — Errata to our paper "Local uniqueness, etc." Amer. J. Math. **81**, 797 (1959).

Budak, B. M.: The straight-line method for certain quasi-linear boundary value problems of parabolic type. (Russian). Z. Vycisl. Mat. Fiz. **1**, 1105–1112 (1961).

— On the method of lines for solving certain quasilinear boundary-value problems of parabolic type with discontinuous data. (Russian). Vestnik Moskov. Univ. Ser. I Mat. Meh. 1961, No. 3, 3–8.

Burton, L. P., Whyburn, W. M.: Minimax solutions of ordinary differential systems. Proc. Amer. Math. Soc. **3**, 794—803 (1952).

Butlewski, Z.: Sur les intégrales d'un système d'équations différentielles linéaires ordinaires. Studia Math. **10**, 40—47 (1948).

— Sur la limitation des solutions d'un système d'équations intégrales de Volterra. Ann. Polon. Math. **6**, 253—257 (1959).

Cafiero, F.: Su due teoremi di confronto relativi ad un'equazione differenziale ordinaria del primo ordine. Boll. Univ. Mat. Ital. (3) **3**, 124—128 (1948).

— Sui teoremi di unicità relativi ad un'equazione differenziale ordinaria del primo ordine. II. Giorn. Mat. Battaglini (4) **2** (78), 193—215 (1949).

Cameron, R. H., Shapiro, J. M.: Nonlinear integral equations. Ann. Math. (2) **62**, 472—497 (1955).

Candirov, G. I.: A generalization of the inequality of Gronwall and its applications. Azerb. Gos. Univ. Uč. Zap. No. 6, 3—11 (1958).

Carathéodory, C.: Vorlesungen über reelle Funktionen. Leipzig: Teubner 1918.

Carslaw, H. S., Jaeger, J. C.: Conduction of heat in solids, 2nd Ed. Oxford: Clarendon Press 1959.

Cesari, L.: Asymptotic behavior and stability problems in ordinary differential equations. Ergebnisse der Mathematik und ihrer Grenzgebiete, N.F., H. 16. Berlin-Göttingen-Heidelberg: Springer 1959.

Chandra, J.: On the asymptotic behaviour of solutions of non-linear differential equations. Proc. Natl. Acad. Sci. India Sect. A **32**, 148—151 (1962).

— On uniqueness of solutions of a non-linear system of differential equations. Proc. Natl. Acad. Sci. India Sect. A **32**, 187—190 (1962a).

Chu, S. C., Diaz, J. B.: The Goursat problem for the partial differential equation $u_{xyz}=f$: A mirage. J. Math. Mech. **16**, 709—713 (1967).

— Metcalf, F. T.: On Gronwall's inequality. Proc. Amer. Math. Soc. **18**, 439—440 (1967).

Ciliberto, C.: Il problema di Darboux per una equazione di tipo iperbolico in due variabili. Ric. Mat. **4**, 15—29 (1955).

— Sul problema di Darboux per l'equazione $s=f(x, y, z, p, q)$. Rend. Accad. Sci. Fis. Mat. Soc. Naz. Sci. Napoli (4) **22**, 221—225 (1956).

— Su alcuni problemi relativi ad una equazione di tipo iperbolico in due variabili. Boll. Univ. Mat. Ital. (3) **11**, 383—393 (1956a).

— Sulle equazioni quasi-lineari di tipo parabolico in due variabili. Ric. Mat. **5**, 97—125 (1956b).

— Nuovi contributi alle teoria dei problemi al contorno per le equazioni paraboliche non lineari in due variabili. Ric. Mat. **5**, 206—225 (1956c).

— Sull'approssimazione delle soluzioni del problema di Darboux per l'equazione $s=f(x, y, z, p, q)$. Ric. Mat. **10**, 106—138 (1961).

— Teoremi di confronto e di unicità per le soluzioni del problema di Darboux. Ric. Mat. **10**, 214—243 (1961a).

— Sul problema di Darboux. Atti Accad. Naz. Lincei. Rend. Classe Sci. Fis. Mat. Nat. (8) **30**, 460—466 (1961b).

— Sull'unicità delle soluzioni di un sistema di equazioni differenziali ordinarie. (English summary). Rend. Accad. Sci. Fis. Mat. Soc. Naz. Sci. Napoli (4) **28**, 187—198 (1961c).

Cinquini, S.: Sopra un teorema di unicità per un sistema di equazioni a derivate parziali non lineari di tipo parabolico. Ann. Mat. Pura Appl. (4) **66**, 117—128 (1964).

Cinquini-Cibrario, M., Cinquini, S.: Equazioni a derivate parziali di tipo iperbolico. Consiglio Nazionale delle Ricerche, Monografie Matematiche Vol 12. Roma: Edizioni Cremonese 1964.

Coddington, E. A., Levinson, N.: Uniqueness and the convergence of successive approximations. J. Indian Math. Soc. (N.S.) **16**, 75—81 (1952).

— Theory of ordinary differential equations. New York-Toronto-London: McGraw-Hill Book Co. 1955.

Collatz, L.: Aufgaben monotoner Art. Arch. Math. **3**, 366—376 (1952).

— Numerische Behandlung von Differentialgleichungen. Berlin-Göttingen-Heidelberg: Springer 1955.

— Fehlerabschätzungen für Näherungslösungen parabolischer Differentialgleichungen. Anais Acad. Brasil. Cienc. **28**, 1—9 (1956).

— Fehlerabschätzungen bei Randwertaufgaben partieller Differentialgleichungen mit unendlichem Grundgebiet. Z. Angew. Math. Phys. **9a**, 118—128 (1958).

— Applications of the theory of monotonic operators to boundary value problems. Boundary problems in differential equations, edit. by E. Langer, p. 35—45. Madison (Wis.): Univ. of Wisconsin Press 1960.

— Schröder, J.: Einschließen der Lösungen von Randwertaufgaben. Num. Math. **1**, 61—72 (1959).

Conlan, J.: The Cauchy problem and the mixed boundary value problem for a non-linear hyperbolic partial differential equation in two independent variables. Arch. Rational Mech. Anal. **3**, 355—380 (1959).

Conlan, J.: An existence theorem for the equation $u_{xyz}=f$. Arch. Rational Mech. Anal. **9**, 64—76 (1962).

— Diaz, J. B.: Existence of solutions of an n-th order hyperbolic partial differential equation. Contrib. Diff. Equations **2**, 277—289 (1963).

Conti, R.: Sul problema di Darboux per l'equazione $z_{xy}=f(x, y, z, z_x, z_y)$. Ann. Univ. Ferrara, Sez. VII. (N.S.) **2**, 129—140 (1953).

— Limitazioni "in ampiezza" delle soluzioni di un sistema di equazioni differenziali e applicazioni. Boll. Univ. Mat. Ital. (3) **11**, 344—349 (1956).

— Sulla prolungabilità delle soluzioni di un sistema di equazioni differenziali ordinarie. Bull. Univ. Mat. Ital. (3) **11**, 510—514 (1956a).

— Sull'equazione integro-differenziale di Darboux-Picard. Matematiche **13**, 30—39 (1958).

Cooper, J. L. B.: The uniqueness of the solution of the equation of heat conduction. J. London Math. Soc. **25**, 173—180 (1950).

Coppel, W. A.: On a differential equation of boundary layer theory. Phil. Trans. Roy. Soc. London, Ser. A **253**, 101—136 (1960).

Corduneanu, C.: Equazioni differenziali negli spazi di Banach, teoremi di esistenza e di prolungabilità. Atti Accad. Naz. Lincei. Rend. Classe Sci. Fis. Mat. Nat. (8) **23**, 226—230 (1957).

— Sur la stabilité asymptotique. An. Şti. Univ. "Al. I. Cuza" Iaşi, N.S. **5**, 37—40 (1959).

— Application of differential inequalities to stability theory. An. Şti. Univ. "Al. I. Cuza" Iaşi, Sect. I, **6**, 47—58 (1960).

— Sur une équation intégrale de la théorie du reglage automatique. Compt. Rend. Acad. Sci. Paris **256**, 3564—3567 (1963).

— Sur une équation intégrale non-linéaire. (Romanian and French summaries). An. Şti. Univ. "Al. I. Cuza" Iaşi Secţ. I (N.S.) **9**, 369—375 (1963a).

— Sur les inégalités différentielles. Mathematica (Cluj) **6** (**29**), 31—33 (1964).

— Problèmes globaux dans la théorie des équations intégrales de Volterra. Ann. Mat. Pura Appl. (4) **67**, 349—363 (1965).

— Sur certaines équations fonctionnelles de Volterra. Funkcial. Ekvac. **9**, 119—127 (1966).

Crank, J.: Simultaneous diffusion and reversible chemical reaction. Phil. Mag. Ser. **7**, 43, 811—825 (1952).

— The mathematics of diffusion. Oxford: Clarendon Press 1956.

Deimling, K.: A Carathéodory theory for systems of integral equations. Ann. Mat. Pura Appl. (to appear) (1969).

— Das Picard-Problem für $u_{xy}=f(x, y, u, u_x, u_y)$ unter Carathéodory-Voraussetzungen. Math. Z. **114**, 303—312 (1970).

De Lucia, P.: Su un problema al contorno per un'equazione di tipo iperbolico di ordine $2m$. Ric. Mat. **16**, 264—302 (1967).

Deo, S. G., Lakshmikantham, V.: On complex differential inequalities. Bol. Soc. Mat. São Paulo **18**,(1963), 3—8 (1966).

Diaz, J. B.: On an analogue of the Euler-Cauchy polygon method for the numerical solution of $u_{xy}=f(x, y, u, u_x, u_y)$. Arch. Rational Mech. Anal. **1**, 357—390 (1958).

— On existence, uniqueness, and numerical evaluation of solutions of ordinary and hyperbolic differential equations. Ann. Mat. Pura Appl. (4) **52**, 163—181 (1960). Also in: Symposium on the numerical treatment of ordinary differential equations, integral and integro-differential equations, pp. 581—602. Basel: Birkhäuser 1960.

— Walter, W.: On uniqueness theorems for ordinary differential equations and for partial differential equations of hyperbolic type. Trans. Amer. Math. Soc. **96**, 90—100 (1960).

Dieudonné, J.: Sur la convergence des approximations successives. Bull. Sci. Math. (2) **69**, 62—72 (1945).
Dinghas, A.: Zur Existenz von Fixpunkten bei Abbildungen vom Abel-Liouvilleschen Typus. Math. Z. **70**, 174—189 (1958).
Dračka, O.: Studium der Kinetik von Elektrodenvorgängen mit Hilfe der Elektrolyse bei konstantem Strom, VI. Collection Czech. Chem. Commun. **26**, 2144—2154 (1961).
Džabbarov, Š. T.: Some integral inequalities. (Russian. Azerbaijani summary). Azerbaĭdžan. Gos. Univ. Učen. Zap. Ser. Fiz.-Mat. i Him. Nauk No. **2**, 13—18 (1964).
Edmunds, D. E.: Sur l'allure asymptotique des solutions d'inégalités paraboliques. Compt. Rend. Acad. Sci. Paris **256**, 5485—5487 (1963).
Eisen, M.: A global existence theorem. Can. Math. Bull. **9**, 519—520 (1966).
Eltermann, H.: Fehlerabschätzung bei näherungsweiser Lösung von Systemen von Differentialgleichungen erster Ordnung. Math. Z. **62**, 469—501 (1955).
Fateeva, G. M.: Boundary value problems for quasilinear degenerating equations of parabolic type. (Russian). Usp. Mat. Nauk **22**, No. 3 (135), 244—245 (1967).
Fedele, N.: Sull' integrale superiore e quello inferiore di alcuni problemi relativi ad equazioni non lineari di tipo iperbolico, in ipotesi di Carathéodory. Matematiche **21**, 167–197 (1966).
Ficken, F. A.: Uniqueness theorems for certain parabolic problems. J. Rational Mech. Anal. **1**, 573—578 (1952).
Fife, P.: Solutions of parabolic boundary problems existing for all time. Arch. Rational Mech. Anal. **16**, 155—186 (1964).
Filippov, A. F.: Sufficient conditions for uniqueness and non-uniqueness of solutions of differential equations. Dokl. Akad. Nauk SSSR (N.S.) **60**, 549—552 (1948).
Fiorenza, R.: Sulla unicità uniforme per una famiglia di equazioni differenziali del primo ordine. Applicazioni a un sistema iperbolico di equazioni a derivate parziali. Ric. Mat. **14**, 71—91 (1965).
Frank, P., v. Mises, R.: Die Differential- und Integralgleichungen der Mechanik und Physik, 2. Aufl., Bd. II. Braunschweig: Fr. Vieweg u. Sohn 1935.
Frasca, M.: Su un problema ai limiti per l'equazione $u_{xyz} = f(x, y, z, u, u_x, u_y, u_z)$. Matematiche (Catania) **21**, 396—412 (1966).
Freud, G.: Über Wärmeleitungs- und Diffusionsprobleme mit zusammengesetzten Randbedingungen. Magy. Tud. Akad. Alkalm. Mat. Int. Közlemen. **3**, 369—394 (1955).
— Eine Eigenschaft der Lösungen parabolischer Differentialgleichungen. Compt. Rend. Acad. Bulgare Sci. **10**, 451—452 (1957).
Friedman, A.: On quasi-linear parabolic equations of the second order. J. Math. Mech. **7**, 793—809 (1958).
— Remarks on the maximum principle for parabolic equations and its applications. Pacific J. Math. **8**, 201—211 (1958a).
— Convergence of solutions of parabolic equations to a steady state. J. Math. Mech. **8**, 57—76 (1959).
— Asymptotic behavior of solutions of parabolic equations. J. Math. Mech. **8**, 387—392 (1959a).
— On the uniqueness of the Cauchy problem for parabolic equations. Amer. J. Math. **81**, 503—511 (1959b).
— Mildly nonlinear parabolic equations with applications to flow of gases through porous media. Arch. Rational Mech. Anal. **5**, 238—248 (1960).
— On quasi-linear parabolic equations of the second order, II. J. Math. Mech. **9**, 539—556 (1960a).

Friedman, A.: A strong maximum principle for weakly subparabolic functions. Pacific J. Math. **11**, 175—184 (1961).
— On integral equations of Volterra type. J. Analyse Math. **11**, 381—413 (1963).
— Partial differential equations of parabolic type. Englewood Cliffs, N.J.: Prentice Hall, Inc., 1964.
— Asymptotic behavior of solutions of parabolic differential equations and of integral equations. Differential Equations and Dynamical Systems (Proc. Intern. Symp., Mayaguez, P.R., 1965), pp. 409—426. New York: Academic Press 1967.
Fulks, W., Maybee, J. S.: A singular non-linear equation. Osaka Math. J. **12**, 1—19 (1960).

Gantmacher, F. R.: Applications of the theory of matrices. New York: Interscience Publishers, Inc. 1959. Or:
— The theory of matrices, Vol. II. New York: Chelsea Publ. Co. 1959.
George, J. H.: On Okamura's uniqueness theorem. Proc. Amer. Math. Soc. **18**, 764—765 (1967).
Gheorghiu, N.: Sur les intégrales des systèmes $dx/dt = A(t)\,x$. (Romanian and Russian summaries). An. Şti. Univ. "Al. I. Cuza" Iaşi Secţ. I a Mat. (N.S.) **11a**, 273—278 (1965).
Giuliano, L.: Sull' unicità della soluzione per una classe di equazioni differenziali alle derivate parziali, paraboliche, non lineari. Atti Accad. Naz. Lincei. Rend. Classe Sci. Fis. Mat. Nat. (8) **12**, 260—265 (1952).
Glick, I. I.: On an analog of the Euler-Cauchy polygon method for the partial differential equation $u_{x_1 \ldots x_n} = f$. Contrib. Diff. Equations **2**, 1—59 (1963).
Gloistehn, H. H.: Monotoniesätze und Fehlerabschätzungen für Anfangswertaufgaben mit hyperbolischer Differentialgleichung. Arch. Rational Mech. Anal. **14**, 384—404 (1963).
Godunova, E. K., Levin, V. I.: Certain qualitative questions of heat conduction. (Russian). Ž. Vyčisl. Mat. i Mat. Fiz. **6**, 1097—1103 (1966).
Görtler, H.: Über die Lösungen nichtlinearer partieller Differentialgleichungen vom Reibungsschichttypus. Z. Angew. Math. Mech. **30**, 265—267 (1950).
— Über nicht-lineare partielle Differentialgleichungen vom Reibungsschichttypus. Atti del Quarto Congresso dell'Unione Matematica Italiana, Taormina 1951, Vol. 2, p. 116—120.
Golomb, M.: Bounds for solutions of nonlinear differential systems. Arch. Rational Mech. Anal. **1**, 272—282 (1958).
Gor'kov, Ju. P.: The behaviour as $t \to \infty$ of the solutions of boundary-value problems for a quasi-linear parabolic equation (Russian). Dokl. Akad. Nauk SSSR **157**, 509—512 (1964).
Goursat, E.: Cours d'analyse mathématique, 4. Ed., t. 3. Paris: Gauthier-Villars 1927.
Gronwall T. H.: Note on the derivatives with respect to a parameter of the solutions of a system of differential equations. Ann. Math. **20**, 292—296 (1918/19).
Guglielmino, F.: Sulla risoluzione del problema di Darboux per l'equazione $s = f(x, y, z)$. Boll. Univ. Mat. Ital. (3) **13**, 308—318 (1958).
— Sul problema di Darboux. Ric. Mat. **8**, 180—196 (1959).
— Sull'esistenza delle soluzioni dei problemi relativi alle equazioni non lineari di tipo iperbolico in due variabili. Matematiche **14**, 67—80 (1959a).
— Sul problema di Goursat. Ric. Mat. **9**, 91—105 (1960).

Haar, A.: Über die Eindeutigkeit und Analyzität der Lösungen partieller Differentialgleichungen. Atti del Congresso Internazionale dei Matematici Bologna 1928. Bd. 3, S. 5—10.

Hahn, W.: Theory and application of Liapunov's direct method. Englewood Cliffs, N. J.: Prentice-Hall 1963.

Halilov, Z. I.: On the investigation of asymptotic stability of solutions of boundary problems for partial differential equations. Dokl. Akad. Nauk Azerb. SSR **12**, 375—378 (1956).

Hartman, P.: Ordinary differential equations. New York-London-Sydney: John Wiley 1964.

— Wintner, A.: On hyperbolic partial differential equations. Amer. J. Math. **74**, 834—864 (1952).

Heins, M.: Some remarks on unicity and continuity theorems for ordinary differential equations. Michigan Math. J. **10**, 85—89 (1963).

Hille, E., Phillips, R. S.: Functional analysis and semigroups. Amer. Math. Soc. Colloquium Publications, Vol. 31 (1957).

Hirschmann Jr., I. I.: A note on the heat equation. Duke Math. J. **19**, 487—492 (1952).

Holmgren, E.: Sur les solutions quasianalytiques de l'équation de la chaleur. Arkiv Mat. Astron. Fys. **18**, 1—5 (1924).

Hopf, E.: Elementare Bemerkungen über die Lösungen partieller Differentialgleichungen zweiter Ordnung vom elliptischen Typus. Sitz.-ber. Preuss. Akad. Wiss. **19**, 147—152 (1927).

— A remark on linear elliptic differential equations of second order. Proc. Amer. Math. Soc. **3**, 791—793 (1952).

Hukuhara, M.: Théorèmes fondamentaux de la théorie des équations différentielles ordinaires, I. Mem. Fac. Sci. Kyūsyū Imp. Univ. A. **1**, 111—127 (1940).

— Théorèmes fondamentaux de la théorie des équations différentielles ordinaires, II. Mem. Fac. Sci. Kyūsyū Imp. Univ. A. **2**, 1—25 (1941).

— Sur le fonction $S(x)$ de M. E. Kamke. Japan. J. Math. **17**, 289—298 (1941a).

— Théorèmes fondamentaux de la théorie des équations différentielles ordinaires dans l'espace vectoriel topologique. J. Fac. Sci. Univ. Tokyo, Sect. I **8**, 111—138 (1959).

— Satō, T.: Theory of differential equations. Modern Math. Series, **11** (1957). Kyōritsu Shuppan Co. Ltd. [Japanese: Reference based on a mimeographed French translation of Chapters 6 to 8 furnished by Professor Satō.]

Ikeda, Y.: On quasi-linear parabolic equations. Nagoya Math. J. **30**, 163—179 (1967).

Il'in, A. M., Kalashnikov, A. S., Oleinik, O. A.: Linear equations of the second order of parabolic type. Usp. Mat. Nauk **17**, 3—146 (1962) [= Russian Math. Surveys **17**, 1—143 (1962)].

Ivanov, I. K.: A relation between the number of changes of sign of the solution of the equation $\partial[a(t,x)\,\partial u/\partial x]/\partial x - \partial u/\partial t = 0$ and the nature of its diminution. (Bulgarian. Russian summary). Godišnik Visš. Tehn. Učebn. Zaved. Mat. **1** (1964), kn. 1, 107—116 (1965).

Iyanaga, S.: Über die Unitätsbedingungen der Lösung der Differentialgleichung $y' = f(x, y)$. Japan. J. Math. **5**, 253—257 (1928).

Jedryka, T. M.: An estimation of the solution of Volterra's integral equation for vector-valued functions. (Polish. English summary). Prace Mat. **9**, 267—271 (1965).

John, F.: On integration of parabolic equations by difference methods. Commun. Pure Appl. Math. **5**, 155—211 (1952).

Kadlec, J., Výborný, R.: Strong maximum principle for weakly nonlinear parabolic equations. Comm. Math. Univ. Carolinae **6**, 19–20 (1965).

Kalašnikov, A. S.: Classes of uniqueness of solution of the Cauchy problem for certain quasi-linear degenerating parabolic equations. (Russian. English summary). Vestn. Mosk. Univ. Ser. I Mat. Meh. **21**, no. 6, 56—60 (1966).

Kamke, E.: Über die eindeutige Bestimmtheit der Integrale von Differentialgleichungen. Math. Z. **32**, 101—107 (1930).

— Über die eindeutige Bestimmtheit der Integrale von Differentialgleichungen II. Sitz.-ber. Heidelberg. Akad. Wiss., Math.-Naturw. Kl., 17. Abhandl. (1930a).

— Zur Theorie der Systeme gewöhnlicher Differentialgleichungen II. Acta Math. **58**, 57—85 (1932).

— Differentialgleichungen reeller Funktionen, 2. Aufl. Leipzig: Akademische Verlagsgesellschaft 1945.

— Das Lebesgue-Stieltjes-Integral. Leipzig: Teubner 1956.

— Differentialgleichungen, Lösungsmethoden und Lösungen, 6. Aufl., Bd. 1. Leipzig: Akademische Verlagsgesellschaft 1959.

— Über die erste Randwertaufgabe bei der Laplace- und der Wärmeleitungs-Differentialgleichung. Jahresber. Deut. Math. Verein **62**, Abt. 1, 1—33 (1959a).

Kamynin, L. I.: On the convergence of the finite-difference process for the heat equation. (Russian). Dokl. Akad. Nauk SSSR **85**, 701–714 (1952).

— On the applicability of a finite-difference method to the solution of the heat equation, II. (Russian). Izv. Akad. Nauk SSSR Ser. Mat. **17**, 249–268 (1953).

— The stability of parabolic difference equations. Soviet Math. Dokl. **2**, 192–196 (1961).

Kanazawa, A., Murakami, H.: Unpublished manuscript (1955). [The remark of Satō and Iwasaki (1955, Footnote 2), that the work appeared in the Proc. Japan. Acad. **31**, (1955) is wrong. Mr. Murakami told the author that it has not yet been published.]

Kanel', Ja. I.: Stabilization of solutions of the Cauchy problem for equations encountered in combustion theory (Russian). Mat. Sb. (N.S.) **59** (**101**), suppl., 245—288 (1962).

— Stabilization of solutions of the Cauchy problem for certain linear parabolic equations (Russian). Usp. Mat. Nauk **18**, no. 2 (110), 127—134 (1963).

Kaplan, S.: On the growth of solutions of quasi-linear parabolic equations. Commun. Pure Appl. Math. **16**, 305—330 (1963).

Kasatkina, N. V.: Integral inequalities for many-dimensional integral equations of Volterra type (Russian). Izv. Vysshikh Uchebn. Zavedenii Mat., no. 2 (45) 77—85 (1965).

— Uniqueness theorems for a system of multidimensional Volterra integral equations (Russian). Differencial'nye Uravnenija **3**, 273—277 (1967).

Kato, T.: On linear differential equations in Banach spaces. Commun. Pure Appl. Math. **9**, 479—486 (1956).

Kikodze, M. S.: On the question of uniqueness of a solution of the Cauchy problem and the convergence of successive approximations (Russian). Differencial'nye Uravnenija **2**, 1553—1560 (1966).

Kisyński, J.: Sur l'existence et l'unicité des solutions des problèmes classiques relatifs à l'équation $s = F(x, y, z, p, q)$. Ann. Univ. Mariae Curie-Sklodowska, Lublin-Polonia, Sect. A, **11**, 73—112 (1957).

— Sur les équations différentielles dans les spaces de Banach. Bull. Acad. Polon. Sci. **7**, 381—385 (1959).

Kisyński, J.: Remarque sur l'existence des solutions en large de l'equation $\partial^2 z/\partial x dy = f(x, y, z, \partial z/\partial x, \partial z/\partial y)$. Ann. Univ. Mariae Curie-Sklodowska, Lublin-Polonia, Sect. A, **13**, 25—32 (1959a).

— Sur la convergence des approximations successives pour l'équation $\frac{\partial^2 z}{\partial x \partial y} = f\left(x, y, z, \frac{\partial z}{\partial x}, \frac{\partial z}{\partial y}\right)$. Ann. Polon. Math. **7**, 233—240 (1960).

— Tym, W.: Sur la convergence des approximations successives pour l'équation $\partial^2 z/\partial x \partial y = f(x, y, z, \partial z/\partial x, \partial z/\partial y)$ (Polish and Russian summaries). Ann. Univ. Mariae Curie-Sklodowska, Lublin-Polonia, Sect. A, **16** (1962), 107—121 (1964).

Kolodner, I. I., Pederson, R. N.: Pointwise bounds for solutions of semilinear parabolic equations. J. Diff. Equations **2**, 353—364 (1966).

Kono, M., Kusano, T.: A boundary value problem for nonlinear parabolic equations of the second order. Comment. Math. Univ. St. Paul. **14**, 85—96 (1966).

Kooi, O.: The method of successive approximations and a uniqueness-theorem of Krasnoselskii and Krein in the theory of differential equations. Ned. Akad. Wetensch. Proc. Ser. A 61 = Indag. Math. **20**, 322—327 (1958).

Kovač, Ju. I.: Application of the theorem of differential inequalities to the Goursat problem for a linear system of partial differential equations (Russian). Differencial'nye Uravnenija **1**, 411—420 (1965).

Krasnosel'skii, M. A., Krein, S. G.: Nonlocal existence theorems and uniqueness theorems for systems of ordinary differential equations. Dokl. Akad. Nauk SSSR (N.S.) **102**, 13—16 (1955).

— — On a class of uniqueness theorems for the equation $y' = f(x, y)$. Usp. Mat. Nauk (N.S.) **11**, No. 1 (67), 209—213 (1956).

Krawczyk, R.: Über Differenzenverfahren bei parabolischen Differentialgleichungen. Arch. Rational Mech. Anal. **13**, 81—121 (1963).

Kružkov, S. N.: A priori estimate for the derivative of a solution of a parabolic equation and certain of its applications (Russian). Dokl. Akad. Nauk SSSR **170**, 501—504 (1966).

Krzyżański, M.: Sur les solutions des équations du type parabolique déterminées dans une région illimitée. Bull. Amer. Math. Soc. **47**, 911—915 (1941).

— Sur les solutions de l'équation linéaire du type parabolique déterminées par les conditions initiales. Ann. Soc. Polon. Math. **18**, 145—156 (1945).

— Sur les solutions de l'équation linéaire du type parabolique déterminées par les conditions initiales. Ann. Soc. Polon. Math. **20**, 7—9 (1948).

— Sur les solutions de l'équation linéaire du type elliptique, discontinues sur la frontière du domaine de leur existence. Studia Math. **11**, 95—125 (1950).

— Sur l'équation aux dérivées partielles de la diffusion. Ann. Soc. Polon. Math. **23**, 95—111 (1950a).

— Sur l'allure asymptotique des solutions d'équation du type parabolique. Bull. Acad. Polon. Sci. Cl. III, **4**, 247—251 (1956).

— Evaluations des solutions de l'équation aux dérivées partielles du type parabolique, déterminées dans un domaine non borné. Ann. Polon. Math. **4**, 93—97 (1957).

— Recherches concernant l'allure des solutions de l'équation du type parabolique lorsque la variable du temps tend vers l'infini. Atti Accad. Naz. Lincei. Rend. Classe Sci. Fis. Mat. Nat. (8) **23**, 28—32 (1957a).

— Certaines inégalités relatives aux solutions de l'équation parabolique linéaire normale. Bull. Acad. Polon. Sci. Sér. Sci. Math. Astron. Phys. **7**, 131—135 (1959).

Krzyżański, M.: Sur l'unicité des solutions des second et troisième problèmes de Fourier relatifs à l'équation linéaire normale du type parabolique. Ann. Polon. Math. **7**, 201—208 (1960).
— Sur l'allure asymptotique des solutions des problèmes de Fourier relatifs à une équation linéaire parabolique. Atti Accad. Naz. Lincei Rend. Classe Sci. Fis. Mat. Nat. (8) **28**, 37—43 (1960a).
— Evaluations des solutions de l'équation linéaire du type parabolique à coefficients non bornés. Ann. Polon. Math. **11**, 253—260 (1962).
— Solutions of a linear second order equation of parabolic type defined in an unbounded domain. Differential equations and their applications (Proc. Conf., Prague, 1962a), pp. 55—63. Prague: Publ. House Czechoslovak Acad. Sci. New York: Academic Press 1963.
— Sur l'unicité et l'existence des solutions des problèmes aux limites relatifs aux équations paraboliques. Conf. Sem. Mat. Univ. Bari No. 97—98—99, 3—11 (1964).
— Sur l'allure asymptotique des solutions des équations paraboliques. Conf. Sem. Mat. Univ. Bari No. 97—98—99, 20—28 (1964a).
Kulikov, N. P.: Interior boundary-value problems without initial conditions for the equation $\Delta u = \partial u/\partial t + qu + F$ (Russian). Izv. Vysshikh Uchebn. Zavedenii Mat. No. 6 (19), 140—149 (1960).
— Existence theorems for solutions of a problem for parabolic systems without initial conditions (Russian). Volž. Mat. Sb. Vyp. **4**, 115—122 (1966).
Kusano, T.: On the maximum principle for quasi-linear parabolic equations of the second order. Proc. Japan Acad. **39**, 211—216 (1963).
— Remarks on some properties of solutions of some boundary value problems for quasi-linear parabolic and elliptic equations of the second order. Proc. Japan Acad. **39**, 217—222 (1963a).
— On the first boundary problem for quasilinear systems of parabolic differential equations in non-cylindrical domains. Funkcial. Ekvac. **7**, 103—118 (1965).
— On bounded solutions of elliptic partial differential equations of the second order. Funkcial. Ekvac. **7**, 1—13 (1965a).
— On bounded solutions of exterior boundary value problems for linear and quasilinear elliptic differential equations. Japan. J. Math. **35**, 31—59 (1965b).
Kvapiš, M. [Kwapisz, M.]. A remark on the boundedness of solutions of a system of Volterra integral equations (Polish. Russian and English summaries). Prace Mat. **7**, 1—6 (1962).
— Existence and uniqueness of the solution and convergence of successive approximations for functional equations of Volterra type in a Banach space (Russian. English summary). Tr. Sem. Teor. Diff. Uravneniĭ s Otklon. Argumentom Univ. Družby Narodov Patrisa Lumumby **4**, 123—138 (1967).
Ladyzenskaja, O. A., Solonnikov, V. A., Ural'ceva, N. N.: Linear and quasilinear equations of parabolic type. Amer. Math. Soc., Transl. Math. Monogr. Vol. 23, Providence, R.I. 1968.
Lakshmikantham, V.: On the boundedness of solutions of nonlinear differential equations. Proc. Am. Math. Soc. **8**, 1044—1048 (1957).
— On the extension of Lyapunov's method to various problems of differential systems. MRC Technical Summary Report No. 277 (1961).
— Uniqueness theorems for ordinary and hyperbolic differential equations. Michigan Math. J. **9**, 161—166 (1962).
— Notes on a variety of problems of differential systems. Arch. Rational Mech. Anal. **10**, 119—126 (1962a).

Lakshmikantham, V.: Properties of solutions of abstract differential inequalities. MRC Technical Summary Report No. 334 (1962b).
— Upper and lower bounds of the norm of solutions of differential equations. Proc. Am. Math. Soc. **13**, 615—616 (1962c).
— Differential equations in Banach spaces and the extension of Lyapunov's method. Proc. London Math. Soc. (3) **14**, 74—82 (1964).
— On the Kamke's function in the uniqueness theorem of ordinary differential equations. Proc. Natl. Acad. Sci. India Sect. A **34**, 11—14 (1964a).
— Parabolic differential equations and Lyapunov like functions. J. Math. Anal. Appl. **9**, 234—251 (1964b).
— Leela, S.: Remarks on minimax solutions. Ann. Polon. Math. **19**, 301—306 (1967).

Landis, E. M.: Some questions in the qualitative theory of elliptic and parabolic equations. Usp. Mat. Nauk (85) **14**, 21—85 (1959); — Transl. in Amer. Math. Soc. Transl., ser. 2, No. 20.
— Some questions in the qualitative theory of second order elliptic equations (case of several variables). Usp. Mat. Nauk **18**, 3—62 (1963); — Transl. in Russian Math. Surv. **18**, 3—61 (1963).
— A property of solutions of a parabolic equation (Russian). Dokl. Akad. Nauk SSSR **169**, 262—265 (1966).

Langenhop, C. E.: Bounds on the norm of a general differential equation. Proc. Amer. Math. Soc. **11**, 795—799 (1960).

LaSalle, J.: Uniqueness theorems and successive approximations. Ann. Math. **50**, 722—730 (1949).

Lasota, A.: Sur l'effet épidermique extérieur et intérieur pour les inégalités différentielles ordinaires. Ann. Polon. Math. **6**, 259—264 (1959).

Leehey, P.: On the existence of not necessarily unique solutions of classical hyperbolic boundary value problems for non-linear second order partial differential equations in two independent variables. Ph. D. Brown University 1950.

Levinson, N.: The asymptotic behavior of a system of linear differential equations. Amer. J. Math. **68**, 1—6 (1946).

Li Mun Su: On the existence and continuability of the solution of a system of integral Volterra equations with a discontinuous operator as the right side. Diff. Uravnenija **1**, 387—392 (1965); — Engl. transl.: Diff. Equations **1**, 294—299 (1965).
— On integral inequalities for a system of Volterra equations whose right-hand side is a discontinuous operator (Russian). Diff. Uravnenija **1**, 1027—1041 (1965a).
— Ragimhanov, R. K.: On the question of Volterra equations with a discontinuous operator in the right-hand side (Russian). Izv. Vysshilch Uchebn. Zavedenii Mat., No. 2 (57), 26—34 (1967).

Lipschitz, R.: Lehrbuch der Analysis, Bd. 2. Bonn: 1880.

Liskovets, O. A.: The method of lines (Review). (Russian). Differentsial'nye Uravneniya **1**, 1662–1678 (1965). Engl. Transl.: Differential Equations **1**, 1308–1323 (1965).

Łojczyk-Królikiewicz, I.: Certaines évaluations des solutions des problèmes de Fourier relatifs à l'équation du type parabolique. Ann. Polon. Math. **13**, 213—219 (1963).
— L'allure asymptotique des solutions des problèmes de Fourier relatifs aux équations linéaires normales du type parabolique dans l'espace $\mathscr{E}^{m+1}$. Ann. Polon. Math. **14**, 1—12 (1963a).

Łojczyk-Królikiewicz, I.: Sur l'unicité et les limitations des solutions des problèmes de Fourier relatifs aux équations paraboliques à coefficients non bornés. Ann. Polon. Math. **15**, 33—41 (1964).

— Sur la stabilité asymptotique de la solution d'un système non linéaire d'équations aux dérivées partielles du type parabolique. Ann. Polon. Math. **18**, 243—255 (1966).

Lu Li-jiang [Lu Li-chiang]: A uniqueness theorem for the first boundary value problems of linear elliptic and parabolic equations with discontinuous boundary values. Acta Math. Sinica **15**, 372—385 (1965) (Chinese); — translated as Chinese Math. Acta **7**, 74—89 (1965).

Luxemburg, W. A. J.: On the convergence of successive approximations in the theory of ordinary differential equations. Can. Math. Bull. **1**, 9—20 (1958).

Malcharek, A.: Sur les solutions bornées de l'équation différentielle $x^{(n)} = f(t, x, x', \ldots \ldots, x^{(n-1)})$. Zeszyty Nauk. Uniw. Jagiel. Prace Mat. No. 10, 61—66 (1965).

Mamedov, Ja. D.: On the theory of solutions of non-linear operator equations of Volterra type (Russian). Sibirsk. Mat. Žh. **5**, 1305—1316 (1964).

Maslennikova, V. N.: A class of systems of quasilinear diffusion equations. Soviet Math. Dokl. **3**, 1300—1303 (1963).

— The first boundary value problem for quasilinear diffusion systems. Soviet Math. Dokl. **4**, 803—806 (1963a).

— On a certain class of diffusion systems. Outlines Joint Sympos. Partial Differential Equations (Novosibirsk, 1963b), pp. 159—162. Moscow: Acad. Sci. USSR Sibirian Branch, 1963.

— The first boundary-value problem for certain quasilinear systems in the mathematical theory of diffusion (Russian). Ž. Vyčisl. Mat. i Mat. Fiz. **3**, 467—477 (1963c).

Matrosov, V. M.: Differential equations and inequalities with discontinuous right member. I (Russian). Diff. Uravnenija **3**, 395—409 (1967).

— Differential equations and inequalities with discontinuous right hand sides. II (Russian). Diff. Uravnenija **3**, 839—848 (1967a).

McNabb, A.: Strong comparison theorems for elliptic equations of second order. J. Math. Mech. **10**, 431—440 (1961).

— Comparison and existence theorems for multicomponent diffusion systems. J. Math. Anal. Appl. **3**, 133—144 (1961a).

Medvedev, N. V.: On bounded solutions of a system of differential equations (Russian). Diff. Uravnenija **1**, 1592—1596 (1965).

Miheeva, E. A.: Uniqueness of solutions of the second and third boundary value problem for a second order elliptic equation. Dokl. Akad. Nauk SSSR **167**, 978—981 (1966a); — Transl. in Soviet Math. **6**, 507—510 (1966).

Mises, R. v.: Bemerkungen zur Hydrodynamik. Z. Angew. Math. Mech. **7**, 425—431 (1927).

Mlak, W.: On the epidermic effect for ordinary differential inequalities of the first order. Ann. Polon. Math. **3**, 37—40 (1956).

— The epidermic effect for partial differential inequalities of the first order. Ann. Polon. Math. **3**, 157—164 (1956a).

— Differential inequalities of parabolic type. Ann. Polon. Math. **3**, 349—354 (1957).

— Remarks on the stability problem for parabolic equations. Ann. Polon. Math. **3**, 343—348 (1957a).

— Limitation of solutions of parabolic equations. Ann. Polon. Math. **5**, 237—245 (1958/59).

— Limitations and dependence on parameter of solutions of non-stationary differential operator equations. Ann. Polon. Math. **6**, 305—322 (1959).

Mlak, W.: Parabolic differential inequalities and Chaplighin's method. Ann. Polon. Math. **8**, 139—153 (1960).

Mohr, E.: Die laminare Strömung längs der Platte und damit verwandte Flüssigkeitsbewegungen. Deut. Math. **4**, 477—513 (1937).

Montaldo, O.: Su un problema di valori al contorno nella teoria diffusiva dei reattori nucleari. Rend. Seminario Fac. Sci. Univ. Cagliari **28**, 118—120 (1958).

Montel, P.: Sur l'intégrale supérieure et l'intégrale inférieure d'une équation différentielle. Bull. Sci. Math. **50**, 205—217 (1926).

Morgenstern, D.: Beiträge zur nichtlinearen Funktionalanalysis. Dissertation, Technische Universität Berlin 1952.

Morozov, V. A.: Differential-difference schemes of second order precision for quasilinear problems of parabolic type with discontinuous elements. (Russian). Vestnik Moskov. Univ. Ser. I Mat. Meh. 1964, No. 2, 12–22.

Moyer, R. D.: A general uniqueness theorem. Proc. Amer. Math.Soc. **17**, 602—607 (1966).

Müller, M.: Über das Fundamentaltheorem in der Theorie der gewöhnlichen Differentialgleichungen. Math. Z. **26**, 619—645 (1926).

— Über die Eindeutigkeit der Integrale eines Systems gewöhnlicher Differentialgleichungen und die Konvergenz einer Gattung von Verfahren zur Approximation dieser Integrale. Sitz.-ber. Heidelberg. Akad. Wiss., Math.-Naturw. Kl. 1927, 9. Abh.

Nagumo, M.: Eine hinreichende Bedingung für die Unität der Lösung von Differentialgleichungen erster Ordnung. Japan. J. Math. **3**, 107—112 (1926).

— Über das Verfahren der sukzessiven Approximation zur Integration gewöhnlicher Differentialgleichungen und die Eindeutigkeit ihrer Integrale. Japan. J. Math. **7**, 143—160 (1930).

— Über die Ungleichung $\frac{\partial u}{\partial x} > f\left(x, y, u, \frac{\partial u}{\partial y}\right)$. Japan. J. Math. **15**, 51—56 (1938).

— Note in "Kansū-Hōteisiki" No. 15 (1939) [Japanese. Reference given in Nagumo-Simoda (1951)].

— Simoda, S.: Note sur l'inégalité différentielle concernant les équations du type parabolique. Proc. Japan. Acad. **27**, 536—539 (1951).

Narasimhan, R.: On the asymptotic stability of solutions of parabolic differential equations. J. Rational Mech. Anal. **3**, 303—313 (1954).

Nickel, K.: Einige Eigenschaften von Lösungen der Prandtlschen Grenzschicht-Differentialgleichungen. Arch. Rational Mech. Anal. **2**, 1—31 (1958).

— Die äußere Randbedingung der Grenzschicht-Differentialgleichung. Z. Angew. Math. Mech. **38**, 400—401 (1958a).

— Fehlerabschätzungen bei parabolischen Differentialgleichungen. Math. Z. **71**, 268—282 (1959).

— Ein Eindeutigkeitssatz für instationäre Grenzschichten. Math. Z. **74**, 209—220 (1960).

— Parabolic equations with applications to boundary layer theory. Partial differential equations and continuum mechanics, p. 319—330. Madison (Wis.): Univ. of Wisconsin Press 1961.

— Fehlerabschätzungs- und Eindeutigkeitssätze für Integro-Differentialgleichungen. Arch. Rational Mech. Anal. **8**, 159—180 (1961a).

— Gestaltaussagen über Lösungen parabolischer Differentialgleichungen. J. Reine Angew. Math. **211**, 78—94 (1962).

— Eine einfache Abschätzung für Grenzschichten. Ing.-Arch. **31**, 85—100 (1962a).

Nickel, K.: Die Prandtlschen Grenzschichtdifferentialgleichungen als asymptotischer Grenzfall der Navier-Stokesschen und der Eulerschen Differentialgleichungen. Arch. Rational Mech. Anal. **13**, 1—14 (1963).

— Ein Eindeutigkeitssatz für instationäre Grenzschichten II. Math. Z. **83**, 1—7 (1964).

Nicolescu, M., Foias, C.: Représentation de Poisson et problème de Cauchy pour l'équation de la chaleur. Atti Accad. Naz. Lincei Rend. Classe Sci. Fis. Mat. Nat. (8) **38**, 466—476 (1965).

— — Représentation de Poisson et problème de Cauchy pour l'équation de la chaleur. II. Atti Accad. Naz. Lincei Rend. Classe Sci. Fis. Mat. Nat. (8) **38**, 621—626 (1965a).

— — Sur l'unicité du problème de Cauchy pour l'équation de la chaleur (Italian summary). Atti Accad. Naz. Lincei Rend. Classe Sci. Fis. Mat. Nat. (8) **40**, 785—791 (1966).

Nirenberg, L.: A strong maximum principle for parabolic equations. Commun. Pure Appl. Math. **6**, 167—177 (1953).

Nitsche, J. C. C.: Ein erweitertes Maximumprinzip für nicht-lineare parabolische Differentialgleichungen. Arch. Math. (Basel) **17**, 65—69 (1966).

— The general maximum principle and removable singularities — a counter-example. Arch. Rational Mech. Anal. **22**, 22—23 (1966a).

Novruzov, G. M.: On the theory of differential equations with a Volterra operator (Russian. Azerbaijani summary). Azerb. Gos. Univ. Vč. Zap. Ser. Fiz.-Mat. Nauk, No. 6, 42—47 (1966).

Oddson, J. K.: Phragmén-Lindelöf and comparison theorems for elliptic equations with mixed boundary conditions. Arch. Rational Mech. Anal. **26**, 316—334 (1967).

Okamura, H.: Condition nécessaire et suffisante remplie par les équations différentielles ordinaires sans points de Peano. Mem. Coll. Sci. Kyoto Imp. Univ. Ser. A, **24**, 21—28 (1942).

Olech, C.: Remarks concerning criteria for uniqueness of solutions of ordinary differential equations. Bull. Acad. Polon. Sci. Sér. Sci. Math. Astron. Phys. **8**, 661—666 (1960).

— On the existence and uniqueness of solutions of an ordinary differential equation in the case of Banach space. Bull. Acad. Polon. Sci. Sér. Sci. Math. Astron. Phys. **8**, 667—673 (1960a).

— Opial, Z.: Sur une inégalité différentielle. Ann. Polon. Math. **7**, 247—254 (1960).

— On a system of integral inequalities. Colloq. Math. **16**, 137—139 (1967).

Oleinik, O. A.: On properties of solutions of certain boundary problems for equations of elliptic type. Mat. Sbornik, N.S., **30**, 695—702 (1952).

— Sur certaines équations paraboliques dégénérescentes de la mécanique. Les Equations aux Dérivées Partielles (Paris, 1962), pp. 175—180. Paris: Editions du Centre National de la Recherche Scientifique 1963.

— On the system of Prandtl equations in boundary-layer theory (Russian). Dokl. Akad. Nauk SSSR **150**, 28—31 (1963).

— On a system of equations in boundary-layer theory (Russian). Ž. Vyčisl. Mat. i Mat. Fiz. **3**, 489—507 (1963a).

— Mathematical problems of boundary layer theory. Lecture Notes, Spring Quarter 1969, University of Minnesota, Minneapolis, Minn. 1969.

Opial, Z.: Sur un système d'inégalités intégrales. Ann. Polon. Math. **3**, 200—209 (1957).

— Sur la dépendence des solutions d'un système d'équations différentielles de leurs second membres. Application aux systèmes prèsque autonomes. Ann. Polon. Math. **8**, 75—89 (1960).

Osgood, W. F.: Beweis der Existenz einer Lösung der Differentialgleichung $dy/dx = f(x, y)$ ohne Hinzunahme der Cauchy-Lipschitzschen Bedingung. Monatsh. Math. Phys. **9**, 331—345 (1898).

Ostrowski, A.: Sur le rayon de convergence de la série de Blasius. Compt. Rend. Acad. Sci. Paris **227**, 580—582 (1948).

Oudart, A.: Publ. sc. et techn. du Ministère de l'air, no. 213, 123—127 (1948). [Reference given in Ostrowski 1948].

Padmavally, K.: On a non-linear integral equation. J. Math. Mech. **7**, 533—555 (1958).

Pagni, M.: Un'osservazione sull'unicità della soluzione del problema di Cauchy per l'equazione $p = f(x, y, z, q)$. Rend. Seminarie Mat. Univ. Padova **20**, 470—474 (1951).

Palczewski, B.: Existence and uniqueness of solutions of the Darboux problem for the equation

$$\frac{\partial^3 u}{\partial x_1\,\partial x_2\,\partial x_3} = f\left(x_1, x_2, x_3, u, \frac{\partial u}{\partial x_1}, \frac{\partial u}{\partial x_2}, \frac{\partial u}{\partial x_3}, \frac{\partial^2 u}{\partial x_1\,\partial x_2}, \frac{\partial^2 u}{\partial x_1\,\partial x_3}, \frac{\partial^2 u}{\partial x_2\,\partial x_3}\right).$$

Ann. Polon. Math. **13**, 267—277 (1963).

— On the uniqueness of solutions and the convergence of successive aproximations in the Darboux problem under the conditions of the Krasnosielski and Krein type. Ann. Polon. Math. **14**, 183—190 (1963/64).

— Pawelski, W.: Some remarks on the uniqueness of solutions of the Darboux problem with conditions of the Krasnosielski-Krein type. Ann. Polon. Math. **14**, 97—100 (1963/64).

Paraf, M. A.: Sur le problème de Dirichlet et son extension au cas de l'équation linéaire générale du second ordre. Ann. Fac. Sci. Toulouse **6**, Fasc. 47—Fasc. 54 (1892).

Payne, L. E., Sather, D.: On a singular hyperbolic operator. Duke Math. J. **34**, 147—162 (1967).

Peano, G.: Démonstration de l'intégrabilité des équations différentielles ordinaires. Math. Ann. **37**, 182—228 (1890).

Pelczar, A.: On the existence and uniqueness of solutions of the Darboux problem for the equation $z_{xy} = f(x, y, z, z_x, z_y)$. Bull. Acad. Polon. Sci. Sér. Sci. Math. Astron. Phys. **12**, 703—707 (1964).

Perov, A. I.: Some remarks on differential inequalities (Russian). Izv. Vysshikh Uchebn. Zavedenii Mat. No. 4 (47), 104—112 (1965).

Perron, O.: Über den Integralbegriff. Sitz.-ber. Heidelberg. Akad. Wiss., Abt. A, 14. Abhandl. (1914).

— Ein neuer Existenzbeweis für die Integrale der Differentialgleichung $y' = f(x, y)$. Math. Ann. **76**, 471—484 (1915).

— Eine neue Behandlung des ersten Randwertproblems für $\Delta u = 0$. Math. Z. **18**, 42—54 (1923).

— Über Ein- und Mehrdeutigkeit des Integrals eines Systems von Differentialgleichungen. Math. Ann. **95**, 98—101 (1926).

Picard, E.: Sur les méthodes d'approximations successives dans la théorie des équations différentielles, Note 1 dans t. 4 de G. Darboux, Leçons sur la théorie générale surfaces, f. 353—367. Paris 1896.

Picone, M.: Sul problema della propagazione del calore in un mezzo privo di frontiera, conduttore, isotropo e omogeneo. Math. Ann. **101**, 701—712 (1929).

— Sull'equazione integrale non lineare di Volterra. Ann. Mat. Pura Appl. (4) **49**, 1—10 (1960).

— Sur les équations intégrales linéaires de deuxième espèce de Volterra avec noyau de translation. Compt. Rend. Acad. Sci. Paris **250**, 46—48 (1960a).

Pini, B.: Sul primo problema di valori al contorno per l'equazione parabolica non lineare del secondo ordine. Rend. Seminarie Mat. Univ. Padova **27**, 149—161 (1957).

Porath, G.: Abschätzungen für Näherungslösungen gewöhnlicher Differentialgleichungen im Banachschen Raum. Math. Nachr. **29**, 333—345 (1965).

Prandtl, L.: Über Flüssigkeitsbewegungen bei sehr kleiner Reibung. Verh. d. III. Int. Math.-Kongr. Heidelberg 1904, S. 484—494. Leipzig: Teubner 1905.

Price, H. S., Varga, R. S., Warren, J. E.: Application of oscillation matrices to diffusion-convection equations. J. Math. Phys. **45**, 301–311 (1966).

Prodi, G.: Questioni di stabilità per equazioni non lineari alle derivate parziali di tipo parabolico. Atti Accad. Naz. Lincei. Rend. Classe Sci. Fis. Mat. Nat. (8) **10**, 365—370 (1951).

Protter, M. H.: A boundary value problem for an equation of mixed type. Trans. Amer. Math. Soc. **71**, 416—429 (1951).

— A maximum principle for hyperbolic equations in a neighborhood of an initial line. Trans. Amer. Math. Soc. **87**, 119—129 (1958).

— Properties of solutions of parabolic equations and inequalities. Can. J. Math. **13**, 331—345 (1961).

— Weinberger, H. F.: Maximum principles in differential equations. Englewood Cliffs. New York: Prentice-Hall, Inc. 1967.

Pucci, C.: Maggiorazione della soluzione di un problema al contorno, di tipo misto, relativo a una equazione a derivate parziali, lineare, del secondo ordine. Atti Accad. Naz. Lincei. Rend. Classe Sci. Fis. Mat. Nat. (8) **13**, 360—366 (1952).

— Bounds for solutions of Laplace's equation satisfying mixed conditions. J. Rational Mech. Anal. **2**, 299—302 (1953).

— Proprietà di massimo e minimo delle soluzioni di equazioni a derivate parziali del secondo ordine di tipo ellitico e parabolico, I. Atti Accad. Naz. Lincei. Rend. Classe Sci. Fis. Mat. Nat. (8) **23**, 370—375 (1957).

— Proprietà di massimo e minimo delle soluzioni di equazioni a derivate parziali del secondo ordine ti tipo ellitico e parabolico, II. Atti Accad. Naz. Lincei. Rend. Classe Sci. Fis. Mat. Nat. (8) **24**, 3—6 (1958).

— Limitazioni per soluzioni di equazioni ellitiche. Ann. Mat. Pura Appl. (4) **74**, 15—30 (1966).

Pulvirenti, G.: Problemi lineari per le equazioni differenziali ordinarie in uno spazio di Banach. Matematiche **15**, 98—107 (1960).

— Il fenomeno di Peano nel problema di Darboux per l'equazione $s = f(x, y, z)$ in ipotesi di Carathéodory. Matematiche **15**, 15—28 (1960a).

— Equazioni differenziali in uno spazio di Banach. Teorema di esistenza e struttura del pennello delle soluzioni in ipotesi di Carathéodory. Ann. Mat. Pura Appl. (4) **56**, 281—300 (1961).

— Il fenomeno di Peano nel problema di Darboux per l'equazione $z_{xy} = f(x, y, z, z_x, z_y)$ in ipotesi di Carathéodory. Ric. Mat. **14**, 9—40 (1965).

Punnis, B.: Zur Differentialgleichung der Plattengrenzschicht von Blasius. Arch. Math. **7**, 165—171 (1956).

Radziszewski, B.: Note on the asymptotic behavior of solutions of a n-th order nonlinear equation. Nonlinear vibration problems, Vol. 6, pp. 346—347. Warsaw: Państw. Wydawn. Nauk. 1966.

Rama Mohana Rao, M.: The local uniqueness and successive approximations (Russian and Romanian summaries). Bul. Inst. Politeh. Iaşi (N.S.) **9** (**13**), fasc. 1—2, 13—18 (1963).

Redheffer, R. M.: Maximum principles and duality. Monatsh. Math. **62**, 56—75 (1958).

Redheffer, R. M.: On the inequality $\Delta u \geqq f(u, |\mathrm{grad}\, u|)$. J. Math. Anal. Appl. **1**, 277—299 (1960).
— Bemerkungen über Monotonie und Fehlerabschätzung bei nichtlinearen partiellen Differentialgleichungen. Arch. Rational Mech. Anal. **10**, 427—457 (1962).
— An extension of certain maximum principles. Monatsh. Math. **66**, 32—42 (1962a).
— Die Collatzsche Monotonie bei Anfangswertproblemen. Arch. Rational Mech. Anal. **14**, 196—212 (1963).
— Elementary remarks on problems of mixed type. J. Math. Phys. **43**, 1—14 (1964).
— Differentialungleichungen unter schwachen Voraussetzungen. Abhandl. Math. Sem. Univ. Hamburg **31**, 33—50 (1967).
— Straus, E. G.: Degenerate elliptic equations. Pacific J. Math. **14**, 265—268 (1964).
Reichert, M.: Über die Fixpunkte einer Klasse singulärer Volterrascher Abbildungen. Dissertation. Veröffentlichungen der Mathematischen Institute der Freien Universität Berlin, Bd. I (1962).
— Eindeutigkeits- und Iterationsfragen bei Volterra-Hammersteinschen Integralgleichungen. J. Reine Angew. Math. **220**, 74—87 (1965).
— Über die Gesamtheit der Lösungen bei hyperbolischen Anfangswertaufgaben. Math. Ann. **182**, 281—308 (1969).
Rektorys, K.: Die Lösung des ersten Randwertproblems „im Ganzen" für eine nichtlineare parabolische Gleichung mit der Netzmethode (Russian summary). Czech. Math. J. **12** (**87**), 69—103 (1962).
Rosenblatt, A.: Über die Existenz von Integralen gewöhnlicher Differentialgleichungen. Arkiv Mat. Astron. Fys. **5**, Nr. 2 (1909).
Rosenbloom, P.: Linear partial differential equations. Surveys in Applied Math., Vol. V. New York: John Wiley 1958.
Rosenbloom, P. C., Widder, D. V.: A temperature function which vanishes initially. Amer. Math. Monthly **65**, 607—609 (1958).

Saks, S.: Theory of the integral. Monografie Matematyczne, Vol. 7. Warsaw 1937.
Sansone, G., Conti, R.: Equazioni differenzali non lineari. Roma: Cremonese 1956.
Santagati, G.: Il problema di Darboux per una equazione del secondo ordine di tipo iperbolico. Matematiche **14**, 115—147 (1959).
— Su alcuni sistemi di equazioni integrodifferenziali (English summary). Ann. Mat. Pura Appl. (4) **63**, 71—121 (1963).
— Sul problema de Picard in ipotesi di Carathéodory (English summary). Ann. Mat. Pura Appl. (4) **75**, 47—94 (1967).
Santoro, P.: Sul problema di Darboux per l'equazione $s = f(x, y, z, p, q)$ e il fenomeno di Peano. Rend. Accad. Naz. dei XL, (4) **10**, 3—17 (1959).
— Sulla dipendenza continua dai dati iniziali delle soluzioni del problema di Darboux. Boll. Un. Mat. Ital. (3) **18**, 148—153 (1963).
Šatalov, Ju. S.: On the question of non-uniqueness of solutions of systems of singular integral equations of Volterra type (Russian). Diff. Uravnenija **3**, 905—911 (1967).
Sather, D.: Maximum and monotonicity properties of initial boundary value problems for hyperbolic equations. Pacific J. Math. **19**, 141—157 (1966).
Satō, T.: Sur l'équation aux dérivées partielles hyperbolique $s = f(x, y, z, p, q)$. I. Mem. Fac. Sci. Kyūsyū Imp. Univ. A. **2**, 107—123 (1941).
— Sur l'équation aux dérivées partielles du type hyperbolique. Rep. Fac. Sc. Kyūsyū Imp. Univ. **1**, 203—249 (1945) [Japanese].
— Sur la limitation des solutions d'un système d'équations intégrales de Volterra. Tohoku Math. J. (2) **4**, 272—274 (1952).
— Sur l'équation intégrale non linéaire de Volterra. Compositio Math. **11**, 271—290 (1953).

Satō, T., Iwasaki, A.: Sur l'équation intégrale de Volterra. Proc. Japan. Acad. **31**, 395—398 (1955).

Schleinkofer, G.: Die erste Randwertaufgabe und das Cauchy-Problem für parabolische Differentialgleichungen mit unstetigen Anfangswerten. Math. Z. **111**, 87—97 (1969).

— Die erste Randwertaufgabe für quasilineare parabolische Differentialgleichungen bei unstetigen Randwerten. Math. Z. **113**, 33—48 (1970).

Schlichting, H.: Boundary layer theory. London: Pergamon Press Ltd. Karlsruhe: G. Braun 1955.

Schröder, J.: Fehlerabschätzungen bei gewöhnlichen und partiellen Differentialgleichungen. Arch. Rational Mech. Anal. **2**, 367—392 (1958/59).

— Vom Defekt ausgehende Fehlerabschätzungen bei Differentialgleichungen. Arch. Rational Mech. Anal. **3**, 219—228 (1959).

— Anwendung von Fixpunktsätzen bei der numerischen Behandlung nichtlinearer Gleichungen in halbgeordneten Räumen. Arch. Rational Mech. Anal. **4**, 177—192 (1959/60).

— Error estimates for boundary value problems using fixed point theorems. Boundary problems in differential equations, edit. by E. Langer, p. 85—96. Madison (Wis.): Univ. of Wisconsin Press 1960.

— Fehlerabschätzungen mit Rechenanlagen bei gewöhnlichen Differentialgleichungen erster Ordnung. Num. Math. **3**, 39—61 (1961).

— Verbesserung einer Fehlerabschätzung für gewöhnliche Differentialgleichungen erster Ordnung. Num Math. **3**, 125—130 (1961a).

Scorza Dragoni, G., Volpato, M.: Un teorema di unicità per le soluzioni di una equazione alle derivate parziali del primo ordine. Rend. Seminario Mat. Univ. Padova **20**, 446—461 (1951).

Seyferth, C.: Die Eindeutigkeit von Lösungen der eindimensionalen Diffusionsgleichung mit konzentrationsabhängigem Diffusionskoeffizienten und die Lage ihrer Extrema. Z. Angew. Math. Mech. **39**, 441—443 (1959).

— Die Eindeutigkeit von Lösungen der eindimensionalen Diffusionsgleichung mit konzentrationsabhängigem Diffusionskoeffizienten. Math. Nachr. **24**, 13—32 (1962).

Shanahan, J. P.: On uniqueness questions for hyperbolic differential equations. Pacific J. Math. **10**, 677—688 (1960).

— On existence in the large of solutions of hyperbolic partial differential equations. Rend. Circ. Mat. Palermo (2) **11** 225—236 (1962).

Shapiro, V. L.: The uniqueness of solutions of the heat equation in an infinite strip. Trans. Amer. Math. Soc. **125**, 326—361 (1966).

Shen Xin-yao [Shen Hsin-yao]: Remarks on the uniqueness of solutions of differential equations (Chinese). Shuxue Jinzhan **6**, 267—271 (1963).

Simoda, S.: Notes pour la théorie des équations aux dérivées partielles du type elliptique. Mem. Osaka Univ. Lir. Arts Ed. Ser. B. mo **5**, 5—15 (1956).

— Nagumo, M.: Sur la solution bornée de l'équation aux dérivées partielles du type elliptique. Proc. Japan. Acad. **27**, 334—339 (1951).

Smirnova, G. N.: Uniqueness classes for the solution of the Cauchy problem for parabolic equations (Russian). Dokl. Akad. Nauk SSSR **153**, 1269—1272 (1963).

— The Cauchy problem for parabolic equations degenerating at infinity (Russian). Mat. Sb. (N.S.) **70** (**112**), 591—604 (1966).

Sternberg, W.: Über die Gleichung der Wärmeleitung. Math. Ann. **101**, 329—398 (1929).

Storm, M. L.: Heat conduction in simple metals. J. Appl. Phys. **22**, 940—951 (1951).

Styś, T.: Hopf's theorem for a certain elliptic system of linear second-order equations (Polish. Russian and English summaries). Prace Mat. **8**, 143—146 (1963/64).

— On the unique solvability of the first Fourier problem for a parabolic system of linear differential equations of second order (Russian). Prace Mat. **9**, 283—289 (1965).

Sugiyama, S.: On a certain functional-differential inequality. Kōdai Math. Sem. Rep. **17**, 273—280 (1965).

Švec, M.: Fixpunktsatz und monotone Lösungen der Differentialgleichung $y^{(n)}+B(x,y,y',\ldots,y^{(n-1)})\,y=0$. Arch. Math. (Brno) **2**, 43—55 (1966).

— L'existence globale et les propriétés asymptotiques des solutions d'une équation différentielle non-linéaire d'ordre n. Arch. Math. (Brno) **2**, 141—151 (1966a).

Szarski, J.: Sur un système d'inégalités différentielles. Ann. Soc. Polon. Math. **20**, 126—134 (1947).

— Sur certains systèmes d'inégalités différentielles aux dérivées partielles du premier ordre. Ann. Soc. Polon. Math. **21**, 7—25 (1948).

— Sur certaines inégalités entre les intégrales des équations différentielles aux dérivées partielles du premier ordre. Ann. Soc. Polon. Math. **22**, 1—34 (1949).

— Sur les systèmes majorants d'équations différentielles ordinaires. Ann. Soc. Polon. Math. **23**, 206—223 (1950).

— Sur les systèmes d'inégalités différentielles ordinaires remplies en dehors de certains ensembles. Ann. Soc. Polon. Math. **24**, Teil II, 1—8 (1951).

— Systèmes d'inégalités différentielles aux dérivées partielles du premier ordre, et leurs applications. Ann. Polon. Math. **1**, 149—165 (1954).

— Sur la limitation et l'unicité des solutions d'un système non-linéaire d'équations paraboliques aux dérivées partielles du second ordre. Ann. Polon. Math. **2**, 237—249 (1955).

— Sur la limitation et l'unicité des solutions des problèmes de Fourier pour un système non linéaire d'équations paraboliques. Ann. Polon. Math. **6**, 211—216 (1959).

— Remarque sur un critère d'unicité des intégrales d'une équation différentielle ordinaire. Ann. Polon. Math. **12**, 203—205 (1962).

— Sur un système non linéaire d'inégalités différentielles paraboliques. Ann. Polon. Math. **15**, 15—22 (1964).

— Differential inequalities. Monografie Matematyczne, Tom 43. Warszawa 1965.

Szmydt, Z.: Sur les systèmes d'équations différentielles dont toutes les solutions sont bornées. Ann. Polon. Math. **2**, 234—236 (1955).

— Sur un nouveau type de problèmes pour un système d'équations différentielles hyperboliques du second ordre à deux variables indépendantes. Bull. Acad. Polon. Sci. Cl. III, **4**, 67—72 (1956).

— Sur une généralisation des problèmes classiques concernant un système d'équations différentielles hyperboliques du second ordre à deux variables indépendantes. Bull. Acad. Polon. Sci. Cl. III, **4**, 579—584 (1956a).

— Sur l'existence de solutions de certains nouveaux problèmes pour un système d'équations différentielles hyperboliques de second ordre à deux variables indépendantes. Ann. Polon. Math. **4**, 40—60 (1957).

— Sur l'existence d'une solution unique de certains problèmes pour un système d'équations différentielles hyperboliques du second ordre à deux variables indépendantes. Ann. Polon. Math. **4**, 165—182 (1958).

Szmydt, Z.: Sur l'existence de solutions de certains problèmes aux limites relatifs à un système d'équations différentielles hyperboliques. Bull. Acad. Polon. Sci. Sér. Sci. Math. Astron. Phys. **6**, 31—36 (1958a).

Tonelli, L.: Sull'unicità della soluzione di un'equazione differenziale ordinarie. Atti Accad. Naz. Lincei. Rend. (6) **1**, 272—277 (1925).

— Sulle equazioni integrali di Volterra. Mem. della Reale Accad. delle Sci. dell'Istituto di Bologna, Classe Sci. Fis. (8) **5**, 59—64 (1927/28).

— Sulle equazioni funzionali del tipo di Volterra. Bull. Calcutta Math. Soc. **20**, 31—48 (1928).

Tychonoff, A. N.: Théorèmes d'unicité pour l'équation de la chaleur. Rec. Math. Moscou **42**, 199—215 (1935).

— Samarski, A. A.: Differentialgleichungen der mathematischen Physik. Berlin: VEB Deutscher Verlag der Wissenschaften 1959.

— — Partial differential equations of mathematical physics (2 Vols.). San Francisco: Holden Day, Inc. 1967.

Uhlmann, W.: Fehlerabschätzungen bei Anfangswertaufgaben gewöhnlicher Differentialgleichungssysteme 1. Ordnung. Z. Angew. Math. Mech. **37**, 88—99 (1957).

— Fehlerabschätzungen bei Anfangswertaufgaben einer gewöhnlichen Differentialgleichung höherer Ordnung. Z. Angew. Math. Mech. **37**, 99—111 (1957a).

Vaghi, C.: Sul comportamento asintotico delle soluzioni di equazioni non lineari di tipo parabolico (English summary). Atti Accad. Naz. Lincei Rend. Classe Sci. Fis. Mat. Nat. (8) **41**, 25—31 (1966).

Velte, W.: Eine Anwendung des Nirenbergschen Maximumprinzips für parabolische Differentialgleichungen in der Grenzschichttheorie. Arch. Rational Mech. Anal. **5**, 420—431 (1960).

— Bemerkungen über Randmaxima bei parabolischen Differentialoperatoren (English and Russian summaries). Z. Angew. Math. Mech. **43**, 421—423 (1963).

Vinokurov, V. R.: Positive and monotone solutions of Volterra integral equations (Russian). Izv. Vysshikh Uchebne Zavedenii Mat., No. 4 (59), 40—46 (1967).

Viswanatham, B.: The general uniqueness theorem and successive approximations. J. Indian Math. Soc. (N.S.) **16**, 69—74 (1952).

— On the structure of the set of solutions of a non-linear differential equation $y' = f(x, y)$. Math. Student **33**, 95—96 (1965).

Volpato, M.: Criteri di confronto e di unicità per le soluzioni dell'equazione $p = f(x, y, z, q)$ coi dati di Cauchy. Rend. Seminario Mat. Univ. Padova **20**, 232—243 (1951).

— Sugli elementi uniti di transformazioni funzionali: un problema ai limiti per una classe di equazioni alle derivate parziali di tipo iperbolico. Ann. Univ. Ferrara **2**, 95—109 (1952/53).

Volterra, V.: Theory of functionals and of integral and integro-differential equations. 1927 [Reprinted New York: Dover Publications Inc. 1959].

Výborný, R.: Über einige Eigenschaften der Lösungen von Randwertaufgaben einer parabolischen partiellen Differentialgleichung (Russian. German summary). Czech. Math. J. **8** (83), 537—551 (1958).

— Über das erweiterte Maximumprinzip (Russian summary). Czech. Math. J. **14** (89), 116—121 (1964).

Wadsworth, M., Wragg, A.: The numerical solution of the heat conduction equation in one dimension. Proc. Cambridge Phil. Soc. **60**, 897–907 (1964).

Wadsworth, M., Wragg, A.: On matrix methods for the solution of partial differential equations. Proc. Cambridge Phil. Soc. **61**, 129—132 (1965).

Walter, W.: Über die Differentialgleichung $u_{xy}=f(x,y,u,u_x,u_y)$, Teil I. Math. Z. **71**, 308—324 (1959).

— Über die Differentialgleichung $u_{xy}=f(x,y,u,u_x,u_y)$, Teil II. Math. Z. **71**, 436—453 (1959a).

— On the existence theorem of Carathéodory for ordinary and hyperbolic equations. Technical Note BN-172, AFOSR (1959b).

— Über die Differentialgleichung $u_{xy}=f(x,y,u,u_x,u_y)$, Teil III. Math. Z. **73**, 268—279 (1960).

— Eindeutigkeitssätze für gewöhnliche, parabolische und hyperbolische Differentialgleichungen. Math. Z. **74**, 191—208 (1960a).

— Fehlerabschätzungen bei hyperbolischen Differentialgleichungen. Arch. Rational Mech. Anal. **7**, 249—272 (1961).

— Fehlerabschätzungen und Eindeutigkeitssätze für gewöhnliche und partielle Differentialgleichungen. Z. Angew. Math. Mech. **42**, T 49—T 62 (1962).

— Bemerkungen zu verschiedenen Eindeutigkeitskriterien für gewöhnliche Differentialgleichungen. Math. Z. **84**, 222—227 (1964).

— Über sukzessive Approximation bei Volterra-Integralgleichungen in mehreren Veränderlichen. Ann. Acad. Sci. Fennicae Ser. A I, **345**, 1—32 (1965).

— On nonlinear Volterra integral equations in several variables. J. Math. Mech. **16**, 967—985 (1967).

— On the non-existence of maximal solutions for hyperbolic differential equations. Ann. Polon. Math. **19**, 307—311 (1967a).

— Die Linienmethode bei nichtlinearen parabolischen Differentialgleichungen. Numer. Math. **12**, 307—321 (1968).

— Ein Existenzbeweis für nichtlineare parabolische Differentialgleichungen aufgrund der Linienmethode. Math. Z. **107**, 173—188 (1968a).

— Wärmeleitung in Systemen mit mehreren Komponenten. ISNM Vol. 9 „Numerische Mathematik, Differentialgleichungen, Approximationstheorie", p. 177—185. Basel: Birkhäuser 1968b.

— Gewöhnliche Differential-Ungleichungen im Banachraum. Arch. Math. **20**, 36—47 (1969a).

— Existenzsätze im Großen für das Cauchyproblem bei nichtlinearen parabolischen Differentialgleichungen mit der Linienmethode. Math. Ann. **183**, 254—274 (1969b).

— Approximation für das Cauchyproblem bei parabolischen Differentialgleichungen mit der Linienmethode. ISNM Vol. 10 "Abstract Spaces and Approximation", p. 139—145. Basel: Birkhäuser 1969c.

— Existence and convergence theorems for the boundary layer equations based on the line method. MRC Technical Summary Report No. 1025. Arch. Rational Mech. Anal. (in print) (1970).

— On the asymptotic behavior of solutions of the Prandtl boundary layer equations. MRC Technical Summary Report No. 1056 (1970a). J. Math. Mech. (in print).

— Über die Eindeutigkeitsbedingung von Krasnosel'skii-Krein bei hyperbolischen Differentialgleichungen. Ann. Polon. Math. XXII, 117—124 (1970b).

Waltman, P.: On the asymptotic behavior of solutions of an nth order equation. Monatsh. Math. **69**, 427—430 (1965).

Ważewski, T.: Sur le domaine d'existence des intégrales de l'équation aux dérivées partielles du premier ordre linéaire. Ann. Soc. Polon. Math. **12**, 81—86 (1933).

Ważewski, T.: Sur l'unicité et la limitation des intégrales des équations aux dérivées partielles du premier ordre. Atti. Accad. Naz. Lincei. Rend. Classe Sci. Fis. Mat. Nat. (6) **18**, 372—376 (1933a).
— Sur la limitation des intégrales des systèmes d'équations différentielles linéaires ordinaires. Studia Math. **10**, 48—59 (1948).
— Systèmes des équations et des inégalités différentielles ordinaires aux deuxièmes membres monotones et leurs applications. Ann. Soc. Polon. Math. **23**, 112—166 (1950).
— Certaines propositions de caractère "epidermique" relatives aux inégalités différentielles. Ann. Soc. Polon. Math. **24**, 1—12 (1952).
— Remarque sur un système d'inégalités intégrales. Ann. Polon. Math. **3**, 210—212 (1957).
— Sur l'existence et l'unicité des intégrales des équations différentielles ordinaires au cas de l'espace de Banach. Bull. Acad. Polon. Sci. Sér. Sci. Math. Astron. Phys. **8**, 301—305 (1960).
— Sur la convergence des approximations successives pour les équations différentielles ordinaires au cas de l'espace de Banach. Ann. Polon. Math. **16**, 231—235 (1965).
Weigel, H.: Randwertaufgaben für parabolische Differentialgleichungen bei unstetigen Randwerten. Math. Z. **111**, 98—120 (1969).
Weinberger, H. F.: A maximum property of Cauchy's problem. Ann. Math. **64**, 503—513 (1956).
Weissinger, J.: Zur Theorie und Anwendung des Iterationsverfahrens. Math. Nachr. **8**, 193—212 (1952).
Westphal, H.: Zur Abschätzung der Lösungen nichtlinearer parabolischer Differentialgleichungen. Math. Z. **51**, 690—695 (1949).
Weyl, H.: On the differential equations of the simplest boundary-layer problems. Ann. Math. **43**, 381—407 (1942).
— Comment on the preceding paper (Levinson, 1946). Amer. J. Math. **68**, 7—12 (1946).
Widder, D. V.: Positive temperatures on an infinite rod. Trans. Amer. Math. Soc. **55**, 85—95 (1944).
Willett, D.: Nonlinear vector integral equations as contracting mappings. Arch. Rational Mech. Anal. **15**, 79—86 (1964).
— A linear generalization of Gronwall's inequality. Proc. Amer. Math. Soc. **16**, 774—778 (1965).
— Wong, J. S. W.: On the discrete analogues of some generalizations of Gronwall's inequality. Monatsh. Math. **69**, 362—367 (1965).
Wilson, A. H.: A diffusion problem in which the amount of diffusing substance is finite. Philos. Mag. Ser. 7, 39, 48—58 (1948).
Wintner, A.: The non-local existence problem of ordinary differential equations. Amer. J. Math. **67**, 277—284 (1945).
— On the convergence of successive approximations. Amer. J. Math. **68**, 13—19 (1946).
— Asymptotic equilibria. Amer. J. Math. **68**, 125—132 (1946a).
— The infinities in the non-local existence problem of ordinary differential equations. Amer. J. Math. **68**, 173—178 (1946b).
— On the local uniqueness of the initial value problem of the differential equation $d^n x/dt^n = f(t, x)$. Boll. Univ. Mat. Ital. (3) **11**, 496—498 (1956).
— On non-constant Lipschitz factors in the uniqueness problem of ordinary differential equations. Arch. Math. **7**, 465—468 (1956a).

Wong, J. S. W.: Remarks on the uniqueness theorem of solutions of the Darboux problem. Can. Math. Bull. **8**, 791—796 (1965).

— On the convergence of successive approximations in the Darboux problem. Ann. Polon. Math. **17**, 329—336 (1966).

Wu, Zhueng-hai [Wu, Chung-hai]: Theorems on differential inequalities for nonlinear Bianchi equations. Acta Math. Sinica **13**, 584—606 (1963) (Chinese); translated as Chinese Math. **4**, 637—660 (1964).

Ziebur, A. D.: Uniqueness and the convergence of successive approximations. Proc. Amer. Math. Soc. **13**, 899—903 (1962).

— Uniqueness and the convergence of successive approximations. II. Proc. Amer. Math. Soc. **16**, 335—340 (1965).

Zitarosa, A.: Sulle equazioni funzionali di Volterra-Tonelli. Ric. Mat. **3**, 108—126 (1954).

— Alcune osservazioni su certi teoremi di compattezza e sul problema di Darboux. Rend. Accad. Sci. Fis. Mat. Napoli (4) **27**, 25—35 (1960).

Zwirner, G.: Criteria di unicità per gli integrali delle equazioni differenziali del primo ordine. Rend. Seminario Mat. Univ. Padova **19**, 273—293 (1950).

— Sull'approssimazione degli integrali del sistema differenziale. $\partial^2 z/\partial x \partial y = f(x, y, z)$, $z(x_0, y) = \psi(y)$, $z(x, y_0) = \varphi(x)$. Ist. Veneto Sci. Lett. Arti. Classe Sci. Mat. Nat. **109**, 219—231 (1951).

— Teoremi di unicità e di confronto per gli integrali di una particolare classe di equazioni differenziali a derivate parziali del secondo ordine. Rend. Seminario Mat. Univ. Padova **20**, 329—345 (1951a).

— Sull'equazione $\frac{\partial^2 z}{\partial x \partial y} = f\left(x, y, z, \frac{\partial z}{\partial y}\right)$. Ann. Univ. Ferrara. Sez. VII (N.S.) **1**, 9—16 (1952).

Zygmund, A.: Trigonometric series, 2. Ed., Vol. 1. Cambridge: University Press 1959.

Subject-Index

Operator equations are listed under integral equations. The phrase "one-dimensional integral equations" or "multidimensional integral equations" means that we are concerned with a problem from Chapter I (one-dimensional domain) or Chapter III (multi-dimensional domain).

Author-Index

Ergebnisse der Mathematik und ihrer Grenzgebiete

---

1. Bachmann: Transfinite Zahlen. DM 38,–; US $ 10.50
2. Miranda: Partial Differential Equations of Elliptic Type. DM 58,–; US $ 16.00
4. Samuel: Méthodes d'Algèbre Abstraite en Géométrie Algébrique. DM 26,–; US $ 7.20
5. Dieudonné: La Géométrie des Groupes Classiques. DM 38,–; US $ 10.50
6. Roth: Algebraic Threefolds with Special Regard to Problems of Rationality. DM 19,80; US $ 5.50
7. Ostmann: Additive Zahlentheorie. 1. Teil: Allgemeine Untersuchungen. DM 38,–; US $ 10.50
8. Wittich: Neuere Untersuchungen über eindeutige analytische Funktionen. DM 28,–; US $ 7.70
11. Ostmann: Additive Zahlentheorie. 2. Teil: Spezielle Zahlenmengen. DM 28,–; US $ 7.70
13. Segre: Some Properties of Differentiable Varieties and Transformations. In preparation
14. Coxeter/Moser: Generators and Relations for Discrete Groups. DM 32,–; US $ 8.80
15. Zeller/Beekmann: Theorie der Limitierungsverfahren. DM 64,–; US $ 17.60
17. Severi: Il teorema di Riemann-Roch per curve-superficie e varietà questioni collegate. DM 23,60; US $ 6.50
18. Jenkins: Univalent Functions and Conformal Mapping. DM 34,–; US $ 9.40
19. Boas/Buck: Polynomial Expansions of Analytic Functions. DM 16,–; US $ 4.40
20. Bruck: A Survey of Binary Systems. DM 36,–; US $ 9.90
21. Day: Normed Linear Spaces. DM 17,80; US $ 4.90
23. Bergman: Integral Operators in the Theory of Linear Partial Differential Equations. DM 36,–; US $ 9.00
25. Sikorski: Boolean Algebras. DM 38,–; US $ 9.50
26. Künzi: Quasikonforme Abbildungen. DM 39,–; US $ 10.80
27. Schatten: Norm Ideals of Completely Continuous Operators. DM 26,–; US $ 7.20
28. Noshiro: Cluster Sets. DM 36,–; US $ 9.90
29. Jacobs: Neuere Methoden und Ergebnisse der Ergodentheorie. DM 49,80; US $ 13.70
30. Beckenbach/Bellman: Inequalities. DM 38,–; US $ 10.50
31. Wolfowitz: Coding Theorems of Information Theory. DM 27,–; US $ 7.50
32. Constantinescu/Cornea: Ideale Ränder Riemannscher Flächen. DM 68,–; US $ 18.70
33. Conner/Floyd: Differentiable Periodic Maps. DM 26,–; US $ 7.20
34. Mumford: Geometric Invariant Theory. DM 22,–; US $ 6.10
35. Gabriel/Zisman: Calculus of Fractions and Homotopy Theory. DM 38,–; US $ 9.50
36. Putnam: Commutation Properties of Hilbert Space Operators and Related Topics. DM 28,–; US $ 7.70
37. Neumann: Varieties of Groups. DM 46,–; US $ 12.60
38. Boas: Integrability Theorems for Trigonometric Transforms. DM 18,–; US $ 5.00
39. Sz.-Nagy: Spektraldarstellung linearer Transformationen des Hilbertschen Raumes. DM 18,–; US $ 5.00
40. Seligman: Modular Lie Algebras. DM 39,–; US $ 9.75
41. Deuring: Algebren. DM 24,–; US $ 6.60
42. Schütte: Vollständige Systeme modaler und intuitionistischer Logik. DM 24,–; US $ 6.60
43. Smullyan: First-Order Logic. DM 36,–; US $ 9.90
44. Dembowski: Finite Geometries. DM 68,–; US $ 17.00

45. Linnik: Ergodic Properties of Algebraic Fields. DM 44,– ; US $ 12.10
46. Krull: Idealtheorie. DM 28,– ; US $ 7.70
47. Nachbin: Topology on Spaces of Holomorphic Mappings. DM 18,– ; US $ 5.00
48. A. Ionescu Tulcea/C. Ionescu Tulcea: Topics in the Theory of Lifting. DM 36.– ; US $ 9.90
49. Hayes/Pauc: Derivation and Martingales. DM 48,– ; US $ 13.20
50. Kahane: Séries de Fourier Absolument Convergentes. DM 44,– ; US $ 12.10
51. Behnke/Thullen: Theorie der Funktionen mehrerer komplexer Veränderlichen. DM 48,– ; US $ 13.20
52. Wilf: Finite Sections of Some Classical Inequalities. DM 28,– ; US $ 7.70
53. Ramis: Sous-ensembles analytiques d'une variété banachique complexe. DM 36,– ; US $ 9.90
54. Busemann: Recent Synthetic Differential Geometry. DM 32,– ; US $ 8.80
55. Walter: Differential and Integral Inequalities. DM 74,– ; US $ 20.40
56. Monna: Analyse non-archimédienne. DM 38,– ; US $ 10.50
57. Alfsen: Compact Convex Sets and Boundary Integrals. In preparation
58. Greco/Salmon: Topics in $\mathfrak{m}$-adic Topologies. In preparation
59. López de Medrano: Involutions on Manifolds. In preparation